An Introduction to Optimization

The primary goal of this text is a practical one. Equipping students with enough knowledge and creating an independent research platform, the author strives to prepare students for professional careers. Providing students with a marketable skill set requires topics from many areas of optimization. The initial goal of this text is to develop a marketable skill set for mathematics majors as well as for students of engineering, computer science, economics, statistics, and business. Optimization reaches into many different fields.

This text provides a balance where one is needed. Mathematics optimization books are often too heavy on theory without enough applications; texts aimed at business students are often strong on applications, but weak on math. The book represents an attempt at overcoming this imbalance for all students taking such a course.

The book contains many practical applications but also explains the mathematics behind the techniques, including stating definitions and proving theorems. Optimization techniques are at the heart of the first spam filters, are used in self-driving cars, play a great role in machine learning, and can be used in such places as determining a batting order in a Major League Baseball game. Additionally, optimization has seemingly limitless other applications in business and industry. In short, knowledge of this subject offers an individual both a very marketable skill set for a wealth of jobs as well as useful tools for research in many academic disciplines.

Many of the problems rely on using a computer. Microsoft's Excel is most often used, as this is common in business, but Python and other languages are considered. The consideration of other programming languages permits experienced mathematics and engineering students to use MATLAB® or Mathematica, and the computer science students to write their own programs in Java or Python.

Jeffrey Paul Wheeler earned his PhD in Combinatorial Number Theory from the University of Memphis by extending what had been a conjecture of Erdős on the integers to finite groups. He has published, given talks at numerous schools, and twice been a guest of Trinity College at the University of Cambridge. He has taught mathematics at Miami University (Ohio), the University of Tennessee-Knoxville, the University of Memphis, Rhodes College, the University of Pittsburgh, Carnegie Mellon University, and Duquesne University. He has received numerous teaching awards and is currently in the Department of Mathematics at the University of Pittsburgh. He also occasionally teaches for Pitt's Computer Science Department and the College of Business Administration. Dr. Wheeler's Optimization course was one of the original thirty to participate in the Mathematical Association of America's NSF-funded PIC Math program.

Textbooks in Mathematics
Series editors:
Al Boggess, Kenneth H. Rosen

An Introduction to Analysis, Third Edition
James R. Kirkwood

Multiplicative Differential Calculus
Svetlin Georgiev, Khaled Zennir

Applied Differential Equations
The Primary Course
Vladimir A. Dobrushkin

Introduction to Computational Mathematics: An Outline
William C. Bauldry

Mathematical Modeling the Life Sciences
Numerical Recipes in Python and MATLAB™
N. G. Cogan

Classical Analysis
An Approach through Problems
Hongwei Chen

Classical Vector Algebra
Vladimir Lepetic

Introduction to Number Theory
Mark Hunacek

Probability and Statistics for Engineering and the Sciences with Modeling using R
William P. Fox and Rodney X. Sturdivant

Computational Optimization: Success in Practice
Vladislav Bukshtynov

Computational Linear Algebra: with Applications and MATLAB® Computations
Robert E. White

Linear Algebra With Machine Learning and Data
Crista Arangala

Discrete Mathematics with Coding
Hugo D. Junghenn

Applied Mathematics for Scientists and Engineers
Youssef N. Raffoul

Graphs and Digraphs, 7th ed
Gary Chartrand, Heather Jordon, Vincent Vatter and Ping Zhang

An Introduction to Optimization: With Applications in Machine Learning and Data Analytics
Jeffrey Paul Wheeler

https://www.routledge.com/Textbooks-in-Mathematics/book-series/CANDHTEXBOOMTH

An Introduction to Optimization

With Applications in Machine Learning and Data Analytics

Jeffrey Paul Wheeler

CRC Press is an imprint of the
Taylor & Francis Group, an **informa** business
A CHAPMAN & HALL BOOK

First edition published 2024
by CRC Press
6000 Broken Sound Parkway NW, Suite 300, Boca Raton, FL 33487-2742

and by CRC Press
4 Park Square, Milton Park, Abingdon, Oxon, OX14 4RN

CRC Press is an imprint of Taylor & Francis Group, LLC

ISBN: 978-0-367-42550-0 (hbk)
ISBN: 978-1-032-61590-5 (pbk)
ISBN: 978-0-367-42551-7 (ebk)

DOI: 10.1201/9780367425517

Typeset in CMR10 font
by KnowledgeWorks Global Ltd.

Publisher's note: This book has been prepared from camera-ready copy provided by the authors.

Contents

Acknowledgments

To love gone sour, suspicion, and bad debt.
– The Clarks

Also, to my immediate family for the support as I hid in the basement working on this book, and to my extended family – especially my parents – for their hard work, sacrifice, and support through all the years in giving me a great education and continually encouraging me along the way. You all have optimized the quality of my life.

List of Figures

List of Tables

List of Algorithms

List of Notation

$GL_n(\mathbb{Z})$	general linear group of matrices	59
$T_n(x)$	n^{th} order Taylor polynomial	141
$R_n(x)$	remainder of $T_n(x)$	142
$f_x = f_x(x,y) = \frac{\partial f(x,y)}{\partial x} = \frac{\partial}{\partial x} f(x,y)$	partial derivative of f(x,y) with respect to x	146
$f_y = f_y(x,y) = \frac{\partial f(x,y)}{\partial y} = \frac{\partial}{\partial y} f(x,y)$	partial derivative of f(x,y) with respect to y	146
$\nabla f(x,y)$	gradient of f(x,y)	147
Hf	Hessian of $f(x_1, x_2, \ldots, x_n)$	147
$L(S)$	linear hull (or linear span) of a set S	210
$aff(S)$	affine hull of a set S	210
$coni(S)$	conical hull of a set S	210
$conv(S)$	convex hull of a set S	210
$H(\mathbf{c}, b)$	hyperplane	218
$A + B$	sumset	221
$epi(f)$	epigraph of $f(x)$; i.e. all points above the graph of $f(X)$	235
$hypo(f)$	hypograph of $f(x)$; i.e. all points below the graph of $f(X)$	236
$\|\|\mathbf{x}\|\|_p$	p norm	246
$\mathbb{R}_{++}$	the set of positive real numbers $(0, \infty)$	248
$\mathbb{R}_{+}$	the set of nonnegative real numbers $[0, \infty)$	248
$n!$	factorial function $n! = n(n-1)(n-2)\cdots 3 \cdot 2 \cdot 1$ with $0! := 1$	268
$P(n,r)$	the number of r-permutations from a set of n objects without repetition	270
$C(n,r)$	the number of r-combinations from a set of n objects without repetition	271
$\binom{n}{r}$	binomial coefficient = C(n,r)	274
$\binom{n}{n_1, n_2, \ldots, n_m}$	multinomial coefficient	277
$\lceil r \rceil$	ceiling of r; the smallest integer $k \geq r$	282
$\deg^-(u)$	in-degree of a vertex u	287
$\deg^+(u)$	out-degree of a vertex u	287
P_n	path of order n	289
C_n	cycle of order n	289
K_n	complete graph of order n	291
$K_{s,t}$	complete bipartite graph with partite sets of order s and t	292

Part I

Preliminary Matters

1

Preamble

1.1 Introduction

As a subject, optimization can be admired because it reaches into many different fields. It borrows tools from statistics, computer science, calculus (analysis), numerical analysis, graph theory, and combinatorics as well as other areas of mathematics. It has applications in economics, computer science and numerous other disciplines as well as being incredibly useful outside academics (this is an understatement!). Its techniques were at the heart of the first spam filters, are used in self-driving cars, play a great role in machine learning and can be used in such places as determining a batting order in a Major League Baseball game. Additionally, it has seemingly limitless other applications in business and industry. In short, knowledge of this subject offers an individual both a very marketable skill set for a wealth of jobs as well as useful tools for research in many academic disciplines.

1.2 Software

Even though much of this text has problems that rely on using a computer, I have stayed away from emphasizing any one particular software. Microsoft's *Excel* is most often used as this is common in business, but Python and other languages are considered. There are two reasons for this:

1. Software changes.

2. In a typical *Introduction to Optimization* class in the Mathematics Department at the University of Pittsburgh, I get students from Mathematics, Engineering, Economics, Computer Science, Statistics, and Business and these students come to the class with different computer experience. I do assign problems that involve using a computer, but I never require using a particular software and this has been very rewarding; especially during student presentations. The reason for this is we all get to experience the

DOI: 10.1201/9780367425517-1

Mathematics and Engineering students using MatLab or Mathematica, the Economics and Business majors using Excel, and the Computer Science students writing their own programs in Java or Python. In short, we all see different ways to do similar problems and are each richer for seeing the different processes.

1.3 About This Book

1.3.1 Presentation

This book is a sampling of optimization techniques and only touches on the surface of each. Most of the chapters in this book could be developed into their own texts, but rather than be an expert exposition on a single topic, my colleagues and I have sought to create a buffet from which students and faculty may sample what they wish. Additionally, we have attempted to offer a healthy combination of applications of techniques as well as some of the underlying mathematics of the techniques. This latter goal may have caused us to lose some readers: "Oh, no, this is going to be dry, boring, and soulless. I just want to know how to solve problems". If this is you, please be patient with us and join us on our journey. We believe that even the smallest understanding of the theory enables one to be a better problem solver and, more importantly, provides one with the tools to effectively diagnose what is not working and which direction to head when the technique fails to work in a particular setting. Others will see this as a poser for a proper mathematics text. There is some truth to this assessment and to those individuals I recommend any one of the many excellent mathematical texts referenced in this work. To the many that may regard the majority of the presentations in this work as not proper or formal enough, I quote a New Yorker article on the superb mathematician Alexander Grothendieck:

"Grothendieck argued that mathematicians hide all of the discovery process, and make it appear smooth and deductive. He said that, because of this, the creative side of math is totally misunderstood." [28]

1.3.2 Contents

The structure of the text is such that most chapters and even some sections can be read independently. A refresher, for example, on Taylor's Theorem for multivariable functions or a crash course on matrix factorization can be easily done by finding the appropriate section in the text.

This text grew out of notes for the class *Mathematical Models for Consultants* I taught three times for the Tepper School of Business at Carnegie

Mellon University, *Applied Optimization and Simulation*, I have taught for the Katz College of Business Administration at the University of Pittsburgh, and my *Introduction to Optimization* class offered regularly in the Department of Mathematics at the University of Pittsburgh. I have taught undergraduate and graduate versions of the class more than a dozen times so far for our department and have even had colleagues attend the course. In a typical semester of my math class, I try to cover Linear Programming, Integer Linear Programming, multiple Geometric Programming techniques, the Fundamental Theorem of Linear Programming, transshipment problems, minimum-weight spanning trees, shortest paths, and the Traveling Salesperson Problem as well as some other topics based upon student interest. My goal for the math class at Pitt is to provide a skill set so that our majors can get a job after graduating, but the course – as previously stated – has ended up attracting many students with diverse backgrounds including Engineering, Computer Science, Statistics, Economics, and Business. This mix of student interests explains the structure of the text: a smorgasbord of topics gently introduced with deeper matters addressed as needed. Students have appreciated sampling different dishes from the buffet, and the ones that wanted to dig further were asked to do a little research on their own and give a short presentation. Some extra credit was used as the carrot to motivate them, but the true reward was much greater. All of us in the audience were treated with learning something new, but the student presenter's true reward was the realization that they had the ability to learn on their own (and show an employer that they can give a great presentation). The educator in me has found this to be the most rewarding of all the classes I teach; feeling that same satisfaction of watching my child happily ride a bike for the first time.

As such, I would change nothing about the structure of this text. It probably has too much mathematics for some and not enough mathematics for others, but that is exactly where I want the text to be. I have taught college mathematics for over 30 years and taught a wide range of courses at different levels. I receive many thank you cards, but I get the most from students that took this course; usually because of the job they have secured because of what they learned in this class. I have also had many students continue to graduate school in a wide range of areas, but none more than in some version of Optimization. Numerous students have obtained interesting jobs as well, including working on self-driving cars, doing analysis for a professional sports team, and being a contributing member of the discussion on how to distribute the first COVID vaccine during the pandemic. In short, this approach to the material has worked very well and given the subject's utility, it is the right time for an undergraduate-level survey text in the material. I hope you enjoy the journey through the material as much as I and my students have.

1.4 One-Semester Course Material

Much of this book is not covered in a typical one-semester course that I teach. In a typical one-semester course, I cover

- Linear Programming (Chapter 6)
- Integer Linear Programming (Chapter 8)
- Geometric Programming, specifically Chapters 11, 14, and 13
- affine, conical, and convex sets as well as the Fundamental Theorem of Linear Programming (Chapters 16 and 17)
- an introduction to Graph Theory (the first two sections of Chapter 21)
- minimum weight spanning trees, shortest paths, networks, and transshipment problems, as well as the Traveling Salesperson Problem (Chapters 23, 24, and 25)

and we will mention

- complexity (Chapter 3) and
- sensitivity analysis (Chapter 7)

and spend some time in these chapters if we need to. We also explore other topics if time permits.

I never cover the Algebra, Matrix Factorization, Calculus, or Combinatorics chapters, but often many of my students need a refresher of these topics or a short introduction. Note also that though matrix factorization and matrix norms other than the Euclidean norm are not used later in the text, though they are important matters to consider if one wants to discuss using a computer to solve important problems. Sometimes, also, these topics come up in class discussions and many students need to know this material for what analytical work they have done after taking the class. In a very real sense this book is less of a textbook and more of a handbook of optimization techniques; the mathematics, computer science, and statistics behind them; and essential background material. As my class draws not just math majors but students from across campus, and having resources of necessary review materials has been key to many of my students' success.

1.5 Acknowledgments

This book would not have been possible without significant contributions from talented professionals with whom I have had the privilege to work in some capacity. I am quite pleased to share that most on this list are former optimization students, and all of the contributors have written from their professional expertise. Arranged by chapter, contributors of all or most of the material in particular chapters to the text are:

- Complexity Classes – Graham Zug (Software Engineer, Longpath Technologies)
- Genetic Algorithms – Andy Walsh, Ph.D. (Chief Science Officer, Health Monitoring)
- Convex Optimization – Jourdain Lamperski, Ph.D. (Pitt Department of Industrial Engineering)
- Traveling Salesperson Problem – Corinne Brucato Bauman (Assistant Professor, Allegheny Campus, Community College of Allegheny County)
- Probability – Joseph Datz (former analyst for the Pittsburgh Pirates; Institute Grant, Research, and Assessment Coordinator, Marywood University) and Joseph "Nico" Gabriel (Research Analyst 2, Skaggs School of Pharmacy and Pharmaceutical Sciences at the University California San Diego).
- Regression Analysis via Least Squares – John McKay (Research Scientist at Amazon Transport Science) with contributions from Suren Jayasuria (Assistant Professor Arts, Media and Engineering Departments, Arizona State University)
- Forecasting – Joseph "Nico" Gabriel (Research Analyst 2, Skaggs School of Pharmacy and Pharmaceutical Sciences at the University of California, San Diego).
- Intro to Machine Learning – Suren Jayasuria, Ph.D. (Assistant Professor Arts, Media and Engineering Departments, Arizona State University) with contributions from John McKay (Research Scientist at Amazon Transport Science).

Memphis student and former AMS Senior Editor Avery Carr shared his expert LaTeX editing skills and saved me from agonizing over details I would have to work to understand. Avery also served as a guest editor for the wonderful 100th *Anniversary Edition* ΠME *Problems Edition.* Pitt student Evan Hyzer also edited and solved many LaTeX mysteries for me. Special appreciation is to be extended to Mohd Shabir Sharif for his extended edits increasing the quality of the text.

Distinguished University Professor Tom Hales of the Department of Mathematics at the University of Pittsburgh was kind enough to read an early version of the text and offer incredibly helpful feedback.

Additionally, Joseph Datz contributed a large number of wonderful exercises throughout the various topics in the text. Former student Graham Zug contributed homework problems and also kindly proofread. The contributors are all former students except Andy Walsh and John McKay (who was a student at Pitt, but never took any of my classes).

My wife and children are to be thanked for tending to home matters and allowing me to disappear to my basement lair undisturbed for hours on end, many days and nights in a row.

Those deserving my greatest appreciation, though, are the many students that endured the evolution of the notes that eventually became this book. They also had to withstand all the classes where I often poorly pieced together seemingly unrelated material. I have been blessed with terrific students who have been open to discussion and exploration as well as being eager to learn. They have molded this material, contributed to its content, and made the journey to this point most enjoyable. Thank you all.

– Jeffrey Paul Wheeler, University of Pittsburgh

2

The Language of Optimization

2.1 Basic Terms Defined

It is most likely the case that anyone reading a book such as this is familiar with basic definitions of what we are studying, yet it is worthwhile in a mathematical context to offer formal definitions.

Definition 2.1.1 (Maximizer, Maximum). *Let $f : D \to \mathbb{R}$ where the domain D of f is some subset of the real numbers*[1]

- global *(or* absolute*)* maximizer *of $f(x)$ over D if $f(x^*) \geq f(x)$ for all $x \in D$;*
- strict global *(or* strict absolute*)* maximizer *of $f(x)$ over D if $f(x^*) > f(x)$ for all $x \in D$ with $x \neq x^*$;*
- local *(or* relative*)* maximizer *of $f(x)$ if there exists some positive number ϵ such that $f(x^*) \geq f(x)$ for all x, where $x^* - \epsilon < x < x^* + \epsilon$ (i.e. in some* neighborhood *of x^*);*
- strict local *(or* strict relative*)* maximizer *of $f(x)$ if there exists some positive number ϵ such that $f(x^*) > f(x)$ for all x, where $x^* - \epsilon < x < x^* + \epsilon$ with $x \neq x^*$.*

The $f(x^*)$ in the above is, respectively, the global (or absolute) maximum, strict global (or absolute) maximum, local (or relative) maximum, *or* strict local (or relative) maximum *of $f(x)$ over D.*

It is important to understand the difference between a maximizer and a maximum.

Highlight 2.1.2. *The maximum $f(x^*)$ is the optimal value of f which, if the optimal value exists, is unique. The maximizer x^* is the location of the optimal value, which is not necessarily unique.*

A slight change in detail will lead to another important concept:

Definition 2.1.3 (Minimizer, Minimum). *Let $f : D \to \mathbb{R}$ where $D \subseteq \mathbb{R}$. A point x^* in D is said to be a*

[1] It should be noted that we have no need to restrict ourselves to the reals and could offer the definition in a more abstract field.

DOI: 10.1201/9780367425517-2

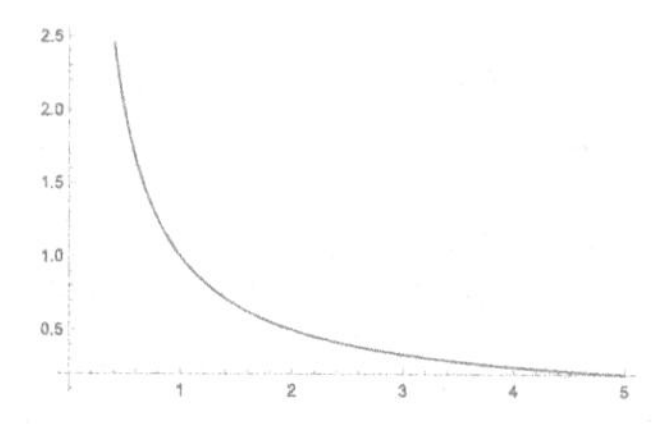

FIGURE 2.1
The graph of $f(x) = \frac{1}{x}$, where $x > 0$.

- global *(or* absolute*)* minimizer *of $f(x)$ over D if $f(x^*) \leq f(x)$ for all $x \in D$;*
- strict global *(or* absolute*)* minimizer *of $f(x)$ over D if $f(x^*) < f(x)$ for all $x \in D$ with $x \neq x^*$;*
- local *(or* relative*)* minimizer *of $f(x)$ if there exists some positive number ϵ such that $f(x^*) \leq f(x)$ for all x, where $x^* - \epsilon \leq x \leq x^* + \epsilon$;*
- strict local *(or* relative*)* minimizer *of $f(x)$ if there exists some positive number ϵ such that $f(x^*) < f(x)$ for all x, where $x^* - \epsilon < x < x^* + \epsilon$ with $x \neq x^*$.*

The $f(x^*)$ in the above is, respectively, the global *(or* absolute*)* minimum*),* strict global minimum, local *(or* relative*)* minimum, *or* strict local *(or* relative*)* minimum *of $f(x)$ over D.*

Note that the plural form of maximum is *maxima* and that the plural form of minimum is *minima*. Together the local and global maxima and minima of a function $f(x)$ are referred to as *extreme values* or *extrema* of $f(x)$. A single maximum or minimum of $f(x)$ is called an *extreme value* or an *extremum* of the function.

We also note that the stated definitions of maximum and minimum are for functions of a single variable, but the definitions[2] are the same for a function of n variables except that $D \subseteq \mathbb{R}^n$ and $\mathbf{x}^* = \langle x_1^*, \ldots, x_n^* \rangle$ would replace x^*.

2.2 When a Max or Min Is Not in the Set

Consider the function $f(x) = \frac{1}{x}$ where $x > 0$. Certainly $f(x)$ is never 0 nor is it ever negative (the graph of $f(x)$ is in Figure 2.1); thus for all $x > 0$, $f(x) > m$ where m is any nonpositive real number. This leads to the following collection of definitions:

[2]There is some concern with how the ϵ interplays with the $x_1, \ldots, x_n$, but the overall idea is the same.

Definition 2.2.1 (Upper and Lower Bounds). *Let f be a function mapping from a set D onto a set R where R is a subset of the real numbers. If there exists a real number M such that $f(x) \leq M$ for all x in D, then f is said to be* bounded from above. *Likewise, if there exists a real number m such that $f(x) \geq m$ for all x in D, then f is said to be* bounded from below. *M is called an* upper bound *of f whereas m is called a* lower bound *of f.*

Example 2.2.2. *The function $f(x) = \frac{1}{x}$ is bounded below by $m = 0$ as well as by $m = -1$. The function is unbounded from above.*

Note that if a function is both bounded above and bounded below, then the function is said to be a *bounded function*; that is

Definition 2.2.3 (Bounded Function). *If there exists a constant M such that $|f(x)| \leq M$ for all x in the domain of f, then f is said to be a bounded function. If no such M exists, then f is said to be* unbounded.

Example 2.2.4. *Since $|\sin x| \leq 1$ for any real number x, $\sin x$ is a bounded function.*

Let us now reconsider f in Example 2.2.2 and observe that $f : D \to R$ where $D = (0, \infty) = R$. As previously observed, $f(x) > 0$ for all x in D, but 0 is not in R. As we can find x that get us as arbitrarily close to 0 as we like, f does not have a minimum, but 0 still plays a special role.

Definition 2.2.5 (Infimum, Supremum). *Let S be a nonempty subset of $\mathbb{R}$. Then b is said to be the* infimum *of S if $b \leq s$ for all $s \in S$ and $b \geq m$ for any lower bound m of S. The infimum of a set is the greatest lower bound of the set and is denoted $b = \inf(S)$. Likewise, a is said to be the* supremum *of S if $a \geq s$ for all $s \in S$ and $a \leq M$ for any upper bound M of S. The supremum of a set is the least upper bound of the set and is denoted $a = \sup(S)$.*

It will come as no surprise that the infimum is also often called the *greatest lower bound (glb)*, and the supremum is referred to as the *least upper bound (lub)*. When an infimum or supremum exists, it is unique. As well, the plural of supremum is *suprema* and infimum has *infima* as its plural.

Example 2.2.6. *For f in Example 2.2.2, $\min f$ does not exist, but for the codomain $\mathbb{R}^+$ (the set of positive real numbers), $\inf R = 0$.*

Thus we see that the infimum (when it exists) can play a role similar to a minimum when a minimum does not exist. An analogous statement can be said of maxima and suprema.

2.3 Solving an Optimization Problem

By "solving" algebraic equations like

$$2x^2 + 2x + 5 = 9 \tag{2.1}$$

we mean "finding the particular values of x that satisfy equation 2.1" (they are -2 and 1). In another circumstance, we may be interested in what a lower bound of the polynomial $2x^2 + 2x + 5$ is (this solution is $9/2$ or anything smaller).

But when solving an optimization problem, we always mean a little more than just some numeric value. For example, consider the classic algebra problem of a farmer having 1000 feet of fence and wanting to know what is the biggest area he can enclose with a rectangular pen for his livestock if he builds the pen adjacent to his barn (big enough that he only needs fence on three sides). If we label the side parallel to the barn y and the other two sides x, then the mathematical model of our problem is

$$\text{Maximize } A = A(x,y) = xy \tag{2.2}$$
$$\text{Subject to } y + 2x = 1000 \tag{2.3}$$
$$\text{with } x, y \geq 0. \tag{2.4}$$

As our goal is to maximize the area, the function representing it, $A(x,y)$, is called the *objective function*. As well, the amount of available fence puts a restriction on the problem, so the corresponding equation $y + 2x = 1000$ is called a *constraint*. As well, we naturally have the *nonnegativity constraints* that $x, y \geq 0$.

Using the constraint to eliminate a variable, the problem simplifies to

$$\text{Maximize } A(x) = 1000x - 2x^2. \tag{2.5}$$

The maximum of this function is 125,000 square feet, and though our farmer will appreciate this knowledge, it is certain he also would like to know what dimensions he needs to make the pen in order to achieve having this maximum area. Thus, by a *solution* to an optimization question, we do not just mean the optimal value of the objective function but also the values of the variables that give the extreme value. Thus we report the answer as $\max A = 125{,}000$ square feet, which occurs when $x = 250$ feet and $y = 500$ feet. We summarize the point of this example in the following Highlight:

Highlight 2.3.1. *A* solution *to an optimization problem is*

1. *the optimal value of the objective function together with*
2. *all possible feasible values of the decision variable(s) that yield the optimal objective value.*

2.4 Algorithms and Heuristics

By an *algorithm* we mean a finite procedure applied to an input with well-defined steps that are repeated to obtain a desired outcome. For example, consider washing your hair. First you wet your hair, then you apply the shampoo and lather, and lastly you rinse. This process may be repeated as many times as you wish to obtain the desired level of cleanliness (read your shampoo bottle; it may have an algorithm written on it). In some sense, an algorithm is a recipe that is repeated.

You may have noticed that we have not offered a formal definition of an algorithm. We are going to avoid unnecessary formality and potential disputes and not offer one all the while noting (modifying Justice Potter Stewart's words in *Jacobellis v. Ohio*:) "I may not know how what the definition of an algorithm is, but I know one when I see it" (Justice Stewart was not addressing algorithms; decency forbids me addressing the matter of that case). It is worthwhile to note that the authoritative text on algorithms – *Algorithms* [11] by Thomas H. Cormen, Charles E. Leiserson, Ronald Rivest, and Clifford Stein – as well does not define the term *algorithm* anywhere in its 1312 pages.

Algorithms have been around for a long time. Perhaps in grade school you learned the *Sieve of Eratosthenes* (circa 3rd century BCE) to find primes. Given a finite list of integers, one circles 2 and crosses out all other multiples of 2. We then proceed to the next available integer, 3, keep it, and cross out all other multiples of 3. We repeat until every integer in our list is circled or crossed out, and what remains are the primes that were in our list of numbers. Algorithms will play a major role in techniques we study iterative methods and combinatorial optimization.

The word *algorithm* has a fascinating origin. It comes from the Latinized ("Algorithmi") version of the Persian name Muḥammad ibn Mūsā al-Khwārizmī whose early 9th century CE book *Al-kitāb al-mukhtaṣar fī ḥisāb al-ğabr wa'l-muqābala* ("The Compendious Book on Calculation by Completion and Balancing") is the first known systematic treatment of algebra as an independent subject. Unlike other early works presenting specific problems and their solution, Al-Khwārizmī's work presents general solution techniques for first- and second-order equations, including completing the square. Al-Khwārizmī can be regarded as the father of algebra, and it is from his text we get the term "algebra" (interestingly, it is also from his name the Spanish and Portuguese get their words for "digit"; see [64]).

Algorithms may produce a globally optimal solution, as we will see in the Simplex Method to solve Linear Programming problems and as well in Kruskal's Algorithm or Prim's Method to find minimum weight spanning trees in a graph. On the other hand, an algorithm may not give a solution but under the right conditions give a good approximation as in Newton's Method.

A *heuristic* is a slightly different monster. A dictionary from before the days of everyday people being familiar with computers would report that "heuristic" is an adjective meaning "enabling a person to discover or learn something for themselves" [15] or "by trial and error" [16]. These days, the word is also regarded as a noun and is most likely shortened from "a heuristic method". When using it as a noun, we mean by *heuristic* a technique that is employed when no method of obtaining a solution (either global or local) is known or a known technique takes too long. It is, in a very true sense, an "educated guess". Consider the *Traveling Salesperson Problem* (TSP) which is introduced in Chapter 25. A salesperson needs to visit a collection of cities and would like to know how to plan her route to minimize distance driven. Unfortunately, there does not yet exist a deterministic-polynomial time algorithm to solve this[3], nor is it known that it is impossible for one to exist ($P = NP$ anyone?), so she can instead do what seems like a good idea: drive to the nearest city and when done with her business there, drive to the nearest city not yet visited, etc. (this is the *Nearest Neighbor Heuristic*[4] that we will see later).

2.5 Runtime of an Algorithm or a Heuristic

A very important matter we will need to consider as we solve our problems is how long it will take a computer to do the work for us. This can be measured in different ways, either time elapsed or the number of operations a computer must perform. Seldom do we use time as the standard in this regard as processor speeds vary and get faster. The standard is to count operations the computer must do, but even this is not precise as sometimes we may count only arithmetic operations performed, but other times we also include calls to memory, etc. This apparent discrepancy is not a large concern as our goal when determining the runtime or *computational complexity*, or simply *complexity*, of an algorithm, heuristic, or computer program is to approximate the amount of effort a computer must put forth. The purpose of these calculations is to compare the runtime efficiency of a given program, algorithm, or heuristic to other known techniques.

Complexity is worthy of its own chapter and is addressed in Chapter 3.

2.6 For Further Study

Parts of these texts have excellent presentations on what we have considered and can be used to deepen one's understanding of the material presented in this chapter:

[3]A brute force algorithm that tests all the paths will give the correct answer and terminate in a finite amount of time. Unfortunately, there are $(n-1)!/2$ possible tours (routes) on n cities, so with 10 cities there are 181,440 possible tours and 20 cities have 60,822,550,204,416,000 possible tours. Hence, though a brute force approach works, we may not live long enough to see the end of the algorithm.

[4]We are referencing specifically the algorithm for "solving" the TSP and not the unrelated algorithm in Machine Learning.

- *An Introduction to Algorithms,* 3rd *edition*, Thomas H. Cormen, Charles E. Leiserson, Ronald L. Rivest, and Clifford Stein; MIT Press (2009). (This is the go-to text for many experts when it comes to the study of algorithms.)

- *Graphs, Algorithms, and Optimization,* 2nd *edition*, William Kocay, Donald L. Kreher, CRC Press (2017)

- *Mathematical Programming An Introduction to Optimization,* Melvyn Jeter, CRC Press (1986)

- *The Mathematics of Nonlinear Programming,* A.L. Peressini, F.E. Sullivan, J.J. Uhl Jr., Springer (1991)

This list is not, of course, an exhaustive list of excellent sources for a general overview of optimization.

2.7 Keywords

(strict) global or absolute maximizer/minimizer, (strict) local or relative maximizer/minimizer, maximum, minimum, infimum, supremum, solution to an optimization problem, algorithm, heuristic, runtime, (computational) complexity.

2.8 Exercises

Exercise 2.1. *State the maximum, minimum, infimum, and supremum (if they exist) of each of the following sets:*

i) $A = \{8, 6, 7, 5, 3, 0, 9\}$,
ii) $B = [a, b)$, *where* $a, b \in \mathbb{R}$,
iii) $C =$ *the range of* $f(x) = 1/(1 - x)$, *where* $x \neq 1$,
iv) $D =$ *the range of* $g(x) = 1/(1 - x)^2$, *where* $x \neq 1$,
v) $E = \{1 + \frac{(-1)^n}{n}\}$, *where* n *is a positive integer,*
vi) $F =$ *the set of prime numbers.*

Exercise 2.2. *Let* $f : \mathbb{R}^n \to \mathbb{R}$ *and* $\mathbf{x}^* = \langle x_1^*, \ldots, x_n^* \rangle \in \mathbb{R}^n$. *Show* $f(\mathbf{x}^*)$ *is a maximum of* f *if and only if* $-f(\mathbf{x}^*)$ *is a minimum of* $-f$.

Exercise 2.3. *Suppose* s_1 *and* s_2 *are suprema of some set* $S \subset \mathbb{R}$. *Prove* $s_1 = s_2$, *thus establishing that the supremum of a set is unique (obviously, a very similar proof shows that, if it exists, the infimum of a set is also unique).*

Exercise 2.4. *Show if* $S \subset \mathbb{R}$ *is a nonempty, closed, and bounded set, then* $\sup(S)$ *and* $\inf(S)$ *both belong to* S.

Exercise 2.5. *Let* $f : (0, \infty) \to \mathbb{R}$ *by* $x \mapsto \ln x$ *(i.e.,* $f(x) = \ln x$*). Prove that* f *is monotonically increasing continuous bijection. [Note: this exercise assumes some familiarity with topics in Calculus/Elementary Real Analysis.]*

Exercise 2.6. *Let* f *be a positive-valued function; that is, its image is a subset of* $(0, \infty)$ *with* $D(f) \subseteq \mathbb{R}$. *Prove that* $f(x^*) = \max_{x \in D(f)}\{f(x)\}$ *if and only if* $\ln f(x^*) = \max_{x \in D(f)}\{\ln f(x)\}$ *(i.e. the max of* f *and* $\ln f$ *occur at the same locations). You may use Exercise 2.5.*

3

Computational Complexity

3.1 Arithmetic Complexity

Many optimization techniques rely on algorithms – especially using a computer – and, as such, it is natural to consider how long an iterative process may take. As processors' speeds vary and technology improves, time is not a good candidate for a metric on "how long". Instead, one approach is to count the number of operations necessary to complete the algorithm. This is illustrated in the following example:

Consider using Gauss-Jordan elimination to solve a system of n equations in n variables as in the following 3×3 case with variables x_1, x_2, and x_3 (for a refresher on this technique, see Section 4.2).

$$\left[\begin{array}{ccc|c} -2 & 3 & 5 & 7 \\ 4 & -3 & -8 & -14 \\ 6 & 0 & -7 & -15 \end{array}\right] \xrightarrow[3R_1+R_3 \to R_3]{2R_1+R_2 \to R_2} \left[\begin{array}{ccc|c} -2 & 3 & 5 & 7 \\ 0 & 3 & 2 & 0 \\ 0 & 9 & 8 & 6 \end{array}\right] \tag{3.1}$$

$$\xrightarrow{-3R_2+R_3 \to R_3} \left[\begin{array}{ccc|c} -2 & 3 & 5 & 7 \\ 0 & 3 & 2 & 0 \\ 0 & 0 & 2 & 6 \end{array}\right] \tag{3.2}$$

We may continue row operations to get the matrix in reduced row echelon form, but this is computationally expensive[1], so we instead back substitute:

$$2x_3 = 6, \text{ so } x_3 = 3; \tag{3.3}$$

$$3x_2 + 2(3) = 0, \text{ thus } x_2 = -2; \text{ and} \tag{3.4}$$

$$-2x_1 + 3(-2) + 5(3) = 7, \text{ hence } x_1 = 1. \tag{3.5}$$

The process of reducing the matrix but stopping short of reaching reduced row echelon form and using back substitution is usually referred to as *Gaussian elimination*.

[1] A good lesson to carry with us as we explore the topics in this text is that it is not always best for a computer to do a problem the same way you and I would solve it on paper. This matter is briefly discussed at the beginning of Section 5.

DOI: 10.1201/9780367425517-3

Counting operations at each step of the Gaussian elimination in our example on 3 variables we have:

step	multiplications	additions	process
3.1	$2(3+1)$	$2(3+1)$	elimination
3.2	$1(2+1)$	$1(2+1)$	elimination
3.3	1	0	back substitution
3.4	2	1	back substitution
3.5	3	2	back substitution

If we consider a system with n variables, then the total number of operations in Gaussian elimination, $G(n)$, is

$$\begin{aligned}
G(n) &:= \#\text{ operations} \\
&= \#\text{ elim mult} + \#\text{ elim add} + \#\text{ back sub mult} + \#\text{ back sub add} \qquad (3.6)\\
&= \sum_{i=1}^{n}(i-1)(i+1) + \sum_{i=1}^{n}(i-1)(i+1) + \sum_{i=1}^{n} i + \sum_{i=1}^{n}(i-1) \qquad (3.7)\\
&= 2\sum_{i=1}^{n}(i^2-1) + 2\sum_{i=1}^{n} i - \sum_{i=1}^{n} 1 \qquad (3.8)\\
&= 2\frac{n(n+1)(2n+1)}{6} - 2n + \frac{n(n+1)}{2} - n \qquad (3.9)\\
&= \frac{(4n^3+6n^2+2n) - 12n + (3n^2+3n) - 6n}{6} \qquad (3.10)\\
&= \frac{4n^3+9n^2-13n}{6} \text{ or roughly } \frac{2n^3}{3}. \qquad (3.11)
\end{aligned}$$

For growing values of n we have

n	# operations	$\frac{2}{3}n^3$	% error
1	0	0.666666	33.3333
2	7	5.333333	23.8095
3	25	18.000000	28.0000
4	58	42.666666	26.4367
5	110	83.333333	24.2424
10	795	666.666666	16.1425
20	5890	5333.333333	9.4510
30	19285	18000.000000	6.6632
40	44980	42666.666666	5.1430
50	86975	83333.333333	4.1870
100	681450	666666.666666	2.1693
500	83707250	83333333.333333	0.4466
10^3	666816645000	666666666.666666	0.2241
10^4	666816645000	666666666666.666666	0.0224
10^5	666681666450000	666666666666666.666666	0.0022
10^6	666668166664500000	666666666666666666.6666	0.0002

The polynomial in 3.11 gives the number of arithmetic operations required in n variables with n unknowns. As we see in the table, as n grows large $\frac{4n^3+9n^2-13n}{6}$ is approximated nicely by $\frac{2n^3}{3}$ thus we say that the *arithmetic complexity* of Gaussian elimination is of the order $\frac{2n^3}{3}$.

It is important to note that there is much more a computer is doing than just arithmetic operations when it does a calculation. One very important process we have ignored in our example is calls to memory and these can be very expensive. Including all that is involved in memory makes a runtime assessment much more difficult and often this component is ignored. The purpose of these calculations is not to get a precise measurement of how long it will take a computer to complete an algorithm, but rather to get a rough idea of all that is involved so that we can compare the algorithm to other techniques that accomplish the same task and therfore have some means to compare which is more efficient. A thorough treatment of all this can be found in the excellent textbook [11].

3.2 Asymptotic Notation

As we saw in the example in the previous section, $\frac{2n^3}{3}$ is a very good approximation for $\frac{4n^3+9n^2-n}{6}$ as n grows large. This agrees with our intuition that as n gets big, the only term that really matters in the polynomial is the leading term. This idea is encapsulated in *asymptotic notation* (think of "asymptotic" as a synonym for "long-run behavior") and, in particular for our purposes, *big O notation.*

Definition 3.2.1 (Big O Notation). *Let g be a real-valued function (though this definition also holds for complex-valued functions). Then*

$$O(g) := \{f \mid \textit{there exist positive constants } C, N \textit{such that } 0 \leq |f(x)| \leq C|g(x)| \textit{ for all } x \geq N\}.$$

Thus $O(g)$ is a family of functions $\mathscr{F}$ for which a constant times $|g(x)|$ is eventually an upper bound for all $f \in \mathscr{F}$. More formally, f being $O(g)$ means that as long as $g(x) \neq 0$ and the limit exists, $\lim_{x\to\infty} |f(x)/g(x)| = C$ or 0 (if g is too big) where C is some positive constant.

Example 3.2.2. *Show that $\frac{4n^3+9n^2-n}{6}$ is $O(\frac{2n^3}{3})$ where $n \in \mathbf{N}$.*

Solution. For $n \geq 1$ (thus $N = 1$),

$$\left|\frac{4n^3 + 9n^2 - n}{6}\right|$$

$$\leq \left|\frac{4n^3}{6}\right| + \left|\frac{9n^2}{6}\right| + \left|\frac{n}{6}\right| \text{ by the Triangle Inequality (Theorem B.2.3)} \tag{3.12}$$

$$\leq \frac{4n^3}{6} + \frac{9n^3}{6} + \frac{n^3}{6} \text{ since } n \geq 1 \tag{3.13}$$

$$= \frac{14n^3}{6} \tag{3.14}$$

$$= \frac{7}{2} \cdot \left|\frac{2n^3}{3}\right| \tag{3.15}$$

establishing $\frac{4n^3+9n^2-n}{6}$ is $O(\frac{2n^3}{3})$ where $C = 7/2$. Notice that

$$\lim_{n\to\infty} \left(\frac{4n^3 + 9n^2 - n}{6}\right) / \left(\frac{2n^3}{3}\right) = 1 \tag{3.16}$$

■

Regarding our work in Example 3.2.2 one usually would not include the constant 2/3 but rather report the answer as $\frac{4n^3+9n^2-n}{6}$ is $O(n^3)$. This is, of course, because the constant does not matter in big O notation. Some Numerical Analysis texts use this problem as an example, though, and include the constant 2/3 when reporting the approximate number of [2]. We have kept the constant to be consistent with those texts and though technically correct, including the constant in the $O(\cdot)$ can viewed as bad form.

It is important to realize that Big O notation gives an asymptotic (long run) upper bound on a function. It should also be noted that when using big O notation, $O(g(x))$ is a set and, as such, one should write "$h(x) \in O(g(x))$". Note, though, that it is standard practice to abuse the notation and state "$h(x) = O(g(x))$" or "$h(x)$ is $O(g(x))$".

One further observation before some examples. We have shown that the number of arithmetic operations in performing Gaussian elimination to solve a linear system in n variables is $G(n) = \frac{4n^3+9n^2-n}{6}$ and that this function is $O(\frac{2}{3}n^2)$. Furthering our work in Example 3.2.2 by picking up in 3.15 we have

$$\frac{7}{2}\left|\frac{2n^3}{3}\right| = \frac{7}{3}n^3 \tag{3.17}$$

$$< n^4 \text{ for } n \geq 3. \tag{3.18}$$

[2]The reason for this is that using *Cholesky Decomposition* (Chapter 5) to solve a system of linear equations is $O(\frac{n^3}{3})$; i.e. twice as fast as Gaussian elimination.

Thus, not only is $G(n) = O(\frac{2}{3}n^3)$, $G(n) = O(n^4)$. In fact,

Observation 3.2.3. *Let x be a positive real number and suppose $f(x)$ is $O(x^k)$. Then for any $l > k$, $f(x)$ is $O(x^l)$.*

We now consider a few more important examples before moving on.

Example 3.2.4. *Let $n \in \mathbf{Z}^+$. Then*

$$1 + 2 + 3 + \cdots + n \leq \overbrace{n + n + n + \cdots + n}^{n \text{ terms}} = n^2 \tag{3.19}$$

and taking $C = N = 1$ we see that the sum of the first n positive integers is $O(n^2)$.

Example 3.2.5. *Let $n \in \mathbf{Z}^+$. Then*

$$n! := n(n-1)(n-2)\cdots 3 \cdot 2 \cdot 1 \leq \overbrace{n \cdot n \cdot n \cdot \cdots \cdot n}^{n \text{ factors}} = n^n \tag{3.20}$$

and taking $C = N = 1$ we see that $n!$ is $O(n^n)$.

Example 3.2.6. *Show that for $n \in \mathbf{N}$, $f(n) = n^{k+1}$ is not $O(n^k)$ for any nonnegative integer k.*

Solution. Let us assume for contradiction that the statement is true, namely there exist positive constants C and N such that $n^{k+1} \leq Cn^k$ for all $n \geq N$. Thus for all $n \geq N$, $n \leq C$, which is absurd, since n is a natural number and therefore unbounded. Thus n^{k+1} cannot be $O(n^k)$. ∎

Growth of basic "orders" of functions are shown in Figure 3.1. Note that though $n!$ is defined for nonnegative integers, Exercises 9.1 and 9.2 show how to extend the factorial function to the real numbers.

Before concluding this section, we mention that other asymptotic notation exists, for example: $o(g(x))$, $\omega(g(x))$, $\Omega(g(x))$, $\Theta(g(x))$, etc., but we do not consider them here[3].

3.3 Intractability

Some problems we will encounter will have solutions that can be reached in theory, but take too much time in practice, are said to be *intractable*. Conversely, any problem that can be solved in practice is said to be *tractable*; that is, "easily worked" or "easily handled or controlled" [16]. The bounds between

[3]The interested reader is encouraged to read the appropriate sections in [11] or [48].

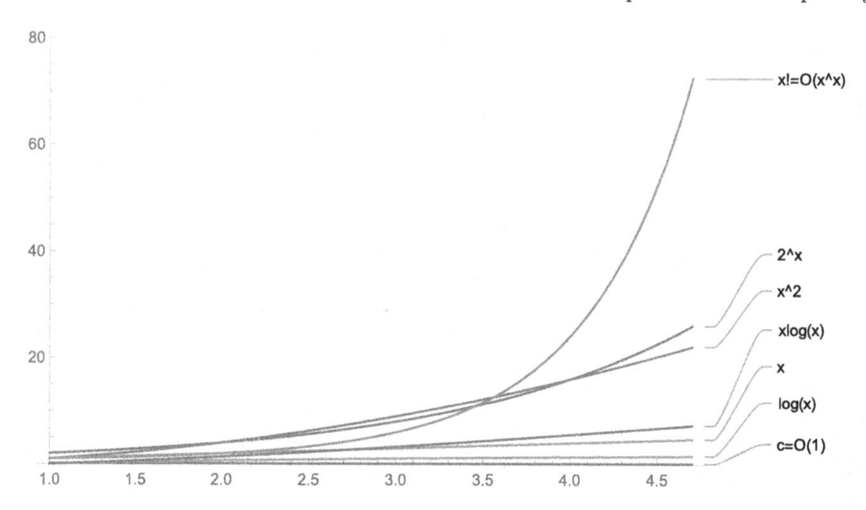

FIGURE 3.1
The growth of functions.

these two are not clearly defined and depend on the situation. Though the discipline lacks a precise definition of both tractable and intractable, their usage is standard and necessary for many situations encountered in Optimization.

For an example, let us revisit using Gauss-Jordan elimination as was considered in Section 3.1. Oak Ridge National Laboratory unveiled in 2018 its supercomputer *Summit* capable of 122.3 petaflops (122.3×10^{15}) calculations per second. Without worrying about the details, let us assume a good PC can do $100{,}000{,}000{,}000 = 10^{11}$ calculations per second (this is a little generous). By our work in Section 3.1, to perform Gaussian elimination on a matrix with 10^6 rows (i.e. a system of equations with 10^6 variables) it would take Summit

$$\frac{2}{3}(10^6)^3/122.3 \times 10^{15} \approx 5.45 \text{ seconds.} \tag{3.21}$$

On a good PC this would take

$$\left(\frac{2}{3}(10^6)^3/10^{11}\right)/86400 \text{ seconds per day } \approx 77 \text{ days.} \tag{3.22}$$

Note that these calculation are not exact as we have not consider calls to memory, etc., but they do illustrate the point.

Spending 77 days to solve a problem is a nuisance, but not an insurmountable situation. Depending on the practice one would have to decide if this amount of time makes the problem intractable or not. But Gauss-Jordan elimination is a polynomial time algorithm, so to better illustrate this point let us now assume we have a program that runs in *exponential time*; say one that has as its runtime 2^n. For $n = 100$ (considerably less than 1,000,000) this program would run on Summit for

$$\left(2^{100}/122.3 \times 10^{15}\right)/31536000 \text{ seconds per year } \approx 3.3 \text{ years!} \tag{3.23}$$

We will not even consider how long this would take on a good PC. If we consider a system with $n = 10^3$ variables, the runtime on Summit becomes 2.7×10^{276} years which is 2×10^{266} times the age of the universe.

We will close this section by noting that most computer security depends on intractability. The most used public key encryption scheme is known as RSA encryption. This encryption scheme uses a 400 digit number that is known to be the product of two primes and it works well since currently factoring algorithms for a number this large are intractable[4]. The intractability of this problem will most likely change with quantum computing.

3.4 Complexity Classes

3.4.1 Introduction

It would be helpful to have a metric by which to classify the computational difficulty of a problem. One such tool is the complexity class. A *complexity class* is a set of problems that can be solved using some model of computation (often this model of computation is given a limited amount of space, time, or some other resource to work out the problem). A *model of computation* is any method of computing an output given an input.

One of the most useful models of computation is the Turing machine. For the sake of our discussion we can consider a "Turing machine" as any algorithm that follows the following method:

0. Initialize an infinite string of "blank symbols". Each symbol can be referred to by its position, starting at position 0. This string of symbols is called the *tape* of the Turing machine.

1. Write a given finite string of symbols, s, to the tape starting at position 0. Each symbol in s must be taken from a given finite set of symbols, A, known as the *alphabet* of the Turing machine. A contains the blank symbol. We often restrict s so that it cannot contain the blank symbol.

2. Choose an arbitrary non-negative integer k and read the kth symbol of the tape. This step is referred to as "moving the head of the Turing Machine to position k".

[4]The nature of primes is also involved here. Although they are the building blocks of all integers, we know little about them, especially how many there are in (large) intervals and where they reside.

3. Change the kth symbol of the tape to any symbol in A.
4. Go to step 2 or step 5.
5. Declare either "accept" or "reject" (This is referred to as "accepting or rejecting the input string".) A Turing machine that reaches this step is said to *halt*.

When we discuss Turing machines we usually just assume step 0 has already occurred but its important to note that Turing machines in concept have an infinite amount of space on the tape to store information and that they look up information by moving the head to its position on the tape. For any position k on the tape we say that position $k - n$ is "n to the left" of position k and the position $k + n$ is "n to the right" of position k. The following algorithm is an example of a Turing Machine that we will call TM:

1. Write a given string of symbols, s, to the beginning of the tape where every symbol of s is in the set $\{0, 1\}$.
2. Move the head to the first blank after s. (We will call this position on the tape p).
3. Write the symbol 0 to position p.
4. Move the head to position 0.
5. If the symbol at the head's position is 1, move the head to position p and change the symbol at position p to a 1 if it is a 0 and to a 0 if it is a 1, then move the head back to where it was when we started this step.
6. Move the head one symbol to the right.
7. If the head is at position p, go to step 8. Otherwise go to step 5.
8. Read the symbol at position p. If it is 1, accept. If it is 0, reject.

TM is a simple Turing machine that accepts all strings that contain an odd number of 1 symbols and rejects all strings that do not. Hence, TM solves a decision problem. A *decision problem* is a problem with an answer than is either "yes" or "no". TM solves or *decides* the decision problem "Does the given string s have an odd number of 1 symbols?" When we say that TM "solves" this problem, we mean:

1. If L is the set of all strings that contain an odd number of 1 symbols, TM will accept a string s if and only if $s \in L$. (This is equivalent to saying that "L is the language of TM")
2. TM halts on every string. (It always reaches step 5 in the original given Turing machine definition).

Note that the second point is not trivial because Turing machines can get stuck in an infinite loop of moving the head and writing symbols. Now that we have a working definition of a Turing machine, we will explore how complexity classes are defined using Turing machines as their model of computation.

3.4.2 Time, Space, and Big O Notation

As previously mentioned, complexity classes usually give their chosen model of computation a limited resource. When defining complexity classes involving Turing Machines we often limit either the amount of *time* or *space* the Turing machine can use to solve the problem. *Time* refers to the amount of head-moving[5] and writing steps the Turing machine can make and *space* refers to the number of positions after the given input string on the tape the Turing machine may use.

In general, we are concerned with how Turing machines run on all inputs they can take. This presents a problem – because how much space or time a Turing machine uses to run can significantly vary based on the size of the input it is given. It is therefore usually the case that we do not place problems in complexity classes based on the absolute amount of time or space a Turing machine that solves the problems will take. Instead we allow Turing machines to use an asymptotically increasing amount of resources relative to the size of their inputs.

As we are thinking asymptotically, it is helpful to use Big O notation when defining complexity classes. Consider a Turing machine T and $h(n)$, where $h(n)$ is a function that maps the integer n to the greatest amount of read and write steps T will take on an input string of size n. If $h(n) \in O(f(n))$ we say that "T runs in $O(f)$ time".

3.4.3 The Complexity Class P

P is the set of all decision problems that can be solved in polynomial time by a Turing machine. More formally, a set of strings L is in P if and only if there exists a Turing machine whose language is L that runs in $O(n^k)$ time where k is some integer (Note that this means P is a set of languages. Not a set of Turing machines). P is sometimes thought of as the set of decision problems whose solutions are tractable. This is not the whole truth – some problems can be bounded above by large polynomials and be in P and some problems can not be asymptotically bound by a polynomial but still be solvable for reasonably sized inputs – but it is very often the case that problems in P are tractable and problems out of P with large inputs are not.

[5]In our given definition of a Turing machine we conceptualized moving the head of the Turing machine to an arbitrary position as "one step" in the algorithm. However – when counting the number of operations a Turing machine takes if a Turing machine moves from position p_1 to position p_2 the Turing machine has taken $|p_1 - p_2|$ steps.

For this reason polynomial time algorithms – especially algorithms that run in $O(n \log n)$ time or quicker – are often the goal when attempting to optimize the speed at which a decision problem is solved by a Turing machine. Though finding such algorithms for some problems is often elusive and sometimes impossible. That being said many practical problems exist in P. Determining whether or not there is a path of length n or less between two cities when given the set of roads between them, determining whether or not an item is on a list, determining whether two integers are coprime, and determining whether an integer is prime are all decision problems that exist in P. (The input strings for these problems are string representations of the integers, paths between cities, etc).

3.4.4 The Complexity Class NP

NP is a complexity class that contains all languages L for which if $s \in L$ there exists a proof that can be verified in polynomial time that s is in L[6]. More precisely, A language L is in NP if there exists a Turing machine, $TM(s, c)$[7] for which the following properties hold:

1. TM always runs in $O(n^k)$ time on any input s, c where n is the length of s and k is some integer.

2. If $s \in L$ – there exists a c such that $TM(s, c)$ accepts. (We say that c is the *certificate* of s).

3. if $s \notin L$ – no c exists such that $TM(s, c)$ accepts.

An example may be enlightening. Consider the subset-sum problem: "Given an integer n and a set of integers S, is there some subset of S that adds to n?" In this instance the certificate that can be verified in polynomial time is a subset of S that adds to n. If given S, n, and a subset of S that adds to n, a deterministic Turing machine could add the given subset and accept if and only if it adds to n. Such a subset will always exist if $(S, n) \in L$ and because we only accept if the subset of S adds to n – this Turing machine will never be "tricked" into accepting a string that is not in L (i.e. the third above premise holds). Hence, if $(S, n) \in L$ there is a proof that can be verified in polynomial time that (S, n) is in L.

NP is an interesting complexity class because many important optimization problems exist in NP. Whether or not a graph is n-colorable (see Chapter 21), determining whether or not there is a Hamiltonian cycle of weight less than n (also introduced in Chapter 21), whether or not a set of Boolean clauses

---

[6]NP is short for "non-deterministic polynomial" which is a description of a more generalized version of a Turing machine. For more information check out Chapter 2 of Computational complexity: a modern approach in the further reading section.

[7]Because Turing machines can take arbitrary strings as arguments – it is conventional to use the notation $TM(s_1, s_2, ...s_n)$ if drawing a distinction between different parts of the input is important.

can be satisfied, determining whether a set of flights can be scheduled without any conflicts, and determining whether k varying sized objects can be packed into n bins are all problems in NP[8]. NP is also mysterious in that it is unknown whether or not $P = NP$ – which means it is possible that there is are efficient algorithms for solving the toughest problems in NP that have not yet been discovered. Much effort has been dedicated to trying to find efficient algorithms for solving NP problems and there are a wide variety of heuristics for doing so that work well but not perfectly. For an example of a heuristic, see the *Nearest Neighbor Algorithm* or the insertion algorithms for solving the *Traveling Salesperson Problem* (Chapter 25).

3.4.5 Utility of Complexity Classes

Complexity classes are a useful tool for classifying the computational difficulty of solving a problem. It is almost universally the case that if a problem is in $EXPTIME$ (the class of problems solvable by Turing machines in exponential time) and not in P – that problem will be more difficult than one in P as inputs get large. If we are trying to determine the computational difficulty of a problem it may therefore be helpful to discover information about which complexity class it belongs to. In particular it may be useful to determine if the problem is in P, NP, or neither.

3.5 For Further Study

Parts of these texts have excellent presentations on what we have considered and can be used to deepen ones understanding of the material presented in this chapter:

- *An Introduction to Algorithms,* 3rd edition, Thomas H. Cormen, Charles E. Leiserson, Ronald L. Rivest, and Clifford Stein; MIT Press (2009). (This is the go-to text for many experts when it comes to the study of algorithms.)
- *Discrete Mathematics and Its Applications,* 8th edition by Kenneth Rosen, McGraw-Hill (2019)
- *Concrete Mathematics a Foundation for Computer Science,* 2nd edition by Ronald Graham, Donald Knuth, and Oren Patashnik, Addison Wesley (1994)

[8]These problems are *NP-complete* meaning that all problems in NP are less difficult than they are. For a more thorough discussion of what that means – see chapter 2 of *Computational complexity: A Modern Approach* listed in the further study section of this chapter.

This list is not, of course, an exhaustive list of excellent sources for a general overview of optimization.

3.6 Keywords

Gauss-Jordan elimination, Gaussian elimination, arithmetic complexity, asymptotic notation, big O notation, intractability, complexity class, model of computation, Turing machine, P, NP.

3.7 Exercises

Exercise 3.1. *In this chapter, we determined that the number of arithmetic operations to perform Gaussian elimination to solve a system in n variables is $G(n) = \frac{4n^3+9n^2-13n}{6} = O(\frac{2}{3}n^3)$. Consider a linear system with 412 variables.*

i) What is the exact number of arithmetic operations to use Guassian elimination to solve this system?
ii) Calculate the value of the order $\frac{2}{3}n^3$ for this system and determine the percent error of this approximation of $G(n)$.
iii) What is the first value of n for which $O(\frac{2}{3}n^3)$ has less than a 1% error in approximating $G(n)$?

Exercise 3.2. *Show that $x^2 + x + 1$ is $O(x^2)$ but not $O(x)$.*

Exercise 3.3. *Let n, k be positive integers. Show $1^k + 2^k + \cdots + n^k$ is $O(n^{k+1})$.*

Exercise 3.4. *Prove Observation 3.2.3.*

Exercise 3.5. *Construct a Turing machine whose alphabet is $\{0, 1\}$ that has the language of all strings that contain more 1s than 0s.*

Exercise 3.6. *Construct a Turing machine whose alphabet is the set of integers between 0 and 100 that sorts the integers in the given string s in ascending order, then accepts.*

Exercise 3.7. *Prove that if L is a language with a finite amount of strings then $L \in P$.*

Exercise 3.8. coP *is the class of decision problems for which if $L \in P$ then $\bar{L} \in coP$. Prove that $P = coP$.*

Exercise 3.9. *Prove that $P \subseteq NP$.*

4

Algebra Review

> *"For the things of this world cannot be made known without a knowledge of mathematics."*
>
> – Roger Bacon

> *"Here arises a puzzle that has disturbed scientists of all periods. How can it be that mathematics, being after all a product of human thought which is independent of experience, is so admirably appropriate to the objects of reality? Is human reason, then, without experience, merely by taking thought, able to fathom the properties of real things?"*
>
> – Albert Einstein

Many techniques used in this text will require some knowledge of various levels of algebra, both elementary algebra and linear algebra. Linear algebra is at the heart of machine learning. This chapter provides a refresher of relevant algebraic techniques and may either be read carefully, skimmed, or visited as needed.

4.1 Systems of Linear Inequalities in Two Variables – Geometric Solutions

The very first Optimization topic we will study will require an understanding of solving systems of linear inequalities. Before we look at those, though, let us make sure we understand the terminology.

Recall that a *linear function* in n variables is a function such that the maximum total degree of each term is 1. The *graph* of a linear function

$$y = c_0 + c_1 x_1 + \cdots + c_n x_n \tag{4.1}$$

in n variables is a hyperplane in $\mathbb{R}^{n+1}$. A point in $\mathbb{R}^{n+1}$ is on this hyperplane if and only if its coordinates satisfy 4.1. Of course, in $\mathbb{R}^2$, this hyperplane is a line and, in $\mathbb{R}^3$, the this hyperplane is a plane.

A *linear inequality* is merely a statement like 4.1 but with the "$=$" replaced by a "$<$","$>$","$\leq$", or "$\geq$". A linear inequality in one variable has as its graph

DOI: 10.1201/9780367425517-4

either a line (if "$\leq$" or "$\geq$") or dotted line (if "$<$" or "$>$") together with the corresponding half-plane (we will see examples soon). Analogous statements hold for hyperplanes and half-spaces in higher dimensions; that is, functions involving more than one variable. A *system of linear inequalities* is a collection of linear inequalities.

The *solution set* to a system of linear inequalities is the collection of all points that satisfy each inequality in the system. Often this is most easily represented by a graph. A solution set is said to be *bounded*[1] if the entire region can be enclosed in a circle in the two dimensional case or a ball in the general case. If the region cannot be entirely enclosed in an arbitrary circle, then the region is said to be *unbounded.* As well, we note that a point in the solution region that is the intersection of two boundary lines is called a *corner point* or *extreme point of a solution set.* When the solution set is for a system of linear inequalities in a linear programming problem (Chapter 6), the solution set is called a *feasible region* and its corner points are, in this context, often referred to as *vertices of the feasible region.* The significance of the corner points of feasible regions will be discussed later.

4.1.1 Examples

We will solve each of the following systems of linear inequalities graphically and state whether the system's solution region is bounded or unbounded. By the *graph of a system of inequalities* we mean "the collection of points that satisfy all the stated inequalities".

Example 4.1.1.

$$2x + y \leq 10 \tag{4.2}$$
$$x + 2y \leq 8 \tag{4.3}$$
$$x \geq 0 \tag{4.4}$$
$$y \geq 0 \tag{4.5}$$

Example 4.1.2.

$$2x + y > 10 \tag{4.6}$$
$$x + 2y \geq 8 \tag{4.7}$$
$$x \geq 0 \tag{4.8}$$
$$y \geq 0 \tag{4.9}$$

[1]A precise mathematical definition of *bounded* exists, but we will be comfortable in this text with some informal definitions.

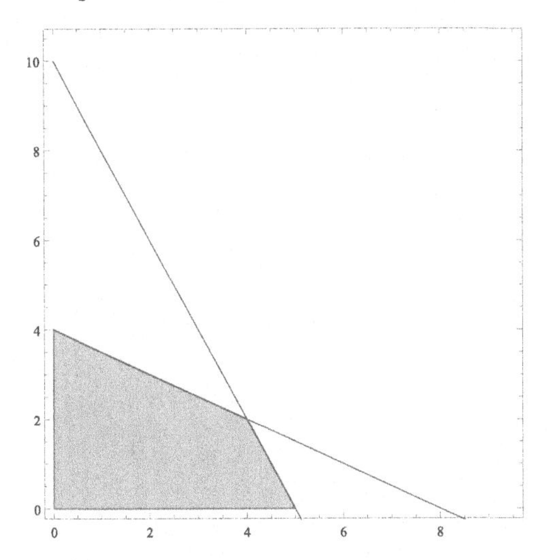

FIGURE 4.1
Example 4.1.1 – bounded solution set.

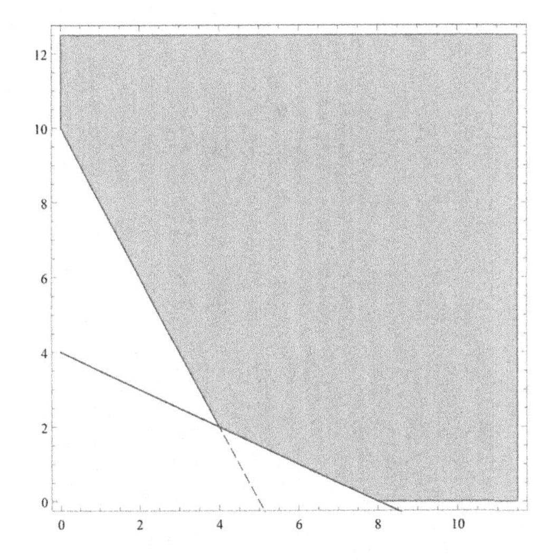

FIGURE 4.2
Example 4.1.2 – unbounded solution set.

Example 4.1.3.

$$x + 4y \leq 32 \tag{4.10}$$

$$3x + y \leq 30 \tag{4.11}$$

$$4x + 5y \geq 51 \tag{4.12}$$

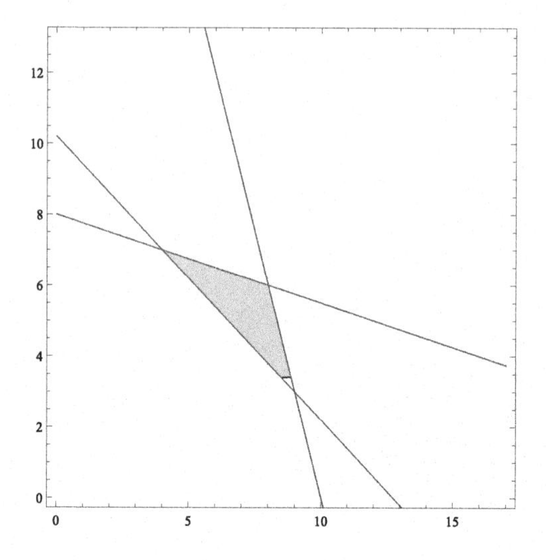

FIGURE 4.3
Example 4.1.3 – bounded solution set.

Example 4.1.4.

$$4x + 3y \leq 48 \tag{4.13}$$
$$2x + y \geq 24 \tag{4.14}$$
$$x \leq 9 \tag{4.15}$$

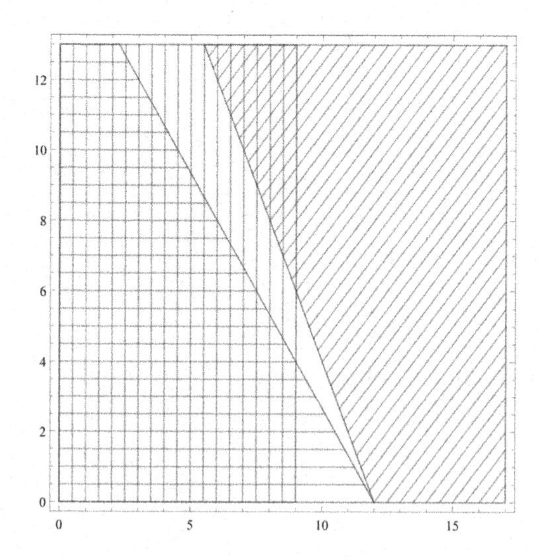

FIGURE 4.4
Example 4.1.4 – empty solution set.

We note that in each of these examples the corner points can be found by solving algebraically the system of equations corresponding to the intersecting lines. For instance, to find the corner point in Example 4.1.1 that is not on either the x or y axis, we solve

$$\begin{cases} 2x + y &= 10 \\ x + 2y &= 8. \end{cases} \tag{4.16}$$

The corner point is $(4, 2)$.

4.2 Solving Systems of Linear Equations Using Linear Algebra

4.2.1 Gauss-Jordan Elimination

Solving a system of linear equations via geometry has the advantage that we have a visual representation of what is taking place. The disadvantage is that the graph may be too complicated to get an understanding of what is going on if there are too many equations. Worse, it may be impossible to draw as is the case of a system with functions having three or more variables.

Suppose we are faced with solving the system

$$\begin{cases} 3x_1 - 5x_2 + x_3 &= 18 \\ -2x_1 + x_2 - x_3 &= 5 \\ x_1 - x_2 + 3x_3 &= -4. \end{cases}$$

We could pick one equation, isolate a variable, then substitute for that variable in the remaining equations. This reduces the system to two equations in two unknowns and then we could go from there, but that is too much work. Instead, we can combine equations to eliminate variables. For example, we can

- multiply equation 3 by -3 and add it to equation 1 and
- multiply equation 3 by 2 and add it to equation 2

to get

$$\begin{cases} -2x_2 - 8x_3 &= 30 \\ -x_2 + 5x_3 &= -3 \\ x_1 - x_2 + 3x_3 &= -4 \end{cases}$$

Let us now

- multiply equation 2 by -1 and
- interchange equations 1 and 3:

$$\begin{cases} x_1 - x_2 + 3x_3 &= -4 \\ 1x_2 - 5x_3 &= 3 \\ -2x_2 - 8x_3 &= 30 \end{cases}$$

Lastly,

- add equation 2 to equation 1 and
- multiply equation 2 by 2 and add to equation 3

to get

$$\begin{cases} x_1 & & -2x_3 & = & -1 \\ & x_2 & -5x_3 & = & 3 \\ & & -18x_3 & = & 36. \end{cases}$$

The third equation gives $x_3 = -2$ and we may use this to *back substitute* into equation 2 to find $x_2 = -7$ and into equation 1 to find $x_1 = -5$. We note that our legal algebraic operations on the equations[2] were:

Observation 4.2.1 (Legal Operations on Systems of Equations).

- *multiply an equation by a nonzero constant,*
- *interchange two equations, and*
- *add a constant times an equation to another equation.*

This is a perfectly legitimate method, but it is too much work and we may want a simpler way to express what we are doing so that a computer can do it for us. Hence we instead introduce the *augmented matrix*:

$$\left[\begin{array}{rrr|r} 3 & -5 & 1 & 18 \\ -2 & 1 & -1 & 5 \\ 1 & -1 & 3 & -4 \end{array}\right]$$

and perform *Gauss-Jordan elimination* (row operations) to get the matrix in reduced form. The legal row operations[3] are called *elementary row operations* and correspond exactly to the algebraic operations on equations used above and appearing in Observation 4.2.1.

Theorem 4.2.2 (Elementary Row Operations). *When using Gauss-Jordan elimination, the row operations that do not change the solution set are*

- *multiply a row by a nonzero constant,*
- *interchange two rows, and*
- *add a constant times a row to another row.*

Note that these row operations, called *elementary row operations*, correspond exactly to the algebraic operations on equations listed previously.

[2]In a typical high school algebra class no explanation is provided as to why these operations are "legal" or what even *legal* means. A little Linear Algebra will soon provide the answers.

[3]The reasons why these are the legal operations will be clear after Theorems 4.3.48 and 4.3.50.

So we can conveniently rewrite our example as

$$\left[\begin{array}{rrr|r} 3 & -5 & 1 & 18 \\ -2 & 1 & -1 & 5 \\ 1 & -1 & 3 & -4 \end{array}\right] \xrightarrow[2R_3+R_2\to R_2]{-3R_2+R_1\to R_1} \left[\begin{array}{rrr|r} 0 & -2 & -8 & 30 \\ 0 & -1 & 5 & -3 \\ 1 & -1 & 3 & -4 \end{array}\right] \tag{4.17}$$

$$\xrightarrow[-R_2\to R_2]{R_1\leftrightarrow R_3} \left[\begin{array}{rrr|r} 1 & -1 & 3 & -4 \\ 0 & 1 & -5 & 3 \\ 0 & -2 & -8 & 30 \end{array}\right] \tag{4.18}$$

$$\xrightarrow[2R_2+R_3\to R_3]{R_2+R_1\leftrightarrow R_1} \left[\begin{array}{rrr|r} 1 & 0 & -2 & -1 \\ 0 & 1 & -5 & 3 \\ 0 & 0 & -18 & 36 \end{array}\right] \tag{4.19}$$

Certainly, we can back substitute (see the next section) from here, but two more steps will nicely put the matrix in *reduced row echelon* form (i.e. at most 1 nonzero entry, a 1, in each column of the left hand side of $|$):

$$\xrightarrow{-\frac{1}{18}R_3\to R_3} \left[\begin{array}{rrr|r} 1 & 0 & -2 & -1 \\ 0 & 1 & -5 & 3 \\ 0 & 0 & 1 & -2 \end{array}\right] \tag{4.20}$$

$$\xrightarrow[2R_3+R_1\to R_1]{5R_3+R_2\to R_2} \left[\begin{array}{rrr|r} 1 & 0 & 0 & -5 \\ 0 & 1 & 0 & -7 \\ 0 & 0 & 1 & -2 \end{array}\right] \tag{4.21}$$

and we now easily see the solution is $x_1 = -5, x_2 = -7, x_3 = -2$.

4.2.2 Gaussian Elimination Compared with Gauss-Jordan Elimination

The previous section introduced the important technique *Gauss-Jordan elimination*. We now introduce a truncated version of this technique which is referred to as *Gaussian elimination*. Gaussian elimination has the advantage of being less computationally expensive than Gauss-Jordan elimination[4]. First, some necessary definitions.

Definition 4.2.3 (Row Echelon Form). *A matrix is in* row echelon form *if it satisfies*

i) the first nonzero entry (called the leading entry*) of any row is* 1,

ii) each leading entry is in a column to the right of the previous row's entry, and

iii) rows consisting entirely of 0s *are at the bottom.*

[4] See [1] or other Numerical Analysis texts.

The matrix in (4.19) is in row echelon form.

Definition 4.2.4 (Reduced Row Echelon Form). *A matrix is in* reduced row echelon *form if*

i) it is in row echelon form and
ii) each column with a leading 1 has zeros in all other entries.

The matrix in (4.21) is in reduced row echelon form.

In the previous section, the solution was reached via Gauss-Jordan elimination; that is the augmented matrix representing the original system of equations is manipulated via row operations until it is in reduced row echelon form. Once the augmented matrix is in reduced row echelon form the solution may be directly obtained. In Gaussian elimination the row operations halt when the augmented matrix is in row echelon form then *back substitution* is used. Solving the previous section's example via Gaussian Elimination involves stopping row operations at (4.19) then using back substitution:

From row 3:

$$-18x_3 = 36$$

thus $x_3 = -2$.

Using this solution with row 2:

$$x_2 = 3 + 5x_3$$

so $x_2 = -7$.

Lastly these solutions together with row 1 give

$$x_1 = -1 + 2x_3$$

and we have $x_1 = -5$.

4.3 Linear Algebra Basics

There is no more a pervasive a tool in optimization and the mechanics of computer usage than linear algebra. It is the skeletal structure of machine learning to which all of its other tools attached. Thus understanding what follows is crucial to understanding the operating procedures of most of what is in this text. We begin by recalling some basic definitions and the algebra of vectors and matrices.

4.3.1 Matrices and Their Multiplication

A *matrix* is a rectangular array of numbers. If the matrix has m rows and n columns, the matrix is refereed to as an $m \times n$ *matrix*. A *submatrix* of matrix

A is a matrix formed from A by deleting any amount of rows or columns of A.

Example 4.3.1. *The matrix* $A = \begin{bmatrix} 1 & 0 & -2 & 3 \\ 0 & 7 & -4 & 5 \\ -2 & 4 & -1 & 6 \end{bmatrix}$ *has* $\begin{bmatrix} 1 & -2 & 3 \\ 0 & -4 & 5 \\ -2 & -1 & 6 \end{bmatrix}$ *and* $\begin{bmatrix} 0 & 3 \\ 4 & 6 \end{bmatrix}$ *as two of its possible submatrices.*

A useful operator on matrices is the notion of its transpose.

Definition 4.3.2 (The Transpose of a Matrix).
Consider the $m \times n$ *(*m *rows,* n *columns) matrix*

$$A = \begin{bmatrix} a_{1,1} & a_{1,2} & \cdots & a_{1,n} \\ a_{2,1} & a_{2,2} & \cdots & a_{2,n} \\ \vdots & \vdots & \ddots & \vdots \\ a_{m,1} & a_{m,2} & \cdots & a_{m,n} \end{bmatrix}. \tag{4.22}$$

The transpose *of* A *is*

$$A^T = \begin{bmatrix} a_{1,1} & a_{2,1} & \cdots & a_{m,1} \\ a_{1,2} & a_{2,2} & \cdots & a_{m,2} \\ \vdots & \vdots & \ddots & \vdots \\ a_{1,n} & a_{2,n} & \cdots & a_{m,n} \end{bmatrix}. \tag{4.23}$$

In shorthand notation, if $A = [a_{i,j}]$[5], then $A^T = [a_{j,i}]$; that is we swap the rows and columns of A to form A^T. Note that if A is in $\mathbb{R}_{m \times n}$, then A^T is in $\mathbb{R}_{n \times m}$.

Matrices that are there own transpose will have some special properties, so we will give these matrices a name (actually, many names). A real matrix A is said to be *Hermitian* or *self-adjoint*[6] if $A = A^T$. We may also refer to such a matrix A as a *symmetric matrix*. Clearly a symmetric matrix must be a square matrix. We will explore one useful property of self-adjoint matrices in the chapter on matrix factorization (Chapter 5).

The reader is hopefully familiar with vectors and we note that in the context of this text we are primarily concerned with seeing a vector as a special matrix and not so much all the wonderful qualities vectors have in Physics. For example, the row vector $\mathbf{x} = \langle x_1, x_2, \ldots, x_n \rangle$ can be seen as the $1 \times n$ matrix $A = [x_1, x_2, \ldots, x_n]$ and the column vector $\mathbf{x}^T$ as the $n \times 1$ matrix A^T.

---

[5] It is often the case that the comma is omitted when writing the index of the entry in a matrix. That is, it is acceptable to write a_{32} for $a_{3,2}$.

[6] A proper full definition of Hermitian matrices involves considering complex entries. When the entries of A can include complex numbers, A is Hermitian if it is equal to its own *conjugate transpose*; that is, $A = \overline{A^T}$ meaning $a_{ij} = \overline{a_{ji}}$. As we will only be considering real matrices, stating the full definition is unnecessarily cumbersome.

Since $\mathbf{v} + \mathbf{v}$ should be $2\mathbf{v}$, multiplication between a number and a vector is defined as:

Definition 4.3.3 (Scalar Multiplication). *Let $c \in \mathbb{R}$ and let $\mathbf{v} = \langle v_1, v_2, \ldots, v_n \rangle$. Then*

$$c\mathbf{v} := \langle cv_1, cv_2, \ldots cv_n \rangle$$

There are two different multiplications defined for vectors: the dot product and the cross product (which is only defined for vectors in three dimensions). We will have no use for the cross product of two vectors, so we remind the reader

Definition 4.3.4 (Dot Product). *Let $\mathbf{a} = \langle a_1, \ldots, a_n \rangle$ and $\mathbf{b} = \langle b_1, \ldots, b_n \rangle$. Then the* dot product *of $\mathbf{a}$ and $\mathbf{b}$ is*

$$\mathbf{a} \cdot \mathbf{b} := a_1 b_1 + a_2 b_2 + \cdots + a_n b_n. \tag{4.24}$$

One use of the dot product is calculating the magnitude (or length) of a vector.

Definition 4.3.5 (Magnitude of a Vector). *For $\mathbf{x} = \langle x_1, x_2, \ldots, x_n \rangle$ in $\mathbb{R}^n$, let $||\mathbf{x}||$ represent the length or* magnitude *of $\mathbf{x}$.*

By the Pythagorean Theorem (Theorem B.2.1), for the vector in the definition, we have

$$||\mathbf{x}|| = \sqrt{x_1^2 + x_2^2 + \cdots + x_n^2}. \tag{4.25}$$

As stated, a useful property of the dot product is

Proposition 4.3.6. *For $\mathbf{u}, \mathbf{v} \in \mathbb{R}^n$,*

$$\mathbf{u} \cdot \mathbf{v} = ||\mathbf{u}|| \; ||\mathbf{v}|| \cos\theta$$

where θ is the angle between $\mathbf{u}$ and $\mathbf{v}$.

From either Definition 4.3.4 or Proposition 4.3.6, it follows easily that

Remark 4.3.7.

$$||\mathbf{v}|| = \sqrt{\mathbf{v} \cdot \mathbf{v}}.$$

We add any vector with magnitude 1 is called a *unit vector*. Note that for any nonzero vector $\mathbf{v}$, $\frac{\mathbf{v}}{||\mathbf{v}||}$ is a unit vector parallel to $\mathbf{v}$. Dividing a vector by its magnitude to create a unit vector that is parallel[7] to the original vector is called *normalizing the vector*.

With the dot product defined, we can refresh our memories as to the product of two matrices:

[7] Recall that $\mathbf{u}$ is parallel to vector $\mathbf{v}$ if there exists a real number c such that $\mathbf{u} = c\mathbf{v}$.

Definition 4.3.8 (Matrix Multiplication). *Let A be the $m \times n$ matrix $A = \begin{bmatrix} \mathbf{a}_1 \\ \vdots \\ \mathbf{a}_m \end{bmatrix}$ consisting of row vectors and B be the $n \times r$ matrix $B = [\mathbf{b}_1, \mathbf{b}_2, \cdots, \mathbf{b}_r]$ consisting of column vectors. Then the matrix product $A \times B$ is the $m \times r$ matrix*

$$C = A \times B = \begin{bmatrix} \mathbf{a}_1 \cdot \mathbf{b}_1 & \mathbf{a}_1 \cdot \mathbf{b}_2 & \cdots & \mathbf{a}_1 \cdot \mathbf{b}_r \\ \mathbf{a}_2 \cdot \mathbf{b}_1 & \mathbf{a}_2 \cdot \mathbf{b}_2 & \cdots & \mathbf{a}_2 \cdot \mathbf{b}_r \\ \vdots & \vdots & \ddots & \vdots \\ \mathbf{a}_m \cdot \mathbf{b}_1 & \mathbf{a}_m \cdot \mathbf{b}_2 & \cdots & \mathbf{a}_m \cdot \mathbf{b}_r \end{bmatrix}; \tag{4.26}$$

that is, $c_{ij} := \mathbf{a}_i \cdot \mathbf{b}_j$.

Example 4.3.9. *Find $C = A \times B$ where $A = \begin{bmatrix} 1 & -1 \\ 0 & 2 \\ -1 & 1 \end{bmatrix}$ and $B = \begin{bmatrix} 1 & 2 & 3 & 4 \\ 0 & 1 & 2 & 3 \end{bmatrix}$.*

Solution. We have

$$c_{1,1} = \mathbf{a}_1 \cdot \mathbf{b}_1 = [1, -1] \cdot [1, 0]^T = 1 \tag{4.27}$$
$$c_{2,1} = \mathbf{a}_2 \cdot \mathbf{b}_1 = [0, 2] \cdot [1, 0]^T = 0 \tag{4.28}$$
$$c_{3,1} = \mathbf{a}_3 \cdot \mathbf{b}_1 = [-1, 1] \cdot [1, 0]^T = -1 \tag{4.29}$$
$$c_{1,2} = \mathbf{a}_1 \cdot \mathbf{b}_2 = [1, -1] \cdot [2, 1]^T = 1 \tag{4.30}$$
$$c_{2,2} = \mathbf{a}_2 \cdot \mathbf{b}_2 = [0, 2] \cdot [2, 1]^T = 2 \tag{4.31}$$
$$c_{3,2} = \mathbf{a}_3 \cdot \mathbf{b}_2 = [-1, 1] \cdot [2, 1]^T = 1 \tag{4.32}$$
$$c_{1,3} = \mathbf{a}_1 \cdot \mathbf{b}_3 = [1, -1] \cdot [3, 2]^T = 1 \tag{4.33}$$
$$c_{2,3} = \mathbf{a}_2 \cdot \mathbf{b}_3 = [0, 2] \cdot [3, 2]^T = 4 \tag{4.34}$$
$$c_{3,3} = \mathbf{a}_3 \cdot \mathbf{b}_3 = [-1, 1] \cdot [3, 2]^T = -1 \tag{4.35}$$
$$c_{1,4} = \mathbf{a}_1 \cdot \mathbf{b}_4 = [1, -1] \cdot [4, 3]^T = 1 \tag{4.36}$$
$$c_{2,4} = \mathbf{a}_2 \cdot \mathbf{b}_4 = [0, 2] \cdot [4, 3]^T = 6 \tag{4.37}$$
$$c_{3,4} = \mathbf{a}_3 \cdot \mathbf{b}_4 = [-1, 1] \cdot [4, 3]^T = -1 \tag{4.38}$$

Thus

$$C = \begin{bmatrix} 1 & 1 & 1 & 1 \\ 0 & 2 & 4 & 6 \\ -1 & 1 & -1 & -1 \end{bmatrix}$$

■

4.3.2 Identity Matrices, Inverses, and Determinants of Matrices

Since we have discussed the process of matrix multiplication, it is natural to consider matrices that behave how a 1 does under multiplication of numbers. For any number a, $a \cdot 1 = a = 1 \cdot a$; that is, 1 is the *multiplicative identity* when multiplying numbers; i.e. multiplying by 1 does not change the number's identity. That is we seek a matrix I such that $AI = I = IA$. Once we have an identity, we may develop the process of undoing multiplication.

Definition 4.3.10 (Identity Matrix). *Let I_n (denoted I when the n is understood) be the $n \times n$ matrix with 1's on the diagonal and 0's elsewhere. That is*

$$I_n = I = \begin{bmatrix} 1 & 0 & \cdots & 0 & 0 \\ 0 & 1 & \cdots & 0 & 0 \\ \vdots & & \ddots & & \vdots \\ 0 & 0 & \cdots & 1 & 0 \\ 0 & 0 & \cdots & 0 & 1 \end{bmatrix}$$

I is not the identity matrix just because we say so, but it is not hard to prove that I serves as an identity matrix and that, for a given n, $I = I_n$ is unique (see Exercise 4.11).

Definition 4.3.11 (Invertible Matrix). *Let A be an $n \times n$ matrix. If there exists an $n \times n$ matrix B such that $AB = I = BA$, then A is said to be invertible with inverse B.*

The analogy of Definition 4.3.11 with real numbers is consider something like $3 \times 1/3 = 1$. In this situation, 3 is invertible with (multiplicative) inverse $1/3$. This is why A's inverse B is denoted by A^{-1}. Note also that the *left inverse* and *right inverse* do not generally have to agree in algebraic structures (i.e. there are times when $ba = 1 = ac$ but $a \neq c$), but for square matrices the left and right inverses always agree (see Exercise 4.12).

Let us consider a 2×2 matrix A and find its inverse (assuming one exists); that is find constants e, f, g, and h such that

$$AA^{-1} = \begin{bmatrix} a & b \\ c & d \end{bmatrix} \begin{bmatrix} e & f \\ g & h \end{bmatrix} = \begin{bmatrix} 1 & 0 \\ 0 & 1 \end{bmatrix}. \tag{4.39}$$

Multiplying the matrices gives the system of equations

$$\begin{cases} ab + bg & = 1 \\ af + bh & = 0 \\ ce + dg & = 0 \\ cf + dh & = 1. \end{cases} \tag{4.40}$$

Treating a, b, c, and d as constants gives the solution

$$e = \frac{d}{ad - bc}, f = \frac{-b}{ad - bc}, g = \frac{-c}{ad - bc}, h = \frac{a}{ad - bc}; \tag{4.41}$$

that is,

$$A^{-1} = \frac{1}{ad - bc}\begin{bmatrix} d & -b \\ -c & a \end{bmatrix}. \tag{4.42}$$

Notice how the term $ad - bc$ is present in the denominator of each term. Moreover, we have just shown that a 2×2 matrix like A will have an inverse exactly when $ad - bc \neq 0$. As such, this value is so important, we give it a name.

Definition 4.3.12 (Determinant of a 2×2 Matrix). *Let* $A = \begin{bmatrix} a & b \\ c & d \end{bmatrix}$. *Then the* determinant *of A is*

$$\det(A) := ad - bc.$$

By this discussion we have also proven the important result

Theorem 4.3.13. *Let A be a 2×2 matrix. Then A has an inverse if and only if* $\det(A) \neq 0$.

We will see very soon that it is not just for 2×2 matrices that invertiblity requires a nonzero determinant. Moreover, since it is that case that invertible matrices must have a nonzero determinant, those that have a determinant of zero are given a special name.

Definition 4.3.14 (Singular Matrix). *A matrix A that has the property that* $\det(A) = 0$ *is said to be a* singular matrix. *Matrices whose determinant is nonzero are said to be* non-singular.

Determinants of higher ordered square matrices are more involved and include the notion of *minor matrices* and *cofactors.*

Definition 4.3.15 (Minor Matrix). *Let A be an $n \times n$ matrix with a_{ij} as its entry in the ith row and j^{th} column; that is*

$$A = \begin{bmatrix} a_{11} & a_{12} & \cdots & a_{1n} \\ \vdots & \vdots & & \vdots \\ a_{n1} & a_{n2} & \cdots & a_{nn} \end{bmatrix}$$

Then the ij^{th} minor matrix of A *is the $n-1 \times n-1$ matrix A_{ij} that is formed by removing the ith row and j^{th} column from A; that is*

$$A_{ij} := \begin{bmatrix} a_{11} & a_{12} & \cdots & a_{1j-1} & a_{1j+1} & \cdots & a_{1n} \\ a_{21} & a_{22} & \cdots & a_{2j-1} & a_{2j+1} & \cdots & a_{2n} \\ \vdots & \vdots & & \vdots & \vdots & & \vdots \\ a_{i-11} & a_{i-22} & \cdots & \cdots & a_{i-1j-1} & a_{i-1j+1} & a_{1n} \\ a_{i+11} & a_{i+22} & \cdots & \cdots & a_{i+1j-1} & a_{i+1j+1} & a_{i+1n} \\ \vdots & \vdots & & \vdots & \vdots & & \vdots \\ a_{n1} & a_{n2} & \cdots & \cdots & a_{nj-1} & a_{nj+1} & a_{nn} \end{bmatrix}.$$

One more piece is needed before defining determinants of a general square matrix.

Definition 4.3.16 (Cofactor of a Square Matrix). *Let A be an $n \times n$ matrix with a_{ij} as its entry in the ith row and j^{th} column. Then the ij^{th}* cofactor *of A is*

$$A'_{ij} := (-1)^{i+j} \det(A_{ij})$$

(as we will see in its definition, we will assume we know how to get the determinant of matrices of smaller order; namely, we state a definition that follows an inductive process).

Now we may finish the goal:

Definition 4.3.17 (Determinant of an $n \times n$ Matrix). *If $A = [a]$ (a matrix with a single entry; a* singleton*), then* $\det A = a$. *The determinant of a 2×2 matrix is given in Definition 4.3.12. Let A be an $n \times n$ matrix with $n \geq 2$. Then*

$$\det(A) := a_{11}A'_{11} + a_{12}A'_{12} + \cdots + a_{1n}A'_{1n}$$

where a_{1j} is the jth entry in the first row of A and A'_{1j} is the $1j^{th}$ cofactor.

Definition 4.3.17 may be extended to a very useful form:

Theorem 4.3.18 (General Expansion by Minors). *Let A be an $n \times n$ matrix with $n \geq 2$ and assume that determinants are defined for square matrices of order less than n. Then*

$$\det(A) = a_{i1}A'_{i1} + a_{i2}A'_{i2} + \cdots + a_{in}A'_{in} \tag{4.43}$$

or

$$\det(A) = a_{1j}A'_{1j} + a_{2j}A'_{2j} + \cdots + a_{nj}A'_{nj} \tag{4.44}$$

where a_{ij} is the entry of A in row i and column j and A'_{ij} is the ij^{th} cofactor.

When finding a determinant by general expansion by minors, (4.43) is referred to as the *expansion of* $\det(A)$ *by minors on the ith row of A* while (4.44) is called the *expansion of* $\det(A)$ *by minors on the jth column of A.*

Example 4.3.19. *Find* $\det(A)$ *by general expansion on minors for*

$$A = \begin{bmatrix} 0 & 0 & 1 & 0 & 0 \\ 8 & 6 & 7 & 5 & 3 \\ 0 & 9 & 9 & 3 & 0 \\ 0 & 0 & 0 & 2 & 0 \\ 6 & 4 & 2 & 1 & 0 \end{bmatrix}$$

Solution. The row that will be used for the expansion by minors has been highlighted (note that the last column would also be a good choice). Thus

$$\begin{aligned}
\det(A) &= (-1)^5(0)A'_{41} + (-1)^6(0)A'_{42} + (-1)^7(0)A'_{43} + (-1)^8(2)A'_{44} \\
&\quad + (-1)^9(0)A'_{45} && (4.45) \\
&= 2 \cdot \det\left(\begin{bmatrix} 0 & 0 & 1 & 0 \\ 8 & 6 & 7 & 3 \\ 0 & 9 & 9 & 0 \\ 6 & 4 & 2 & 0 \end{bmatrix}\right) && (4.46) \\
&= 2 \cdot \left[(-1)^5(0)A'_{14} + (-1)^6(3)A'_{24} + (-1)^7(0)A'_{34} + (-1)^8(0)A'_{44}\right] && (4.47) \\
&= 6 \cdot \det\left(\begin{bmatrix} 0 & 0 & 1 \\ 0 & 9 & 9 \\ 6 & 4 & 2 \end{bmatrix}\right) && (4.48) \\
&= 6 \cdot \left[(-1)^2(0)A'_{11} + (-1)^3(0)A'_{12} + (-1)^4(1)A'_{13}\right] && (4.49) \\
&= 6 \cdot \det\left(\begin{bmatrix} 0 & 9 \\ 6 & 4 \end{bmatrix}\right) && (4.50) \\
&= 6(0 \cdot 4 - 6 \cdot 9) = -324. && (4.51)
\end{aligned}$$

■

Since we now have determinants for higher order matrices, we may extend Theorem 4.3.13.

Theorem 4.3.20. *Let A be a square matrix. Then A^{-1} exists if and only if* $\det(A) \neq 0$.

We will not offer a proof of Theorem 4.3.20 but will note it is part of the larger Fundamental Theorem of Invertible Matrices (Theorem 4.3.54). A proof can be found in [44].

Also, now that we are equipped with determinants, we may state a formula for the inverse of a 3×3 matrix:

Highlight 4.3.21 (Inverse of a 3×3 Matrix).

Let $A = \begin{bmatrix} a_{11} & a_{12} & a_{13} \\ a_{21} & a_{22} & a_{23} \\ a_{31} & a_{32} & a_{33} \end{bmatrix}$ *be an invertible matrix. Then*

$$A^{-1} = \frac{1}{\det(A)} \begin{bmatrix} A'_{11} & A'_{21} & A'_{31} \\ A'_{12} & A'_{22} & A'_{32} \\ A'_{13} & A'_{23} & A'_{33} \end{bmatrix}$$

where A'_{ij} is the ij^{th} cofactor of A as in Definition 4.3.16.

Of course, one may also use the following technique.

Highlight 4.3.22 (Inverse of an $n \times n$ Invertible Matrix).
If A is an invertible matrix, then A^{-1} may be obtained using elementary row operations (Theorem 4.2.2):

$$[A|I] \xrightarrow{\text{row operations}} [I|A^{-1}].$$

We will close the section by listing some of the useful properties of determinants.

Theorem 4.3.23 (Properties of Determinants). *Let A and B be square matrices of order n, $I = I_n$ the $n \times n$ identity matrix, and c a constant. Then*

1. $\det(I) = 1$
2. $\det(A^T) = \det(A)$
3. $\det(A^{-1}) = [\det(A)]^{-1} = \frac{1}{\det(A)}$
4. $\det(AB) = \det(A)\det(B)$
5. $\det(cA) = c^n \det(A)$ *and*
6. *if A is an* upper triangular matrix, *that is entry $a_{ij} = 0$ whenever $i < j$, then*
$$\det(A) = a_{11}a_{22}\cdots a_{nn};$$
thus the determinant of an upper triangular matrix is just the product of its diagonal entries.

The proof of these properties is left as Exercise 4.14.

4.3.3 Solving Systems of Linear Equations via Cramer's Rule

Determinants turn out to be incredibly useful as we will see in the following theorem.

Theorem 4.3.24 (Cramer's Rule). *Let A be an $n \times n$ invertible matrix with $\mathbf{x} = [x_1, x_2, \ldots, x_n]^T$ and $\mathbf{b} = [b_1, b_2, \ldots, b_n]^T$. Let A_i represent the matrix formed by replacing the ith column of A with $\mathbf{b}$. Then if the system of linear equations $A\mathbf{x} = \mathbf{b}$ has a unique solution, the solution can be found by*

$$x_i = \det(A_i)/\det(A)$$

for $i = 1, \ldots, n$.

The proof of Cramer's Rule is not difficult, but either involves introducing linear transformations or the adjoint of a matrix. As we wish to be a text on optimization and not linear algebra, we will skip the proof and the curious reader can consult any linear algebra text.

Example 4.3.25. *Use Cramer's Rule to solve the system of equations*

$$\begin{array}{llll} 2x_1 & -3x_2 & +4x_3 & = 19 \\ 4x_1 & -3x_2 & & = 17 \\ 2x_1 & -6x_2 & -x_3 & = 1 \end{array}.$$

Solution. Putting the problem into the form $A\mathbf{x} = \mathbf{b}$, we have $A = \begin{bmatrix} 2 & -3 & 4 \\ 4 & -3 & 0 \\ 2 & -6 & -1 \end{bmatrix}$, $\mathbf{x} = [x_1, x_2, x_3]^T$, and $\mathbf{b} = [19, 17, 1]^T$. Expanding by minors (Theorem 4.3.18), we obtain

$$\det(A) = (-1)^4(4)\begin{bmatrix} 4 & -3 \\ 2 & -6 \end{bmatrix} + 0 + (-1)^6(-1)\begin{bmatrix} 2 & -3 \\ 4 & -3 \end{bmatrix} \tag{4.52}$$

$$= 4(-24+6) - (-6+12) \tag{4.53}$$

$$= -78. \tag{4.54}$$

As $\det(A) \neq 0$, by Theorem 4.3.20, A is invertible and we may use Cramer's Rule. We have

$$A_1 = \begin{bmatrix} 19 & -3 & 4 \\ 17 & -3 & 0 \\ 1 & -6 & -1 \end{bmatrix}, A_2 = \begin{bmatrix} 2 & 19 & 4 \\ 4 & 17 & 0 \\ 2 & 1 & -1 \end{bmatrix}, \text{ and } A_3 = \begin{bmatrix} 2 & -3 & 19 \\ 4 & -3 & 17 \\ 2 & -6 & 1 \end{bmatrix}$$

which leads to $\det(A_1) = -390$, $\det(A_2) = -78$, and $\det(A_3) = -234$. Thus

$$x_1 = \det(A_1)/\det(A) = 5, x_2 = \det(A_2)/\det(A) = 1,$$
$$\text{and } x_3 = \det(A_3)/\det(A) = 3.$$

■

One can see that using Cramer's Rule to solve systems of equations involves calculating many determinants. Thus a program running a naive implementation of this process will be computationally much more expensive than Gauss-Jordan Elimination (see Sections 3.1 and 4.2.2). This means Cramer's Rule does not have much use in practice, but it does have significant theoretical value as we will see in Sections 4.4.2 and 24.2.

In addition to added runtime, Cramer's rule has the additional disadvantage of more numerical instability than Gauss-Jordan elimination. To understand numerical instability, we must first understand matrix norms.

4.3.4 Vector and Matrix Norms

We discussed the magnitude of a vector in Definition 4.3.5 and may do the same for matrices. First, we need some machinery.

Definition 4.3.26 (Vector Norm). *Let* $\mathbf{u}$ *and* $\mathbf{v}$ *be any vectors in* $\mathbb{R}^n$. *A* vector norm $||\cdot||$ *is any mapping from* $\mathbb{R}^n$ *to* $\mathbb{R}$ *satisfying*

i. $||\mathbf{v}|| \geq 0$ *with equlity if and only if* $\mathbf{v} = \mathbf{0}$,
ii. $||c\mathbf{v}|| = |c|||\mathbf{v}||$ *for any scalar* c, *and*
iii. $||\mathbf{u} + \mathbf{v}|| \leq ||\mathbf{u}|| + ||\mathbf{v}||$.

A vector norm generalizes the notion of the magnitude of a vector and is a specific instance of the more general mathematical notion of a *distance* (or a *metric*) and, as such, the first two properties are very natural (see Definition B.1.4). The third property is also a natural requirement for anything behaving like a distance and is commonly called the *triangle inequality* (see Theorem B.2.3).

We did use $|| \cdot ||$ to denote a generic norm in Definition 4.3.26 though we used the same symbol to a specific norm when we addressed the length of a vector in Definition 4.3.5. You will understand our choice to do this when you see the next example. Note that the specific norm in Definition 4.3.5 is called the *Euclidean norm*. We are now ready to introduce other vector norms.

Theorem 4.3.27 (Some Vector Norms)**.** *Let* $\mathbf{v} = \langle v_1, v_2, \ldots, v_n \rangle$ *be any vector in* $\mathbb{R}^n$. *Then each of the following is a norm for the vector* $\mathbf{v}$*:*

$$||\mathbf{v}||_1 := \sum_{i=1}^{n} |v_i| \qquad \text{the } \ell_1 \text{ norm,} \tag{4.55}$$

$$||\mathbf{v}||_2 := \left(\sum_{i=1}^{n} |v_i|^2 \right)^{1/2} \qquad \text{the } \ell_2 \text{ or Euclidean norm, and} \tag{4.56}$$

$$||\mathbf{v}||_\infty := \max_{1 \leq i \leq n} |v_i| \qquad \text{the } \ell_\infty \text{ norm.} \tag{4.57}$$

The proof that these are norms is Exercise 4.15. An example of the norms is not to be excluded, though.

Example 4.3.28. *Let* $\mathbf{v} = [4, 0, -4, 1]^T$ *be a vector in* $\mathbb{R}^4$. *Then*

$$||\mathbf{v}||_1 = |4| + |0| + |-4| + |1| = 9, \tag{4.58}$$

$$||\mathbf{v}||_2 = \sqrt{4^2 + 0^2 + (-4)^2 + 1^2} = \sqrt{33}, \text{ and} \tag{4.59}$$

$$||\mathbf{v}||_\infty = \max\{|4|, |0|, |-4|, |1|\} = 4. \tag{4.60}$$

Also, let $\mathbf{u} = [5, 2, -1, 1]^T$. *Then*

$$||\mathbf{u}||_1 = |5| + |2| + |-1| + |1| = 9, \tag{4.61}$$

$$||\mathbf{u}||_2 = \sqrt{5^2 + 2^2 + (-1)^2 + 1^2} = \sqrt{31}, \text{ and} \tag{4.62}$$

$$||\mathbf{u}||_\infty = \max\{|5|, |2|, |-1|, |1|\} = 5. \tag{4.63}$$

Note that many other vector norms exist and will be introduced as needed. An important theorem for us regarding vector norms is

Theorem 4.3.29 (Equivalence of Vector Norms). *Let $||\cdot||$ and $||\cdot||_*$ be vector norms for any vector in a given finite dimensional vector space*[8] *V. Then there exist constants c and C such that*

$$c||\mathbf{v}|| \leq ||\mathbf{v}||_* \leq C||\mathbf{v}|| \tag{4.64}$$

for all $\mathbf{v}$ in V.

In the language of the discipline, Numerical Analysts summarize Theorem 4.3.29 by saying *"all vector norms are equivalent"*. This, of course, does not mean that all norms give the same value or even that relative magnitudes are preserved (for instance, in Example 4.3.28, $||\mathbf{v}||_1 = ||\mathbf{u}||_1$, but $||\mathbf{v}||_2 > ||\mathbf{u}||_2$, and $||\mathbf{v}||_\infty < ||\mathbf{u}||_\infty$). What is meant by stating Theorem 4.3.29 as *"all vector norms are equivalent"* is that *if a sequence of vectors from a finite dimensional vector space converges under any vector norm $||\cdot||$, then the sequence also converges under any other vector norm $||\cdot||_*$.*

It is also possible to consider the "magnitude" of a matrix.

Definition 4.3.30 (Matrix Norm). *Let A and B be any matrices in $\mathbb{R}_{m\times n}$ and c any scalar. A* matrix norm *$||\cdot||$ is any mapping from $\mathbb{R}_{m\times n}$ to $\mathbb{R}$ satisfying*

i. $||A|| \geq 0$ *with equality if and only if* $a_{ij} = 0$ *for all i and j,*
ii. $||cA|| = |c|||A||$ *for any scalar c,*
iii. $||A + B|| \leq ||A|| + ||B||$.

Theorem 4.3.31 (Some Matrix Norms). *Let A be any matrix in $\mathbb{R}_{m\times n}$. Then each of the following is a norm for the matrix A:*

$$||A||_1 := \max_{1\leq j\leq n} \sum_{i=1}^{m} |a_{ij}| \quad \text{the } \ell_1 \text{ norm (max column sum of magnitudes),} \tag{4.65}$$

$$||A||_F := \sqrt{\sum_{j=1}^{n}\sum_{i=1}^{m} a_{ij}^2} \quad \text{the Frobenius norm, and} \tag{4.66}$$

$$||A||_\infty := \max_{1\leq i\leq n} \sum_{j=1}^{m} |a_{ij}| \quad \text{the } \ell_\infty \text{ norm (max row sum of magnitudes).} \tag{4.67}$$

The proof that the examples in Theorem 4.3.31 are matrix norms is left to Exercise 4.16. Note that $||A||_F$ is not the ℓ_2 matrix norm. The ℓ_2 matrix

[8]We discuss vector spaces in the next section.

norm requires more machinery and will not be discussed here. Also, as with vectors, many matrix norms exist and we will also introduce those as needed.

There is plenty to say about the relationship between matrix and vector norms, but first an example.

Example 4.3.32. *Let* $A = \begin{bmatrix} 1 & -1 & 3 & 4 \\ 2 & 1 & -5 & 3 \\ -2 & -2 & -8 & 7 \end{bmatrix}$. *Then*

$$||A||_1 = \max_{1\leq j\leq 4} \sum_{i=1}^{3} |a_{ij}| \tag{4.68}$$

$$= \max\{|1|+|2|+|-2|, |-1|+|1|+|-2|, |3|+|-5|+|-8|, |4|+|3|+|7|\} \tag{4.69}$$

$$= \max\{5, 4, 16, 14\} = 16, \tag{4.70}$$

$$||A||_F = \sqrt{1^2+(-1)^2+3^2+4^2+2^2+1^2+(-5)^2+3^3+(-2)^2+(-2)^2+(-8)^2+7^2} \tag{4.71}$$

$$= \sqrt{187} \text{ and} \tag{4.72}$$

$$||A||_\infty = \max_{1\leq i\leq 3} \sum_{j=1}^{4} |a_{ij}| \tag{4.73}$$

$$= \max\{|1|+|-1|+|3|+|4|, |2|+|1|+|-5|+|3|, |-2|+|-2|+|-8|+|7|\} \tag{4.74}$$

$$= \max\{9, 11, 19\} = 19. \tag{4.75}$$

As with vector norms, matrix norms are equivalent.

Theorem 4.3.33 (Equivalence of Matrix Norms)**.** *Let* $||\cdot||$ *and* $||\cdot||_*$ *be matrix norms for any finite dimensional matrix space*[9] M. *Then there exist constants* c *and* C *such that*

$$c||A|| \leq ||A||_* \leq C||A|| \tag{4.76}$$

for all A *in* M.

As this chapter can be skimmed, we will restate what was said regarding the equivalent theorem for vectors. Numerical Analysts summarize Theorem 4.3.33 by saying *"all matrix norms are equivalent"*. This does not mean that all matrix norms yield the same value for a given matrix or even that relative magnitudes between matrices are preserved. What is meant by stating Theorem 4.3.33 as *"all matrix norms are equivalent"* is that *if a sequence of matrices from a finite dimensional space converges under any matrix norm* $||\cdot||$, *then the sequence also converges under any other matrix norm* $||\cdot||_*$.

A special relationship exists between matrix and vector norms. Let A be a matrix in $\mathbb{R}_{m\times n}$ and $\mathbf{v}$ a vector in $\mathbb{R}^n$; that is, an $n \times 1$ matrix. Their product

[9]By *matrix space* we mean "think of each matrix as a big vector".

$A\mathbf{v} = \mathbf{b}$ is a vector in $\mathbb{R}^m$ (i.e. a $m \times 1$ matrix). Suppose $||\cdot||_{(n)}$ is a vector norm in $\mathbb{R}^n$ and $||\cdot||_{(m)}$ is a vector norm in $\mathbb{R}^m$. We may introduce a matrix norm based on $||\cdot||_{(n)}$ and $||\cdot||_{(m)}$; that is,

$$||A|| := \sup_{\substack{\mathbf{v}\in\mathbb{R}^n \\ \mathbf{v}\neq\mathbf{0}}} \frac{||A\mathbf{v}||_{(m)}}{||\mathbf{v}||_{(n)}} \tag{4.77}$$

where the $\mathbf{v}$ range over all nonzero vectors in $\mathbb{R}^n$.

Normalizing $\mathbf{v}$ and using property *ii)* of vector norms (Definition 4.3.26), 4.77 becomes

$$\begin{aligned} ||A|| := \sup_{\substack{\mathbf{v}\in\mathbb{R}^n \\ \mathbf{v}\neq\mathbf{0}}} \frac{||A\mathbf{v}||_{(m)}}{||\mathbf{v}||_{(n)}} &= \sup_{\substack{\mathbf{v}\in\mathbb{R}^n \\ \mathbf{v}\neq\mathbf{0}}} \frac{||\frac{A\mathbf{v}}{||\mathbf{v}||_{(n)}}||_{(m)}}{||\mathbf{v}||_{(n)}/||\mathbf{v}||_{(n)}} = \sup_{\substack{\mathbf{x}\in\mathbb{R}^n \\ ||x||_{(n)}=1}} \frac{||A\mathbf{x}||_{(m)}}{||\mathbf{x}||_{(n)}} \\ &= \sup_{\substack{\mathbf{x}\in\mathbb{R}^n \\ ||x||_{(n)}=1}} ||A\mathbf{x}||_{(m)}. \end{aligned} \tag{4.78}$$

Thus we have the following

Definition 4.3.34 (Induced Matrix Norm). *Let $||\cdot||$ be a vector norm over $\mathbb{R}^n$ and let A be an $n \times n$ matrix. Then the* matrix norm induced by $||\cdot||$ *is*

$$||A|| := \sup_{\substack{\mathbf{x}\in\mathbb{R}^n \\ ||x||_{(n)}=1}} ||A\mathbf{x}||.$$

Induced matrix norms are also commonly called *subordinate matrix norms* and have additional properties over non-induced matrix norms (though some non-induced norms satisfy some of these):

Remark 4.3.35. *Let $||\cdot||$ be an induced matrix norm and A and B two matrices subject to the norm and $\mathbf{v}$ any vector subject to the original vector norm. Then*

i) $||I|| = 1$,
ii) $||A\mathbf{v}|| \leq ||A||\ ||\mathbf{v}||$ *(i.e. the norm is* scalable*), and*
iii) $||AB|| \leq ||A||\ ||B||$ *(i.e. the norm is* multiplicative*).*

Let us now determine an induced matrix norm from a given vector norm.

Example 4.3.36. *Let $\mathbf{x} = \langle x_1, x_2, \ldots, x_n\rangle$ be a vector in $\mathbb{R}^n$ and let A be an $m \times n$ matrix whose colums are $\mathbf{a}_1, \ldots, \mathbf{a}_n$. Put $M = \max_{1\leq j\leq m}\{\sum_{i=1}^n |a_{ij}|\}$; that is, M is the max column sum of absolute values of the entries in A. Then* *the matrix norm induced by the ℓ_1 vector norm is*

$$||A|| := \sup_{\substack{\mathbf{x}\in\mathbb{R}^n \\ ||x||_{(n)}=1}} ||A\mathbf{x}||_1 \tag{4.79}$$

$$= \sup_{\substack{\mathbf{x}\in\mathbb{R}^n \\ ||x||_{(n)}=1}} ||x_1\mathbf{a}_1 + x_2\mathbf{a}_2 \cdots + x_n\mathbf{a}_n||_1 \tag{4.80}$$

$$\leq \sup_{\substack{\mathbf{x}\in\mathbb{R}^n \\ ||x||_{(n)}=1}} |x_1|||\mathbf{a}_1||_1 + |x_2|||\mathbf{a}_2||_1 \cdots + |x_n|||\mathbf{a}_n||_1 \tag{4.81}$$

$$\leq \sup_{\substack{\mathbf{x}\in\mathbb{R}^n \\ ||x||_{(n)}=1}} |x_1|M + |x_2|M \cdots + |x_n|M \tag{4.82}$$

$$= \sup_{\substack{\mathbf{x}\in\mathbb{R}^n \\ ||x||_{(n)}=1}} (|x_1| + |x_2| \cdots + |x_n|)M \tag{4.83}$$

$$= \sup_{\substack{\mathbf{x}\in\mathbb{R}^n \\ ||x||_{(n)}=1}} 1 \cdot M \tag{4.84}$$

$$= M = ||A||_1 \tag{4.85}$$

since the maximum is obtained by letting $\mathbf{x} = \mathbf{e}_k$ *where* k *is a column* $\mathbf{a}_k$ *whose sum of absolute values is* M.

Showing that the vector norm $||\cdot||_\infty$ induces the ℓ_∞ matrix norm is the same argument as in Example 4.3.36 except M is the maximum row sum of absolute values. Note that the matrix norm induced by the ℓ_2 (Euclidean) vector norm is *not* the Frobenius norm, but rather the ℓ_2 or *spectral matrix norm.* We will introduce the spectral norm as needed as it involves more machinery than we currently have.

Equipped with matrix norms, we may now introduce the important concept of the *condition number* of a matrix.

Definition 4.3.37 (Condition Number of a Matrix). *Let* A *be an invertible matrix and* $||\cdot||$ *any induced matrix norm or the Frobenius norm*[10]. *Then the* condition number *of* A *is*

$$\kappa(A) := ||A||\ ||A^{-1}||.$$

The reason this is called a condition number is as follows: suppose we are solving a linear system $A\mathbf{x} = \mathbf{b}$, with A invertible, and supposed $\mathbf{x}$ is slightly *perturbed*, that is, it experiences a small change, say $\Delta\mathbf{x}$. Then to find the perturbation of $\mathbf{b}$ (call it $\Delta\mathbf{b}$)

$$A(\mathbf{x} + \Delta\mathbf{x}) = A\mathbf{x} + A\Delta\mathbf{x} = \mathbf{b} + \Delta\mathbf{b} \tag{4.86}$$

[10]The condition number of a matrix can be defined using any matrix norm but the conditional number is only useful if satisfies property *ii*) in Remark 4.3.35. It can be shown that the Frobenius norm satisfies property *ii*) – see Exercise 4.18.

and since $\mathbf{b} = A\mathbf{x}$, we get

$$A\Delta\mathbf{x} = \Delta\mathbf{b}. \tag{4.87}$$

Thus $\Delta\mathbf{x} = A^{-1}\Delta\mathbf{b}$ and for any induced matrix norm or the Frobenius norm [or any norm satisfying property *ii*) in Remark 4.3.35]

$$||\Delta\mathbf{x}|| = ||A^{-1}\Delta\mathbf{b}|| \leq ||A^{-1}||\ ||\Delta\mathbf{b}||. \tag{4.88}$$

Moreover, since $\mathbf{b} = A\mathbf{x}$ we again have by property *ii*)

$$||\mathbf{b}|| = ||A\mathbf{x}|| \leq ||A||\ ||\mathbf{x}|| \tag{4.89}$$

giving

$$\frac{1}{||\mathbf{x}||} \leq \frac{||A||}{||\mathbf{b}||}. \tag{4.90}$$

Combining 4.88 and 4.90 gives

$$\frac{||\Delta\mathbf{x}||}{||\mathbf{x}||} \leq ||A||\ ||A^{-1}||\frac{||\Delta\mathbf{b}||}{||\mathbf{b}||} = \kappa(A)\frac{||\Delta\mathbf{b}||}{||\mathbf{b}||}. \tag{4.91}$$

Notice that what 4.91 tells us is that ratio of the perturbation of the change in the constants on the right hand side of $A\mathbf{x} = \mathbf{b}$ can lead to a perturbation of the solution $\mathbf{x}$ by up to a factor of $\kappa(A)$; that is, any matrix with a large condition number can have a big change in the solution $\mathbf{x}$ given a small change in the right hand side constants $\mathbf{b}$. This translates to computer usage in that a small round-off error could lead to a serious error in the solution of a linear system. Such matrices are said to be *ill-conditioned* and, as a rule of thumb,:

Highlight 4.3.38 (Rule of Thumb for the Condition Number of a Matrix). *If the condition number of a matrix $A = \kappa(A) = 10^s$, then one can expect to loose up to k digits of precision in the computer solution of $A\mathbf{x} = \mathbf{b}$.*

Matrices that have small condition numbers are said to be *well-conditioned* matrices.

Example 4.3.39 (Ill-conditioned Matrix). *For a positive integer n, consider the nth* Hilbert matrix

$$H_n := \begin{bmatrix} 1 & \frac{1}{2} & \frac{1}{3} & \frac{1}{4} \\ \frac{1}{2} & \frac{1}{3} & \frac{1}{4} & \frac{1}{5} \\ \frac{1}{3} & \frac{1}{4} & \frac{1}{4} & \frac{1}{6} \\ \frac{1}{4} & \frac{1}{4} & \frac{1}{6} & \frac{1}{7} \end{bmatrix}.$$

H_n turns out to be an ill-conditioned matrix. An easy way to see this is to solve the matrix equation $H_4X = I$. We note that when a computer does arithmetic, it does not store a fraction and only has a finite amount of memory to store a decimal. Moreover, it sees each number in scientific notation, as in 2.2598×10^7.[11] *If we assume the computer is has five digits in its significand, it will store* 1/3 *as* 0.33333 *and the calculation* $12.154 + 0.011237$ *as* 12.165.[12]

[11]Of course, the computer is seeing the number in binary, but we are not going to worry about that detail here.

[12]The real situation is a little more involved than this. See [1], for example, and *IEEE single precision floating point representation.*

If we assume five digits in the significand, then the calculated solution is

$$X = H_n^{-1} I = \hat{H}_n^{-1} = \begin{bmatrix} 16.248 & -122.72 & 246.49 & -144.20 \\ -122.72 & 1229.9 & -2771.3 & 1726.1 \\ 246.49 & -2771.3 & 6650.1 & -4310.0 \\ -144.20 & 1726.1 & -4310.0 & 2871.1 \end{bmatrix}.$$

The true solution is

$$X = H_n^{-1} I = H_n^{-1} = \begin{bmatrix} 16 & -120 & 240 & -140 \\ -120 & 1200 & -2700 & 1680 \\ 240 & -2700 & 6480 & -4200 \\ -140 & 1680 & -4200 & 2800 \end{bmatrix}.$$

Calculating the condition number using the ℓ_1 norm

$$\begin{aligned} \kappa(H_4) = ||H_4||_1 ||H_4^{-1}||_1 &= \left(1 + \frac{1}{2} + \frac{1}{3} + \frac{1}{4}\right)(240 + 2700 + 6480 + 4200) \\ &= 28375 \approx 10^{4.4529}. \end{aligned} \tag{4.92}$$

As $s \approx 4$ for H_4, by Highlight 4.3.38, we can expect a loss of accuracy in up to four of the significant digits (from the right) of some of the calculated values. For H_4, this occurs three times and is illustrated with the highlight entries of the calculate $\hat{H}_4^{-1}$ and the true H_4^{-1}.

4.3.5 Vector Spaces

We have discussed the basic algebra of vectors. The collections of these objects often behave nicely and lead to the following definition.

Definition 4.3.40 (Vector Space). *A set of vectors V together with a set of scalars (numbers) S is a* **vector space** *if for all $\mathbf{u}, \mathbf{v}, \mathbf{w} \in V$ and $a, b \in S$*

1. $\mathbf{u} + \mathbf{v}$ *is in V (i.e. V is* ***closed under vector addition****);*
2. $\mathbf{u} + \mathbf{v} = \mathbf{v} + \mathbf{u}$ *(****vector addition is commutative****);*
3. $\mathbf{u} + (\mathbf{v} + \mathbf{w}) = (\mathbf{u} + \mathbf{v}) + \mathbf{w}$ *(****vector addition is associative****);*
4. *there exist a vector $\mathbf{0}$ in V such that $\mathbf{u} + \mathbf{0} = \mathbf{u}$ (****existence of an additive vector identity****);*
5. *there exist a vector $-\mathbf{u}$ in V such that $-\mathbf{u} + \mathbf{u} = \mathbf{0}$ (****existence of additive vector inverses****);*
6. $a\mathbf{u}$ *is in V (****closure under scalar multiplication****);*
7. $1\mathbf{u} = \mathbf{u}$ *(****existence of a scalar multiplicative identity****);*
8. $a(\mathbf{u} + \mathbf{v}) = a\mathbf{u} + a\mathbf{v}$ *(****scalar multiplication distributes over vector addition****);*
9. $(a + b)\mathbf{u} = a\mathbf{u} + b\mathbf{u}$ *(****scalar multiplication distributes over scalar addition****); and*
10. $a(b\mathbf{u}) = (ab)\mathbf{u}$ *(****associativity of scalar multiplication****).*

A subset of a vector space which is itself a vector space is called a ***subspace***.

A natural next step is to introduce the essence of a vector space, that is the important idea of a *basis* of a vector space. For this we need a few more definitions.

Definition 4.3.41 (Linearly Independent). *A nonempty set of vectors* $\{\mathbf{v}_1, \mathbf{v}_2, \dots \mathbf{v}_n\}$ *in a vector space* V *is* ***linearly independent*** *if and only if the only coefficients satisfying*

$$c_1\mathbf{v}_1 + c_2\mathbf{v}_2 + \cdots + c_n\mathbf{v}_n = \mathbf{0}$$

are $c_1 = c_2 = \cdots = c_n = 0$*. Vectors that are not linearly independent are said to be* ***linearly dependent****.*

Example 4.3.42. *In* $\mathbb{R}^2$*, the vectors* $\mathbf{u}_1 = \langle 1, 1\rangle$ *and* $\mathbf{u}_2 = \langle 2, -1\rangle$ *are linearly independent as the only solution to* $c_1\mathbf{u}_1 + c_2\mathbf{u}_2 = 0$ *is* $c_1 = c_2 = 0$ *(try it!). The set of vectors* $\{\mathbf{u}_1, \mathbf{u}_2.\mathbf{u}_3\}$ *where* $\mathbf{u}_3 = \langle 1, -2\rangle$ *is a set of linearly dependent vectors as* $1\mathbf{u}_1 - 1\mathbf{u}_2 + 1\mathbf{u}_3 = 0$*.*

Definition 4.3.43 (Span). *For vectors* $\mathbf{v}_1, \mathbf{v}_2, \dots \mathbf{v}_n$ *in a vector space* V *with scalars* S*, the* ***span*** *of* $\mathbf{v}_1, \mathbf{v}_2, \dots \mathbf{v}_n$ *is the set of all linear combinations of* $\mathbf{v}_1, \mathbf{v}_2, \dots \mathbf{v}_n$ *with coefficients from* S*. That is,*

$$span(\mathbf{v}_1, \mathbf{v}_2, \dots \mathbf{v}_n) = \{c_1\mathbf{v}_1 + c_2\mathbf{v}_2 + \dots c_n\mathbf{v}_n \mid c_i \in S\}.$$

Example 4.3.44. *The vectors* $\mathbf{u}_1 = \langle 1, 1\rangle$ *and* $\mathbf{u}_2 = \langle 2, -1\rangle$ *span* $\mathbb{R}^2$ *as any arbitrary vector* $\langle x, y\rangle$ *of* $\mathbb{R}^2$ *can be expressed as a linear combination of* $\mathbf{u}_1$ *and* $\mathbf{u}_2$ *by* $\langle x, y\rangle = \frac{x+2y}{3}\langle 1, 1\rangle + \frac{x-y}{3}\langle 2, -1\rangle$*.*

Definition 4.3.45 (Basis). *A set of vectors* U *in a vector space* V *is a basis of* V *if*

1. *U spans V and*
2. *U is a linearly independent set of vectors.*

Example 4.3.46. *From Example 4.3.42 and Example 4.3.44, we see that the vectors* $\mathbf{u}_1 = \langle 1, 1\rangle$ *and* $\mathbf{u}_2 = \langle 2, -1\rangle$ *form a basis of* $\mathbb{R}^2$*.*

Note that $\langle 1, 0\rangle$ and $\langle 0, 1\rangle$ also form a basis of $\mathbb{R}^2$ and thus a basis of a vector space need not be unique.

Armed with the definitions we may now state some very important theorems. The proofs of these theorems can be found in any Linear Algebra text.

Theorem 4.3.47. *If a vector space has a finite basis, then all bases of the vector space have the same number of vectors.*

Because of this theorem, the number of vectors in a basis for a vector space V is said to be the ***dimension*** of V, written $dim(V)$.

We note that as a matter of convention, we define $\dim(\emptyset) = -1$.

Let matrix A have real entries in m rows and n columns. We refer to A as an $m \times n$ matrix. The span of the columns of A is a subspace of $\mathbb{R}^m$ and is called the ***column space*** of A which is usually denoted $col(A)$. Similarly $row(A)$ is the ***row space*** of A, that is, the subspace of $\mathbb{R}^n$ spanned by the rows of A. The solution set to the homogeneous system of equations $A\mathbf{x} = \mathbf{0}$ is a subspace of $\mathbb{R}^n$ and is called the ***null space*** of A and is denoted $null(A)$. Note that $\mathbf{0}$ is in the null space of A for every matrix A.

Equipped with these, we may state

Theorem 4.3.48. *The elementary row operations (listed in Theorem 4.2.2) do not change the row or null space of a matrix.*

Note well that the previous theorem did not mention the column space of a matrix. The reason for its absence is that elementary row operations may change the column space of a matrix; as seen in the next example.

Example 4.3.49 (Row Operations can change the Column Space).
Consider the matrix

$$A = \begin{bmatrix} 1 & 0 & 1 \\ 0 & 1 & 0 \\ 1 & 1 & 1 \end{bmatrix}.$$

To determine the column space of A let a, b, and c be arbitrary scalars. Then

$$a \begin{bmatrix} 1 \\ 0 \\ 1 \end{bmatrix} + b \begin{bmatrix} 0 \\ 1 \\ 1 \end{bmatrix} + c \begin{bmatrix} 1 \\ 0 \\ 1 \end{bmatrix} = \begin{bmatrix} a+c \\ b \\ a+b+c \end{bmatrix}. \tag{4.93}$$

Thus

$$Col(A) = \left\{ \begin{bmatrix} s \\ t \\ s+t \end{bmatrix} : s, t \in \mathbb{R} \right\} \tag{4.94}$$

(so $s = a + c$ and $b = t$). Thus, for example, $[1, 2, 3]^T \in Col(A)$.

The standard row reductions on A give

$$A \xrightarrow{-R_1 - R_2 + R_3 \to R_3} \begin{bmatrix} 1 & 0 & 1 \\ 0 & 1 & 0 \\ 0 & 0 & 0 \end{bmatrix} =: A^* \tag{4.95}$$

but now we have

$$Col(A^*) = \left\{ \begin{bmatrix} s \\ t \\ 0 \end{bmatrix} : s, t \in \mathbb{R} \right\} \tag{4.96}$$

and this time $[1, 2, 3]^T \notin Col(A^*)$.

We note that it is easy to see that $Row(A^*) = \{[s,t,s|s,t \in \mathbb{R}]$. *If we multiply arbitrary scalars times the rows of* A *we get*

$$a[1,0,1] + b[0,1,0] + c[1,1,1] = [a+c, b+c, a+c] \tag{4.97}$$

and we can get $[s,t,s]$ *by putting* $a = s$, $b = t$, *and* $c = 0$. *This shows* $Row(A) = Row(A^*)$.

A remarkable fact about any matrix is

Theorem 4.3.50. *The row space and column space of a matrix have the same dimension.*

The common dimension of the row and column spaces of a matrix A is called the ***rank*** of A and written $rank(A)$. We then have

Corollary 4.3.51. *For any matrix* A, $rank(A^T) = rank(A)$.

Proof.

$$rank(A^T) = dim(row(A^T)) = dim(col(A)) = rank(A).$$

□

The dimension of the null space of A is also named and is called the ***nullity*** of A and is denoted $nullity(A)$.

And lastly the beautiful

Theorem 4.3.52 (The Rank Theorem)**.** *For any* $m \times n$ *matrix* A,

$$rank(A) + nullity(A) = n.$$

If our matrix A is the coefficient matrix of a system of linear equations, the *Rank Theorem* tells us how many free variables and how many parameters the system's solution has. We illustrate this with an example.

Example 4.3.53. *Find all solutions to the following system:*

$$\begin{cases} x_1 & & +x_3 & +2x_4 & = & 13 \\ 5x_1 & +x_2 & +7x_3 & +14x_4 & = & 79 \\ 12x_1 & +2x_2 & +19x_3 & +38x_4 & = & 196. \end{cases}$$

Solution. The system has as its augmented matrix $[A|b]$

$$\left[\begin{array}{cccc|c} 1 & 0 & 1 & 2 & 13 \\ 5 & 1 & 7 & 14 & 79 \\ 12 & 2 & 19 & 38 & 196 \end{array}\right]$$

and using Gauss-Jordan Elimination

$$\left[\begin{array}{cccc|c} 1 & 0 & 1 & 2 & 13 \\ 5 & 1 & 7 & 14 & 79 \\ 12 & 2 & 19 & 38 & 196 \end{array}\right] \xrightarrow[-5R_1+R_2\to R2]{-12R_1+R_3\to R_3} \left[\begin{array}{cccc|c} 1 & 0 & 1 & 2 & 13 \\ 0 & 1 & 2 & 4 & 14 \\ 0 & 2 & 7 & 14 & 40 \end{array}\right] \quad (4.98)$$

$$\xrightarrow{-2R_2+R_3\to R_3} \left[\begin{array}{cccc|c} 1 & 0 & 1 & 2 & 13 \\ 0 & 1 & 2 & 4 & 14 \\ 0 & 0 & 3 & 6 & 12 \end{array}\right] \quad (4.99)$$

$$\xrightarrow{\frac{1}{3}R_3\to R_3} \left[\begin{array}{cccc|c} 1 & 0 & 1 & 2 & 13 \\ 0 & 1 & 2 & 4 & 14 \\ 0 & 0 & 1 & 2 & 4 \end{array}\right] \quad (4.100)$$

$$\xrightarrow[-R_3+R_1\to R_1]{-2R_3+R_2\to R_2} \left[\begin{array}{cccc|c} 1 & 0 & 0 & 0 & 9 \\ 0 & 1 & 0 & 0 & 6 \\ 0 & 0 & 1 & 2 & 4 \end{array}\right]. \quad (4.101)$$

The rows of A are (as vectors) $[1,0,0,0], [0,1,0,0], [0,0,1,2]$ which are clearly linearly independent. Hence $dim[row(A)] = 3$. The columns of A are (again, as vectors) $[1,0,0]^T, [0,1,0]^T, [0,0,1]^T, [0,0,2]^T$ but by Theorem 4.3.50, $dim[col(A)] = 3$ therefore this set is not linearly independent. Removing $[0,0,1]^T$ or $[0,0,2]^T$ fixes this, so we choose $[1,0,0]^T, [0,1,0]^T, [0,0,1]^T$ as a basis for $col(A)$. By the Rank Theorem, since $rank(A) = 3$, $null(A) = 4 - 3 = 1$. Thus we have one free variable and require one parameter. For convenience, it is easiest to choose x_4 as the free variable thus we put $x_4 = t$. Thus the solution set for the system is

$$x_1 = 9, x_2 = 6, x_3 = 4 - 2t, x_4 = t \text{ where } t \text{ is any real number.}$$

■

As if the Rank Theorem is not beautiful enough, we offer the following theorem that brings together much of what we have just discussed.

Theorem 4.3.54 (The Fundamental Theorem of Invertible Matrices). *If A is an $n \times n$ matrix, then the following are equivalent:*

i. *A is invertible.*
ii. *$A\mathbf{x} = \mathbf{b}$ has a unique solution for every $\mathbf{b}$ in $\mathbb{R}^n$.*
iii. *$A\mathbf{x} = \mathbf{0}$ has only the trivial solution.*
iv. *The reduced row echelon form of A is I_n.*
v. *A is the product of elementary matrices.*
vi. *$rank(A) = n$*
vii. *$nullity(A) = 0$*
viii. *The column vectors of A are linearly independent.*
ix. *The column vectors of A span $\mathbb{R}^n$.*
x. *The column vectors of A are form a basis for $\mathbb{R}^n$.*
xi. *The row vectors of A are linearly independent.*

xii. *The row vectors of* A *span* $\mathbb{R}^n$.
xiii. *The row vectors of* A *are form a basis for* $\mathbb{R}^n$.
xiv. $\det(A) \neq 0$
xv. 0 *is not an eigenvalue of* A.

We could also include in the theorem results involving linear transformations, but as they are not relevant to this text we have chosen to leave them out. A result involving eigenvalues has been included, which we discuss next.

4.4 Matrix Properties Important to Optimization

4.4.1 Eigenvalues

We have seen that matrix multiplication can be quite cumbersome where scalar multiplication is quite kind. Would it not be wonderful if matrix multiplication could be replaced by scalar multiplication? Indeed this would be a dream, and – in general – is asking too much, but there are situations where this dream is a reality and the properties that follow are almost magical.

Let A be an $n \times n$ matrix with real entries. A scalar λ satisfying $A\mathbf{x} = \lambda\mathbf{x}$ for some nonzero $\mathbf{x}$ in $\mathbb{R}^n$ is called an *eignenvalue* of A with $\mathbf{x}$ called an *eigenvector* of A corresponding to λ.

Example 4.4.1. *Consider the matrix* $A = \begin{bmatrix} 2 & 3 \\ 2 & 1 \end{bmatrix}$.

Since

$$\begin{bmatrix} 2 & 3 \\ 2 & 1 \end{bmatrix} \begin{bmatrix} 3 \\ 2 \end{bmatrix} = \begin{bmatrix} 12 \\ 8 \end{bmatrix} = 4 \begin{bmatrix} 3 \\ 2 \end{bmatrix}, \tag{4.102}$$

$\lambda = 4$ *is an eigenvalue of* A *and* $\mathbf{x} = [3, 2]^T$ *is an eigenvector of* A *corresponding to the eigenvalue* $\lambda = 4$.

If $\mathbf{x} \neq \mathbf{0}$ is an eigenvector of the $n \times n$ matrix A corresponding to the eigenvalue λ, then

$$A\mathbf{x} = \lambda\mathbf{x} \Longleftrightarrow [A - \lambda I]\,\mathbf{x} = \mathbf{0} \tag{4.103}$$

which means[13] that eigenvalues λ of A must satisfy $\det[A - \lambda I] = 0$.

Example 4.4.2. *Let* A *be the matrix in Example 4.4.1. Since*

$$\det \begin{bmatrix} 2-\lambda & 3 \\ 2 & 1-\lambda \end{bmatrix} = \lambda^2 - 3\lambda - 4 = 0 \tag{4.104}$$

has $\lambda = 4, -1$ *as solutions, the eigenvalues of* A *are* $\lambda = 4$ *and* $\lambda = -1$.

[13] For the details missing here, see any Linear Algebra text.

Though finding eigenvalues will be enough for our work, for the sake of completeness, let us find the eigenvectors corresponding to the eigenvalue $\lambda = -1$ for A in the example.

Example 4.4.3. *Since $A\mathbf{x} = \lambda\mathbf{x}$, we have the matrix equation $(A - \lambda I)\mathbf{x} = \mathbf{0}$. With the A from Example 4.4.1 and its eigenvalue $\lambda = -1$, this gives a linear system which we solve by Guassian elimination:*

$$\left[\begin{array}{cc|c} 2-\lambda & 3 & 0 \\ 2 & 1-\lambda & 0 \end{array}\right] = \left[\begin{array}{cc|c} 3 & 3 & 0 \\ 2 & 2 & 0 \end{array}\right] \tag{4.105}$$

$$\longrightarrow \left[\begin{array}{cc|c} 1 & 1 & 0 \\ 1 & 1 & 0 \end{array}\right] \tag{4.106}$$

$$\longrightarrow \left[\begin{array}{cc|c} 1 & 1 & 0 \\ 0 & 0 & 0 \end{array}\right], \tag{4.107}$$

This means that $x_1 + x_2 = 0$, thus $x_2 = -x_1$ and therefore any vector of the form $[c, -c]$ where c is any constant is an eigenvector corresponding to the eigenvalue $\lambda = -1$ for the matrix A (try it!).

The example was not an anomaly; for a given eigenvalue there is always a corresponding set of associated eigenvectors.

Though, as stated, we will have no use for eigenvectors, we will see how nicely eigenvalues can be used to quickly solve optimization questions in Chapter 12.

4.4.2 Unimodular Matrices

We now explore a type of matrix will be important to our study of network flow problems, especially after we understand how difficult integer linear programming is.

Definition 4.4.4 (Unimodular Matrices)**.** *A unimodular matrix of order n is an $n \times n$ matrix with determinant ± 1.*

Example 4.4.5. *The matrix $A = \begin{bmatrix} \frac{5}{3} & \frac{1}{4} \\ \frac{7}{2} & \frac{9}{8} \end{bmatrix}$ is unimodular since* $\det(A) = \frac{5}{3} \cdot \frac{9}{8} - \frac{7}{2} \cdot \frac{1}{4} = 1$.

To see how wonderful unimodular matrices turn out to be, let us consider solving a 2×2 system of linear equations $A\mathbf{x} = \mathbf{b}$ where A is unimodular and of full rank (i.e. the row vectors and column vectors of A are linearly independent and thus there exists a unique solution). Moreover, let us further suppose A and $\mathbf{b}$ have integer entries.

Example 4.4.6. *Suppose for the 2×2 linear system $A\mathbf{x} = \mathbf{b}$ that A is of full rank, unimodular, and that A and $\mathbf{b}$ have integer entries. We then have*

$$\begin{bmatrix} a & b \\ c & d \end{bmatrix} \cdot \begin{bmatrix} x_1 \\ x_2 \end{bmatrix} = \begin{bmatrix} b_1 \\ b_2 \end{bmatrix}$$

and by Cramer's Rule (Theorem 4.3.24)

$$x_1 = \det(A_i)/\det(A) = \pm(b_1 d - b_2 b) \tag{4.108}$$

$$x_2 = \det(A_i)/\det(A) = \pm(a b_2 - c b_1) \tag{4.109}$$

where $\det(A) = \pm 1$ *because A is unimodular. Since a, b, c, d, b_1, and b_2 are all integers, x_1 and x_2 must be integers as well.*

If we had introduced the notion of adjoint (or adjugate) matrices, it would not be hard to prove by Cramer's Rule that if A is a unimodular matrix with integer entries then A^{-1} is also unimodular and has integer entries. In fact, it is the case that for a given order n, the collection of all $n \times n$ unimodular matrices with integer entries form a group[14], the *general linear group*, which is denoted $GL_n(\mathbb{Z})$. More relevant to our work, though, are the conditions which will guarantee integer solutions to a linear system.

Things worked very nicely in Example 4.4.6 but this was because we had a system that was 2×2. Larger ordered systems will require more work, so we need a stronger requirement to guarantee integer solutions. By Cramer's Rule (Theorem 4.3.24), we can solve a linear system using determinants and by Expansion by Minors (Theorem 4.3.18) we can reduce the determinant of any matrix down to linear combinations of determinants of 2×2 submatrices. Thus a way to get integer solutions to a larger system it would be a wonderful thing if every square submatrix of A had a determinant of 0 or ± 1. This leads to the following definition.

Definition 4.4.7 (Totally Unimodular Matrices). *An $m \times n$ matrix is* totally unimodular *if the determinant of any square submatrix is* 0, 1, *or* -1.

The definition of a totally unimodular matrix can be stated in an alternate form that is worth emphasizing.

Highlight 4.4.8. *We may alternatively state that a matrix is totally unimodular if every square, non-singular submatrix is unimodular.*

Example 4.4.9. *Clearly any identity matrix is totally unimodular. For a nontrivial example, let us show* $A = \begin{bmatrix} 1 & -1 & 0 \\ 0 & 1 & -1 \\ 1 & 0 & 1 \end{bmatrix}$ *is totally unimodular.*

Solution. By Exercises 20.9 and 20.10, A has $\binom{6}{3} = 20$ square submatrices. We will not be concerned with the empty square submatrix and clearly the nine 1×1 matrices all have the proper determinant. Checking the remaining 10:

[14] A group is a set with a binary operation defined on its elements. A group has the properties that it is closed under the operation (matrix multiplication in this case), associative, there is an identity, and every element has an inverse. This is addressed in Exercise 4.34.

$$\det\left(\begin{bmatrix}1 & -1\\0 & 1\end{bmatrix}\right) = 1, \det\left(\begin{bmatrix}0 & -1\\1 & 1\end{bmatrix}\right) = 1, \det\left(\begin{bmatrix}0 & 1\\1 & 0\end{bmatrix}\right) = -1, \quad (4.110)$$

$$\det\left(\begin{bmatrix}-1 & 0\\0 & 1\end{bmatrix}\right) = -1, \det\left(\begin{bmatrix}1 & 0\\1 & 1\end{bmatrix}\right) = 1, \det\left(\begin{bmatrix}1 & -1\\1 & 0\end{bmatrix}\right) = 1, \quad (4.111)$$

$$\det\left(\begin{bmatrix}-1 & 0\\1 & -1\end{bmatrix}\right) = 1, \det\left(\begin{bmatrix}1 & 0\\0 & -1\end{bmatrix}\right) = -1, \det\left(\begin{bmatrix}1 & -1\\0 & 1\end{bmatrix}\right) = 1, \text{ and} \quad (4.112)$$

$$\left(\begin{bmatrix}1 & -1\\0 & 1\end{bmatrix}\right) = 1 \quad (4.113)$$

$$\det(A) = 1 \cdot \det\left(\begin{bmatrix}1 & -1\\0 & 1\end{bmatrix}\right) - 1 \cdot \det\left(\begin{bmatrix}0 & -1\\1 & 1\end{bmatrix}\right) + 0 = 1 - 1 + 0 = 0 \quad (4.114)$$

Since all square submatrices of A have determinant 0,1, or -1, A is totally unimodular. ■

We had to check 10 nontrivial matrices in the example. A 4×4 matrix will have $\binom{8}{4} - 16 - 1 = 53$ nontrivial square submatrices (though a totally unimodular matrix need not be square) and a square matrix of order 5 will have $\binom{10}{5} - 25 - 1 = 226$. Clearly a brute force approach to checking total unimodularity is undesirable. Fortunately, we have this result due to I. Heller and C.B. Tompkins from 1956 [33]:

Theorem 4.4.10. *Let A be an $m \times n$ matrix with rows that can be partitioned into two disjoint sets R_1 and R_2. Then if*

- *every entry in A is 0, 1, or -1,*
- *every column of A contains at most two nonzero entries,*
- *two nonzero entries of a column of A have the same sign, then the row containing one of these entries is in R_1 and the other entry's row is in R_2, and*
- *two nonzero entries of a column of A have opposite signs, then the rows of both of these entries are both in R_1 or both in R_2*

then A is totally unimodular.

In the light of this theorem we can now see easily that the matrix in Example 4.4.9 is totally unimodular by putting row 1 in R_1, row 3 in R_2 with row 2 of the matrix belonging to either of R_1 or R_2. We note that P.D. Seymour's 1980 paper [52] gives a complete characterization of totally unimodular matrices, thought the sufficient conditions in the previous theorem are enough for us to do our work.

As mentioned in the paragraph preceding Definition 4.4.7, the importance of a totally unimodular matrix A is that it guarantees integer solutions to $A\mathbf{x} = \mathbf{b}$ when A and $\mathbf{b}$ have integer entries. The significance of this will be more clear once we understand how much more difficult integer linear programming is than linear programming. We will explore all of this in more detail in Section 24.2.

4.5 Keywords

linear function, linear inequality, system of linear equations, solution set of a system of linear inequalities, (un)bounded set, graph, corner/extreme point, back substitution, Gauss-Jordan elimination, elementary row operations, row echelon form versus reduced row echelon form, matrix, submatrix, $m \times n$ matrix, transpose, symmetric matrix, Hermitian matrix, self-adjoint matrix, (row/column) vector, scalar multiplication, dot product, magnitude of a vector, unit vector, normalizing a vector, matrix multiplication, (multiplicative) identity, inverse, identity matrix, vector norm, matrix norm, induced matrix norm, condition number $\kappa(A)$ of a matrix A, ill-conditioned matrix, vector space, subspace, linearly independent, span, basis, dimension, column space, row space, null space, rank, nullity, eigenvalue, eigenvector.

4.6 Exercises

Exercise 4.1. *Convert the following system of linear equations to a matrix and solve:*

$$x + 2y + 3z = 4$$
$$4x + 3y + z = 40$$
$$2x + 3y + z = 10$$

Exercise 4.2. *Use a graph to show why there is no solution to the system of equations:*

$$x + 0y = 2$$
$$x + y = 1$$
$$-1x + y = 3$$

Exercise 4.3. *Let*

$$A = \begin{bmatrix} 1 & -1 & 0 \\ 0 & 1 & 2 \\ 3 & -1 & 4 \end{bmatrix}. \tag{4.115}$$

Let A^* be the reduce echelon form of A. As in Example 4.3.49, show that $Row(A) = Row(A^*)$ but $Col(A) \neq Col(A^*)$. Provide an example of a vector that is in $Col(A)$ but not in $Col(A^*)$.

Exercise 4.4. *We can associate the polynomial*

$$f(x) = 5x^3 + 2x^2 - 3x + 2$$

with the vector $\mathbf{v} = [5, 2, -3, 2]^T$. Find a matrix A such that $A\mathbf{v} = f'(x)$.

Exercise 4.5. *Let A be an $m \times n$ matrix and $\mathbf{e}_i$ the ith standard basis (column) vector of $\mathbb{R}^n$. Show that*

$$A\mathbf{e}_i = \mathbf{a}_i \tag{4.116}$$

where $\mathbf{a}_i$ is the ith column of A.

Exercise 4.6. *Show that for all $\mathbf{u}$ and $\mathbf{v}$ in $\mathbb{R}^n$*

$$\mathbf{u} \cdot \mathbf{v} = \frac{1}{4}||\mathbf{u} + \mathbf{v}||^2 - \frac{1}{4}||\mathbf{u} - \mathbf{v}||^2 \tag{4.117}$$

where $||\cdot||$ is the Euclidean norm.

Exercise 4.7. *Prove the following properties of the dot product for $\mathbf{a}, \mathbf{b}, \mathbf{c} \in \mathbb{R}^n$ with c a scaler.*

i) $\mathbf{a} \cdot \mathbf{a} = ||\mathbf{a}||^2$ *where $||\cdot||$ is the Euclidean norm*
ii) $\mathbf{a} \cdot \mathbf{b} = \mathbf{b} \cdot \mathbf{a}$
iii) $\mathbf{a} \cdot (\mathbf{b} + \mathbf{c}) = \mathbf{a} \cdot \mathbf{b} + \mathbf{a} \cdot \mathbf{c}$
iv) $(c\mathbf{a}) \cdot \mathbf{b} = c(\mathbf{a} \cdot \mathbf{b}) = \mathbf{a} \cdot (c\mathbf{b})$
v) $\mathbf{0} \cdot \mathbf{a} = 0$

Exercise 4.8. *Prove Proposition 4.3.6 (HINT: use the Law of Cosines and properties of the dot product).*

Exercise 4.9. *Find the inverse of* $\begin{bmatrix} 2 & 3 \\ 4 & 5 \end{bmatrix}$.

Exercise 4.10. *Find the inverse of* $\begin{bmatrix} 1 & -1 & 2 \\ 2 & -3 & 4 \\ 3 & -3 & 7 \end{bmatrix}$.

Exercise 4.11. *Let A be an $n \times n$ matrix, $A \neq \mathbf{0}$, and consider the identity matrix $I = I_n$ as in Definition 4.3.10. Show*

i) $AI = A = IA$ *and*

ii) if $AI = A = IA$ and $AI' = A = I'A$, then $I = I'$.

(This exercise shows that i) I serves as an identity and ii) that identity matrices are unique up to their order.)

Exercise 4.12. *Let A be an $n \times n$ and suppose there exist B and B' such that $AB = I = B'A$. Show that $B = B'$. (Here $B = A^{-1}$ and this exercise shows that inverse matrices are unique.)*

Exercise 4.13. *Use the definition of a determinant (Definition 4.3.17) to prove that one can find determinants by the General Expansion by Minors (Theorem 4.3.18).*

Exercise 4.14. *Prove the properties of determinants in Theorem 4.3.23.*

Exercise 4.15. *For $\mathbf{v}$ in $\mathbb{R}^n$, show that $||\mathbf{v}||_1$, $||\mathbf{v}||_2$, and $||\mathbf{v}||_\infty$ from Theorem 4.3.27 are vector norms.*

Exercise 4.16. *For A in $\mathbb{R}_{m\times n}$, show that $||A||_1$, $||A||_F$, and $||A||_\infty$ from Theorem 4.3.31 are matrix norms.*

Exercise 4.17. *Prove Remark 4.3.35.*

Exercise 4.18. *Show that the Frobenius matrix norm $||\cdot||_F$ introduced in Theorem 4.3.31 satisfies property ii) of Remark 4.3.35.*

Exercise 4.19. *Let $A = \begin{bmatrix} 2 & -1 & 0 \\ -1 & 2 & -1 \\ 0 & -1 & 2 \end{bmatrix}$. Determine the condition number of A, $\kappa(A)$, under $||\cdot||_1$, $||\cdot||_F$, and $||\cdot||_\infty$.*

Exercise 4.20. *Let $B = \begin{bmatrix} 0 & 1 & 0 \\ 1 & 1 & 1 \\ 0 & 0 & 1 \end{bmatrix}$. Determine the condition number of B, $\kappa(B)$, under $||\cdot||_1$, $||\cdot||_F$, and $||\cdot||_\infty$.*

Exercise 4.21. *Rework Example 4.3.39 but for H_3. Your response should include the true H_3^{-1}, the calculated $\hat{H}_3^{-1}$ assuming five digits in the significand, and $\kappa(H_3)$. Note the significant differences between $\hat{H}_3^{-1}$ and H_3^{-1} and how this relates to $\kappa(H_3)$.*

Exercise 4.22. *Give an informal explanation as to why the length of a list of linearly independent vectors must be less than or equal to length of a spanning list of vectors.*

Exercise 4.23. *Show why or why not the set*

$$\{\mathbf{v} \in \mathbb{R}^2 \quad | \quad v_1, v_2 \geq 0\}$$

where v_1 and v_2 are individual components of a vector, is a vector space.

Exercise 4.24. *Prove that the intersection of vector spaces is also a vector space.*

Exercise 4.25. *Show that* $\{[1,2]^T, [-2,3]^T\}$ *is a linearly independent set of vectors.*

Exercise 4.26. *Consider the following set of vectors:* $\{[1,1,0]^T, [1,0,1]^T, [0,1,a]^T\}$.

i) *Show that the set is linearly independent if and only if* $a \neq -1$.

ii) *Since the set is linearly dependent if* $a = -1$, *write* $[0,1,-1]^T$ *as a linear combination of* $[1,1,0]^T$ *and* $[1,0,1]^T$.

Exercise 4.27. *Allow* $\mathbf{v} \in \mathbf{R^n}$ *to be a vector. Show that* $A = \mathbf{v}\mathbf{v}^T$ *must be* positive semidefinite *(see Definition 12.1.2); that is, both* $A = A^T \geq 0$ *and* $\mathbf{x}^T A\mathbf{x} \geq 0$ *for all* $\mathbf{x} \in \mathbf{R^n}$.

Exercise 4.28. *Suppose we think of a matrix* T *as a function* $T : R^n \to R^m$. *If* $n > m$, *state whether you believe* T *would be injective or surjective (or both) and give an informal reason as to why. Give another statement with explanation for* $n < m$.

Exercise 4.29. *Let* V *be a vector space and* $S \subseteq V$. *Show that* S *is also a vector space if and only if* $a\mathbf{x} + b\mathbf{y} \in S$ *whenever* $\mathbf{x}, \mathbf{y} \in S$ *and* a *and* b *are any scalars; that is* S *is a subspace of the vector space* V *if and only if* S *is closed under linear combinations.*

Exercise 4.30. *Let* S *be a subspace of* $\mathbb{R}^n$, $B = \{\mathbf{v}_1, \ldots, \mathbf{v}_m\}$ *a basis for* S, *and* $\mathbf{u}$ *an arbitrary vector in* S. *Show there exists a unique set of scalars* $\{c_1, \ldots, c_m\}$ *such that* $\mathbf{u} = c_1\mathbf{v}_1 + \cdots + c_m\mathbf{v}_m$.

Exercise 4.31. *Prove Theorem 4.3.50.*

Exercise 4.32. *Let* $A = \begin{bmatrix} 1 & 1 \\ 4 & -2 \end{bmatrix}$.

i) *Find the eigenvalues of* A.
ii) *For each eigenvalue of* A, *find the set of corresponding eigenvectors.*

Exercise 4.33. *Let* $B = \begin{bmatrix} 1 & 0 & 0 \\ 0 & -1 & 0 \\ 1 & 1 & 0 \end{bmatrix}$.

i) *Find the eigenvalues of* A.
ii) *For each eigenvalue of* A, *find the set of corresponding eigenvectors.*

Exercise 4.34. *In Abstract Algebra, a* group *is a set* G *together with a binary operation* $\star$ *such that for any* a, b, *and* c *in* G

i) $a \star b \in G$ *(*G *is* closed *under* $\star$*),*

ii) $(a \star b) \star c = a \star (b \star c)$ *($\star$ is* associative*),*

iii) *there exists an $e \in G$ such that $a \star e = a = e \star a$ (there exist an* identity element*), and*

iv) *there exists for each $g \in G$ elements h and h' such that $g \star h = e = h' \star g$ and that $h = h'$ (every element has a left and right inverse and this inverse is the same).*

Show that the set of all unimodular matrices with integer entries $GL_n(\mathbb{Z})$ – the general linear group – is, in fact, a group.

Exercise 4.35. *Show each of the following matrices is unimodular (from [61]).*

i) $\begin{bmatrix} 2 & 3 & 5 \\ 3 & 2 & 3 \\ 9 & 5 & 7 \end{bmatrix}.$

ii) $\begin{bmatrix} 16 & 3 & 4 \\ 8 & 2 & 3 \\ 9 & 1 & 1 \end{bmatrix}.$

iii) $\begin{bmatrix} 48 & 5 & 8 \\ 24 & 2 & 3 \\ 25 & 2 & 3 \end{bmatrix}.$

iv) $\begin{bmatrix} 2 \cdot 4n(n+1) & 2n+1 & 4n \\ 4n(n+1) & n+1 & 2n+1 \\ 4n(n+1)+1 & n & 2n-1 \end{bmatrix}$ *for any n.*

Exercise 4.36. *Show that* $\begin{bmatrix} 1 & 0 & 1 \\ 0 & 1 & -1 \\ -1 & 0 & 1 \end{bmatrix}$ *is totally unimodular.*

5

Matrix Factorization

In Section 4.2, the reader was reminded of Gaussian Elimination to solve systems of linear equations. A large advantage of this technique is that it is easy to instruct a computer to efficiently carry out these steps with good stability[1] and, as such, this technique is important for numerical solutions to systems of equations. Heavy computer usage to solve systems of equations as well as their role in machine learning techniques are the motivation for this section. We will consider *LU factorization*, *QR factorization*, and *Cholesky decomposition* but note that there are other factorizations that are not included here. **It is important to realize that the reason for writing matrices in these forms is that it is not always best for a computer to solve things the way you and I would solve them on paper.** Various matrix factorizations exist to make computer run time tractable as well as to minimize round-off errors.

5.1 *LU* Factorization

Let us return to the example from Section 4.2 but rewrite it so that it is obvious that row swaps are not needed.

Example 5.1.1. *Use Gaussian Elimination to solve*

$$\begin{cases} x_1 - x_2 + 3x_3 & = -4 \\ 3x_1 - 5x_2 + x_3 & = 18 \\ -2x_1 + x_2 - x_3 & = 5 \end{cases}$$

Solution.

$$\left[\begin{array}{rrr|r} 1 & -1 & 3 & -4 \\ 3 & -5 & 1 & 18 \\ -2 & 1 & -1 & 5 \end{array}\right] \xrightarrow[2R_1+R_3\to R_3]{-3R_1+R_2\to R_2} \left[\begin{array}{rrr|r} 1 & -1 & 3 & -4 \\ 0 & -2 & -8 & 30 \\ 0 & -1 & 5 & -3 \end{array}\right] \tag{5.1}$$

$$\xrightarrow{-\frac{1}{2}R_2+R_3\to R_3} \left[\begin{array}{rrr|r} 1 & -1 & 3 & -4 \\ 0 & -2 & -8 & 30 \\ 0 & 0 & 9 & -18 \end{array}\right] \tag{5.2}$$

[1]See the very last paragraph on well- and ill-conditioned matrices in the Vector and Matrix Norms section of the last chapter (Section 4.3.4).

DOI: 10.1201/9780367425517-5

Back substitution will once again led to the desired solution. ■

The elementary row operations used in the solution of Example 5.1.1 can be expressed as the *elementary matrices*

$$E_1 = \begin{bmatrix} 1 & 0 & 0 \\ 0 & 1 & 0 \\ 2 & 0 & 1 \end{bmatrix}, \quad E_2 = \begin{bmatrix} 1 & 0 & 0 \\ -3 & 1 & 0 \\ 0 & 0 & 1 \end{bmatrix}, \quad E_3 = \begin{bmatrix} 1 & 0 & 0 \\ 0 & 1 & 0 \\ 0 & -\frac{1}{2} & 1 \end{bmatrix}.$$

Let A be the matrix representing the coefficients in the original system of equations, $\mathbf{x} = [x_1, x_2, x_3]^T$, and $\mathbf{b}$ the constants. Then (5.1.1) can be written $A\mathbf{x} = \mathbf{b}$. Notice the left hand side of the augmented matrix in 5.2 is an *upper triangular matrix* (i.e. the entries below the diagonal are each 0) which we will denote U.

With this notation we may write the row operations in the Gaussian Elimination as

$$E_3 E_2 E_1 A = U \tag{5.3}$$

and thus

$$A = {E_1}^{-1}{E_2}^{-1}{E_3}^{-1}U. \tag{5.4}$$

Then

$$A = LU \tag{5.5}$$

where

$$L = {E_1}^{-1}{E_2}^{-1}{E_3}^{-1} = \begin{bmatrix} 1 & 0 & 0 \\ 3 & 1 & 0 \\ -2 & \frac{1}{2} & 1 \end{bmatrix} \tag{5.6}$$

with L a *unit lower triangular matrix* (i.e. the diagonal entries are 1 and the entries above the diagonal are each 0) and from line 5.2

$$U = \begin{bmatrix} 1 & -1 & 3 \\ 0 & -2 & -8 \\ 0 & 0 & 9 \end{bmatrix} \tag{5.7}$$

where U is upper-triangular. That is,

$$A = \begin{bmatrix} 1 & -1 & 3 \\ 3 & -5 & 1 \\ -2 & 1 & -1 \end{bmatrix} = \begin{bmatrix} 1 & 0 & 0 \\ 3 & 1 & 0 \\ -2 & \frac{1}{2} & 1 \end{bmatrix} \begin{bmatrix} 1 & -1 & 3 \\ 0 & -2 & -8 \\ 0 & 0 & 9 \end{bmatrix} = LU. \tag{5.8}$$

When a square matrix A can be factored into the product of a lower triangular matrix L as its first factor and an upper triangular matrix U as its second factor, we say A has an *LU factorization*. We will only concern ourselves with square matrices in this text, but will mention that the notion of LU factorization can be extended to nonsquare matrices by requiring U to be in row echelon form.

A matrix does not always have an LU factorization and plenty is known about the necessary conditions, but we will be satisfied with the following condition:

Theorem 5.1.2. *If A is a square matrix requiring no row swaps in reducing it to row echelon form, then A has an LU factorization.*

You may have noticed that the numbers in L look pretty similar to the constants used in the row operations. This is no accident.

Theorem 5.1.3. *If A has an LU factorization and $R_j - m_{ij}R_i \to R_j$ is a row operation used to row reduce A, then m_{ij} is entry l_{ij} in matrix L.*

That is, for $i > j$, the ij^{th} entry in L is the multiplier used to eliminated entry a_{ij} when row reducing A; so $l_{ij} = m_{ji}$.

Moreover, regarding the L and the U we have

Theorem 5.1.4. *Suppose $A = LU$ where L is unit lower triangular and U is upper triangular. Then*

- *L is invertible,*
- *U is invertible if and only if A is invertible,*
- *L^{-1} is also lower triangular*
- *when it exists, U^{-1} is upper triangular, and*
- *the LU factorization is unique.*

It turns out that to determine a matrix's LU factorization, we do not need to work as hard as we did in Example 5.1.1 and the work that followed. In particular, we do not need to perform Gaussian Elimination to determine the multipliers that make up L.

Theorem 5.1.5. *(Dolittle's Method) Suppose an $n \times n$ matrix A has an LU factorization, that is*

$$\begin{bmatrix} a_{11} & \cdots & a_{1j} & \cdots & a_{1n} \\ a_{21} & \cdots & a_{2j} & \cdots & a_{2n} \\ \vdots & & \vdots & & \vdots \\ a_{i1} & \cdots & a_{ij} & \cdots & a_{in} \\ \vdots & & \vdots & & \vdots \\ a_{n1} & \cdots & a_{nj} & \cdots & a_{nn} \end{bmatrix} = \begin{bmatrix} 1 & 0 & \cdots & 0 & \cdots & 0 \\ m_{21} & 1 & \cdots & 0 & \cdots & 0 \\ \vdots & & \vdots & & \vdots & \\ m_{i1} & m_{i2} & \cdots & 1 & \cdots & 0 \\ \vdots & & \vdots & & \vdots & \\ m_{n1} & m_{n2} & \cdots & m_{nj} & \cdots & 1 \end{bmatrix}$$
$$\times \begin{bmatrix} u_{11} & u_{12} & \cdots & u_{1n-1} & u_{1n} \\ 0 & u_{22} & \cdots & u_{2n-1} & u_{2n} \\ \vdots & \vdots & & \vdots & \vdots \\ \vdots & \vdots & & \vdots & \vdots \\ 0 & 0 & \cdots & u_{n-1n-1} & u_{n-1n} \\ 0 & 0 & \cdots & 0 & u_{nn} \end{bmatrix}$$

Then

$$\begin{array}{ccccc}
u_{11}=a_{11}, & u_{12}=a_{12}, & u_{13}=a_{13}, & \cdots & u_{1n}=a_{1n} \\
a_{21}=m_{21}u_{11}, & a_{22}=m_{21}u_{12} & a_{23}=m_{21}u_{13} & \cdots & a_{2n}=m_{21}u_{1n} \\
 & +u_{22}, & +u_{23}, & & +u_{2n} \\
a_{31}=m_{31}u_{11}, & a_{32}=m_{31}u_{12} & a_{33}=m_{31}u_{13}+m_{32}u_{23} & \cdots & a_{3n}=m_{31}u_{1n}+m_{32}u_{2n} \\
 & +m_{32}u_{22}, & +u_{33}, & & +u_{3n} \\
\vdots & \vdots & \vdots & \vdots & \vdots
\end{array}$$

LU factorization can be used to solve a system of linear equations (see, e.g., [1]). If $A\mathbf{x} = \mathbf{b}$ is a system of equations and A is an $n \times n$ matrix with an LU factorization, then it can be shown that the operations count in solving the system of equations is of the order $\frac{1}{3}n^3$ which is of the same order as Gaussian Elimination, but slightly quicker for large n. The LU factorization technique has an advantage with memory storage as a computer can be programmed to rewrite A as the entries of L and U are calculated. Furthermore, if A is *sparse* – that is, many entries are 0 – then the LU factorization technique is much quicker than Gaussian Elimination.

5.2 Cholesky Decomposition

Recall that a real matrix A is Hermitian if $A = A^T$ and positive-semidefinite if $\mathbf{x}^T A\mathbf{x} > 0$ for every nonzero column vector $\mathbf{x}$. These matrices have a very special property.

Theorem 5.2.1 (Cholesky Decomposition).
When A is a real Hermitian (symmetric) positive-definite matrix, there exist a lower triangular matrix $\tilde{L}$ such that

$$A = \tilde{L}\tilde{L}^T.$$

Moreover, this factorization[2] *is unique.*

We will use a constructive proof to establish Cholesky Decomposition thus, in addition to establishing the result, we will obtain an algorithm to calculate $\tilde{L}$ for a given A.

[2]As stated when the definition of Hermitian was introduced, a little more is needed if A can have complex values. In this case, the decomposition is $A = LL^*$ where L^* is the conjugate transpose of L. As in the real case, this L is unique.

Proof.
Let A be real symmetric positive definite $n \times n$ matrix. Suppose $A = \tilde{L}\tilde{L}^T$ where $\tilde{L}\tilde{L}^T$ is lower triangular. Then

$$\begin{bmatrix} a_{11} & a_{12} & a_{13} & \cdots & a_{1n} \\ a_{21} & a_{22} & a_{23} & \cdots & a_{2n} \\ a_{31} & a_{32} & a_{33} & \cdots & a_{3n} \\ \vdots & \vdots & \vdots & \ddots & \vdots \\ a_{n1} & a_{n2} & a_{n3} & \cdots & a_{nn} \end{bmatrix} = \begin{bmatrix} l_{11} & 0 & 0 & \cdots & 0 \\ l_{21} & l_{22} & 0 & \cdots & 0 \\ l_{31} & l_{32} & l_{33} & \cdots & 0 \\ \vdots & \vdots & \vdots & \ddots & \vdots \\ l_{n1} & l_{n2} & l_{n3} & \cdots & l_{nn} \end{bmatrix} \times \begin{bmatrix} l_{11} & l_{21} & l_{31} & \cdots & l_{n1} \\ 0 & l_{22} & l_{32} & \cdots & l_{n2} \\ 0 & 0 & l_{33} & \cdots & l_{n3} \\ \vdots & \vdots & \vdots & \ddots & \vdots \\ 0 & 0 & 0 & \cdots & l_{nn} \end{bmatrix} \quad (5.9)$$

By direct calculation we have

$$a_{11} = l_{11}^2; \quad (5.10)$$

$$a_{1i} = l_{11}l_{i1} = a_{i1} \quad \text{for } i = 2, 3, \ldots, n; \quad (5.11)$$

$$\begin{aligned} a_{ii} &= [l_{i1}, l_{i2}, \ldots, l_{ii}, 0, \ldots, 0] \cdot [l_{i1}, l_{i2}, \ldots, l_{ii}, 0, \ldots, 0] \\ &= l_{i1}^2 + l_{i2}^2 + \cdots + l_{ii}^2 \quad \text{for } i = 2, 3, \ldots, n; \text{ and} \end{aligned} \quad (5.12)$$

$$\begin{aligned} a_{ij} &= [l_{i1}, l_{i2}, \ldots, l_{ij}, \ldots, l_{ii}, 0, \ldots, 0] \cdot [l_{j1}, l_{j2}, \ldots, l_{jj}, 0, \ldots, 0] \\ &= l_{i1}l_{j1} + l_{i2}l_{j2} + \cdots + l_{ij}l_{jj} \quad \text{for } j = 2, \ldots, n-1 \text{ and } j < i \leq n. \end{aligned} \quad (5.13)$$

Thus

$$l_{11} = \sqrt{a_{11}}; \quad (5.14)$$

$$l_{i1} = \frac{a_{i1}}{l_{11}} = \frac{a_{1i}}{l_{11}} \quad \text{for } i = 2, 3, \ldots, n; \quad (5.15)$$

$$l_{ii} = \sqrt{a_{ii} - \sum_{k=1}^{i-1} l_{ik}^2} \quad \text{for } i = 2, 3, \ldots, n; \text{ and} \quad (5.16)$$

$$l_{ij} = \left(a_{ij} - \sum_{k=1}^{j-1} l_{ik}l_{jk} \right) / l_{jj} \quad \text{for } j = 2, \ldots, n-1 \text{ and } j < i \leq n. \quad (5.17)$$

Without providing the details, we mention that an induction argument using the positive definiteness of A guarantees that all the square roots exist (and the entries of L are therefore unique). □

Example 5.2.2. *It can be shown that the matrix* $A = \begin{bmatrix} 1 & -2 & -1 \\ -2 & 13 & 2 \\ -1 & 2 & 2 \end{bmatrix}$ *is positive-definite. As it is clearly symmetric, it has a Cholesky decomposition. Using the proof of Theorem 5.2.1 we have*

$$l_{11} = \sqrt{a_{11}} = \sqrt{1} = 1, \tag{5.18}$$

$$l_{21} = \frac{a_{21}}{l_{11}} = \frac{-2}{1} = -2, \tag{5.19}$$

$$l_{22} = \sqrt{a_{22} - l_{21}^2} = \sqrt{13 - (-2)^2} = \sqrt{9} = 3, \tag{5.20}$$

$$l_{31} = \frac{a_{31}}{l_{11}} = \frac{-1}{1} = -1, \tag{5.21}$$

$$l_{32} = \frac{a_{32} - l_{31}l_{21}}{l_{22}} = \frac{2 - (-1)(-2)}{3} = 0, \text{ and} \tag{5.22}$$

$$l_{33} = \sqrt{a_{33} - l_{31}^2 - l_{32}^2} = \sqrt{2 - (-1)^2 - 0^2} = 1. \tag{5.23}$$

Thus

$$\begin{bmatrix} 1 & -2 & -1 \\ -2 & 13 & 2 \\ -1 & 2 & 2 \end{bmatrix} = \begin{bmatrix} 1 & 0 & 0 \\ -2 & 3 & 0 \\ -1 & 0 & 1 \end{bmatrix} \begin{bmatrix} 1 & -2 & -1 \\ 0 & 3 & 0 \\ 0 & 0 & 1 \end{bmatrix}.$$

Since $A = \tilde{L}\tilde{L}^T$, Cholesky decomposition can be viewed as "finding the square root of a matrix". More importantly, it can be shown that the run-time to solve the system $A\mathbf{x} = \mathbf{b}$, when A is symmetric positive-definite and Cholesky decomposition can be used, is $O(\frac{1}{3}n^3)$, which is twice as fast as Gaussian elimination.

5.3 Orthogonality

The reader is most likely familiar with the notion of *perpendicular* and we now discuss this property and its applications in the context of vectors. Without worrying about subtle mathematical details, let us regard *orthogonal* and *perpendicular* to be synonymous.

Definition 5.3.1 (Orthogonal Vectors, Orthogonal Set)**.** *Let* $\mathbf{u}$ *and* $\mathbf{w}$ *be vectors. If* $\mathbf{u} \cdot \mathbf{w} = 0$*, then* $\mathbf{u}$ *and* $\mathbf{w}$ *are said to be* orthogonal*. A collection* $\{\mathbf{v}_1, \mathbf{v}_2, \ldots, \mathbf{v}_n\}$ *is an* orthogonal set *if the vectors are pairwise orthogonal; that is if* $\mathbf{v}_i \cdot \mathbf{v}_j = 0$ *for* $1 \leq i < j \leq n$*.*

Of course, things are just not perpendicular because we say so. Suppose $\mathbf{u}, \mathbf{w} \neq \mathbf{0}$ and that the vectors are orthogonal. Let θ be the angle between the vectors. Then, by Proposition 4.3.6, $0 = ||\mathbf{u}|| \, ||\mathbf{w}|| \cos\theta$, but since neither vector is $\mathbf{0}$, $\cos\theta = 0$, giving $\theta = \frac{\pi}{2}$ and $\mathbf{u} \perp \mathbf{w}$.

Example 5.3.2. *The standard basis vectors* $\mathbf{e}_1 = [1,0,0]^T$, $\mathbf{e}_2 = [0,1,0]^T$, *and* $\mathbf{e}_3 = [0,0,1]^T$ *of* $\mathbb{R}^3$ *form an orthogonal set.*

Orthogonal sets are wonderful things; exactly because of the following property.

Theorem 5.3.3. *The vectors* $\{\mathbf{v}_1, \mathbf{v}_2, \dots, \mathbf{v}_n\}$ *of an orthogonal set of nonzero vectors are linearly independent.*

Proof. Suppose $\{\mathbf{v}_1, \mathbf{v}_2, \dots, \mathbf{v}_n\}$ is an *orthogonal set* of nonzero vectors and that

$$c_1\mathbf{v}_1 + c_2\mathbf{v}_2 + \cdots + c_n\mathbf{v}_n = \mathbf{0} \tag{5.24}$$

for some collection of scalars $c_1, c_2, \dots, c_n$. Thus for any i, $1 \leq i \leq n$

$$\begin{aligned} 0 = \mathbf{0} \cdot \mathbf{v}_i &= (c_1\mathbf{v}_1 + c_2\mathbf{v}_2 + \cdots + c_n\mathbf{v}_n) \cdot \mathbf{v}_i && (5.25) \\ &= c_1(\mathbf{v}_1 \cdot \mathbf{v}_i) + c_2(\mathbf{v}_2 \cdot \mathbf{v}_i) + \cdots + c_n(\mathbf{v}_n \cdot \mathbf{v}_i) && (5.26) \\ &= c_1 \cdot 0 + \cdots c_i(\mathbf{v}_i \cdot \mathbf{v}_i) + \cdots + c_n \cdot 0 && (5.27) \\ &= c_i(\mathbf{v}_i \cdot \mathbf{v}_i) && (5.28) \end{aligned}$$

since the vectors form an orthogonal set. Moreover, since $\mathbf{v}_i \neq \mathbf{0}$, $c_i = 0$. This holds for each i, thus

$$c_1 = c_2 = \cdots = c_i = \cdots = c_n = 0 \tag{5.29}$$

and the vectors $\{\mathbf{v}_1, \mathbf{v}_2, \dots, \mathbf{v}_n\}$ form a linearly indpendent set. □

As the vectors of on orthogonal set are linearly independent, it is natural to discuss having these vectors as a basis.

Definition 5.3.4 (Orthogonal Basis). *An* orthogonal basis *for a subspace* S *of* $\mathbb{R}^n$ *is a basis of* S *that is an orthogonal set.*

Example 5.3.5. *Let* $V = \{[2,1,-1]^T, [1,-1,1]^T, [0,1,1]^T\}$. *Since*

$$\begin{bmatrix} 2 \\ 1 \\ -1 \end{bmatrix} \cdot \begin{bmatrix} 1 \\ -1 \\ 1 \end{bmatrix} = 0, \begin{bmatrix} 1 \\ -1 \\ 1 \end{bmatrix} \cdot \begin{bmatrix} 0 \\ 1 \\ 1 \end{bmatrix} = 0, \text{ and } \begin{bmatrix} 2 \\ 1 \\ -1 \end{bmatrix} \cdot \begin{bmatrix} 0 \\ 1 \\ 1 \end{bmatrix} = 0,$$

V *is an orthogonal set of vectors and by Theorem 5.3.3 the vectors in* V *are linearly independent. Thus by the Fundamental Theorem of Invertible Matrices (Theorem 4.3.54),* V *is a basis of* $\mathbb{R}^3$.

Since are considering a basis

Theorem 5.3.6. *Suppose* S *is a subspace of* $\mathbb{R}^n$, *that* $\mathbf{u}$ *is in* S, *and that* $V = \{\mathbf{v}_1, \dots, \mathbf{v}_m\}$ *is an orthogonal basis for* S. *Then there exist unique scalars* $c_1 \dots, c_m$ *such that*

$$\mathbf{u} = c_1\mathbf{v}_1 + \cdots + c_m\mathbf{v}_m$$

where for each i, $1 \leq i \leq m$,

$$c_i = \frac{\mathbf{u} \cdot \mathbf{v}_i}{\mathbf{v}_i \cdot \mathbf{v}_i} = \frac{\mathbf{u} \cdot \mathbf{v}_i}{||\mathbf{v}_i||}.$$

Proof. Since V is a basis for S, by Exercise 4.30 there exist unique scalars $c_1 \dots, c_m$ such that $\mathbf{u} = c_1\mathbf{v}_1 + \cdots + c_m\mathbf{v}_m$. Thus for any i

$$\begin{aligned}
\mathbf{u}\cdot\mathbf{v}_i &= (c_1\mathbf{v}_1 + \cdots + c_m\mathbf{v}_m)\cdot\mathbf{v}_i && (5.30)\\
&= c_1(\mathbf{v}_1\cdot\mathbf{v}_i) + \cdots + c_m(\mathbf{v}_m\cdot\mathbf{v}_i) && (5.31)\\
&= c_i(\mathbf{v}_i\cdot\mathbf{v}_i) && (5.32)
\end{aligned}$$

where the last equality follows because V is an orthogonal set. As V is an orthogonal set, every $\mathbf{v}_i$ is nonzero, thus $c_i = (\mathbf{u}\cdot\mathbf{v}_i)/(\mathbf{v}_i\cdot\mathbf{v}_i)$ for each i. □

Example 5.3.7. *Write $\mathbf{u} = [4, 6, -4]^T$ as a linear combination of the vectors in V from Example 5.3.5.*

Solution. By Example 5.3.5, V is an orthogonal basis for $\mathbb{R}^3$. Thus from Theorem 5.3.6

$$c_1 = \frac{\mathbf{u}\cdot\mathbf{v}_1}{\mathbf{v}_1\cdot\mathbf{v}_1} = \frac{18}{6} = 3, c_2 = \frac{\mathbf{u}\cdot\mathbf{v}_2}{\mathbf{v}_2\cdot\mathbf{v}_2} = \frac{-6}{3} = -2, c_3 = \frac{\mathbf{u}\cdot\mathbf{v}_3}{\mathbf{v}_3\cdot\mathbf{v}_3} = \frac{2}{2} = 1 \tag{5.33}$$

so $\mathbf{u} = 3\mathbf{v}_1 - 2\mathbf{v}_2 + 1\mathbf{v}_3$. ■

Finding the coefficients via a direct calculation – that is solving $\mathbf{u} = a\mathbf{v}_1 + b\mathbf{v}_2 + c\mathbf{v}_3$ – would be more tedious; hence the value of orthogonal bases.

Let us take a look at unit vectors parallel to the vectors in V from Example 5.3.5.

$$\begin{aligned}
\frac{\mathbf{v}_1}{||\mathbf{v}_1||} &= \left[\frac{\sqrt{6}}{3}, \frac{\sqrt{6}}{6}, -\frac{\sqrt{6}}{6}\right]^T,\\
\frac{\mathbf{v}_2}{||\mathbf{v}_2||} &= \left[\frac{\sqrt{3}}{3}, -\frac{\sqrt{3}}{3}, \frac{\sqrt{3}}{3}\right]^T,\\
\text{and}\quad \frac{\mathbf{v}_3}{||\mathbf{v}_3||} &= \left[0, \frac{\sqrt{2}}{2}, \frac{\sqrt{2}}{2}\right]^T.
\end{aligned}$$

Thus

$$V_* = \left\{\left[\frac{\sqrt{6}}{3}, \frac{\sqrt{6}}{6}, -\frac{\sqrt{6}}{6}\right]^T, \left[\frac{\sqrt{3}}{3}, -\frac{\sqrt{3}}{3}, \frac{\sqrt{3}}{3}\right]^T, \left[0, \frac{\sqrt{2}}{2}, \frac{\sqrt{2}}{2}\right]^T\right\}$$

is an orthogonal set of unit vectors. We call such a set an *orthonormal set*. As V was shown to be an orthogonal basis for $\mathbb{R}^3$, it is easy to see that V_* is an *orthonormal basis* for $\mathbb{R}^3$; that is, a basis of unit orthogonal vectors.

Much like Theorem 5.3.6, expressing a vector in terms of an orthonormal basis is easy.

Theorem 5.3.8. *Suppose S is a subspace of $\mathbb{R}^n$, that $\mathbf{u}$ is in S, and that $V_* = \{\mathbf{v}_1, \ldots, \mathbf{v}_m\}$ is an orthogonal basis for S. Then*

$$\mathbf{u} = (\mathbf{u} \cdot \mathbf{v}_1)\mathbf{v}_1 + \cdots + (\mathbf{u} \cdot \mathbf{v}_m)\mathbf{v}_m$$

,

Proof. Since V is a basis for S, by Exercise 4.30 there exist unique scalars $c_1 \ldots, c_m$ such that $\mathbf{u} = c_1\mathbf{v}_1 + \cdots + c_m\mathbf{v}_m$. Thus for any i

$$\mathbf{u} \cdot \mathbf{v}_i = (c_1\mathbf{v}_1 + \cdots + c_m\mathbf{v}_m) \cdot \mathbf{v}_i \tag{5.34}$$

$$= c_1(\mathbf{v}_1 \cdot \mathbf{v}_i) + \cdots + c_m(\mathbf{v}_m \cdot \mathbf{v}_i) \tag{5.35}$$

$$= c_i(\mathbf{v}_i \cdot \mathbf{v}_i) = c_i \tag{5.36}$$

where the last two equalities hold because V_* is an orthonormal set. □

Example 5.3.9. *Write $\mathbf{u} = [4, 6, -4]^T$ as a linear combination of the vectors in the orthonormal basis of $\mathbb{R}^3$*

$$V_* = \left\{ \left[\frac{\sqrt{6}}{3}, \frac{\sqrt{6}}{6}, -\frac{\sqrt{6}}{6}\right]^T, \left[\frac{\sqrt{3}}{3}, -\frac{\sqrt{3}}{3}, \frac{\sqrt{3}}{3}\right]^T, \left[0, \frac{\sqrt{2}}{2}, \frac{\sqrt{2}}{2}\right]^T \right\}.$$

Solution. By Theorem 5.3.8,

$$\begin{bmatrix} 4 \\ 6 \\ -4 \end{bmatrix} = \left(\begin{bmatrix} 4 \\ 6 \\ -4 \end{bmatrix} \cdot \begin{bmatrix} \frac{\sqrt{6}}{3} \\ \frac{\sqrt{6}}{6} \\ -\frac{\sqrt{6}}{6} \end{bmatrix} \right) \begin{bmatrix} \frac{\sqrt{6}}{3} \\ \frac{\sqrt{6}}{6} \\ -\frac{\sqrt{6}}{6} \end{bmatrix} + \left(\begin{bmatrix} 4 \\ 6 \\ -4 \end{bmatrix} \cdot \begin{bmatrix} \frac{\sqrt{3}}{3} \\ -\frac{\sqrt{3}}{3} \\ \frac{\sqrt{3}}{3} \end{bmatrix} \right) \begin{bmatrix} \frac{\sqrt{3}}{3} \\ -\frac{\sqrt{3}}{3} \\ \frac{\sqrt{3}}{3} \end{bmatrix}$$

$$+ \left(\begin{bmatrix} 4 \\ 6 \\ -4 \end{bmatrix} \cdot \begin{bmatrix} 0 \\ \frac{\sqrt{2}}{2} \\ \frac{\sqrt{2}}{2} \end{bmatrix} \right) \begin{bmatrix} 0 \\ \frac{\sqrt{2}}{2} \\ \frac{\sqrt{2}}{2} \end{bmatrix} \tag{5.37}$$

$$= 3\sqrt{6} \begin{bmatrix} \frac{\sqrt{6}}{3} \\ \frac{\sqrt{6}}{6} \\ -\frac{\sqrt{6}}{6} \end{bmatrix} - 2\sqrt{3} \begin{bmatrix} \frac{\sqrt{3}}{3} \\ -\frac{\sqrt{3}}{3} \\ \frac{\sqrt{3}}{3} \end{bmatrix} + \sqrt{2} \begin{bmatrix} 0 \\ \frac{\sqrt{2}}{2} \\ \frac{\sqrt{2}}{2} \end{bmatrix}. \tag{5.38}$$

■

5.4 Orthonormal Matrices

Hopefully the reader understands the value of a basis of a vector space; namely, since we can express any vector in the space as a unique linear combination of the basis vectors, we obtain results about the entire space by only considering the basis vectors. Being able to form an orthogonal basis only sweetens the deal.

In this section, we present an easy way to construct an orthonormal basis and – as a bonus – also get a very useful factorization for matrices.

Definition 5.4.1 (Orthonormal Matrix). *A square matrix Q whose columns form an orthonormal set is called an* orthonormal matrix.[3]

Example 5.4.2. *Since the vectors in V_* from Example 5.3.7 form an orthonormal set, the matrix*

$$Q = \begin{bmatrix} \frac{\sqrt{6}}{3} & \frac{\sqrt{3}}{3} & 0 \\ \frac{\sqrt{6}}{6} & -\frac{\sqrt{3}}{3} & \frac{\sqrt{2}}{2} \\ -\frac{\sqrt{6}}{6} & \frac{\sqrt{3}}{3} & \frac{\sqrt{2}}{2} \end{bmatrix}$$

is an orthonormal matrix.

Though we are only concerned with square matrices in this text, we state a more general version of the following theorem.

Theorem 5.4.3. *An $m \times n$ matrix Q is an orthonormal matrix if and only if $Q^TQ = I_n$.*

Proof. We establish the theorem if we show that the appropriate entry of Q^TQ is 0 or 1; that is

$$(Q^TQ)_{ij} = \begin{cases} 0 & \text{if } i \neq j \\ 1 & \text{if } i = j. \end{cases} \tag{5.39}$$

Let $\mathbf{q}_i$ be the ith column of Q. By matrix multiplication (see Definition 4.3.8),

$$(Q^TQ) = \mathbf{q}_i \cdot \mathbf{q}_j. \tag{5.40}$$

But Q is orthogonal if and only if

$$\mathbf{q}_i \cdot \mathbf{q}_j = \begin{cases} 0 & \text{if } i \neq j \\ 1 & \text{if } i = j \end{cases} \tag{5.41}$$

and we have the desired result by 5.40 and 5.41. □

Theorem 5.4.3 has a very useful corollary:

Corollary 5.4.4. *A square matrix Q is orthogonal if and only if $Q^{-1} = Q^T$.*

---

[3]Older textbooks refer to these as *orthogonal matrices*, which they are. They are very much more than orthogonal, though, and orthonormal is the most appropriate term to name them.

Example 5.4.5. *The matrix Q in Example 5.4.2 is orthonormal, hence*

$$Q^T Q = \begin{bmatrix} \frac{\sqrt{6}}{3} & \frac{\sqrt{6}}{6} & -\frac{\sqrt{6}}{6} \\ \frac{\sqrt{3}}{3} & -\frac{\sqrt{3}}{3} & \frac{\sqrt{3}}{3} \\ 0 & \frac{\sqrt{2}}{2} & \frac{\sqrt{2}}{2} \end{bmatrix} \begin{bmatrix} \frac{\sqrt{6}}{3} & \frac{\sqrt{3}}{3} & 0 \\ \frac{\sqrt{6}}{6} & -\frac{\sqrt{3}}{3} & \frac{\sqrt{2}}{2} \\ -\frac{\sqrt{6}}{6} & \frac{\sqrt{3}}{3} & \frac{\sqrt{2}}{2} \end{bmatrix} = \begin{bmatrix} 1 & 0 & 0 \\ 0 & 1 & 0 \\ 0 & 0 & 1 \end{bmatrix}.$$

Moreover, by Corollary 5.4.4, we see that

$$Q^{-1} = Q^T = \begin{bmatrix} \frac{\sqrt{6}}{3} & \frac{\sqrt{6}}{6} & -\frac{\sqrt{6}}{6} \\ \frac{\sqrt{3}}{3} & -\frac{\sqrt{3}}{3} & \frac{\sqrt{3}}{3} \\ 0 & \frac{\sqrt{2}}{2} & \frac{\sqrt{2}}{2} \end{bmatrix}.$$

Example 5.4.6.
Show

$$Q_1 = \begin{bmatrix} 0 & 1 & 0 \\ 0 & 0 & 1 \\ 1 & 0 & 0 \end{bmatrix}, Q_2 = \begin{bmatrix} \cos\theta & -\sin\theta \\ \sin\theta & \cos\theta \end{bmatrix}$$

are orthogonal matrices and find their inverses.

Solution. The columns of Q_1 are the standard basis vectors of $\mathbb{R}^3$ and are clearly orthonormal. Thus by Corollary 5.4.4

$$Q_1^{-1} = Q_1^T = \begin{bmatrix} 0 & 0 & 1 \\ 1 & 0 & 0 \\ 0 & 1 & 0 \end{bmatrix}. \tag{5.42}$$

Regarding Q_2

$$Q_2^T Q_2 = \begin{bmatrix} \cos\theta & \sin\theta \\ -\sin\theta & \cos\theta \end{bmatrix} \begin{bmatrix} \cos\theta & -\sin\theta \\ \sin\theta & \cos\theta \end{bmatrix} \tag{5.43}$$

$$= \begin{bmatrix} \cos^2\theta + \sin^2\theta & -\cos\theta\sin\theta + \sin\theta\cos\theta \\ -\sin\theta\cos\theta + \cos\theta\sin\theta & \cos^2\theta + \sin^2\theta \end{bmatrix} \tag{5.44}$$

$$= \begin{bmatrix} 1 & 0 \\ 0 & 1 \end{bmatrix} \tag{5.45}$$

thus by Theorem 5.4.3, Q_2 is orthonormal, and by Corollary 5.4.4,

$$Q_2^{-1} = \begin{bmatrix} \cos\theta & \sin\theta \\ -\sin\theta & \cos\theta \end{bmatrix}.$$

■

A few poignant comments regarding the matrices of the previous example are in order. First, for any $3 \times n$ matrix A, the product $Q_1 A$ is just A under the row permutations $(1, 3, 2)$; that is, $R_1 \to R_3$, $R_3 \to R_2$, and $R_2 \to R_1$ (try it!). Such a matrix Q_1 is called a *permutation matrix* and permutation matrices are always orthonormal (see Exercise 5.7). Secondly, for any vector $\mathbf{v}$ in $\mathbb{R}^2$, the product $Q_2\mathbf{v}$ is $\mathbf{v}$ rotated counterclockwise in $\mathbb{R}^2$ through an angle θ. Since multiplication by Q_2 is just a rotation, $||Q_2\mathbf{v}|| = ||\mathbf{v}||$; that is, it is a *length preserving transformation* which is called an *isometry*. Since Q_1 is just a permutation of the rows of I, it is not surprising that its multiplication with a vector preserves the length of the vector. It turns out this property is true of all orthonormal matrices and, though not needed for our study, the following theorem is too pretty not to state.

Theorem 5.4.7. *Let Q be an $n \times n$ matrix. Then the following are equivalent:*

i. Q is orthonormal.
ii. For all $\mathbf{u}$ and $\mathbf{v}$ in $\mathbb{R}^n$,

$$Q\mathbf{u} \cdot Q\mathbf{v} = \mathbf{u} \cdot \mathbf{v}.$$

iii. For all $\mathbf{v}$ in $\mathbb{R}^n$,

$$||Q\mathbf{v}|| = ||\mathbf{v}||.$$

Before we begin the proof, we remind the reader that if $\mathbf{u}$ and $\mathbf{v}$ are both column (or row) vectors in $\mathbb{R}^n$, then $\mathbf{u}^T\mathbf{v} = \mathbf{u} \cdot \mathbf{v}$.

Proof. We show that *i.* and *ii.* are a biconditional as well as *ii.* and *iii.*.
(*i.* $\Rightarrow$ *ii.*) Suppose Q is orthonormal, then by Theorem 5.4.3 $Q^TQ = I$ and thus

$$Q\mathbf{u} \cdot Q\mathbf{v} = (Q\mathbf{u})^T Q\mathbf{v} = (\mathbf{u}^T Q^T) Q\mathbf{v} = \mathbf{u}^T (Q^T Q)\mathbf{v} = \mathbf{u}^T I \mathbf{v} = \mathbf{u}^T \mathbf{v} = \mathbf{u} \cdot \mathbf{v}. \tag{5.46}$$

(*ii.* $\Rightarrow$ *i.*) Now suppose that for all $\mathbf{u}$ and $\mathbf{v}$ in $\mathbb{R}^n$, $Q\mathbf{u} \cdot Q\mathbf{v} = \mathbf{u} \cdot \mathbf{v}$ and let $\mathbf{e}_i$ be the ith standard basis vector. Then by Exercise 4.5

$$\mathbf{q}_i \cdot \mathbf{q}_j = Q\mathbf{e}_i \cdot Q\mathbf{e}_j = \mathbf{e}_i \cdot \mathbf{e}_j = \begin{cases} 0 & \text{if } i \neq j \\ 1 & \text{if } i = j \end{cases} \tag{5.47}$$

thus showing Q is orthonormal.
(*ii.* $\Rightarrow$ *iii.*) Again suppose that for all $\mathbf{u}$ and $\mathbf{v}$ in $\mathbb{R}^n$, $Q\mathbf{u} \cdot Q\mathbf{v} = \mathbf{u} \cdot \mathbf{v}$. Thus by Remark 4.3.7

$$||Q\mathbf{v}|| = \sqrt{Q\mathbf{v} \cdot Q\mathbf{v}} = \sqrt{\mathbf{v} \cdot \mathbf{v}} = ||\mathbf{v}||. \tag{5.48}$$

(*iii.* $\Rightarrow$ *ii.*) Here assume $||Q\mathbf{v}|| = ||\mathbf{v}||$ for all $\mathbf{v}$ in $\mathbb{R}^n$. By uusing Exercise 4.6 twice

$$\begin{aligned}
\mathbf{u} \cdot \mathbf{v} &= \frac{1}{4}\left(||\mathbf{u}+\mathbf{v}||^2 - ||\mathbf{u}-\mathbf{v}||^2\right) && (5.49)\\
&= \frac{1}{4}\left(||Q(\mathbf{u}+\mathbf{v})||^2 - ||Q(\mathbf{u}-\mathbf{v})||^2\right) && (5.50)\\
&= \frac{1}{4}\left(||Q\mathbf{u}+Q\mathbf{v}||^2 - ||Q\mathbf{u}-Q\mathbf{v}||^2\right) && (5.51)\\
&= Q\mathbf{u} \cdot Q\mathbf{v}. && (5.52)
\end{aligned}$$

□

For the sake of completeness, we add one more collection of beautiful properties of orthonormal matrices.

Theorem 5.4.8. *Let Q be an orthonormal matrix. Then*

i. *Q^{-1} is orthonormal;*
ii. $\det Q = \pm 1$*;*
iii. *if λ is an eigenvalue of Q, then $|\lambda| = 1$; and*
iv. *if Q_1 and Q_2 are orthonormal $n \times n$ matrices, then so is $Q_1 Q_2$.*

5.5 The Gram-Schmidt Process

One of the more useful matrix factorizations relies on constructing an orthogonal basis for a given space. To build the idea, we could discuss the notion of orthogonal complements and orthogonal projections as well as the Orthogonal Decomposition Theorem, but as the aim of this text is not to teach a linear algebra course, we refer the reader interested in such matters to any linear algebra text and choose rather to just state the desired process.

One technique to build an orthogonal (or eventually orthonormal) basis is the *Gram-Schmidt Process.* It is not without its shortcomings, which we will discuss later.

Theorem 5.5.1 (Gram-Schmidt Process). *Let $\{\mathbf{v}_1, \mathbf{v}_2, \ldots, \mathbf{v}_k\}$ be a basis for a subspace V of $\mathbb{R}^n$. Define*

$$\mathbf{u}_1 := \mathbf{v}_1 \qquad \text{with } U_1 := span(\mathbf{v}_1), \tag{5.53}$$

$$\mathbf{u}_2 := \mathbf{v}_2 - \frac{\mathbf{u}_1 \cdot \mathbf{v}_2}{\mathbf{u}_1 \cdot \mathbf{u}_1}\mathbf{u}_1 \qquad \text{with } U_2 := span(\mathbf{v}_1, \mathbf{v}_2), \tag{5.54}$$

$$\mathbf{u}_3 := \mathbf{v}_3 - \frac{\mathbf{u}_1 \cdot \mathbf{v}_3}{\mathbf{u}_1 \cdot \mathbf{u}_1}\mathbf{u}_1 - \frac{\mathbf{u}_2 \cdot \mathbf{v}_3}{\mathbf{u}_2 \cdot \mathbf{u}_2}\mathbf{u}_2 \qquad \text{with } U_3 := span(\mathbf{v}_1, \mathbf{v}_2, \mathbf{v}_3), \tag{5.55}$$

$$\vdots \qquad\qquad \vdots$$

$$\mathbf{u}_k := \mathbf{v}_k - \frac{\mathbf{u}_1 \cdot \mathbf{v}_k}{\mathbf{u}_1 \cdot \mathbf{u}_1}\mathbf{u}_1 - \frac{\mathbf{u}_2 \cdot \mathbf{v}_k}{\mathbf{u}_2 \cdot \mathbf{u}_2}\mathbf{u}_2 \qquad \text{with } U_k := span(\mathbf{v}_1, \ldots, \mathbf{v}_k).$$

$$- \cdots - \frac{\mathbf{u}_{k-1} \cdot \mathbf{v}_k}{\mathbf{u}_{k-1} \cdot \mathbf{u}_{k-1}}\mathbf{u}_{k-1} \tag{5.56}$$

Then for each i, $1 \leq i \leq k$, $\{\mathbf{u}_1, \ldots \mathbf{u}_i\}$ is an orthogonal basis for U_i.

Note that in particular, Theorem 5.5.1 says that the constructed collection of vectors $\{\mathbf{u}_1, \ldots \mathbf{u}_k\}$ is an orthogonal basis for the subspace $V = U_i$. We obtain an orthonormal basis for V by following the same construction of vectors and then normalize them; that is, if $\{\mathbf{u}_1, \ldots \mathbf{u}_k\}$ is an orthogonal basis for a subspace V, then $\{\mathbf{q}_1, \ldots \mathbf{q}_k\}$ where $\mathbf{q}_i := \mathbf{u}_i / ||\mathbf{u}_i||$ is an orthonormal basis for V.

The proof of Theorem 5.5.1 is not difficult can be done by Induction. Unfortunately, the standard proof requires an understanding of orthogonal projections and orthogonal complements as well as the Orthogonal Decomposition Theorem and as these have not been introduced we omit the proof.

Before we begin the example, a helpful observation. If c is a nonzero scalar and $\mathbf{v}$ is any vector, we know $c\mathbf{v}$ is parallel to $\mathbf{v}$ and thus multiplying any vector in an orthogonal set preserves orthogonality.

Observation 5.5.2. *If* $\{\mathbf{u}_1, \dots \mathbf{u}_k\}$ *is an orthogonal set, then so is* $\{c_1\mathbf{u}_1, \dots c_k\mathbf{u}_k\}$ *for any nonzero scalars* $\{c_1, \dots, c_k\}$.

Example 5.5.3. *Use the Gram-Schmidt Process to construct an orthonormal basis for the space* $V = span(\mathbf{v}_1, \mathbf{v}_2, \mathbf{v}_3)$ *where*

$$\mathbf{v}_1 = \begin{bmatrix} 1 \\ 0 \\ -1 \\ 2 \end{bmatrix}, \mathbf{v}_2 = \begin{bmatrix} 2 \\ -1 \\ 0 \\ 1 \end{bmatrix}, \mathbf{v}_3 = \begin{bmatrix} 1 \\ -2 \\ -1 \\ 0 \end{bmatrix}.$$

Solution. By Theorem 5.5.1,

$$\mathbf{u}_1 = \mathbf{v}_1 = \begin{bmatrix} 1 \\ 0 \\ -1 \\ 2 \end{bmatrix} \tag{5.57}$$

and

$$\mathbf{u}_2 = \mathbf{v}_2 - \frac{\mathbf{u}_1 \cdot \mathbf{v}_2}{\mathbf{u}_1 \cdot \mathbf{u}_1}\mathbf{u}_1 = \begin{bmatrix} 2 \\ -1 \\ 0 \\ 1 \end{bmatrix} - \frac{2}{3}\begin{bmatrix} 1 \\ 0 \\ -1 \\ 2 \end{bmatrix} = \begin{bmatrix} \frac{4}{3} \\ -1 \\ \frac{2}{3} \\ -\frac{1}{3} \end{bmatrix}. \tag{5.58}$$

Since we are doing this by hand, we will take advantage of Observation 5.5.2 and use

$$\mathbf{u}_2^* = 3\mathbf{u}_2 = \begin{bmatrix} 4 \\ -3 \\ 2 \\ -1 \end{bmatrix} \tag{5.59}$$

thus

$$\begin{aligned} \mathbf{u}_3 &:= \mathbf{v}_3 - \frac{\mathbf{u}_1 \cdot \mathbf{v}_3}{\mathbf{u}_1 \cdot \mathbf{u}_1}\mathbf{u}_1 - \frac{\mathbf{u}_2^* \cdot \mathbf{v}_3}{\mathbf{u}_2^* \cdot \mathbf{u}_2^*}\mathbf{u}_2^* \\ &= \begin{bmatrix} 1 \\ -2 \\ -1 \\ 0 \end{bmatrix} - \frac{1}{3}\begin{bmatrix} 1 \\ 0 \\ -1 \\ 2 \end{bmatrix} - \frac{4}{15}\begin{bmatrix} 4 \\ -3 \\ 2 \\ -1 \end{bmatrix} = \begin{bmatrix} -\frac{2}{5} \\ -\frac{6}{5} \\ -\frac{6}{5} \\ -\frac{2}{5} \end{bmatrix}. \end{aligned} \tag{5.60}$$

Though not necessary, we will go ahead and scale $\mathbf{u}_3$

$$\mathbf{u}_3^* = -\frac{5}{2}\mathbf{u} = \begin{bmatrix} 1 \\ 3 \\ 3 \\ 1 \end{bmatrix}.$$

Thus by Gram-Schmidt

$$\left\{ \begin{bmatrix} 1 \\ 0 \\ -1 \\ 2 \end{bmatrix}, \begin{bmatrix} 4 \\ -3 \\ 2 \\ -1 \end{bmatrix}, \begin{bmatrix} 1 \\ 3 \\ 3 \\ 1 \end{bmatrix} \right\}$$

is an *orthogonal* basis for V. Normalizing the vectors gives the *orthonormal* basis

$$\left\{ \begin{bmatrix} \frac{\sqrt{6}}{6} \\ 0 \\ -\frac{\sqrt{6}}{6} \\ \frac{\sqrt{6}}{3} \end{bmatrix}, \begin{bmatrix} \frac{2\sqrt{30}}{15} \\ -\frac{\sqrt{30}}{10} \\ \frac{\sqrt{30}}{15} \\ -\frac{\sqrt{30}}{30} \end{bmatrix}, \begin{bmatrix} \frac{\sqrt{5}}{10} \\ \frac{3\sqrt{5}}{10} \\ \frac{3\sqrt{5}}{10} \\ \frac{\sqrt{5}}{10} \end{bmatrix} \right\}.$$

■

5.6 *QR* Factorization

Let A be an $m \times n$ matrix, $m \geq n$, with full rank; that is, the n column vectors $\{\mathbf{a}_1, \mathbf{a}_2, \ldots, \mathbf{a}_n\}$ of A are linearly independent. Then by the Gram-Schmidt Process (Theorem 5.5.1) with normalizing the vectors, there exists vectors $\{\mathbf{q}_1, \mathbf{q}_2, \ldots, \mathbf{q}_i\}$ that form an orthonormal basis for $span\{\mathbf{a}_1, \mathbf{a}_2, \ldots, \mathbf{a}_i\}$ for each $1 \leq i \leq n$. Thus there exists scalars $r_1, r_2, \ldots r_n$ such that

$$\mathbf{a}_1 = r_{11}\mathbf{q}_1, \tag{5.61}$$

$$\mathbf{a}_2 = r_{12}\mathbf{q}_1 + r_{22}\mathbf{q}_2, \tag{5.62}$$

$$\vdots \tag{5.63}$$

$$\mathbf{a}_n = r_{1n}\mathbf{q}_1 + r_{2n}\mathbf{q}_2 + \cdots + r_{nn}\mathbf{q}_n. \tag{5.64}$$

We can write our work in matrix form as

$$A = [\mathbf{a}_1, \mathbf{a}_2, \ldots, \mathbf{a}_n] = [\mathbf{q}_1, \mathbf{q}_2, \ldots, \mathbf{q}_n] \begin{bmatrix} r_{11} & r_{12} & \cdots & r_{1n} \\ 0 & r_{22} & \cdots & r_{2n} \\ \vdots & \vdots & \ddots & \vdots \\ 0 & 0 & \cdots & r_{nn} \end{bmatrix} \tag{5.65}$$

or $A = QR$ where Q is an orthogonal matrix and R is upper triangular. This is the inspiration behind the next theorem.

Theorem 5.6.1 (QR Matrix Factorization). *Let A be a $m \times n$ matrix with linearly independent columns (i.e. $rank(A) = n \leq m$). Then the exist matrices Q and R such that*

$$A = QR$$

where

- *Q is $m \times n$ orthogonal matrix and*
- *R and invertible upper triangular matrix.*

Proof. The discussion before the theorem establishes there are matrices Q and R such that $A = QR$ with Q orthogonal and R upper triangluar. All that remains is showing that R is indeed invertible. To reach this end, suppose $R\mathbf{x} = \mathbf{0}$ where $\mathbf{x} = [x_1, \ldots, x_n]^T$. Then

$$A\mathbf{x} = QR\mathbf{x} = Q\mathbf{0} = \mathbf{0}. \tag{5.66}$$

But

$$A\mathbf{x} = \begin{bmatrix} a_{11}x_1 + a_{12}x_2 + a_{13}x_3 + \cdots + a_{1n}x_n \\ a_{21}x_1 + a_{22}x_2 + a_{23}x_3 + \cdots + a_{2n}x_n \\ \vdots \\ a_{n1}x_1 + a_{n2}x_2 + a_{n3}x_3 + \cdots + a_{nn}x_n \end{bmatrix} \tag{5.67}$$

$$= \mathbf{a}_1 x_1 + \mathbf{a}_2 x_2 + \cdots + \mathbf{a}_n x_n \tag{5.68}$$

i.e. $A\mathbf{x}$ is a linear combination of the column vectors of A. Since these are linearly independent, if follows from $A\mathbf{x} = \mathbf{0}$ that $\mathbf{x} = \mathbf{0}$. Thus by The Fundamental Theorem of Invertible Matrices (Theorem 4.3.54), R is invertible. □

Example 5.6.2. *Find the QR factorization of*

$$A = \begin{bmatrix} 1 & 2 & 1 \\ 0 & -1 & -2 \\ -1 & 0 & -1 \\ 2 & 1 & 0 \end{bmatrix}$$

Solution. The columns of A are just the vectors of Example 5.5.3. Thus, by the by the Gram-Schmidt Process as in the example

$$Q = \begin{bmatrix} \frac{\sqrt{6}}{6} & \frac{2\sqrt{30}}{15} & \frac{\sqrt{5}}{10} \\ 0 & -\frac{\sqrt{30}}{10} & \frac{3\sqrt{5}}{10} \\ -\frac{\sqrt{6}}{6} & \frac{\sqrt{30}}{15} & \frac{3\sqrt{5}}{10} \\ \frac{\sqrt{6}}{3} & -\frac{\sqrt{30}}{30} & \frac{\sqrt{5}}{10} \end{bmatrix}.$$

Hence we have

$$A = \begin{bmatrix} 1 & 2 & 1 \\ 0 & -1 & -2 \\ -1 & 0 & -1 \\ 2 & 1 & 0 \end{bmatrix} = \begin{bmatrix} \frac{\sqrt{6}}{6} & \frac{2\sqrt{30}}{15} & \frac{\sqrt{5}}{10} \\ 0 & -\frac{\sqrt{30}}{10} & \frac{3\sqrt{5}}{10} \\ -\frac{\sqrt{6}}{6} & \frac{\sqrt{30}}{15} & \frac{3\sqrt{5}}{10} \\ \frac{\sqrt{6}}{3} & -\frac{\sqrt{30}}{30} & \frac{\sqrt{5}}{10} \end{bmatrix} \begin{bmatrix} r_{11} & r_{12} & r_{13} \\ 0 & r_{22} & r_{23} \\ 0 & 0 & r_{33} \end{bmatrix} = QR. \tag{5.69}$$

Thus

$$1 = \frac{\sqrt{6}}{6} r_{11} \tag{5.70}$$

giving $r_{11} = \sqrt{6}$. From the second row of Q

$$a_{22} = -1 = -\frac{\sqrt{30}}{10} r_{22} \tag{5.71}$$

so $r_{22} = \frac{10}{\sqrt{30}} = \frac{\sqrt{30}}{3}$ and by the third row of Q

$$a_{32} = 0 = -\frac{\sqrt{6}}{6} r_{12} + \frac{\sqrt{30}}{15} r_{22} \tag{5.72}$$

giving

$$r_{12} = \frac{6}{\sqrt{6}} \cdot \frac{\sqrt{30}}{15} \cdot \frac{\sqrt{30}}{3} = \frac{2\sqrt{6}}{3}. \tag{5.73}$$

Now we pause and point out that in spite of how clever we have been thus far, we are working way too hard. Remember, in Mathematics, theorems are our friends. By Theorem 5.4.3

$$Q^T A = Q^T QR = IR = R \tag{5.74}$$

thus

$$\begin{aligned} R &= \begin{bmatrix} \frac{\sqrt{6}}{6} & 0 & -\frac{\sqrt{6}}{6} & \frac{\sqrt{6}}{3} \\ \frac{2\sqrt{30}}{15} & -\frac{\sqrt{30}}{10} & \frac{\sqrt{30}}{15} & -\frac{\sqrt{30}}{30} \\ \frac{\sqrt{5}}{10} & \frac{3\sqrt{5}}{10} & \frac{3\sqrt{5}}{10} & \frac{\sqrt{5}}{10} \end{bmatrix} \begin{bmatrix} 1 & 2 & 1 \\ 0 & -1 & -2 \\ -1 & 0 & -1 \\ 2 & 1 & 0 \end{bmatrix} \\ &= \begin{bmatrix} \sqrt{6} & \frac{2\sqrt{6}}{3} & \frac{\sqrt{6}}{3} \\ 0 & \frac{\sqrt{30}}{3} & \frac{4\sqrt{30}}{15} \\ 0 & 0 & -\frac{4\sqrt{5}}{5} \end{bmatrix} \end{aligned} \tag{5.75}$$

Thus

$$\begin{aligned} A &= \begin{bmatrix} 1 & 2 & 1 \\ 0 & -1 & -2 \\ -1 & 0 & -1 \\ 2 & 1 & 0 \end{bmatrix} = \begin{bmatrix} \frac{\sqrt{6}}{6} & \frac{2\sqrt{30}}{15} & \frac{\sqrt{5}}{10} \\ 0 & -\frac{\sqrt{30}}{10} & \frac{3\sqrt{5}}{10} \\ -\frac{\sqrt{6}}{6} & \frac{\sqrt{30}}{15} & \frac{3\sqrt{5}}{10} \\ \frac{\sqrt{6}}{3} & -\frac{\sqrt{30}}{30} & \frac{\sqrt{5}}{10} \end{bmatrix} \begin{bmatrix} \sqrt{6} & \frac{2\sqrt{6}}{3} & \frac{\sqrt{6}}{3} \\ 0 & \frac{\sqrt{30}}{3} & \frac{4\sqrt{30}}{15} \\ 0 & 0 & -\frac{4\sqrt{5}}{5} \end{bmatrix} \\ &= QR \end{aligned} \tag{5.76}$$

■

The technique can be summarized as

Highlight 5.6.3. *(QR Factorization) Suppose A is an $m \times n$ matrix whose n column vectors are linearly independent. Then $A = QR$ where*

- *Q is found by the Gram-Schmidt process (Theorem 5.5.1) and*
- *$R = Q^T A$.*

Some comments regarding QR factorization and the Gram-Schmidt Process are in order. First, if Q is a large matrix then it is very computationally expensive to form Q^T so finding R as we did in (5.75) would be very computationally expensively and possibly intractable by this technique. Thus, if we are writing a program we may prefer to follow what was done in (5.70) through (5.73). This would work, but since we are using Gram-Schmidt to find Q, we actually can work calculating the r_{ij} intro the Gram-Schmidt Process and take care of two jobs at once. There is a concern with Gram-Schmidt, though, in that it is very *unstable*; that is, a small round-off error in the calculation can lead to a large error in the solution. As such, many good programs use what is called the *Modified Gram-Schmidt Process.* A great presentation of what we have addressed in this paragraph appears in [58]. As an alternate to the Gram-Schmidt Process, one could use what is known as a *Householder Transformation* to efficiently compute QR. A nice introduction to this technique can be found in [44].

5.7 Keywords

lower triangular matrix, upper triangular matrix. LU factorization, Cholesky Decomposition, constructive proof, orthogonal vectors, orthogonal basis, orthonormal vectors, orthonormal basis, orthonormal matrix, Gram-Schmidt Process, QR factorization.

5.8 For Further Study

There is a concern with Gram-Schmidt, though, in that it is very *unstable*; that is, a small round-off error in the calculation can lead to a large error in the solution. As such, many good programs use what is called the *Modified Gram-Schmidt Process.* A great presentation of what we have addressed in this paragraph appears in [58]. As an alternate to the Gram-Schmidt Process, one could use what is known as a *Householder Transformation* to efficiently compute QR. A nice introduction to this technique can be found in [44].

5.9 Exercises

Exercise 5.1. *Let L be an arbitrary 3×3 lower triangular matrix. Assuming L^{-1} exists, show that L^{-1} is also lower triangular.*

Exercise 5.2. *Let U be an arbitrary 3×3 upper triangular matrix. Assuming U^{-1} exists, show that U^{-1} is also upper triangular.*

Exercise 5.3. *Show that in Cholesky decomposition $\tilde{L} = LD^{1/2}$ where the L and D are as in the LDL^T factorization and $D^{1/2}$ means the matrix whose entries are the square roots of the entries in the matrix D.*

Exercise 5.4. *Let* $A = \begin{bmatrix} 1 & 0 & 2 & 0 \\ 0 & 4 & 0 & 4 \\ 2 & 0 & 13 & 0 \\ 0 & 4 & 0 & 5 \end{bmatrix}$. *Find A's LU, LDL^T, and Cholesky factorizations.*

Solution: $L = \begin{bmatrix} 1 & 0 & 0 & 0 \\ 0 & 1 & 0 & 0 \\ 2 & 0 & 1 & 0 \\ 0 & 1 & 0 & 1 \end{bmatrix}$, $U = \begin{bmatrix} 1 & 0 & 2 & 0 \\ 0 & 4 & 0 & 4 \\ 0 & 0 & 9 & 0 \\ 0 & 0 & 0 & 1 \end{bmatrix}$, $D = \begin{bmatrix} 1 & 0 & 0 & 0 \\ 0 & 4 & 0 & 0 \\ 0 & 0 & 9 & 0 \\ 0 & 0 & 0 & 1 \end{bmatrix}$,

and $\tilde{L} = \begin{bmatrix} 1 & 0 & 0 & 0 \\ 0 & 2 & 0 & 0 \\ 2 & 0 & 3 & 0 \\ 0 & 2 & 0 & 1 \end{bmatrix}$.

Exercise 5.5. *Prove that if Q is an orthonormal matrix its rows form an orthonormal set.*

Exercise 5.6. *Let $V_* = \{\mathbf{v}_1, \dots, \mathbf{v}_m\}$ be an orthonormal set in $\mathbb{R}^n$ and let $\mathbf{u}$ be a vector in $\mathbb{R}^n$. Prove* Bessel's Inequality

$$||\mathbf{u}||^2 \geq |\mathbf{u} \cdot \mathbf{v}_1|^2 + \cdots + |\mathbf{u} \cdot \mathbf{v}_m|^2$$

with equality if and only if $\mathbf{u} \in span(V_*)$.

Exercise 5.7. *A* permutation matrix *is any matrix formed by permuting (swapping) the rows of I. Show that any permutation matrix is necessarily orthogonal.*

Exercise 5.8. *Use the Gram-Schmidt Process to construct an orthonormal basis for the space $V = span(\mathbf{v}_1, \mathbf{v}_2, \mathbf{v}_3)$ where*

$$\mathbf{v}_1 = [3, 0, 4], \mathbf{v}_2 = [-1, 0, 7], \mathbf{v}_3 = [2, 9, 11].$$

Note that a solution is $[3, 0, 4], [-4, 0, 3], [0, 9, 0]$.

Part II

Linear Programming

6

Linear Programming

After optimizing a function of single or multiple variables (considered for review in Chapter 9) , the least sophisticated optimization problems are concerned with either minimizing or maximizing a linear function which satisfy conditions modeled by linear inequalities. There are numerous practical problems which fall into this category and for these reasons we begin our study of Optimization here.

6.1 A Geometric Approach to Linear Programming in Two Dimensions

To illustrate the material, let us consider the following example:

6.1.1 Example

Example 6.1.1 (Lincoln Outdoors). Lincoln Outdoors, *a camping merchandise manufacturer, makes two types of sleeping bags: the Cabin Model for light camping and the Frontier Model for more rugged use. Each Cabin sleeping bag requires 1 labor-hour from the cutting department and 2 labor hours from the assembly department whereas each Frontier model requires 2 labor-hour from the cutting department and 3 labor hours from the assembly department. The per day maximum amount of labor hours for the cutting department is 40 labor-hours where the assembly department has 72 labor-hours available per day. The company makes a profit of $60 on each Cabin model it sells and a profit of $90 on each Frontier model sold. Assuming that all sleeping bags that are manufactured will sell, how many bags of each type should* Lincoln Outdoors *manufacture per day in order to maximize the total daily profit?*

Solution. We may summarize the information as follows:

The first step will be to identify the ***decision variables*** for the model. In this situation, let us put

DOI: 10.1201/9780367425517-6

TABLE 6.1
Manufacturing Data for Lincoln Outdoors in Example 6.1.1

Labor-Hours	Cabin Model	Frontier Model	Max Hours per Day
Cutting Dept.	1	2	40
Assembly Dept.	2	3	72
Profit per Bag	\$60	\$90	

$x_1 =$ the number of Cabin Model sleeping bags manufactured per day and

$x_2 =$ the number of Frontier Model sleeping bags manufactured per day.

We next form our ***objective function***, that is the function we wish to optimize. In this situation our objective function is

$$P(x_1, x_2) = 60x_1 + 90x_2. \tag{6.1}$$

According to this function, profit can be made arbitrarily large by letting x_1 or x_2 grow without bound. Unfortunately, the situation has restrictions due to the amount of available labor-hours. Thus we have the following ***problem constraints***:

$$\text{Cutting Department Constraints:} \quad x_1 + 2x_2 \leq 40 \quad \text{and} \tag{6.2}$$

$$\text{Assembly Department Constraints:} \quad 2x_1 + 3x_2 \leq 72. \tag{6.3}$$

As well, we have the ***non-negativity constraints***[1] that

$$x_1 \geq 0 \text{ and} \tag{6.4}$$

$$x_2 \geq 0. \tag{6.5}$$

These constraints are usually expressed with the single statement

$$x_1, x_2 \geq 0.$$

Thus the mathematical model for the problem we are considering is

$$\text{Maximize: } P(x_1, x_2) = 60x_1 + 90x_2 \tag{6.6}$$

$$\text{Subject to: } \quad x_1 + 2x_2 \leq 40 \tag{6.7}$$

$$2x_1 + 3x_2 \leq 72 \tag{6.8}$$

$$x_1, x_2 \geq 0. \tag{6.9}$$

The graph of this system of linear inequalities given by the constraints is known as the ***feasible region***.

[1]Certainly we also have the natural constraint that the number of sleeping bags be integer-valued, but this is a matter for Chapter 8.

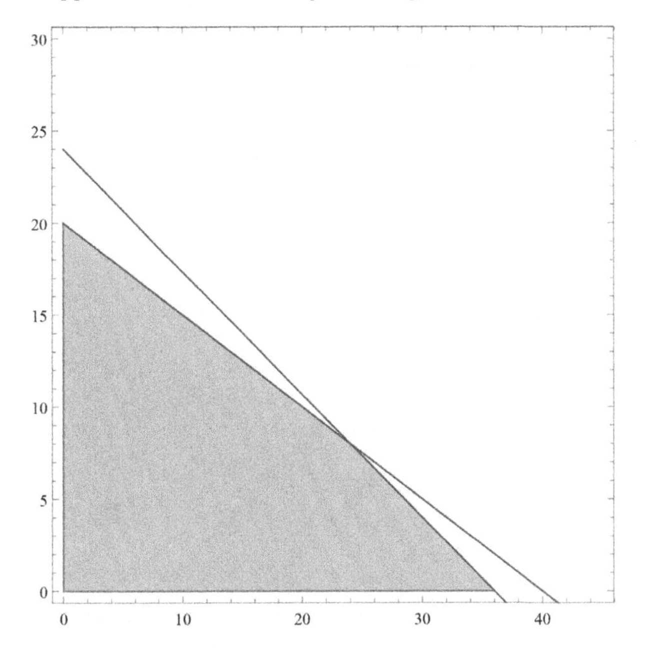

FIGURE 6.1
Feasible region for Lincoln Outdoors.

Now it is a wonderful thing that we are able to graph the feasible region and thus know the set of solutions to the system of linear inequalities, but which pair maximizes the profit function? This is a daunting task as there are infinitely many possible points (unless we only consider integer solutions; more on this later).

Our aim is to maximize $P(x_1, x_2) = 60x_1 + 90x_2$ so let us consider this function. If we fix a value for the profit, call it K, we then have a linear equation in two variables. In particular, if we solve for x_2 we have

$$x_2 = -\frac{2}{3}x_1 + \frac{K}{90}. \tag{6.10}$$

Notice that as K increases, this line moves further away from the origin (see Figure 6.2).

We wish to increase K as much as possible, but recall there are restrictions. Specifically, the line representing the profit must intersect the feasible region. So to maximize profit but also satisfy the constraints, we move the line as far away from the origin as possible but still have at least one point on the line in the feasible region. By this reasoning, we may conclude that an optimal solution to a linear programming problem occurs at a corner point (as mentioned in Section 4.1, a corner point of feasible regions is often called a *vertex* of the feasible region). ■

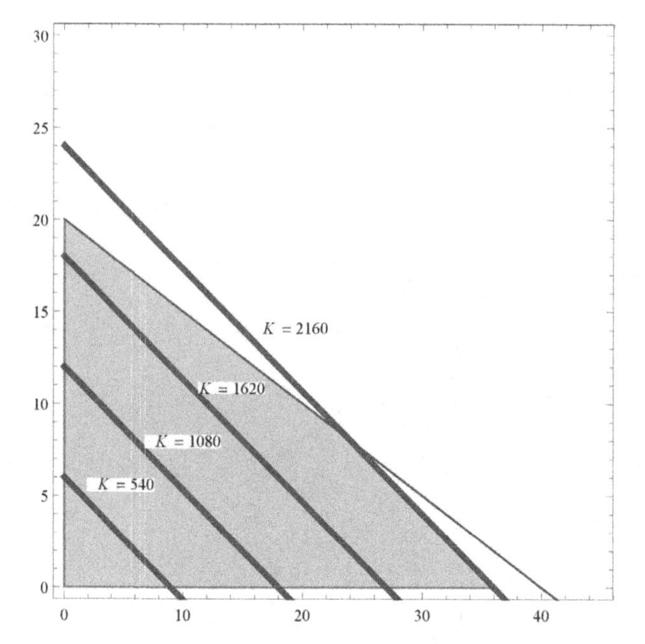

FIGURE 6.2
Graphs of the objective function for Lincoln Outdoors.

Though we have not formally proved[2] it, we have

Theorem 6.1.2 (The Fundamental Theorem of Linear Programming[3]). *If the optimal value of the objective function in a linear programming problem exists, then that value (known as the* optimal solution*) must occur at one or more of the corner points of the feasible region.*

Also from our exploration we have

Remark 6.1.3. *The possible classifications of the solutions to Linear Programming problems are:*

- *If the feasible region of a linear programming problem is bounded, then there exists a maximum and a minimum value for the objective function.*
- *If the feasible region of a linear programming problem is unbounded and if the coefficients of the objective function are positive[4], then there exists a minimum value for the objective function, but there does not exist a maximum value for this function. (An analogous exists for a feasible region not bounded below and the associated min and max... but these are seldom encountered in applications.)*

[2]Formal proofs of the theorems in this section will be offered in Chapter 17.
[3]The formal statement of this theorem is given in Theorem 17.2.2.
[4]Scenarios with negative coefficients are vary rare in applications.

- *If the feasible region is empty, then there does not exist a maximum or a minimum value for the objective function.*

Given the Fundamental Theorem of Linear Programming, we may now answer the question we considered in Example 6.1.1. We accomplish this by evaluating the objective function at each corner point of the feasible region and the work is shown in Table 6.2. We see that profit is maximized at $2160 which occurs when either 24 Cabin Model and 8 Frontier Model bags are made or when 36 Cabin Model and no Frontier Model bags are produced. This leads to

TABLE 6.2
$P(x_1, x_2)$ Evaluated at Corner Points

Corner Point (x_1, x_2)	$P(x_1, x_2)$
$(0, 0)$	0
$(0, 20)$	1800
$(24, 8)$	2160
$(36, 0)$	2160

Remark 6.1.4. *Note that it is possible that the optimal solution occurs at more than one corner point. If this situation occurs, then* any point of the line segment joining the corner points is also an optimal solution.

The situation addressed in Remark 6.1.4 is what we have in the Lincoln Outdoors example and hence we conclude that the objective function $P = 60x_1 + 90x_2$ subjected to the given constraints is maximized at all points on the line segment connecting the points $(24, 8)$ and $(36, 0)$, that is the maximum of $2160 occurs over $\{(t, -\frac{2}{3}t + 24) \mid 24 \leq t \leq 36\}$. This solution set is represented in Figure 6.2 where the objective function P intersects the constraint boundary $2x_1 + 3x_2 = 72$.

Of course, in this situation, we will be concerned with integer solutions and the multiple integer solutions for Lincoln Outdoors are $(24, 8)$, $(27, 6)$, $(30, 4)$, $(33, 2)$, and $(36, 0)$ which appear in Figure 6.3. In Chapter 8 we will explore how to find integer solutions and in Chapter 7 we will show how to use Excel to find other possible solutions. In the meantime, we can be aware that multiple solutions exist not only from the corner point analysis but also by the observation that the objective function $P(x_1, x_2) = 60x_1 + 90x_2$ is parallel to the assembly constraint $2x_1 + 3x_2 \leq 72$ (as seen in Figure 6.2).

6.1.2 Summary

We may summarize our techniques as follows:

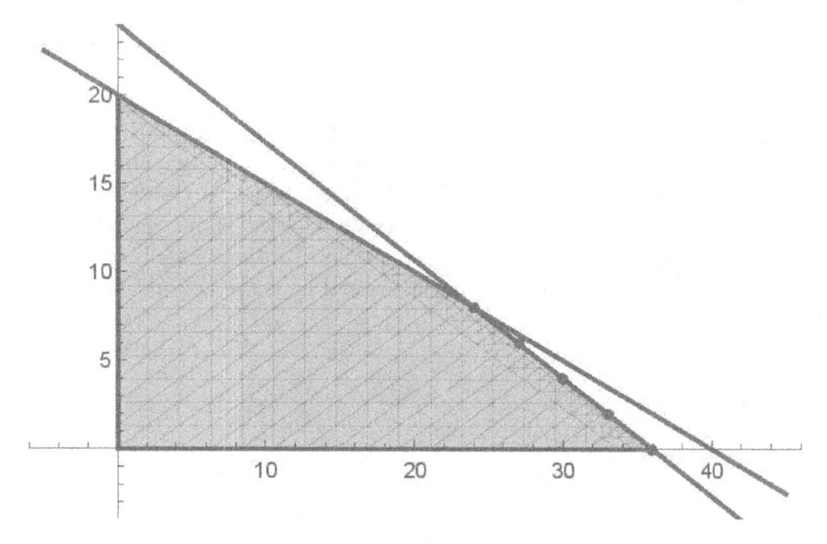

FIGURE 6.3
The multiple integer solutions for Lincoln Outdoors.

1. Summarize the data in table form (see Table 6.1).
2. Form a mathematical model for the problem by
    - introducing decision variables,
    - stating the objective function,
    - listing the problem constraints, and
    - writing the nonnegative constraints.
3. Graph the feasible region.
4. Make a table listing the value of the objective function at each corner point.
5. The optimal solutions will be the largest and smallest values of the points in this table.

6.1.3 Keywords

- *solution set*
- *bounded/unbounded*
- *corner point*
- *decision variables*
- *objective function*
- *problem constraints*
- *nonnegative constraints*
- *feasible region*
- *optimal solution*

6.2 The Simplex Method: Max LP Problems with Constraints of the Form $\leq$

6.2.1 Introduction

We have thus far looked at a geometric means of solving a linear programming problem. Our previous method works fine when we have two unknowns and are thus working with a feasible region that is a subset of the real plane (in other words, is a two-dimensional object). This will also work with three variables, but the graphs of the feasible regions are much more complicated (they would be three dimensional objects). Of course, things get terribly messy if we have four or more unknowns (we could not draw their graphs). As such we must develop another means of tackling linear programming problems.

The technique introduce in this section is due to George Bernard Dantzig (b. November 8, 1914, d. May 13, 2005). Dantzig developed it while on leave from his Ph.D. at Berkeley working for the Army Air Force during World War II. His algorithm was developed while serving and kept secret until it was published in 1951.

The approach begins with examining what we did in Section 4.2 when we considered a system of linear *equalities* of more than one variable. In particular, we were faced with solving the system

$$\begin{cases} 3x_1 - 5x_2 + x_3 & = 18 \\ -2x_1 + x_2 - x_3 & = 5 \\ x_1 - x_2 + 3x_3 & = -4 \end{cases}$$

We considered isolating a variable, substituting, and working to reduce the system to two equations in two unknowns, but that was too much work. Instead, we introduced an augmented matrix

$$\left[\begin{array}{rrr|r} 3 & -5 & 1 & 18 \\ -2 & 1 & -1 & 5 \\ 1 & -1 & 3 & -4 \end{array}\right]$$

and used Gauss-Jordan Elimination (row operations) to get the matrix in reduced form. It is this method that we adapt to solve linear programming problems.

6.2.2 Slack Variables

The careful reader may have observed that in linear programming problems the constraints are expressed as *inequalities* and not *equalities*. To modify the problem from one we do not know how to solve into one we do know how to solve, we introduce **slack variables**.

Recall that in Example 6.1.1 we had the problem constraints that

$$1x_1 + 2x_2 \leq 40, \tag{6.11}$$
$$2x_1 + 3x_2 \leq 72, \text{ and} \tag{6.12}$$
$$x_1, x_2 \geq 0. \tag{6.13}$$

We now introduce <u>nonnegative</u> ***slack variables*** s_1 and s_2 to "pick up the slack", i.e.

$$1x_1 + 2x_2 + s_1 = 40, \tag{6.14}$$
$$2x_1 + 3x_2 + s_2 = 72, \text{ and} \tag{6.15}$$
$$x_1, x_2, s_1, s_2 \geq 0. \tag{6.16}$$

Unfortunately, we are now in a situation where we have four unknowns and two equations. This means that we now have infinitely many solutions to the system (we do know something about these solutions, though; namely that we can fix two variables[5] and express the other two variables as a function of the fixed variables). We get around this little problem of an infinite number of solutions by introducing *basic* and *nonbasic* variables.

Definition 6.2.1 (Basic and Nonbasic Variables).
Basic variables *are chosen arbitrarily but with the restriction that there are exactly the same number of basic variables as there are constraint equations (we will see in Highlight 6.2.2 that there is a clever way to choose which variables are basic). We then say that the remaining variables are* **nonbasic variables**.

We may now present the idea of a basic solution. We **put the nonbasic variables equal to 0** and then the solution of the resulting system of linear equations is called a ***basic solution***. A basic solution is said to be a ***basic feasible solution*** if it lies within the feasible region of the problem (i.e. satisfies all constraints).

Let us revisit Example 6.1.1 with the inclusion of the slack variables. We then have the linear programming problem:

$$\text{Maximize:} \quad P(x_1, x_2) = 60x_1 + 90x_2 \tag{6.17}$$
$$\text{Subject to:} \quad x_1 + 2x_2 + s_1 \qquad = 40 \tag{6.18}$$
$$2x_1 + 3x_2 \qquad + s_2 = 72 \tag{6.19}$$
$$x_1, x_2, s_1, s_2 \geq 0. \tag{6.20}$$

With reference to our original linear programming problem, we may refer to this as the ***modified linear programming problem***.

[5]In this statement we are assuming the constraints are linearly independent, which is almost always the case in applications.

6.2.3 The Method

We begin by writing the model as an augmented matrix with the objective function as the last line:

$$\left[\begin{array}{ccccc|c} x_1 & x_2 & s_1 & s_2 & P & \\ 1 & 2 & 1 & 0 & 0 & 40 \\ 2 & 3 & 0 & 1 & 0 & 72 \\ \hline -60 & -90 & 0 & 0 & 1 & 0 \end{array}\right] \tag{6.21}$$

where the last line refers to rewriting the objective function as $-60x_1 + -90x_2 + P = 0$.[6] Recall from Definition 6.2.1 that there are to be as many basic variables as there are equations. Hence in the example we are to have three basic variables which leaves two to be nonbasic variables. By definition these may be chosen arbitrarily, but some careful thought will make the technique more friendly to use. Note specifically the last three columns of 6.21; they are the vectors $[1, 0, 0]^T$, $[0, 1, 0]^T$, and $[0, 0, 1]^T$. Realizing this matrix has rank 3 we see that our choice of vectors will nicely serve as a basis for the column space of the matrix. For this reason we choose these vectors to be our three basic variables (and this is, in fact, why they are called basic). Hence the decision variables x_1 and x_2 are left to be nonbasic variables. By convention, these are set equal to 0 which means row 1 of our augmented matrix tells us $0 + 0 + s_1 + 0 + 0 = 40$, i.e. $s_1 = 40$. Likewise the second row gives $s_2 = 72$ and the last row $P = 0$.

We now rewrite 6.21 to reflect this:

$$\left[\begin{array}{cccccc|c} (basic) & x_1 & x_2 & s_1 & s_2 & P & \\ (s_1) & 1 & 2 & 1 & 0 & 0 & 40 \\ (s_2) & 2 & 3 & 0 & 1 & 0 & 72 \\ \hline (P) & -60 & -90 & 0 & 0 & 1 & 0 \end{array}\right] \tag{6.22}$$

This form of the matrix representing our linear programming problem is called a ***tableau***.

Highlight 6.2.2 (Selecting Basic Variables). *The variables represented by columns with exactly one nonzero entry (always a 1) are selected to be the basic variables.*

Note in 6.22 we have $\{s_1, s_2, P\}$ as basic variables and $\{x_1, x_2\}$ as nonbasic, hence this corresponds to the solution $x_1 = 0, x_2 = 0, s_1 = 40, s_2 = 72, P = 0$

[6] Note that others choose to put the objective function in the first row of their matrix representing the Linear Programming problem. Also, instead of $-60x_1 + -90x_2 + P = 0$ some choose to write $60x_1 + 90x_2 - P = 0$, thus their row representing the objective function will have positive coefficients and their method is the corresponding adapted version of what follows. Hence other explanations of the Simplex Method may appear different than what we are doing, but really are the same procedure just written differently. This subject is a relatively young one extending over multiple disciplines and notation and procedures are not yet standardized.

or, specifically, $x_1 = 0, x_2 = 0, P = 0$. Though this does not optimize P, it does satisfy the constraints and is therefore feasible. Hence we have that $(0, 0, 40, 72, 0)$ is a basic feasible solution and we now refer to 6.22 as the ***initial Simplex tableau***.

Definition 6.2.3 (Initial or Canonical Simplex Tableau). *If the augmented matrix representing the modified linear programming problem has a solution that is feasible, it is called the* ***initial or canonical Simplex tableau.***

This is all very well and good, but our aim is to maximize profit. Where to go next? Well, since the Frontiersman model produces the most per-unit profit, it seems reasonable that letting x_2 be as large as possible will lead to maximizing the profit. x_2 corresponds to the second column and notice that this column has the largest negative entry in last row (the row corresponding to the objective function). We will choose to work with this column and refer to it as the ***pivot column***.

Highlight 6.2.4 (Selecting the PIVOT COLUMN). *To select the* pivot column, *choose the column with the largest negative entry in the bottom row. If there is a tie, choose either column.* ***If there are no negative entries, we are done and an optimal solution has been found.***

Some reflection will reveal how we know in Highlight 6.2.4 that if a pivot column cannot be selected then the optimal value as been obtained. No decision variable having a negative coefficient in the final row of the corresponding Simplex tableau means that there is no decision variable that can be increased and result in a larger objective function. Thus no changes in any decision variable (unless we leave the feasible region) will lead to a more optimal value of the objective function. Hence the maximum over the feasible region has been obtained and the process terminates.

Once a pivot column has been selected, what is to be done next? Since we are focusing on making as many Frontiersman models as possible (this choice over the Cabin model increases the profit the most), let us recall the constraints. In particular, we know that the cutting department needs 2 labor-hours to cut each sleeping bag and that the assembly department needs 3 labor-hours to assemble a sleeping bag. The cutting department only has 40 labor-hours available which means they can cut for at most 20 sleeping bags. The assembly department has only 72 labor-hours available, so they can assemble at most 24 sleeping bags. Hence we have the restriction that we can make at most 20 Frontiersman models of the sleeping bag in a single day. Hence we choose the first row as the ***pivot row***. Notice that these restrictions correspond to *dividing the each value in the last column by the corresponding value in the pivot column and then selecting the smallest positive ratio.*

Highlight 6.2.5 (Selecting the PIVOT ROW). *To select the* pivot row, *choose the row with the smallest positive ratio of the entry in the last column divided by the corresponding entry in the pivot column. If there is a tie, choose either* *row.* ***If there are no positive entries in the pivot column above the last row, the linear program has no optimal solution and we are done.***

We refer to the element of the tableau that is in the pivot row and the pivot column as the ***pivot element***. In our example, the pivot element is 2. Our job is to now perform legal row operations to make the pivot element 1 and every other entry in the pivot column 0. These row operations are commonly called ***pivot operations***.

Highlight 6.2.6 (PIVOT OPERATIONS). *There are two:*

- *Multiply the pivot row by the reciprocal of the pivot element. This transforms the pivot element into a 1. Symbolically, if k is the pivot element and R_t is the pivot row:* $\frac{1}{k}R_t \to R_t$.
- *Add multiples of the pivot row to all other rows in the tableau in order to annihilate (transform to 0) all other entries in the pivot column. Symbolically, something like* $aR_s + R_t \to R_t$.

Again, some thought sheds light on why not being able to select a pivot row in Highlight 6.2.5 leads to a linear programming problem not having a max. If no entry in the selected pivot column has a positive coefficient, then increasing the corresponding decision variable will lead to a *decrease* in the left-hand side of the constraint that the row represents. As the constraint is of the form $\leq$, the decision variable can be made arbitrarily large – thus increasing the objective function without end – and still satisfy the constraint. Hence when a pivot row cannot be selected, it is the case that the feasible region is unbounded.

Back to the example, we have decided that the second column (the column representing the variable x_2) is the pivot column and the first row (currently representing the basic variable s_1) is the pivot row. Hence we should begin our row operations by doing

$$\frac{1}{2}R_1 \to R_1$$

which gives us

$$\left[\begin{array}{cccccc|c} (basic) & x_1 & x_2 & s_1 & s_2 & P & \\ (s_1) & \frac{1}{2} & 1 & \frac{1}{2} & 0 & 0 & 20 \\ (s_2) & 2 & 3 & 0 & 1 & 0 & 72 \\ \hline (P) & -60 & -90 & 0 & 0 & 1 & 0 \end{array}\right].$$

Next, we do the row operations

$$-3R_1 + R_2 \to R_2 \text{ and } 90R_1 + R_3 \to R_3$$

which gives

$$\left[\begin{array}{cccccc|c} (basic) & x_1 & x_2 & s_1 & s_2 & P & \\ (x_2) & \frac{1}{2} & 1 & \frac{1}{2} & 0 & 0 & 20 \\ (s_2) & \frac{1}{2} & 0 & -\frac{3}{2} & 1 & 0 & 12 \\ \hline (P) & -15 & 0 & 45 & 0 & 1 & 1800 \end{array}\right].$$

Observe that we now have that the basic variables are given by the set $\{x_2, s_2, P\}$ and the non-basic variables are $\{x_1, s_1\}$. It is also worthwhile to point out that we currently have $x_1 = 0$ (since it is currently a non-basic variable) and $x_2 = 20$. This corresponds to a corner point in the graph of the feasible solution and, in fact, **the Simplex Method after one iteration has moved us from the origin to the point** $(0, 20)$ where P has increased from \$0 to \$1800.

Since we still have a negative entry in the bottom row, we repeat the process. Column 1 will now be the pivot column. As well, $20/\frac{1}{2} = 40$ and $12/\frac{1}{2} = 24$, hence we choose the second row as the pivot row. Since the pivot element is $\frac{1}{2}$ we initially perform the row operation

$$2R_2 \to R_2$$

to get

$$\left[\begin{array}{lcccccc|c} (basic) & x_1 & x_2 & s_1 & s_2 & P & \\ (x_2) & \frac{1}{2} & 1 & \frac{1}{2} & 0 & 0 & 20 \\ (s_2) & 1 & 0 & -3 & 2 & 0 & 24 \\ \hline (P) & -15 & 0 & 45 & 0 & 1 & 1800 \end{array}\right].$$

then the row operations

$$-\frac{1}{2}R_2 + R_1 \to R_1 \text{ and } 15R_2 + R_3 \to R_3$$

which gives us

$$\left[\begin{array}{lcccccc|c} (basic) & x_1 & x_2 & s_1 & s_2 & P & \\ (x_2) & 0 & 1 & 2 & -1 & 0 & 8 \\ (x_1) & 1 & 0 & -3 & 2 & 0 & 24 \\ \hline (P) & 0 & 0 & 0 & 30 & 1 & 2160 \end{array}\right].$$

Again, notice that x_1 has now entered as a basic variable (selected column 1 as the pivot column) while s_2 has excited the set of basic variables (selected row 2 as the pivot row). As well, since there are no more negative entries in the bottom row, the Simplex Method terminates and we have found the optimal solution; in particular that producing 24 Weekend model tents (which is what the variable x_1 represents) and producing 8 Backcountry model tents (represented by x_2) produces a maximum daily profit of \$2160. Specifically, **the Simplex Method during the second iteration moved us from the** $(0, 20)$ **to** $(24, 8)$ and P has increased from \$1800 to \$2160 and any other move will not increase P.

It is worthwhile to emphasize a very important property of the Simplex Method.

Discussion 6.2.7. *The Simplex method is designed in such a way that as long as the method begins at a feasible solution (hence the importance of the* initial simplex tableau *over just a tableau), are bounded, and we do not have* *degeneracy (see Section 6.5), the algorithm will efficiently move from corner point to corner point until it terminates at the optimal value. This is important because it guarantees*

1. *the process always terminates and*
2. *the process always produces a feasible solution.*

6.2.4 Summary

To summarize the Simplex Method,

1. Introduce slack variables into the mathematical model and write the initial tableau.
2. Are there any negative entries in the bottom row?
   - Yes – go to step 3.
   - No – the optimal solution has been found.
3. Select the pivot column.
4. Are there any positive elements above the last row (above the solid line)?
   - Yes – go to step 5.
   - No – no optimal solution exists (the feassible region is unbounded).
5. Select the pivot row and thus the pivot element. Perform the appropriate pivot operations then return to step 2.

Example 6.1.1 summarized:

$$\begin{aligned} \text{Maximize: } & P(x_1, x_2) = 60x_1 + 90x_2 \\ \text{Subject to: } & x_1 + 2x_2 + s_1 = 40 \\ & 2x_1 + 3x_2 + s_2 = 72 \\ & x_1, x_2, s_1, s_2 \geq 0. \end{aligned}$$

$$\overbrace{\left[\begin{array}{ccccccc|c} (basic) & x_1 & x_2 & s_1 & s_2 & P & \\ (s_1) & 1 & \boxed{2} & 1 & 0 & 0 & 40 \\ (s_2) & 2 & 3 & 0 & 1 & 0 & 72 \\ \hline (P) & -60 & -90 & 0 & 0 & 1 & 0 \end{array}\right]}^{Notes\ i,\ ii.}$$

$$\xrightarrow{\frac{1}{2}R_1 \to R_1} \left[\begin{array}{cccccc|c} & x_1 & x_2 & s_1 & s_2 & P & \\ (s_1) & \frac{1}{2} & 1 & \frac{1}{2} & 0 & 0 & 20 \\ (s_2) & 2 & 3 & 0 & 1 & 0 & 72 \\ \hline (P) & -60 & -90 & 0 & 0 & 1 & 0 \end{array}\right]$$

$$\xrightarrow[90R_1+R_3\to R_3]{-3R_1+R_2\to R_2} \overbrace{\left[\begin{array}{cccccc|c} & x_1 & x_2 & s_1 & s_2 & P & \\ (x_2) & \frac{1}{2} & 1 & \frac{1}{2} & 0 & 0 & 20 \\ (s_2) & \boxed{0.5} & 0 & -\frac{3}{2} & 1 & 0 & 12 \\ \hline (P) & -15 & 0 & 45 & 0 & 1 & 1800 \end{array}\right]}^{Note\ iii.}$$

$$\xrightarrow{2R_2\to R_2} \left[\begin{array}{cccccc|c} & x_1 & x_2 & s_1 & s_2 & P & \\ (x_2) & \frac{1}{2} & 1 & \frac{1}{2} & 0 & 0 & 20 \\ (s_2) & \boxed{1} & 0 & -3 & 2 & 0 & 24 \\ \hline (P) & -15 & 0 & 45 & 0 & 1 & 1800 \end{array}\right]$$

$$\xrightarrow[15R_2+R_3\to R_3]{-\frac{1}{2}R_2+R_1\to R_1} \overbrace{\left[\begin{array}{cccccc|c} & x_1 & x_2 & s_1 & s_2 & P & \\ (x_2) & 0 & 1 & 1 & -\frac{3}{2} & 0 & 8 \\ (x_1) & 1 & 0 & -3 & 2 & 0 & 24 \\ \hline (P) & 0 & 0 & 0 & 30 & 1 & 2160 \end{array}\right]}^{Note\ iv.}.$$

Notes:

i. As this matrix's solution set $x_1 = 0, x_2 = 0, s_1 = 40, s_2 = 72$ satisfies all constraints, the solution set is in the feasible region and thus we have an *initial Simplex tableau.*

ii. Since $|-90| > |-60|$ and $0 < \frac{40}{2} < \frac{72}{3}$, the 2 is the pivot element.

iii. Since the first column is the only column with a negative entry and since $0 < 12/\frac{1}{2} < 20/\frac{1}{2}$, the circled $\frac{1}{2}$ is the pivot element.

iv. As there are no negative entries in the bottom row, the process terminates and we have an optimal value.

6.2.5 Keywords

- *slack variables*
- *basic variables*
- *nonbasic variables*
- *basic solution*
- *basic feasible solution*
- *pivot row*
- *pivot column*
- *pivot element*
- *pivot operation*
- *Simplex Method*
- *Simplex Tableau*
- *Initial Simplex Tablea*

6.3 The Dual: Minimization with Problem Constraints of the Form ≥

6.3.1 How It Works

Instead of maximizing a particular objective function, let us are now consider a situation where we want to minimize an objective function, e.g. minimizing costs. These problems are of the form:

$$\begin{aligned}
\text{Minimize: } & C(y_1, y_2) = 40y_1 + 72y_2 && (6.23)\\
\text{Subject to: } & y_1 + 2y_2 \geq 60 && (6.24)\\
& 2y_1 + 3y_2 \geq 90 && (6.25)\\
& y_1, y_2 \geq 0. && (6.26)
\end{aligned}$$

As we will see in Section 6.3.2, each problem of this form can be associated with a corresponding maximization problem which we refer to as the ***dual problem***.

Our first step in forming the dual problem will be to form a matrix from the problem constraints and the objective function. The appropriate matrix for our example is

$$A = \left[\begin{array}{cc|c} 1 & 2 & 60 \\ 2 & 3 & 90 \\ \hline 40 & 72 & 1 \end{array}\right].$$

Please note that A is <u>not the matrix associated with the initial simplex tableau</u> as in Example 6.1.1. We now consider A^T, the ***transpose*** of matrix A. Hence, for our example:

$$A^T = \left[\begin{array}{cc|c} 1 & 2 & 40 \\ 2 & 3 & 72 \\ \hline 60 & 90 & 1 \end{array}\right].$$

Given A^T we may now form the dual problem, namely we now have a maximization problem with constraints of the form $\leq$. In particular, we have the dual problem:

$$\begin{aligned}
\text{Maximize: } & P(x_1, x_2) = 60x_1 + 90x_2 && (6.27)\\
\text{Subject to: } & x_1 + 2x_2 \leq 40 && (6.28)\\
& 2x_1 + 3x_2 \leq 72 && (6.29)\\
& x_1, x_2 \geq 0. && (6.30)
\end{aligned}$$

Look familiar? This is the example we have considered previously. This time, though, let us name our slack variables y_1 and y_2. Our initial Simplex tableau is

$$\left[\begin{array}{ccccc|c} x_1 & x_2 & y_1 & y_2 & P & \\ 1 & 2 & 1 & 0 & 0 & 40 \\ 2 & 3 & 0 & 1 & 0 & 72 \\ \hline -60 & -90 & 0 & 0 & 1 & 0 \end{array}\right]. \tag{6.31}$$

and the Simplex method gives us as the final tableau:

$$\left[\begin{array}{ccccc|c} x_1 & x_2 & y_1 & y_2 & P & \\ 0 & 1 & 1 & -\frac{3}{2} & 0 & 8 \\ 1 & 0 & -3 & 2 & 0 & 24 \\ \hline 0 & 0 & 0 & 30 & 1 & 2160 \end{array}\right]. \tag{6.32}$$

We note that the bottom row of the simplex tableau gives us the solution to our minimization problem, namely that $y_1 = 0$ and $y_2 = 30$ minimizes $C(y_1, y_2) = 60y_1 + 90y_2$ under the stated constraints. I.e.

Highlight 6.3.1. *An optimal solution to a minimization problem is obtained from the bottom row of the final simplex tableau for the dual maximization problem.*

One word of caution, though.

Warning!! 6.3.2. *Never multiply the inequality representing a problem constraint in a maximization problem by a number if that maximization problem is being used to solve a corresponding minimization problem.*

Example 6.3.3.

6.3.2 Why It Works

The type of problems considered in Section 6.2 were problems of the form

$$\begin{array}{ll} \text{maximize} & P(\mathbf{x}) = \mathbf{c^T x} \\ \text{subject to} & A\mathbf{x} \leq \mathbf{b} \\ & \mathbf{x} \geq \mathbf{0}. \end{array} \tag{6.33}$$

Here we have abbreviated the statement of the problem by using notation from Linear Algebra. By writing vectors together, e.g. $\mathbf{uv}$, we mean to multiply the vectors using the dot product. By the inequality of (6.33) we mean that for each component of the resulting vectors we have $a_{i1}x_1 + \cdots + a_{in}x_n \leq b_i$. Linear programming problems the form (6.33) ("**maximize**" together with "$A\mathbf{x} \leq \mathbf{b}$") are said to be of ***first primal form***, which was the class of problems we studied in Section 6.2. The ***dual of the first primal form*** is

$$\begin{array}{ll} \text{minimize} & C(\mathbf{y}) = \mathbf{b^T y} \\ \text{subject to} & A^T\mathbf{y} \geq \mathbf{c} \\ & \mathbf{y} \geq \mathbf{0} \end{array} \tag{6.34}$$

where this class of problems is characterized by "**minimize**" together with "$A\mathbf{y} \le \mathbf{c}$".

These two forms are joined by the following theorem:

Theorem 6.3.4 (The Fundamental Principle of Duality). *A minimization problem has a solution if and only if the corresponding dual maximization problem has a solution.*

More precisely

Theorem 6.3.5 (Weak Duality). *If* $\mathbf{x}$ *satisfies the constraints of a linear programming problem in first primal form and if* $\mathbf{y}$ *satisfies the constraints of the corresponding dual, then*

$$\mathbf{c}^T\mathbf{x} \le \mathbf{b}^T\mathbf{y}. \tag{6.35}$$

Proof. Since we have a problem of first primal form, we know that $\mathbf{x} \ge \mathbf{0}$ and $A\mathbf{x} \le \mathbf{b}$. Likewise, by the dual $\mathbf{y} \ge \mathbf{0}$ and $A^T\mathbf{y} \ge \mathbf{c}$ are satisfied. Hence

$$\mathbf{c}^T\mathbf{x} \le (A^T\mathbf{y})^T\mathbf{x} = \mathbf{y}^T A\mathbf{x} \le \mathbf{y}^T\mathbf{b} = \mathbf{b}^T\mathbf{y} \tag{6.36}$$

where the last equality holds because both expressions are dot products. □

Showing that $max\{\mathbf{c}^T\mathbf{x}\}$ and $min\{\mathbf{b}^T\mathbf{y}\}$ exist and are equal in this case is known as *Strong Duality* and this proof is left as an exercise.

The significance of 6.36 in Theorem 6.3.2 is clear once we realize we are trying to maximize the LHS and minimize the RHS. Thus we have a solution to each problem exactly when $\mathbf{c}^T\mathbf{x} = \mathbf{b}^T\mathbf{y}$; in other words if $\mathbf{x}$ is a solution to the first primal LP problem, then $\mathbf{y}$ is a solution to its dual and vice versa.

6.3.3 Keywords

- *dual problem*
- *the Fundamental Theorem of Duality*

6.4 The Big M Method: Max/Min LP Problems with Varying Constraints

6.4.1 Maximization Problems with the Big M Method

We motivate a technique for solving maximization problems with mixed constraints by considering the following example:

$$\text{Maximize: } P(x_1, x_2) = 2x_1 + x_2 \tag{6.37}$$
$$\text{Subject to: } x_1 + x_2 \leq 10 \tag{6.38}$$
$$-x_1 + x_2 \geq 2 \tag{6.39}$$
$$x_1, x_2 \geq 0. \tag{6.40}$$

As before, since the first inequality involves a $\leq$, we introduce a slack variable s_1:

$$x_1 + x_2 + s_1 = 10. \tag{6.41}$$

Note that this slack variable is necessarily nonnegative.

We need the second inequality to be an equality as well, so we introduce the notion of a *surplus variable*, s_2 (remember, the left hand side exceeds the 2, so the s_2 makes up for the difference). It would be natural to make this variable nonpositive, but, to be consistent with the other variables, let us require the surplus variable s_2 to as well be nonnegative. Hence the second inequality is rewritten

$$-x_1 + x_2 - s_2 = 2. \tag{6.42}$$

Hence the modified problem is:

$$\text{Maximize: } P(x_1, x_2) = 2x_1 + x_2 \tag{6.43}$$
$$\text{Subject to: } x_1 + x_2 + s_1 = 10 \tag{6.44}$$
$$-x_1 + x_2 - s_2 = 2 \tag{6.45}$$
$$x_1, x_2, s_1, s_2 \geq 0. \tag{6.46}$$

and the preliminary tableau for the Simplex Method is

$$\left[\begin{array}{lrrrrr|r} (basic) & x_1 & x_2 & s_1 & s_2 & P & \\ (s_1) & 1 & 1 & 1 & 0 & 0 & 10 \\ (s_2) & -1 & 1 & 0 & -1 & 0 & 2 \\ \hline (P) & -2 & -1 & 0 & 0 & 1 & 0 \end{array}\right]. \tag{6.47}$$

which has as its basic solution

$$x_1 = 0, x_2 = 0, s_1 = 10, s_2 = -2. \tag{6.48}$$

Unfortunately, this is not feasible (s_2 fails to satisfy the nonnegativity constraint). Hence this cannot be the initial Simplex tableau[7] and some work most be done to modify the problem so that we have an initial feasible basic solution (if the basic solution is not feasible, we eventually reach a step in the

[7] Some literature refers to an initial tableau which gives a basic feasible solution as a **canonical Simplex tableau**.

process where we do not yet have the optimal value for P, but we are unable to select a next pivot element – try it!).

We thus introduce an *artificial variable* for each surplus variable. These artificial variables are included in the constraint equations involving the respective surplus variable with the purpose of the surplus variable can now be nonnegative and the artificial variable will pick up the slack:

$$-x_1 + x_2 - s_2 + a_1 = 2. \tag{6.49}$$

The introduction of the artificial variable is very clever, but it is "artificial", so we do not want it to be part of an optimal solution. In particular, we want the artificial variable to be 0 in the solution. As such, we introduce a very large *penalty* into the objective function if the, artificial variable is anything other than 0 namely:

$$P(x_1, x_2) = 2x_1 + x_2 - Ma_1 \tag{6.50}$$

where M is an arbitrary number (hence the name "Big M" Method).

Hence the modified problem becomes:

$$\begin{aligned} \text{Maximize:} \quad & P(x_1, x_2) = 2x_1 + x_2 - Ma_1 && (6.51)\\ \text{Subject to:} \quad & x_1 + x_2 + s_1 \qquad\qquad = 10 && (6.52)\\ & -x_1 + x_2 \qquad - s_2 + a_1 = 2 && (6.53)\\ & \qquad\qquad x_1, x_2, s_1, s_2, a_1 \geq 0. && (6.54) \end{aligned}$$

and the tableau is

$$\left[\begin{array}{lcccccc|c} (basic) & x_1 & x_2 & s_1 & s_2 & a_1 & P & \\ (s_1) & 1 & 1 & 1 & 0 & 0 & 0 & 10 \\ (s_2) & -1 & 1 & 0 & -1 & 1 & 0 & 2 \\ \hline (P) & -2 & -1 & 0 & 0 & M & 1 & 0 \end{array}\right] \tag{6.55}$$

which has as its basic solution

$$x_1 = 0, x_2 = 0, s_1 = 10, s_2 = -2, a_1 = 0. \tag{6.56}$$

Note we still have the same problem... a nonfeasible basic solution and therefore the tableau in 6.55 is not the initial Simplex tableau. To remedy this, let us make a_1 a basic variable:
The row operation $(-M)R_2 + R_3 \to R_3$ gives us

$$\left[\begin{array}{lcccccc|c} (basic) & x_1 & x_2 & s_1 & s_2 & a_1 & P & \\ (s_1) & 1 & 1 & 1 & 0 & 0 & 0 & 10 \\ (a_1) & -1 & 1 & 0 & -1 & 1 & 0 & 2 \\ \hline (P) & M-2 & -M-1 & 0 & M & 0 & 1 & -2M \end{array}\right]. \tag{6.57}$$

which has as its basic solution

$$x_1 = 0, x_2 = 0, s_1 = 10, s_2 = 0, a_1 = 2, \tag{6.58}$$

which is feasible.

Now we are able to employ the Simplex method where M is some large *fixed* positive number. The Simplex Method yields:

$$\left[\begin{array}{ccccccc|c} (basic) & x_1 & x_2 & s_1 & s_2 & a_1 & P & \\ (x_1) & 1 & 0 & \frac{1}{2} & \frac{1}{2} & -\frac{1}{2} & 0 & 4 \\ (x_2) & 0 & 1 & \frac{1}{2} & -\frac{1}{2} & \frac{1}{2} & 0 & 6 \\ \hline (P) & 0 & 0 & \frac{3}{2} & \frac{1}{2} & M-\frac{1}{2} & 1 & 14 \end{array}\right]. \tag{6.59}$$

which has as its basic solution

$$x_1 = 4, x_2 = 6, s_1 = 0, s_2 = 0, a_1 = 0, P = 14, \tag{6.60}$$

which is feasible.

To summarize:

THE BIG M METHOD

SET-UP:

1. Multiply any constraints that have negative constants on the right by -1 (this is so that the notions of slack and surplus variables will be consistent).
2. Introduce a slack variable for every constraint that has a $\leq$.
3. Introduce a surplus variable and an artificial variable for every constraint that has a $\geq$.
4. For every artificial variable a_i that has been introduced, add $-Ma_i$ to the objective function.

SOLUTION:

1. Form the preliminary tableau to use the Simplex Method on the modified problem.
2. Do the necessary row operations to make each artificial variable a basic variable (i.e. make sure each column that represents an artificial variable has exactly 1 nonzero entry).
3. Apply the Simplex method to obtain an optimal solution to the modified problem.
4. a) If the modified problem has no optimal solution, then the original problem has no solution.

   b) If all artificial variables are 0, an optimal solution to the original problem has been found.

   c) If any artificial variable is nonzero, then the original problem has no optimal solution.

6.4.2 Minimization Problems with the Big M Method

In Exercise 2.2, you showed that for a real-valued function $f(\mathbf{x})$, $f(\mathbf{x}^*)$ is a maximum of f if and only if $-f(\mathbf{x}^*)$ is a minimum of $-f$. In other words, minimizing $f(x)$ is the same as maximizing $-f(x)$; they can be different values, but the location of $maxf(\mathbf{x})$ and $minf(\mathbf{x})$ are the same: $\mathbf{x}^*$. Therefore if one is to minimize $C(x_1, x_2, ..., x_n)$, merely apply the above procedures to maximize $-C(x_1, x_2, ..., x_n)$.

Highlight 6.4.1. *Note that when using the Big M Method to solve a minimization problem, the penalty $+Ma_1$ (add since we are minimizing) is to be introduced **before** negating the objective function C.*

Example 6.4.2.

$$\begin{aligned} \textit{Minimize:} \quad & C = 5x_1 + 3x_2 && (6.61)\\ \textit{Subject to:} \quad & 3x_1 + 4x_2 \geq 12 && (6.62)\\ & 2x_1 + 5x_2 \leq 20 && (6.63)\\ & x_1, x_2 \geq 0. && (6.64) \end{aligned}$$

Thus the modified problem is

$$\begin{aligned} \textit{Minimize:} \quad & C = 5x_1 + 3x_2 + Ma_1 && (6.65)\\ \textit{Subject to:} \quad & 3x_1 + 4x_2 - s_1 \quad + a_1 = 12 && (6.66)\\ & 2x_1 + 5x_2 \quad + s_2 \quad = 20 && (6.67)\\ & x_1, x_2 \geq 0 && (6.68) \end{aligned}$$

with surplus variable s_1, slack variable s_2, and artificial variable a_1. Note that we have added the penalty Ma_1 because this is a minimization problem and we want a_1 having any value above 0 to take us away from the minimum.

As the Big M Method is designed for maximization problems, we adapt the given problem in the following way:

$$\begin{aligned} \textit{Maximize:} \quad & P = -C = -5x_1 - 3x_2 - Ma_1 && (6.69)\\ \textit{Subject to:} \quad & 3x_1 + 4x_2 - s_1 \quad + a_1 = 12 && (6.70)\\ & 2x_1 + 5x_2 \quad + s_2 \quad = 20 && (6.71)\\ & x_1, x_2, s_1, s_2 \geq 0. && (6.72) \end{aligned}$$

The corresponding tableau is

$$\left[\begin{array}{ccccccc|c} (basic) & x_1 & x_2 & s_1 & s_2 & a_1 & P & \\ (s_1) & 3 & 4 & -1 & 0 & 1 & 0 & 12 \\ (s_2) & 2 & 5 & 0 & 1 & 0 & 0 & 20 \\ \hline (P) & 5 & 3 & 0 & 0 & M & 1 & 0 \end{array}\right]. \qquad (6.73)$$

which has $x_1 = x_2 = a_1 = 0$ and $s_1 = -12, s_2 = 20$ as its basic solution.

Note this is not feasible, so we perform the row operation $-MR_1 + R_3 \to R_3$ to obtain

$$\left[\begin{array}{ccccccc|c} (basic) & x_1 & x_2 & s_1 & s_2 & a_1 & P & \\ (a_1) & 3 & 4 & -1 & 0 & 1 & 0 & 12 \\ (s_2) & 2 & 5 & 0 & 1 & 0 & 0 & 20 \\ \hline (P) & 5-3M & 3-4M & M & 0 & 0 & 1 & -12M \end{array}\right]. \qquad (6.74)$$

TABLE 6.3
Summary of Applying the Simplex Method to LP Problems

LP	Constraints	RHS Constants	Coeff. of P	Solution
Max	$\leq$	nonnegative	any	Simplex w/ slack variables
Min	$\geq$	any	nonnegative	the dual
Max	$\geq$ or mixed	nonnegative	any	Big M
Min	$\leq$ or mixed	nonnegative	any	Big M max negative obj. func.

which has $x_1 = x_2 = s_1 = 0$ and $a_1 = 12, s_2 = 20$ as its basic solution. Since this is feasible, we may now proceed with the Simplex Method.

$$\left[\begin{array}{ccccccc|c} (basic) & x_1 & x_2 & s_1 & s_2 & a_1 & P & \\ (a_1) & 3 & \boxed{4} & -1 & 0 & 1 & 0 & 12 \\ (s_2) & 2 & 5 & 0 & 1 & 0 & 0 & 20 \\ \hline (P) & 5-3M & 3-4M & M & 0 & 0 & 1 & -12M \end{array}\right] \tag{6.75}$$

Using row operations $\frac{1}{4}R_1 \to R_1$, $-\frac{5}{4}R_1 + R_2 \to R_2$, and $\frac{4M-3}{4}R_1 + R_3 \to R_3$, we obtain:

$$\left[\begin{array}{ccccccc|c} & x_1 & x_2 & s_1 & s_2 & a_1 & P & \\ (x_2) & \frac{3}{4} & 1 & -\frac{1}{4} & 0 & \frac{1}{4} & 0 & 3 \\ (s_2) & -\frac{7}{4} & 0 & \frac{5}{4} & 1 & -\frac{5}{4} & 0 & 5 \\ \hline (P) & \frac{11}{4} & 0 & \frac{3}{4} & 0 & \frac{4M-3}{4} & 1 & -9 \end{array}\right] \tag{6.76}$$

As there are no negative entries in the columns involving the variables, the process terminates and we see that max $P = -C - 9$ therefore min $C = 9$ and this occurs at $x_1 = 0, x_2 = 3$.

6.5 Degeneracy and Cycling in the Simplex Method

We have seen that the notion of a basic feasible solution is fundamental to the Simplex Method. The name *basic* comes from the fact that the column vectors representing the basic variables form a basis for the column space ofthe linear programming problem's corresponding matrix. The Simplex Method can fail when at some iteration in the process a basic variable is 0. The pivot that introduces the basic variable into the feasible solution is referred to as a *degenerate pivot* and we refer to the method in this case as being *degenerate.*

Degenerate pivots can happen and not be a death blow to the algorithm accomplishing its task. Unfortunately, there are times when a degenerate pivot leads to further application of the algorithm causing old decision variables to leave the set of basic variables and new ones enter, but the value of the objective function does not change. In this situation, it is possible for the Simplex Method to cycle through these sets in some maddening infinite loop and the process thus not terminate. Such a situation is referred to as *cycling* and causes a problem only in that the algorithm will not stop. Including an anti-cycling rule such as *Bland's Smallest Subscript Rule* (or just *Bland's Rule*) [3] when employing the Simplex Method guarantees that the algorithm will terminate.

6.6 Exercises

For Exercises 6.1 through 6.5, do each of the following:

I. Solve the following linear programming problems by graphing the feasible region then evaluating the objective function at each corner point. "Solve" means state the optimal value of the objective function and ***all points*** in the feasible region at which this optimal value occurs.
II. Solve each problem using the Simplex Method, it's dual, or the Big M Method. Your work should contain a clear statement of the model after the introduction of slack, surplus, and artificial variables. You may use a calculator or computer to do the row operations, but write down the obtained simplex tableau after each iteration of the method. At each iteration identify the pivot element.
III. Check your work using a software package of your choice (Solver, Matlab, etc.). Print and submit your answer screen and please make clear what software you have used.

Exercise 6.1.

$$\begin{aligned}
\textit{Minimize } P(x, y) &= 5x + 2y \\
\textit{Subject to } x + y &\geq 2 \\
2x + y &\geq 4 \\
x, y &\geq 0
\end{aligned}$$

For this question only *(i.e. Exercise 6.1), when the Simplex method (part II above), at each iteration state which variables are basic and which are non-basic. Also, at each iteration state the value of the objective function.*

Exercise 6.2.

$$\begin{aligned} &\textit{Maximize } P(x,y) = 5x + 2y \\ &\textit{Subject to } x + y \geq 2 \\ &\qquad 2x + y \geq 4 \\ &\qquad x, y \geq 0 \end{aligned}$$

Exercise 6.3.

$$\begin{aligned} &\textit{Maximize } P(x,y) = 20x + 10y \\ &\textit{Subject to } x + y \geq 2 \\ &\qquad x + y \leq 8 \\ &\qquad 2x + y \leq 10 \\ &\qquad x, y \geq 0 \end{aligned}$$

Exercise 6.4.

$$\begin{aligned} &\textit{Maximize } P(x,y) = 20x + 10y \\ &\textit{Subject to } 2x + 3y \geq 30 \\ &\qquad 2x + y \leq 26 \\ &\qquad -2x + 5y \leq 34 \\ &\qquad x, y \geq 0 \end{aligned}$$

Exercise 6.5.

$$\begin{aligned} &\textit{Minimize } P(x,y) = 20x + 10y \\ &\textit{Subject to } 2x + 3y \geq 30 \\ &\qquad 2x + y \leq 26 \\ &\qquad -2x + 5y \leq 34 \\ &\qquad x, y \geq 0 \end{aligned}$$

Exercise 6.6. *In this problem, there is a tie for the choice of the first pivot column. When you do your work using the simplex method use the method twice to solve the problem two different ways; first by choosing column 1 as the first pivot column and then for your second solution effort, solve by choosing column 2 as the first pivot column. You may use a computer or calculator to perform the Simplex Method, but do write down the results of each iteration.*

$$\begin{aligned} \textit{Maximize } & P(x, y) = x + y \\ \textit{Subject to } & 2x + y \leq 16 \\ & x \leq 6 \\ & y \leq 10 \\ & x, y \geq 0 \end{aligned}$$

Exercise 6.7. *This problem has multiple parts.*

1. *Solve the following by using the dual of the Simplex Method. Your work should contain each the details of each iteration.*

$$\begin{aligned} \textit{Minimize } & C(x_1, x_2, x_3) = 40x_1 + 12x_2 + 40x_3 \\ \textit{Subject to } & 2x_1 + x_2 + 5x_3 \geq 20 \\ & 4x_1 + x_2 + x_3 \geq 30 \\ & x_1, x_2, x_3 \geq 0 \end{aligned}$$

2. *The dual problem has as its first constraint*

$$2y_1 + 4y_2 \leq 40. \tag{6.77}$$

Replace this constraint by its simplified version

$$y_1 + 2y_2 \leq 20 \tag{6.78}$$

then proceed with the Simplex Method.

3. *Compare your answers from the first two parts. Why are they different?*

7

Sensitivity Analysis

7.1 Motivation

Recall Lincoln Outdoors manufacturing sleeping bags from Example 6.1.1 with the details of the situation summarized in Table 7.1.

We sought to help Lincoln Outdoors by determining what manufacturing levels maximized profit and modeled the situation with the linear programming problem:

$$\begin{aligned} \text{Maximize:} \quad & P(x_1, x_2) = 60x_1 + 90x_2 && (7.1)\\ \text{Subject to:} \quad & x_1 + 2x_2 = 40 && (7.2)\\ & 2x_1 + 3x_2 = 72 && (7.3)\\ & x_1, x_2 \geq 0. && (7.4) \end{aligned}$$

We found that any of $(24, 8)$, $(27, 6)$, $(30, 4)$, $(33, 2)$, or $(36, 0)$ would give a maximum profit of \$2,160.

In many real life situations, the vales used for the coefficients of the constraints and the objective functions are often (hopefully) good estimates of what is taking place (or even a good prediction of what will take place) and, as such, management may be skeptical of the analysis. Also, situations may change in that the cost of materials might fluctuate as may the efficiency of our machines or labor force. For example, it may be that due to a change in the cost of materials the profit realized on the production and sale of a Cabin Model sleeping bag is \$58.77 instead of \$60 or by an employee's clever suggestion assembling a Frontier Model sleeping bag only now takes 2.67 hours instead of 3. *Sensitivity analysis* can help convince someone that the model is reliable as well as give insight into the how a solution may change if any of the contributing factors vary slightly.

7.2 An Excel Example

We will illustrate the ideas of this chapter by considering the Lincoln Outdoors problem. We can enter it in an Excel worksheet as in Figure 7.1.

DOI: 10.1201/9780367425517-7

TABLE 7.1
Manufacturing Data for Lincoln Outdoors in Example 6.1.1

Labor-Hours	Cabin Model	Frontier Model	Max Hours per Day
Cutting Dept.	1	2	40
Assembly Dept.	2	3	72
Profit per Bag	\$60	\$90	

Cabin Model				
Frontier Model				
		Cabin	Frontier	Available
Profit	0	60	90	
Cutting	0	1	2	40
Assembly	0	2	3	72

FIGURE 7.1
The Lincoln Outdoors problem in Excel.

Cabin Model	24			
Frontier Model	8			
		Cabin	Frontier	Available
Profit	2160	60	90	
Cutting	40	1	2	40
Assembly	72	2	3	72

FIGURE 7.2
An Excel solution for Lincoln Outdoors.

Using Excel's add-in Solver and selecting "Simplex LP" as the Solving Method option gives the result shown in Figure 7.2.

Before accessing the result, Solver permits accessing some optional reports as in Figure 7.3. As we are exploring sensitivity analysis, we are going to want Solver to show us an Answer Report, Sensitivity Report, and Limits Report and we get these options by clicking on the appropriate words in the upper right of the window.

After highlighting the Answer, Sensitivity, and Limits reports, select the "OK" tab and Solver returns to the workbook displaying the solution with tabs at the bottom of the sheet. Sheet 1 (or whatever name you change it to) displays Figure 7.4.

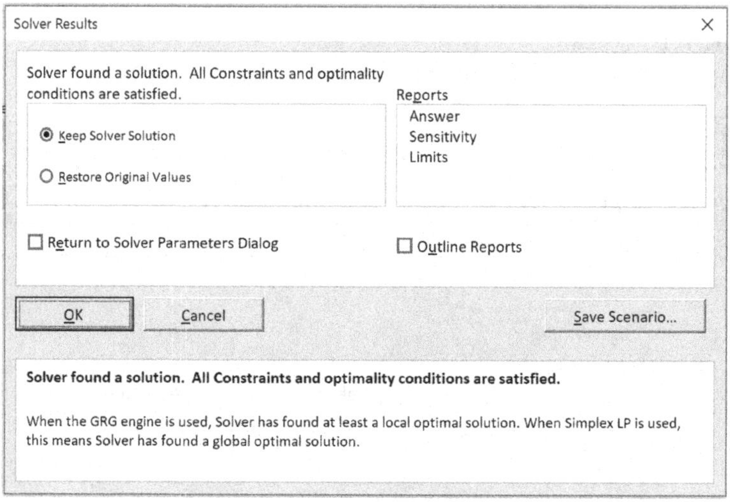

Solver Results

Solver found a solution. All Constraints and optimality conditions are satisfied.

Keep Solver Solution

Restore Original Values

Reports
Answer
Sensitivity
Limits

Return to Solver Parameters Dialog

Outline Reports

OK Cancel Save Scenario...

Solver found a solution. All Constraints and optimality conditions are satisfied.

When the GRG engine is used, Solver has found at least a local optimal solution. When Simplex LP is used, this means Solver has found a global optimal solution.

FIGURE 7.3
Options in Excel's solution for Lincoln Outdoors.

	A	B	C	D	E	F	G
1							
2		Cabin Model	0				
3		Frontier Model	0				
4				Cabin	Frontier	Available	
5		Profit	0	60	90		
6							
7		Cutting	0	1	2	40	
8		Assembly	0	2	3	72	
9							
10							

Answer Report 1 | Sensitivity Report 1 | Limits Report 1 | **Sheet1**

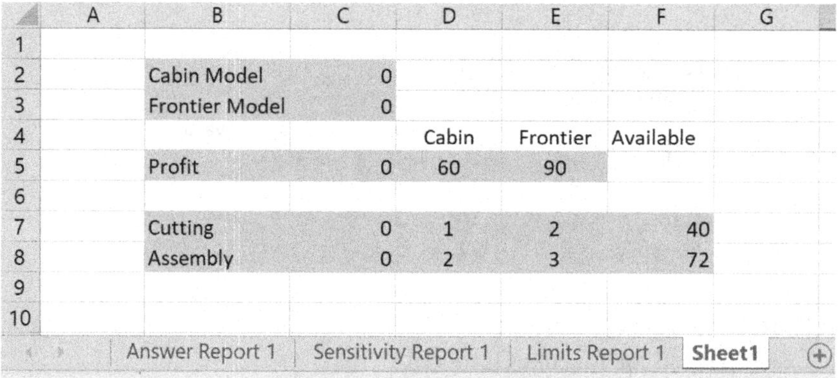

FIGURE 7.4
Report options displayed in Excel's solution for Lincoln Outdoors.

7.2.1 Solver's Answer Report

Let us select the Answer Report tab. Excel displays what is shown in Figure 7.5. The Objective Cell (Max) part of the report tells us that our Profit cell (C5) started with a value of 0 (the default value for a blank cell) and reached its max at $2,160. When using the Simplex LP solution option, the starting value should be at the origin as the technique is designed to start there, though in practice any value should work as the search is very quick and (barring

Objective Cell (Max)

Cell	Name	Original Value	Final Value
$C$5	Profit	0	2160

Variable Cells

Cell	Name	Original Value	Final Value	Integer
$C$2	Cabin Model	0	24	Contin
$C$3	Frontier Model	0	8	Contin

Constraints

Cell	Name	Cell Value	Formula	Status	Slack
$C$7	Cutting	40	$C$7<=$F$7	Binding	0
$C$8	Assembly	72	$C$8<=$F$8	Binding	0

FIGURE 7.5
The answer report for Lincoln Outdoors.

irregularities that seldom occur in application) the process is guaranteed to converge.

The next part of the report, Variable Cells, gives us the initial and solution values of the decision variables as well as the additional information that we found the solution over the real numbers ("Contin" = "Continuous" as compared to an integer or binary solution).

Lastly the Constraints section reports the value of the constraints when the solution is reached. For the situation when $x = \textit{Cabin Models} = 24$ and $y = \textit{Frontier Models} = 8$, the Cutting constraint has a value of 40 while the Assembly constraint has a value of 72; both of which are at their respective limits hence the report that for each variable the Slack is 0 and therefore each constraint is "Binding" meaning there is no room to move either constraint any higher.

The previous report serves its purpose, but for the sake of this chapter we want to focus on the reports in the second and third tabs, namely the Sensitivity Report (Figure 7.6) and the Limits Report (Figure 7.8).

7.2.2 Solver's Sensitivity Report

In the Sensitivity Report (Figure 7.6), the Variable Cells section refers to the decision variables. The Final Value column displays the value of the decision variables returned by Solver for yields the optimal solution. Recall that in this example there are multiple solutions: $(24, 8)$, $(27, 6)$, $(30, 4)$, $(33, 2)$, and $(36, 0)$. Solver returns $(24, 8)$ because it used the Simplex Method which begins

Variable Cells

Cell	Name	Final Value	Reduced Cost	Objective Coefficient	Allowable Increase	Allowable Decrease
$C$2	Cabin Model	24	0	60	0	15
$C$3	Frontier Model	8	0	90	30	0

Constraints

Cell	Name	Final Value	Shadow Price	Constraint R.H. Side	Allowable Increase	Allowable Decrease
$C$7	Cutting	40	0	40	8	4
$C$8	Assembly	72	30	72	8	12

FIGURE 7.6
The sensitivity report for Lincoln Outdoors.

at $(0, 0)$, one iteration takes us to $(0, 20)$ and the next iteration to $(24, 8)$. As no selection of a pivot column would improve the objective function's value (no negative values in the bottom row as per Highlight 6.2.4), the algorithm terminates here. The Reduced Cost requires a little bit of work and we will address that after we understand Shadow Prices. The Objective Coefficient is (as the name states) the coefficient of the decision variable in the objective function. The Allowable Increase and Allowable Decrease columns tell us how much range we have for a change in the particular coefficient to not affect the solution **provided all other coefficients remain unchanged**. For example, in this situation, if we think of the objective function as being $P(x_1, x_2) = Ax_1 + 90x_2$ then the stated global solution [the $(24, 8)$; *not the others!*] will remain the same as long as $45 \leq A \leq 60$. Likewise, $P(x_1, x_2) = 60x_1 + Bx_2$ has also has the solution unchanged as long as $90 \leq B \leq 120$.

You are encouraged to experiment with the ranges of the decision variables in a spreadsheet using Solver. Geometrically, Figure 7.7 shows the idea behind these numbers: for instance, once $A < 45$, the furthest corner point the objective function would intersect as it leaves the origin (as in Figure 6.2) would be $(0, 20)$. If the A goes above 60, the objective function becomes steep enough that the last corner point it would intersect leaving the feasible region would be $(36, 0)$. Each of these situations is illustrated in Figure 7.7. Again, it is important to restate that this range of values only holds if the other decision variable's constant – the 90 – remains unchanged.

In situations where other solutions of a Linear Programming problem exist, just like in Lincoln Outdoors, Solver's Sensitivity Report has another use.

Highlight 7.2.1 (Other LP Solutions in Solver). *Apart from degenerate cases, when using Solver to find the optimum in a Linear Programming problem, a* 0 *in either the Allowable Increase or Allowable Decrease in the Variable Cells section of the Sensitivity Report signals that other solutions exists.*

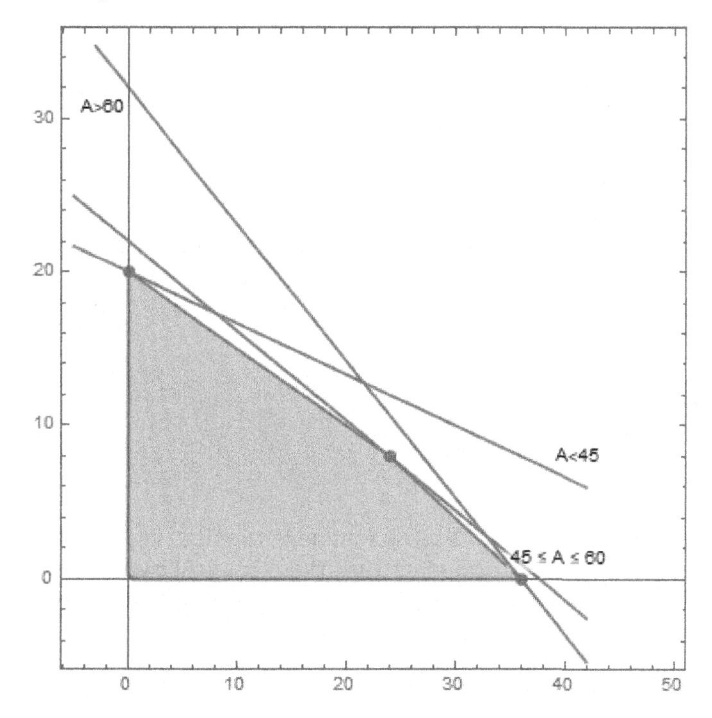

FIGURE 7.7
Solution changes for different A in the objective function $P = Ax_1 + 90x_2$ with Lincoln Outdoors.

In this situation we can use Solver to attempt to find other solutions by using Algorithm 7.2.1.

Algorithm 7.2.1 Finding Additional Non-Degenerate LP Solutions Using Solver.

Input: Solved LP problem in Solver with decision variables $x_1, \ldots, x_n$.
1: Add a constraint to the model that holds the objective function at the optimal value.
2: **for** $i = 1$ to k **do**
3:     **if** Allowable Decrease $= 0$ for x_i **then**
4:         run Solver to minimize x_i
5:     **end if**
6:     **if** Allowable Increase $= 0$ **then**
7:         run Solver to maximize x_i
8:     **end if**
9: **end for**
Output: Additional non-degenerate LP solutions via Solver (if they exist).

Cell	Objective Name	Value
$C$5	Profit	2160

Cell	Variable Name	Value	Lower Limit	Objective Result	Upper Limit	Objective Result
$C$2	Cabin Model	24	0	720	24	2160
$C$3	Frontier Model	8	0	1440	8	2160

FIGURE 7.8
The limits report for Lincoln Outdoors.

Note that in Algorithm 7.2.1, we minimize the decision variable if its Allowable Decrease $= 0$ due to the fact that there is an Allowable Increase and a move in that direction would not change the current optimal solution (hence an "*allowable*" increase). Likewise, if the Allowable Increase $= 0$ we would not want to minimize as there is a positive allowable decrease that would not change the stated solution.

In the Lincoln Outdoors example, this would mean we add $2160 = 60x_1 + 90x_2$ to the constraints. Then if we choose to work with the decision variable representing the number of Cabin Models, x_1, we would make the objective function $Max\ x_1$; or, if we rather choose to focus on the number of Frontier Models, we would make the objective function $Min\ x_2$. The models to explore would then be

$$\text{Maximize: } f(x_1, x_2) = x_1 \tag{7.5}$$
$$\text{Subject to: } 60x_1 + 90x_2 = 2160 \tag{7.6}$$
$$x_1 + 2x_2 = 40 \tag{7.7}$$
$$2x_1 + 3x_2 = 72 \tag{7.8}$$
$$x_1, x_2 \geq 0 \tag{7.9}$$

and

$$\text{Minimize: } f(x_1, x_2) = x_2 \tag{7.10}$$
$$\text{Subject to: } 60x_1 + 90x_2 = 2160 \tag{7.11}$$
$$x_1 + 2x_2 = 40 \tag{7.12}$$
$$2x_1 + 3x_2 = 72 \tag{7.13}$$
$$x_1, x_2 \geq 0. \tag{7.14}$$

Next we turn to the Constraints section of the Sensitivity Report. The Final Value is the value of the constraint when it is evaluated at the point given in the solution and the Constraint R.H. Side is the constant on the right of the constraint. Note that in the Lincoln Outdoors example these columns agree for both constraints. This is exactly because both constraints are binding and there are no labor-hours available for either at the production level given in the solution.

The Shadow Price returns the marginal value the problem constraints (usually resources). Specifically, this value is the amount that the objective function will change with a unit increase in the constant on the right hand side of the constraint; a positive shadow price means an increase in the objective function's optimal value where a negative shadow price means the optimal value of the objective function would decrease. For example, the shadow price of the assembly constraint in the example is 30; this means that if we make one more labor-hour available in the assembly department, the objective function will increase by $30 (note *this is a theoretical return as 1 more assembly hour does not return a full sleeping bag and we will only be selling an integer amount of sleeping bags*). The shadow price of 0 for the cutting constraint means a unit increase in the labor-hours available for cutting will result in no change in the optimal value of the objective function. The Allowable Increase and Allowable Decrease columns that follow provide a range of the changes in the constraint's bound, i.e. the constant on the right hand side of the constraint, for which these shadow prices hold. For example, in the Lincoln Outdoors solution any amount of cutting room labor hours from 36 to 48 would still have a shadow price (marginal value) of 0.

Now that we understand Shadow Price, we can address Reduced Cost. The Reduced Cost of a variable is calculated by

$$\begin{aligned}\textsf{Reduced Cost} &= \textsf{coefficient of variable in objective function}\\ &\quad - \textsf{value per unit of resources used}\end{aligned} \tag{7.15}$$

where *the resources are valued at their shadow price.*

For example, the reduced cost of the Cabin Model for Lincoln Outdoors is (all values are *per unit*)

$$\begin{aligned}&\text{contribution to objective function} - \text{cutting hours} \cdot \text{shadow price}\\ &\quad - \text{assembly hours} \cdot \text{shadow price}\\ &= 60 - 1 \cdot 0 - 2 \cdot 30 = 0.\end{aligned}$$

As well, the reduced cost of the Frontier Model is

$$90 - 2 \cdot 0 - 3 \cdot 30 = 0.$$

It is not a coincidence that both of these values are 0, as the only time Reduced Cost of a decision variable is non-zero is if the variable is at either its lower or upper bound of the feasible region. For example, based on the cutting and assembly constraints we will only be able to produce between 0 and 36 Cabin Model sleeping bags. The solution $x_1 = 24$ is easily within this range.

7.3 Exercises

Exercise 7.1. *Solve the linear programming problem in Exercise 6.1 using Solver. Provide the Answer Report and Limits Report generated by Solver. Explain all the details given in these reports.*

Exercise 7.2. *Explain why Highlight 7.2.1 is true for non-degenerate Linear Programming problems with multiple solutions.*

8

Integer Linear Programming

8.1 Introduction

Many applications of Linear Programming do not welcome fractional solutions. For example, why would anyone want to spend resources to produce $\frac{3}{8}$ of a blender? If a Linear Programming model requires integer solutions, it is natural to assume that we may use the techniques previously learned to solve the problem over the real numbers, then round the solution. Unfortunately, this does not always yield a feasible solution. Consider the following example:

Example 8.1.1. *[Anna's Cozy Home Furnishings]* Anna's Cozy Home Furnishings *(ACHF) handcrafts two kinds of quality hardwood tables for local furniture stores. The company offers a simple Farmhouse model and an elegant Designer model. It takes 1 labor-hour to cut each Farmhouse model whereas each Designer table takes 2 labor-hours to cut. Assembly and finishing time for the Farmhouse model is 3 labor-hours and 5 for the Designer. ACHF assembles and finishes the tables and has a total of 71 labor-hours available per week but contracts the cutting externally and by contract must place an order requiring at least 30 hours per week of cutting. ACHF rents its assembly and finishing facility to another company on the weekends and, as such, they can leave no materials or unfinished tables on site. ACHF makes a profit of* \$100 *per farmhouse table sold and* \$250 *per designer model. Assuming all tables ACHF produces will sell, what should ACHF weekly production be in order to maximize profit?*

Letting x represent the number of Farmhouse tables produced and y the number of Designer tables, the Linear Programming model for this problem is

$$\begin{aligned} \text{Maximize } P &= 100x + 250y && (8.1)\\ \text{Subject to } \quad 3x + 5y &\le 71, && \text{(assembly and finishing constraint)}\\ 1x + 2y &\ge 30, && \text{(cutting constraint)}\\ x, y &\ge 0 \text{ and integer-valued.} \end{aligned}$$

The graph of the feasible region is

DOI: 10.1201/9780367425517-8

The optimal solution as a linear programming problem is to produce 0 Farmhouse models and 14.2 Designer models, but this is not realistic as ACHF cannot have unfinished tables around. If we consider rounding, 0 Farmhouse models and 14 Designer models is certainly feasible, but this may not be optimal. Note that making 1 Farmhouse model and 14 Designer models is not feasible. As we will see shortly, the optimal integer solution is actually to make 3 Farmhouse models and 13 Designer models.

The points in an n dimensional rectangular coordinate system with all coordinates integer are called *lattice points*. As we saw in the discussion at the beginning of this section, considering lattice points neighboring the LP solution over the reals (i.e. rounding) does not necessarily yield the ILP solution.

We present two integer linear programming techniques whose approaches both involve whittling away parts of the feasible region until an integer-valued solution is obtained.

8.2 Dakin's Branch and Bound

Our first solution technique was introduced by R.J. Dakin in 1964 [13] which improves on a technique introduced by A.H. Land and A.G. Doig in 1960 [41]. This approach solves the LP problem over the reals, then introduces bounds in hope of obtaining a solution. It should be noted that this process is not guaranteed to terminate or to give a globally optimal solution.

Algorithm 8.2.1 Dakin's Branch and Bound Algorithm for ILP.

Input: LP problem to be solved over the integers.

1: Solve the linear programming problem over the real numbers.
2: Identify a decision variable x_i whose solution is a non-integer value b_i.
3: Branch the LP problem into two new problems by introducing into one subproblem the constraint $x_i \leq \lfloor b_i \rfloor$ and into the other subproblem introduce the constraint $x_i \geq \lceil b_i \rceil$.
4: Solve each branch as an LP problem over the reals. If either has a lattice point as its solution, stop. Else repeat step 1 for both branches.

Output: A local solution for an Integer Linear Programming problem.

Note that Algorithm 8.2.1 can be repeated as needed in hopes of getting a local solution closer to a global solution. As we will see in Example 8.2.2, we may not want to terminate our search at the first local solution.

The first set of subproblems of Algorithm 8.2.1 is illustrated in the following tree (note: we are free to choose the x_i with which we wish to work):

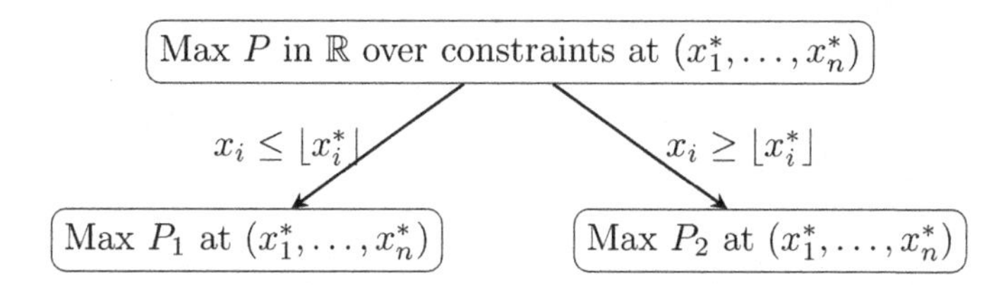

Before continuing, an important observation.

Observation 8.2.1. *Each branch of Dakin's method introduces a new constraint which shrinks the feasible region hence the optimal value at each branch can be no better than the value before the new constraints were introduced.*

In other words, if we seek to maximize a function P and have found an LP solution P^*, branching will lead to values P_1^* and P_2^* where $P_1^* \leq P^*$ and $P_2^* \leq P^*$. Whereas if we seek to minimize a function C and have found an LP solution C^*, branching will lead to values C_1^* and C_2^* such that $C_1^* \geq C^*$ and $C_2^* \geq C^*$. This is illustrated in returning to *Anna's Cozy Home Furniture* (Example 8.2.2) in the following example:

Example 8.2.2 (ACHF Tables using Dakin's Branch and Bound). *Solving over the reals gives a maximum profit of* $P = \$3550$ *when producing* $x = 0$ *Farmhouse tables and* $y = 14.2$ *Designer tables. The problem now branches into two LP problems with the first branch having the additional constraint that* $y \leq 14$ *and the second branch* $y \geq 15$*. Branch 1 yields* $P^* = \$3550$ *at* $(0.5, 14)$ *where Branch 2 has an empty feasible region. We then further split Branch 1 into Branches 3 and 4 by respectively introducing the constraints* $x \leq 0$ *(i.e.* $x = 0$*) and* $x \geq 1$*. Branch 3 yields* $P^* = 3500$ *at* $(0, 14)$ *where Branch 4 gives* $P^* = 3550$ *at* $(1, 13.8)$*. Branch 3 has an integer solution, but we may be able to do better than its profit of* \$3500*. Branch 4 further subdivides by introducing the constraints* $y \leq 13$ *(Branch 5) and* $y \geq 14$ *(Branch 6). Branch 6 has an empty feasible region whereas Branch 5 produces* $P^* = \$3550$ *at* $(3, 13)$*. Continuing down any branch would only lead to smaller values for* P*, thus we terminate the process.*

The example is summarized in the following tree and in Table 8.1 (note that the table does not include the nonnegativity constraints):

TABLE 8.1
LP Branches of Dankin's Method for Example 8.2.2, ACHF Tables

Branch	Introduce	Constraints	Removing Redundancies	Relaxed Solution over $\mathbb{R}$
		$2x+5y \leq 71$, $x+3y \geq 30$		3550 at $(0, 14.2)$
1	$y \leq 14$	$2x+5y \leq 71$, $x+3y \geq 30$, $y \leq 14$		3550 at $(0.5, 14)$
2	$y \geq 15$	$2x+5y \leq 71$, $x+3y \geq 30$, $y \geq 15$		Not feasible
3	$x \leq 0$	$2x+5y \leq 71$, $x+3y \geq 30$, $y \leq 14$, $x=0$		3500 at $(0, 14)$
4	$x \geq 1$	$2x+5y \leq 71$, $x+3y \geq 30$, $y \leq 14$, $x \geq 1$		3550 at $(1, 13.8)$
5	$y \leq 13$	$2x+5y \leq 71$, $x+3y \geq 30$, $y \leq 14$,$x \geq 1$,$y \leq 13$	$2x+5y \leq 71$, $x+3y \geq 30$, $y \leq 13$, $x \geq 1$	3550 at $(3, 13)$
6	$y \geq 14$	$2x+5y \leq 71$, $x+3y \geq 30$, $y \leq 14$, $x \geq 1$, $y \geq 14$	$2x+5y \leq 71$, $x+3y \geq 30$, $y=14$, $x \geq 1$	Not Feasible

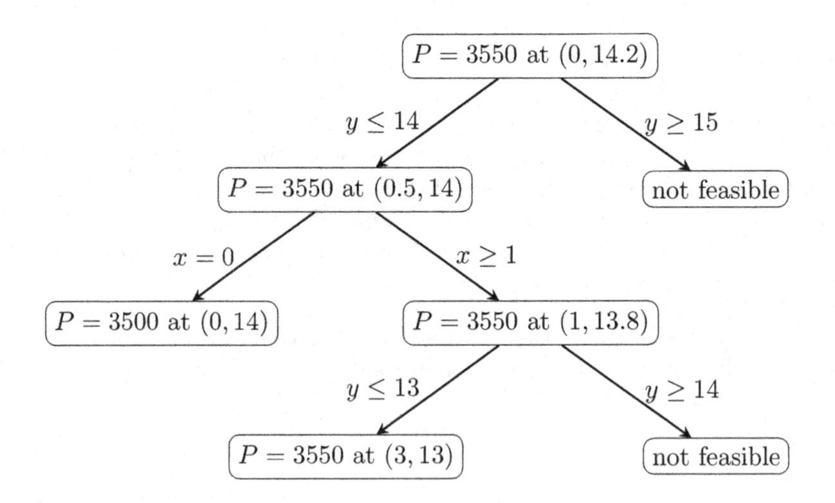

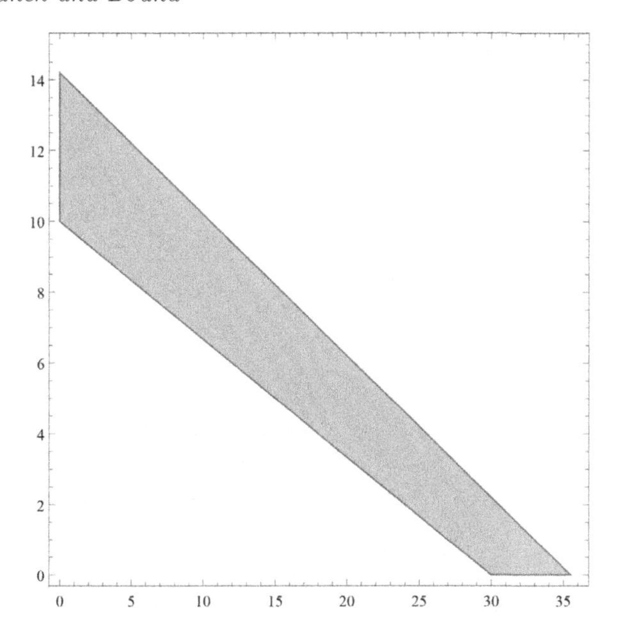

FIGURE 8.1
Feasible region for Anna's tables.

Discussion 8.2.3 (When to terminate Dakin's Branch and Bound Algorithm (Algorithm 8.2.1)). *Our stated algorithm for Dankin's Branch and Bound method was intentionally not precise and, in particular, we stated that the algorithm terminates when an integer solution is found. This was not how we proceeded in Example 8.2.2, namely we kept going even though we found an integer solution in Branch 3 (Note:* ***there is no need to continue further along this branch as we have an integer solution and further work along this branch would decrease*** *P). This is typically how the procedure is implemented and, unlike with the Simplex Method, using Algorithm 8.2.1 (or even a more precise version) on a bounded feasible region* **will not** *guarantee that the procedure will terminate at the globally optimal value.* Excel 2016's Solver, *for example, knows what the globally optimal value is over the reals (by the Simplex Method) and will branch and bound until an integer solution is within a certain tolerance of the global solution over the reals.*

We will close this section by illustrating the cuts in the feasible region in Example 8.2.2 that occurred by following Branches 1, 4, and 5. Recall that the original feasible region is illustrated in Figure 8.1.

Before introducing a second technique, let us consider a slightly more involved example.

Example 8.2.4 (Soylent Foods). *Soylent Foods, a local farm-to-table food service, makes two natural foods from soy and lentil beans and – as part of*

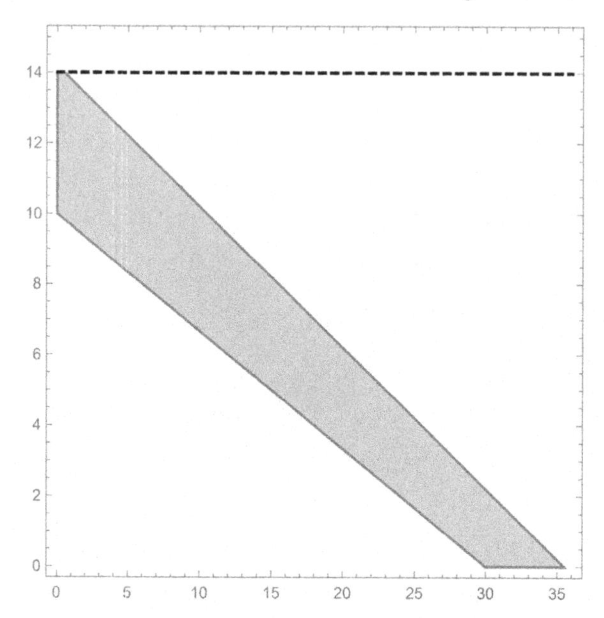

FIGURE 8.2
Feasible region for Anna's tables after Branch 1 cut.

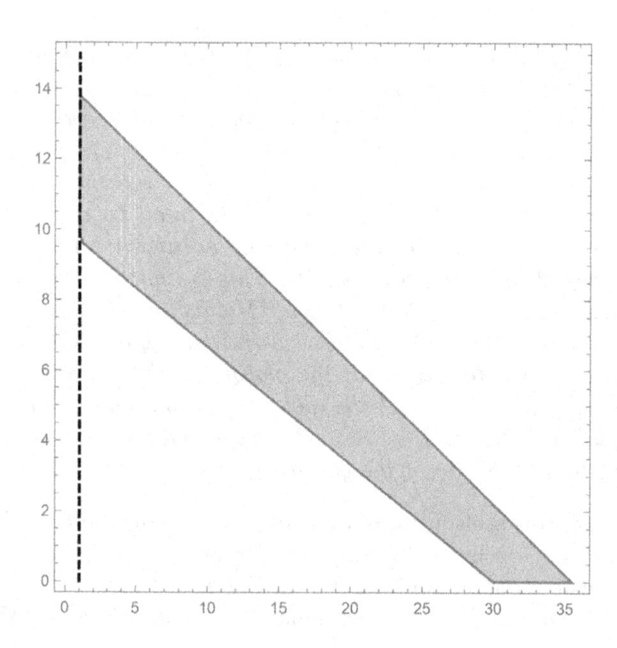

FIGURE 8.3
Feasible region for Anna's tables after Branch 4 cut.

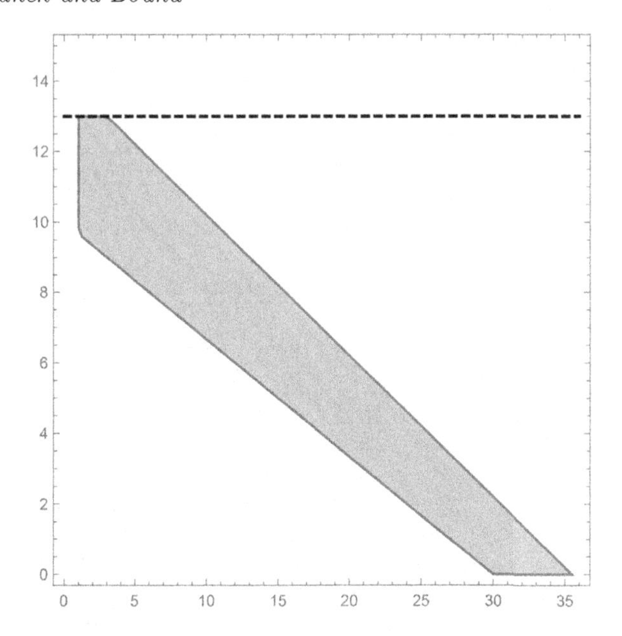

FIGURE 8.4
Feasible region for Anna's tables after Branch 5 cut.

their marketing campaign – the foods are named after the hues "Blue" and "Yellow" and are sold in bulk. Each may be consumed individually or mixed to form the popular "Green". Every unit of "Blue" requires 3lbs of soy beans and 7lbs of lentil beans. The "Yellow" units require 5lbs of soy and 8lbs of lentil beans. Soylent Foods has available each week 200lbs of soy beans and 350lbs of lentil and makes a profit of $62 on each unit of "Blue" sold and $71 on each unit of "Yellow" sold. How many units of each food type should Soylent Foods produce in order to maximize weekly profit?

We (again) assume that Solyent Foods will sell all units they produce, that only whole units will be produced, and we let x and y respectively represent the number of units of "Blue" and "Yellow" produced. Thus the model for this problem is:

$$\begin{aligned}
\text{Maximize: } & P(x,y) = 62x + 71y && (8.2)\\
\text{Subject to: } & 3x + 5y \leq 200 \text{ (soy beans)} && (8.3)\\
& 7x + 8y \leq 350 \text{ (lentil beans)} && (8.4)\\
& x, y \geq 0 \text{ and integer-valued.} && (8.5)
\end{aligned}$$

The process tree for Dankin's Branch and Bound method is (note that all numbers are truncated at the hundredths place):

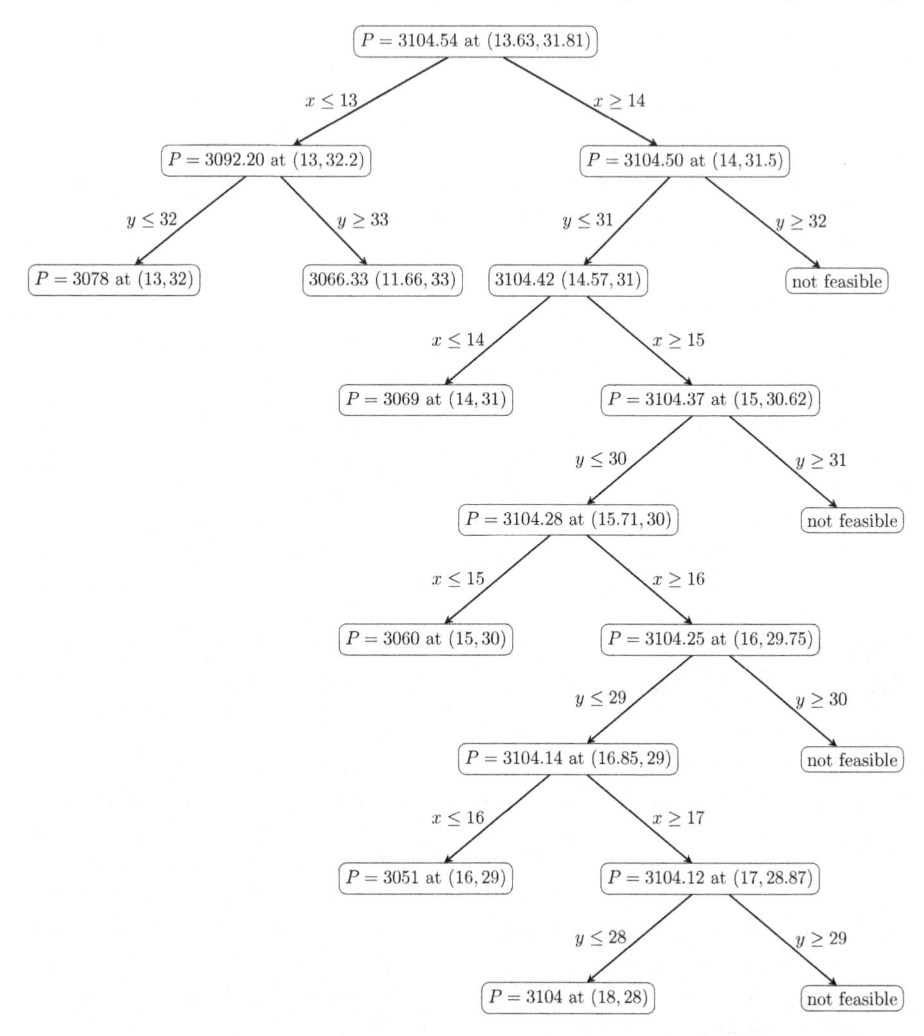

8.3 Gomory Cut-Planes

The next technique we consider for solving ILP problems is from R.E. Gomory in 1958[29]. Gomory's method is similar to Dankin's in that each iteration cuts pieces from the feasible region.

Suppose we wish to solve

$$\text{Maximize:} \quad P(x, y, z) = 5x + 3y + 2z \tag{8.6}$$

$$\text{Subject to:} \quad x + 2y + 3z \leq 35 \tag{8.7}$$

$$2x + y + 4z \leq 33 \tag{8.8}$$

$$3x + 2y + 2z \leq 56 \tag{8.9}$$

$$x, y, z \geq 0 \text{ and integer-valued.}$$

As in Dankin's Branch and Bound technique, we relax the integer requirement and solve the LP problem. Two iterations of the Simplex Method gives

$$\left[\begin{array}{ccccccc|c} x & y & z & s_1 & s_2 & s_3 & P & \\ 1 & 2 & 3 & 1 & 0 & 0 & 0 & 35 \\ 2 & 1 & 4 & 0 & 1 & 0 & 0 & 33 \\ 3 & 2 & 2 & 0 & 0 & 1 & 0 & 56 \\ \hline -5 & -3 & -2 & 0 & 0 & 0 & 1 & 0 \end{array}\right]$$

$$\rightarrow \left[\begin{array}{ccccccc|c} x & y & z & s_1 & s_2 & s_3 & P & \\ 0 & 1 & \frac{2}{3} & \frac{2}{3} & -\frac{1}{3} & 0 & 0 & \frac{37}{3} \\ 1 & 0 & \frac{5}{3} & -\frac{1}{3} & \frac{2}{3} & 0 & 0 & \frac{31}{3} \\ 0 & 0 & -\frac{13}{3} & -\frac{1}{3} & -\frac{4}{3} & 1 & 0 & \frac{1}{3} \\ \hline 0 & 0 & \frac{25}{3} & \frac{1}{3} & \frac{7}{3} & 0 & 1 & \frac{266}{3} \end{array}\right] \tag{8.10}$$

which has as its solution

$$x = \frac{37}{3}, y = \frac{31}{3}, z = 0 \text{ and } P = \frac{266}{3}.$$

The relaxed LP solution does not meet the integrality condition of the original ILP problem, so we must do more. Our approach this time will be to focus on the first row of the final Simplex tableau in 8.10 and note that this row corresponds to the equation:

$$y + \frac{2}{3}z + \frac{2}{3}s_1 - \frac{1}{3}s_2 = \frac{37}{3}. \tag{8.11}$$

Solving for y and writing the right side in the form `constant term - variable terms` (note: each variable written with the operation subtraction) we get

$$y = \overbrace{\frac{37}{3}}^{constant} \overbrace{-\frac{2}{3}z - \frac{2}{3}s_1 - \left(-\frac{1}{3}s_2\right)}^{-variables} \tag{8.12}$$

(the reasons for what we do will be clear shortly).

We now write each number of the right side as an `integer` + `positive fraction` which gives

$$y = 12 + \frac{1}{3} - \left(0 + \frac{2}{3}\right)z - \left(0 + \frac{2}{3}\right)s_1 - \left(-1 + \frac{2}{3}\right)s_2. \tag{8.13}$$

Collecting integer and fractional parts on the right:

$$y = 12 + 1s_2 + \left[\frac{1}{3} - \frac{2}{3}z - \frac{2}{3}s_1 - \frac{2}{3}s_2\right]. \tag{8.14}$$

Since y, 12, and s_2 are integers, $\left[\frac{1}{3} - \frac{2}{3}z - \frac{2}{3}s_1 - \frac{2}{3}s_2\right]$ **must be an integer**. Moreover, since z, s_1, and s_2 are nonnegative, this expression has a **maximum of** $\frac{1}{3}$. Thus

$$\frac{1}{3} - \frac{2}{3}z - \frac{2}{3}s_1 - \frac{2}{3}s_2 \leq 0 \tag{8.15}$$

which gives the equality constraint with a new surplus variable

$$\frac{2}{3}z + \frac{2}{3}s_1 + \frac{2}{3}s_2 - t_1 = \frac{1}{3}. \tag{8.16}$$

Thus we now have a new constraint (8.16) that we can introduce into the tableau solving the last iteration of the relaxed ILP problem. This new constraint is called a *Gomory cut* or a *cut-plane*.

Introducing 8.16 into the second tableau of 8.10 gives

$$\left[\begin{array}{cccccccc|c} x & y & z & s_1 & s_2 & s_3 & t_1 & P & \\ 0 & 1 & \frac{2}{3} & \frac{2}{3} & -\frac{1}{3} & 0 & 0 & 0 & \frac{37}{3} \\ 1 & 0 & \frac{5}{3} & -\frac{1}{3} & \frac{2}{3} & 0 & 0 & 0 & \frac{31}{3} \\ 0 & 0 & -\frac{13}{3} & -\frac{1}{3} & -\frac{4}{3} & 1 & 0 & 0 & \frac{1}{3} \\ 0 & 0 & -\frac{2}{3} & -\frac{2}{3} & -\frac{2}{3} & 0 & 1 & 0 & -\frac{1}{3} \\ \hline 0 & 0 & \frac{25}{3} & \frac{1}{3} & \frac{7}{3} & 0 & 0 & 1 & \frac{266}{3} \end{array}\right] \tag{8.17}$$

Notice the basic solution to this system with the new constraint has $t = -\frac{1}{3}$, which is not feasible, therefore the Simplex Method performed on the first primal form is not guaranteed to work. We must turn to the Big M Method (noting that t_1 is a surplus variable):

$$\left[\begin{array}{ccccccccc|c} x & y & z & s_1 & s_2 & s_3 & t_1 & a_1 & P & \\ 0 & 1 & \frac{2}{3} & \frac{2}{3} & -\frac{1}{3} & 0 & 0 & 0 & 0 & \frac{37}{3} \\ 1 & 0 & \frac{5}{3} & -\frac{1}{3} & \frac{2}{3} & 0 & 0 & 0 & 0 & \frac{31}{3} \\ 0 & 0 & -\frac{13}{3} & -\frac{1}{3} & -\frac{4}{3} & 1 & 0 & 0 & 0 & \frac{1}{3} \\ 0 & 0 & \frac{2}{3} & \frac{2}{3} & \frac{2}{3} & 0 & -1 & 1 & 0 & \frac{1}{3} \\ \hline 0 & 0 & \frac{25}{3} & \frac{1}{3} & \frac{7}{3} & 0 & 0 & M & 1 & \frac{266}{3} \end{array}\right] \tag{8.18}$$

We now repeat the procedure and hope for convergence. To summarize,

---

Algorithm 8.3.1 Gomory Cuts for ILP.

Input: LP problem to be solved over the integers.

1: Solve the linear programming problem over the real numbers.

2: Select any row from the final simplex tableau except the row corresponding to the objective function. Let

$$x, c_1t_1, c_2t_2, \ldots, c_kt_k, b$$

represent the selected row where x and the t_i are variables and the c_i and b are constants.

3: Write the equation represented by the selected row; that is

$$x + c_1t_1 + c_2t_2 + \cdots + c_kt_k = b.$$

4: Solve for one of the variables and rewrite the right-hand side of the solution in the form CONSTANT − VARIABLES.

$$x = b - c_1t_1 - c_2t_2 - \cdots c_kt_k.$$

5: After factoring out a negative in each variable term on the right-hand side, write each number factored number that remains as an INTEGER + a POSITIVE FRACTION.

$$x = (I_b + f_b) - (I_{c_1} + f_{c_1})t_1 - (I_{c_2} + f_{t_2})t_2 - \cdots (I_{c_k} + f_{c_k})t_k$$

where the I_i are integers and the f_k are positive fractions.

6: Collect all the integer and fraction parts on the right-hand side.

$$x = \overbrace{(I_b - I_{c_1}t_1 - \cdots I_{c_k}t_k)}^{integers} + \overbrace{(-f_b - f_{c_1}t_1 - \cdots - f_{c_k}t_k)}^{fractions}.$$

7: Since $x, I_b, I_{c_1}t_{c_1}, \ldots I_{c_k}t_{c_k}$ are all integers, $-f_b - f_{c_1}t_1 - \cdots - f_{c_k}t_k$ must be an integer. Moreover, it is nonnegative.

8: Thus $-f_b - f_{c_1}t_1 - \cdots - f_{c_k}t_k < 0$, which is the new constraint introduced into the most recent Simplex tableau solving the LP problem. This cut does not remove a location of the globally optimal solution.

9: Repeat.

Remark 8.3.1. *Answering an integer LP question exclusively using Gomory Cuts can be incredibly tedious as a tremendous quantity of cuts must be made to obtain some convergence. These cuts were initially seen as not effective in practice until the work of Géard Cornuéols and colleagues of the Tepper School of Business at Carnegie Mellon University [12] showed they can be quite* *effective when combined with branch and bound techniques. At the writing of this text, most software uses a hybrid technique which is a combination of Dankin's Branch and Bound and Gomory Cuts.*

8.4 For Further Study

There related problems that involve different solution techniques. Two important examples are *Balas' Method* (or related methods) for solving *Binary Programming* questions (the decision variables take on the values of 0 or 1) and *Mixed-Integer Programming* where we require only some of the decision variables to be integer valued.

8.5 Exercises

Exercise 8.1. *Solve the following problem by incrementally using Dakin's Branch and Bound Method:*

$$\textit{Maximize} \quad P(x_1, x_2, x_3) = 4x_1 + 3x_2 + 3x_3$$

Subject to:

$$4x_1 + 2x_2 + x_3 \leq 10 \tag{8.19}$$

$$3x_1 + 4x_2 + 2x_3 \leq 14 \tag{8.20}$$

$$2x_1 + x_2 + 3x_3 \leq 7 \tag{8.21}$$

where x_1, x_2, x_3 *are nonnegative integers.*

Draw a decision tree with your answers to the subproblems as in Example 8.2.4. As well, for each iteration of the particular path you are following, please clearly state the LP problem you are answering. You may use Solver (or a program of your choice) to answer each of these individual LP problems.

Exercise 8.2. *Use Solver (or a program of your choice) to answer the above question as a LP problem. Observe the computation time (should be incredibly fast). Return to the original problem, solve it a second time but choose "integer" as a constraint for each of the decision variables. Do this a third time but change the tolerance under "Options" to 0.1% or less. Observe the computation time for the IP problem. This should be a little longer, but for this small of a problem the difference in computation time is probably not very noticeable. Submit the Answer Report for the third run of the problem (if you are using software or a program other than Solver, submit a screenshot of the program's output).*

Exercise 8.3. *Solve the following problem by hand incrementally using Cut-Planes:*

$$\textit{Minimize} \quad C(x, y) = x - y$$

Subject to:

$$3x + 4y \leq 6 \tag{8.22}$$
$$x - y \leq 1 \tag{8.23}$$

where x and y are nonnegative integers.

1. *State the modified Linear Programming problem (what we have after introducing slack, surplus, and artificial variables, etc.);*
2. *graph the feasible region;*
3. *provide the initial simplex tableau;*
4. *begin iteratively introducing Gomory cuts until an integer solution is attained where for each iteration:*
    (a) *clearly show work supporting why you have introduced a particular cut-plane (i.e. the new constraint),*
    (b) *write down the new LP problem (the one from the previous iteration plus the new constraint) and the new initial Simplex Tableau,*
    (c) *provide a diagram showing the feasible region for the decision variables and*
    (d) *use any computer resource to find the final simplex tableau for this iteration's LP and provide the final Simplex tableau (a screen shot is acceptable).*

Exercise 8.4. *Solve the following problem by hand incrementally using Gomory Cut-Planes:*

$$\textit{Maximize} \quad P(x_1, x_2, x_3) = 4x_1 + 3x_2 + 3x_3$$

Subject to:

$$4x_1 + 2x_2 + x_3 \leq 10 \tag{8.24}$$
$$3x_1 + 4x_2 + 2x_3 \leq 14 \tag{8.25}$$
$$2x_1 + x_2 + 3x_3 \leq 7 \tag{8.26}$$

where x_1, x_2, x_3 are nonnegative integers.

1. *State the modified Linear Programming problem (what we have after introducing slack, surplus, and artificial variables, etc.);*
2. *provide the initial simplex tableau;*
3. *begin iteratively introducing Gomory cuts until an integer solution is attained where for each iteration:*

(a) *clearly show work supporting why you have introduced a particular cut-plane (i.e. the new constraint),*

(b) *write down the new LP problem (the one from the previous iteration plus the new constraint) and the new initial Simplex Tableau, and*

(c) *use any computer resource to find the final simplex tableau for this iteration's LP and provide the final Simplex tableau (a screen shot is acceptable).*

Exercise 8.5. *Use Solver (or a program of your choice) to answer the following IP question (you are advised to not answer this by hand!). Do note that solving the problem requires making certain decision variables binary (i.e. 0–1) variables. If you are running Solver, there is an option in the drop-down list box for the binary constraint (this is the box with "<=", etc.).*

(From Ragsdale [47]) In his position as vice president of research and development (R&D) for CRT Technologies, Mark Schwartz is responsible for evaluating and choosing which R&D projects to support. The company received 18 R&D proposals from its scientists and engineers and identified six projects as being consistent with the company's mission. However, the company does not have the funds available to undertake all six projects, so Mark must determine which projects to select. The funding requirements for each project are summarized in the following table along with the NPV (Net Present Value; let's not worry about what that means) the company expects each project to generate.

Project	*Expected NPV*	*Year 1 CR*	*Year 2 CR*	*Year 3 CR*	*Year 4 CR*	*Year 5 CR*
1	$141	$75	$25	$20	$15	$10
2	$187	$90	$35	$0	$0	$30
3	$121	$60	$15	$15	$15	$15
4	$83	$30	$20	$10	$5	$5
5	$262	$100	$25	$20	$20	$20
6	$127	$50	$20	$10	$30	$40

KEY: NPV = Net Present Value; CR = Capital Required.
NOTE: all dollar values are in $1,000's
(So think of NPV as revenue and CR as costs.)

The company currently has $250,000 available to invest in new projects. It has budgeted $75,000 for continued support for these projects in year 2 and $50,000/year for years 3,4,and 5. Surplus funds in any year are reappropriated for other uses within the company and may not be carried over to future years (note: this actually makes the problem easier).

So, what projects should the company select in order to maximize NPV? (Note to all future analysts; please begin your solution by clearly stating what your decision variables are and what they represent and clearly state the model.) Submit your model and the answer report.

Part III

Nonlinear (Geometric) Programming

9

Calculus Review

Many techniques used in this text will require some knowledge of various levels of Calculus, especially multivariable Calculus. This chapter provides a refresher of relevant Calculus topics and may either be read carefully, skimmed, or visited as needed.

Certainly we can work in the algebraic completion of the reals, namely do our analysis in $\mathbb{C}^n$, or even a more abstract space; but as Optimization very often involves applications to real-world problems, we choose to do our work in this text in $\mathbb{R}^n$.

9.1 Derivatives and Continuity

No ϵ - δ proofs will be done in this text, but limits will be used and the reader should be familiar with their properties, including the notion of directional limits. We remind the reader that the *derivative* of a function (if it exists) at the point $x = a$ gives the slope of the tangent line at $x = a$. Recall also that a tangent line is the result of a limiting process on secant lines. More formally,

Definition 9.1.1 (The Derivative). *Let $f(x)$ be a function where x is a real-valued variable. For variables x, y, the* derivative *of $f(x)$ is defined to be*

$$f'(x) := \lim_{h\to 0} \frac{f(x+h) - f(x)}{h} = \lim_{y\to x} \frac{f(y) - f(x)}{y - x}. \tag{9.1}$$

$f(x)$ is said to be differentiable at a *if $f'(a)$ exists (i.e. if the limit in 9.1 exists). $f(x)$ is differentiable over an interval I if it is differentiable at every point in I.*

Other notations for $f'(x)$ are

$$\frac{df}{dx}, \frac{d}{dx}f(x), Df(x), \text{ and } D_x f(x).$$

DOI: 10.1201/9780367425517-9

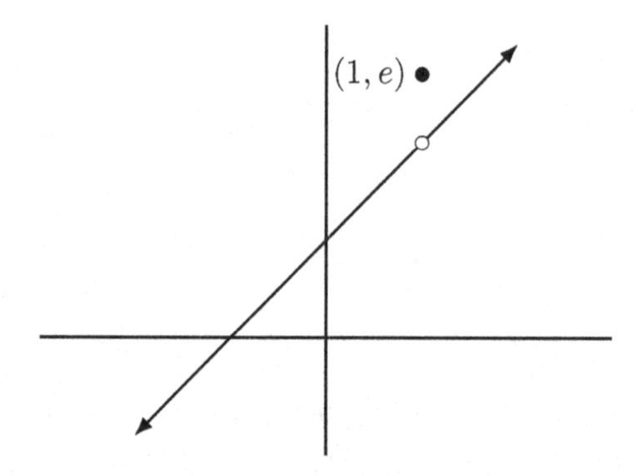

FIGURE 9.1
The discontinuous function in Example 9.1.3.

Another important use of limits is qualifying a valuable property of functions:

Definition 9.1.2 (Continuity). *A function $f(x)$ is said to be* continuous *at $x = a$ if*

$$\lim_{x \to a} f(x) = f(a).$$

Definition 9.1.2 agrees very well with our intuition. Consider

Example 9.1.3. *The piecewise function*

$$f(x) = \begin{cases} x + 1 \text{ for } x \neq 1 \\ e \text{ for } x = 1 \end{cases}$$

is not continuous at $x = 1$ as

$$\lim_{x \to 1} f(x) = 2 \neq e = f(1).$$

We will assume familiarity with derivative rules and integration techniques of single-variable functions and remind the reader of some important theorems involving derivatives, integrals, and continuity.

Theorem 9.1.4 (The Intermediate Value Theorem). *Suppose $f(x)$ is continuous on $[a, b]$ and that N be any number between $f(a)$ and $f(b)$ where $f(a) \neq f(b)$. Then there is a number c in (a, b) such that $f(c) = N$.*

Though the result of the Intermediate Value Theorem seems intuitively obvious, its proof relies on the Completeness Property of the Real Numbers and can be found in a beginning Real Analysis text (see [59], for example).

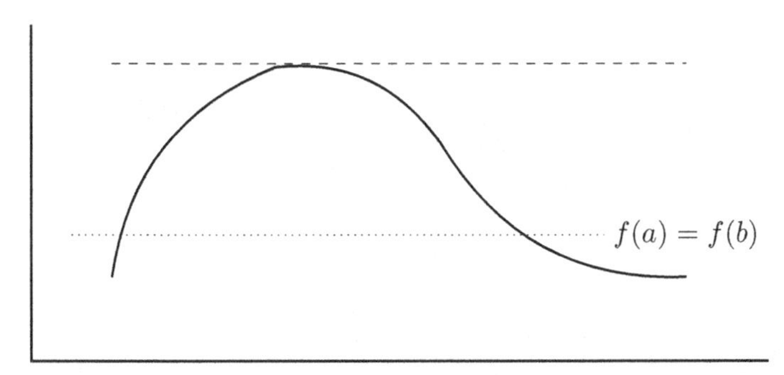

FIGURE 9.2
A case illustrating Rolle's theorem.

There also is a nice relationship between differentiation and continuity:

Theorem 9.1.5. *If $f(x)$ is differentiable at a, then it is continuous at a.*

Next we consider how derivatives and integrals are involved in averages. An important tool for these is

Theorem 9.1.6 (Rolle's Theorem). *Suppose $f(x)$ is continuous on $[a, b]$ and differentiable on (a, b) with $f(a) = f(b)$. Then there is a number c in (a, b) such that $f'(c) = 0$.*

Rolle's Theorem is the mathematical equivalent of "What goes up must come down" as shown in the following figure (and, of course, Rolle's Theorem also means "What goes down must come up"...)

Rolle's Theorem is an important part of establishing

Theorem 9.1.7 (The Mean Value Theorem). *Let $f(x)$ be continuous on $[a, b]$ and differentiable on (a, b) with $a \neq b$. Then there is a c in (a, b) such that $f'(c) = \frac{f(b)-f(a)}{b-a}$.*

There is a geometric interpretation of the Mean Value Theorem, namely that somewhere in the interval there is a place where the tangent at that point is parallel to the secant line joining $(a, f(a))$ to $(b, f(b))$.

Note that this geometric interpretation helps us to remember the result.

The Mean Value Theorem is amazingly useful. Most of the important results that are studied at the end of a typical Calculus 1 class hold because of this theorem (e.g., the Mean Value Theorem is the mathematical reason behind why there is a $+C$ when we do indefinite integrals; see exercise 9.4).

An incredibly useful extension of the Mean Value Theorem is

Theorem 9.1.8 (Extended Mean Value Theorem). *Let $f(x)$ be a twice differentiable function on the interval (s, t) with a and b in (s, t). Then there is a c between a and b where*

$$f(b) = f(a) + f'(a)(b-a) + \frac{1}{2}f''(c)(b-a)^2. \tag{9.2}$$

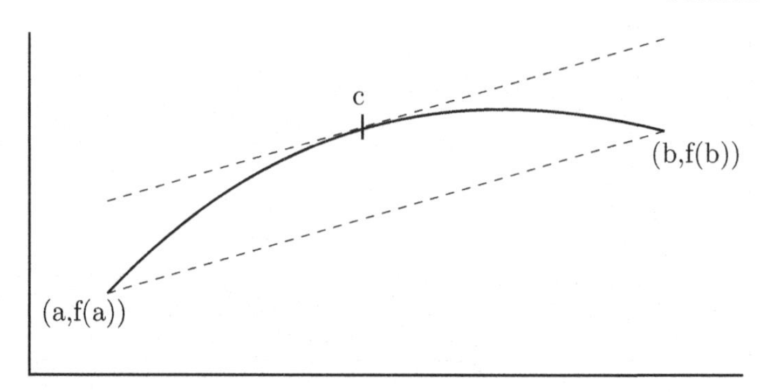

FIGURE 9.3
Illustrating the mean value theorem.

It is the case that 9.2 is a merely a special case of Taylor's Theorem (to come in Theorem 9.2.3), but it is important enough to merit its own name. Additionally, it can be proven without Taylor's Theorem via either integration by parts together with the Mean Value Theorem for Integrals (see below) or by a direct calculation using Rolle's Theorem.

Not to be left out, integrals also have a theorem about averages.

Theorem 9.1.9 (The Mean Value Theorem for Integrals). *Let $f(x)$ be continuous on $[a, b]$. Then there is a number c in $[a, b]$ such that*

$$f(c)(b-a) = \int_a^b f(x)\ dx. \tag{9.3}$$

The geometric meaning of the Mean Value Theorem for Integrals is that if $f(x)$ is a positive function that is continuous over $[a, b]$, then there is place c in the interval where the area under the curve is equal to the rectangle formed by the interval and $f(c)$. Moreover, if we rewrite 9.3 as

$$f(c) = \frac{1}{b-a}\int_a^b f(x)\ dx = \text{ average value of } f(x) \text{ over } [a, b]$$

we see that the Mean Value Theorem for Integrals tells us that there is a place in the interval, c, where the value of $f(x)$ at c equals the average value of the function over the interval.

Theorem 9.1.9 can be generalized to

Theorem 9.1.10 (Generalized Mean Value Theorem for Integrals). *Let $g(x)$ be continuous over $[a, b]$ and $h(x)$ integrable over $[a, b]$ with $h(x) \neq 0$ for all x in $[a, b]$*[1]*. Then there is a c in (a, b) such that*

$$\int_a^b g(x)h(x)\ dx = g(c)\int_a^b h(x)\ dx.$$

[1] $h(x)$ integrable means it is continuous and thus if it is never 0, $h(x)$ will have the same sign over the interval.

Note that it is common to refer to either Theorem 9.1.9 or Theorem 9.1.10 as the *Mean Value Theorem for Integrals.* Moreover, Theorem 9.1.10 reduces to Theorem 9.1.9 by putting $h \equiv 1$.

Lastly, it would a crime against nature to not also state the Fundamental Theorem of Calculus, part of which is directly derived from the Mean Value Theorem.

Theorem 9.1.11 (The Fundamental Theorem of Calculus). *Let $f(x)$ be continuous over $[a, b]$. Then*

1. *If $g(x) = \int_a^x f(t)\,dt$, then $g'(x) = f(x)$.*
2. *$\int_a^b f(x)\,dx = F(b) - F(a)$ where $F(x)$ is any antiderivative of $f(x)$.*

9.2 Taylor Series for Functions of a Single Variable

An incredibly valuable application of derivatives is using them to approximate a function (provided the right amount of derivatives exist).

Early in Calculus 1 one learns that the equation of the line tangent for a given function at a point. That is, if $f(x)$ is differentiable at x_0, then the line tangent to $f(x)$ at $x_0 = a$ is given by

$$y = f'(a)x + [f(a) - af'(a)] = f(a) + f'(a)(x - a) \tag{9.4}$$

Example 9.2.1. *To find equation of the line tangent to $f(x) = x^2 + 3$ at $x = 1$, we have $f'(x) = 2x$. Thus by (9.4) the desired tangent line is*

$$y = f'(1)x + [f(1) - 1f'(1)] = 2x + 2.$$

This example is illustrated in Figure 9.4.

Notice that the tangent line can serve as an approximation of the curve and that this approximation works nicely for values near $x = 1$, but we get a better approximation if we use higher degree polynomials instead of a linear function. Hence

Definition 9.2.2 (Taylor Polynomial). *Let $f(x)$ be a function with an nth order derivative in some open interval containing x_0. The nth* ***order Taylor polynomial*** *T_n is given by*

$$T_n(x) = \sum_{k=0}^{n} \frac{f^{(k)}(x_0)}{k!}(x - x_0)^k$$

where $f^{(k)}$ is the kth derivative of $f(x)$ with $f^{(0)}(x) = f(x)$ and $k! = k(k-1)\cdots 1$ with $0! := 1$.

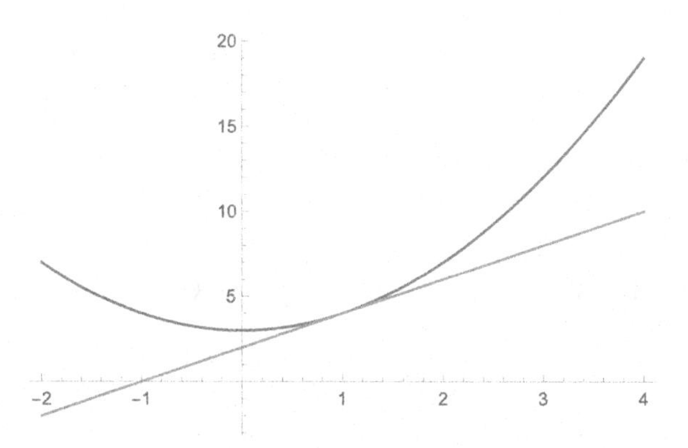

FIGURE 9.4
$f(x) = x^2 + 3$ and its tangent line at $x = 1$.

We may know how good of an approximation we have by

Theorem 9.2.3 (Taylor's Theorem). *Suppose that $f(x)$ has $n+1$ derivatives in some open interval containing the value a and let $R_n(x)$ be the difference (i.e.* remainder*) between $f(x)$ and the nth order Taylor polynomial $T_n(x)$ at $x = a$, then*

$$R_n(x) = \int_a^x \frac{(x-t)^n}{n!} f^{(n+1)}(t)\, dt.$$

We may state the remainder without involving an integral via

Theorem 9.2.4 (Lagrange's Remainder Theorem). *Let $f(x)$ be a function over some open interval I that has $n+1$ derivatives. Let $R_n(x)$ be the remainder for $T_n(x)$ and suppose $a \in I$. Then for every x in I there exists a value c between x and a such that*

$$R_n(x) = \frac{f^{(n+1)}(c)}{(n+1)!}(x-a)^{n+1}.$$

If we wish not to have a remainder,

Definition 9.2.5 (Taylor Series). *Let $f(x)$ be a function with derivatives of all orders at a. Then the* ***Taylor series of*** *$f(x)$* ***at*** *a is*

$$\sum_{k=0}^{\infty} \frac{f^{(k)}(a)}{k!}(x-a)^k. \tag{9.5}$$

A Taylor series is just a name attached to the process of representing a function as a *power series* where a **power series** is an expression of the form $\sum_{k=0}^{\infty} c_k(x-a)^k$. It is well-known that a power series converges either only at its center a, over some interval $(a-R, a+R)$ centered at a, or over all the real numbers. The *interval of convergence*, I, of a power series and its *radius of convergence*, R, can be found using the convergence tests discussed in any Calculus text. Regarding rates of convergence, one should consult a Numerical Analysis text.

Example 9.2.6. *Find the Taylor series expansion of $f(x) = e^x$ at $a = 0$.*

Solution.

k	$f^{(k)}(x)$	$\frac{f^{(k)}(0)}{k!}$
0	e^x	1
1	e^x	1
2	e^x	$\frac{1}{2!}$
3	e^x	$\frac{1}{3!}$
$\vdots$	$\vdots$	$\vdots$
n	e^x	$\frac{1}{n!}$
$\vdots$	$\vdots$	$\vdots$

Thus

$$e^x = 1 + x + \frac{x^2}{2!} + \frac{x^3}{3!} + \cdots + \frac{x^n}{n!} + \cdots$$

and we can obtain T_n for any nonnegative integer by truncating the series at the desired n. It is a short argument using the Ratio Test to show that the stated infinite series converges for all real numbers. ∎

We will close this section by stating a quite useful result related to the previous theorems. This result is known as *Taylor's Formula.*

Theorem 9.2.7 (Taylor's Formula). *Suppose f, f', and f'' all exist on a closed interval I and that x and x^* are distinct points in I. Then there exists a c strictly between x and x^* such that*

$$f(x) = f(x^*) + f'(x^*)(x - x^*) + \frac{f''(c)}{2}(x - x^*)^2. \tag{9.6}$$

Taylor's Formula is quite useful in Optimization. If $f'(x^*) = 0$ (we will see in Chapter 10 that such an x^* is called a *critical point*), and $f''(x) > 0$ for all $x \neq x^*$ in I, then by 9.6

$$f(x) = f(x^*) + 0 + \text{ something positive;}$$

i.e.

$$f(x^*) < f(x) \text{ for all } x \neq x^* \text{ in } I.$$

That is, x^* is a strict global minimizer for f over I. Similarly, if $f''(x) < 0$ for all $x \neq x^*$ in I, then x^* is a strict global maximizer for f over I.

9.3 Newton's Method

We may use the ideas of the previous section to find approximations of zeros of a function. We motivate the idea by trying to find a real root of $f(x) = x^4 - x^2 - 2$. Since $f(1) = -2$, $f(2) = 10$, and f is continuous, we know by the Intermediate Value Theorem that $f(x)$ has a root in the interval $(1, 2)$. Let us now approximate the curve of $f(x)$ by using its tangent line at $x = 2$. By (9.4) we have that the tangent line at this point is $y = 28x - 46$ and finding its zero is easy: $x = \frac{46}{28} = \frac{23}{14}$. Our strategy will be to find the line tanget to f at this new point $x = \frac{23}{14}$ and repeat the process.

In general, we are finding the root of the tangent line $y = f(a) + f'(a)(x - a)$ at $x_0 = a$ to approximate a root of $f(x)$. That is, solving

$$0 = f(x_0) + f'(x_0)(x - x_0) \tag{9.7}$$

which gives (as long as $f'(x_0) \neq 0$)

$$x = x_1 = x_0 - \frac{f(x_0)}{f'(x_0)}. \tag{9.8}$$

We then repeat the process now using the tangent line at x_1. This gives the iteration

$$x_{n+1} = x_n - \frac{f(x_n)}{f'(x_n)} \tag{9.9}$$

and this process is known as **Newton's Method** or the **Newton-Raphson Method**.

Table 9.1 gives the values of the first few iterations of this method for our example $f(x) = x^4 - x^2 - 2$. From this we see that $x_n \to 1.414213562373$ and note that $\sqrt{2}$ is a root of $x^4 - x^2 - 2$.

We should address that not only will Newton's method fail if $f'(x_n) = 0$, but will occasionally fail when $f'(x_n)$ is close to 0 (Newton's method works nicely for $f(x) = x^2$, but this is not always the case). Also, there is the issue of convergence. These matters are addressed in any good Numerical Analysis text.

9.4 Vectors

The language of higher dimensions – functions of multiple variables – is spoken in vectors. Recall that a *vector* in $\mathbb{R}^n$ is an n-tuple $\langle x_1, x_2, \ldots, x_n \rangle$ subject to the rules

1. $\langle x_1, x_2, \ldots, x_n \rangle + \langle y_1, y_2, \ldots, y_n \rangle = \langle x_1 + y_1, x_2 + y_2, \ldots, x_n + y_n \rangle$ (*component-wise addition* of vectors) and

TABLE 9.1
Iterative Values of Newton's Method for $x^4 - x^2 - 2$ with $x_0 = 2$ (Truncated)

n	x_n
0	2
1	1.64285714285714
2	1.46393444394806
3	1.41718633854642
4	1.41422496580582
5	1.41421356254167
6	1.41421356237309
7	1.41421356237310
$\vdots$	$\vdots$
$\sqrt{2}$	1.41421356237309504

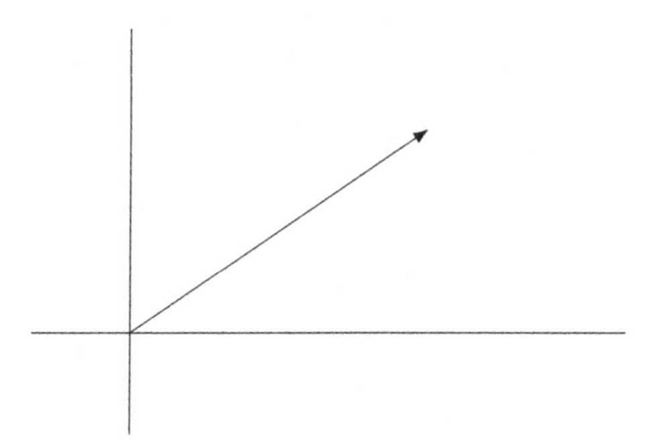

FIGURE 9.5
The vector $\langle 3, 2\rangle$ drawn in standard position.

2. for any real number c, $c\langle x_1, x_2, \ldots, x_n\rangle = \langle cx_1, cx_2, \ldots, cx_n\rangle$ (*scalar multiplication*).

Though it may not be obvious from the stated definition, vectors relay two key physical properties: *magnitude* and *direction.* As such, any two vectors having the same magnitude and direction are regarded as being equal in spite of their physical location. Thus we may always write a vector $\langle x_1, x_2, \ldots, x_n\rangle$ in *standard position*: that is, position it so that its *initial point* is at the origin and its *terminal point* is at $(x_1, x_2, \ldots, x_n)$. For example, the vector $\langle 3, 2\rangle$ in $\mathbb{R}^2$ when written in standard position has its terminal point at $(3, 2)$ (see Figure 9.5). Thus there is a one-to-one correspondence (which is also an isometry) between vectors drawn in standard position in $\mathbb{R}^n$ and points in $\mathbb{R}^n$ and we will henceforth loosely treat the two as the same.

9.5 Partial Derivatives

We now briefly consider some of the Calculus of multivariable functions. It must be noted that there are concerns with derivatives of multivariable functions existing and considering *directional derivatives* instead of just focusing on partials with respect to particular variables; but we wish to not be a Calculus text. There is an excellent treatment of this in [57].

Definition 9.5.1 (Partial Derivatives – two variables)**.** *If $f(x,y)$ is a function of two variables, then the* **partial derivatives** f_x **and** f_y *are defined by*

$$f_x = f_x(x,y) = \frac{\partial f(x,y)}{\partial x} = \frac{\partial}{\partial x} f(x,y) = \lim_{h \to 0} \frac{f(x+h,y) - f(x,y)}{h}$$

and

$$f_y = f_y(x,y) = \frac{\partial f(x,y)}{\partial y} = \frac{\partial}{\partial y} f(x,y) = \lim_{h \to 0} \frac{f(x,y+h) - f(x,y)}{h}.$$

The nuts and bolts of taking partial derivatives is given by

Rule 9.5.2. *To find $f_x(x,y)$, regard y as a constant and differentiate with respect to x. To find $f_y(x,y)$, regard x as a constant and differentiate with respect to y.*

As in single variable Calculus, higher derivatives are done by differentiating the previous derivative. Care is needed, though, as $f_{xy}(x,y)$ means to first differentiate $f(x,y)$ with respect to x then differentiate that derivative with respect to y where $f_{yx}(x,y)$ means to first differentiate $f(x,y)$ this time with respect to y then differentiate that derivative with respect to x. That is

$$f_{xy} = f_{xy}(x,y) = \frac{\partial}{\partial y}\left(\frac{\partial f(x,y)}{\partial x}\right) \tag{9.10}$$

and

$$f_{yx} = f_{yx}(x,y) = \frac{\partial}{\partial x}\left(\frac{\partial f(x,y)}{\partial y}\right). \tag{9.11}$$

Example 9.5.3. *For $f(x,y) = x^2y^3 \sin y$, we have*

$$f_x = 2xy^3 \sin y \textit{ thus } f_{xy} = 2x(y^3 \cos y + 3y^2 \sin y) \tag{9.12}$$

and

$$f_y = x^2(y^3 \cos y + 3y^2 \sin y) \textit{ thus } f_{yx} = 2x(y^3 \cos y + 3y^2 \sin y). \tag{9.13}$$

One may have noticed the nice coincidence in this example that $f_{xy} = f_{yx}$. This was no accident as there are many times when the order in which the partials are done is irrelevant. The details are

Theorem 9.5.4 (Two-variable Clairaut's Theorem). *Let D be any disk over which f is defined and $(a, b) \in D$. If f_{xy} and f_{yx} are both continuous on D, then $f_{xy}(x, y) = f_{yx}(x, y)$.*

Note that partial derivatives and Clairaut's Theorem both translate nicely to functions of more than two variables. Care is needed with higher order versions of Clairaut's Theorem though, in that we must be certain that *all variants* of the mixed partials are continuous. That is, for example, that Clairaut's Theorem does not say that if f_{xyy} and f_{yxy} are continuous on some D that $f_{xyy} = f_{yxy}$ but rather that if f_{xyy}, f_{yxy}, and f_{yyx} are continuous on some D then $f_{xyy} = f_{yxy} = f_{yyx}$.

Now that we have partial derivatives, we may define the multivariable version of the first derivative:

Definition 9.5.5 (The Gradient of a Function of Two Variables). *Let $f(x, y)$ be a function of two variables. The **gradient** of f is*

$$\nabla f(x, y) := \langle f_x(x, y), f_y(x, y)\rangle.$$

Example 9.5.6. *The gradient of the function given in Example 9.5.3 is*

$$\nabla f(x, y) = \langle 2xy^3 \sin y, x^2(y^3 \cos y + 3y^2 \sin y)\rangle.$$

In general, we have

Definition 9.5.7. *[The Gradient of a Function of Multiple Variables] Let $f(x_1, x_2, \ldots, x_n)$ be a multi-variable function mapping from $\mathbb{R}^n$ to $\mathbb{R}$. The **gradient** of f is*

$$\nabla f(x_1, x_2, \ldots, x_n) := \left\langle \frac{\partial f}{\partial x_1}, \frac{\partial f}{\partial x_2}, \ldots, \frac{\partial f}{\partial x_n} \right\rangle.$$

Just like the derivative of a single-variable function, the gradient gives the slope of the tangent hyperplane to the graph of a function.

Also as the gradient is the "first derivative" of a multivariable function, there is convenient notation for the "second derivative" of a multivariable function:

Definition 9.5.8. *[Hessian] Let $f(x_1, x_2, \ldots, x_n)$ be a multi-variable function mapping from $\mathbb{R}^n$ to $\mathbb{R}$. The **Hessian** of f is*

$$Hf = \begin{bmatrix} \frac{\partial^2 f}{\partial x_1^2} & \frac{\partial^2 f}{\partial x_1 \partial x_2} & \cdots & \frac{\partial^2 f}{\partial x_1 \partial x_n} \\ \frac{\partial^2 f}{\partial x_2 \partial x_1} & \frac{\partial^2 f}{\partial x_2^2} & \cdots & \frac{\partial^2 f}{\partial x_2 \partial x_n} \\ \vdots & \vdots & \ddots & \vdots \\ \frac{\partial^2 f}{\partial x_n \partial x_1} & \frac{\partial^2 f}{\partial x_n \partial x_2} & \cdots & \frac{\partial^2 f}{\partial x_n^2} \end{bmatrix}. \tag{9.14}$$

So far we have considered *scalar valued* functions; that is functions whose output is a number. We now consider functions whose output is a vector. Specifically, the function f in Definition 9.5.7 takes an n-dimensional input and gives a single output, that is $f : \mathbb{R}^n \to \mathbb{R}$. In some cases, we may want something like the gradient for a function whose output is also multi-dimensional, say $f : \mathbb{R}^n \to \mathbb{R}^m$ where $m > 1$.

In this situation we have, say,

$$f(x_1, x_2, x_3, \ldots, x_n) = f(\mathbf{x}) = \begin{bmatrix} f_1(\mathbf{x}) \\ f_2(\mathbf{x}) \\ \vdots \\ f_m(\mathbf{x}) \end{bmatrix}. \tag{9.15}$$

With such a function, we will find it useful to define the *Jacobian* of f:

$$\left[\frac{\partial f}{\partial x_1}, \frac{\partial f}{\partial x_2}, \ldots, \frac{\partial f}{\partial x_n}\right] = \begin{bmatrix} \frac{\partial f_1}{\partial x_1} & \frac{\partial f_1}{\partial x_2} & \cdots & \frac{\partial f_1}{\partial x_n} \\ \frac{\partial f_2}{\partial x_1} & \frac{\partial f_2}{\partial x_2} & \cdots & \frac{\partial f_2}{\partial x_n} \\ \vdots & \vdots & \ddots & \vdots \\ \frac{\partial f_m}{\partial x_1} & \frac{\partial f_m}{\partial x_2} & \cdots & \frac{\partial f_m}{\partial x_n} \end{bmatrix}. \tag{9.16}$$

9.6 The Taylor Series of a Function of Two Variables

Taylor's Theorem (Theorem 9.2.3) and its related theorems all extend nicely to multivariable functions, though we will not restate each of them here. We will, however, give attention to a function of two variables $f(x, y)$ that has as its Taylor Series at a point (a, b) and state a multivariable version of Taylor's Formula (Theorem 9.2.7).

Definition 9.6.1 (Two-Variable Taylor Series)**.** *Let $f(x, y)$ be a function with derivatives of all orders at (a, b). Then the* **Taylor series** *of $f(x, y)$ centered at (a, b) is*

$$\begin{aligned} f(x,y) = {} & f(a,b) + f_x(a,b)\cdot(x-a) + f_y(a,b)\cdot(y-b) + \frac{1}{2}f_{xx}(a,b)\cdot(x-a)^2 + \\ & + \frac{1}{2}f_{xy}(a,b)\cdot(x-a)(y-b) + \frac{1}{2}f_{yx}(a,b)(x-a)(y-b) + \frac{1}{2}f_{yy}(a,b)(y-b)^2 + \\ & + \frac{1}{3!}f_{xxx}(a,b)\cdot(x-a)^3 + \frac{1}{3!}f_{xxy}(a,b)\cdot(x-a)^2(y-b) + \cdots. \end{aligned} \tag{9.17}$$

Example 9.6.2. *Find the* 3rd *order Taylor polynomial of $f(x, y) = x \sin y$ at $(0, 0)$.*

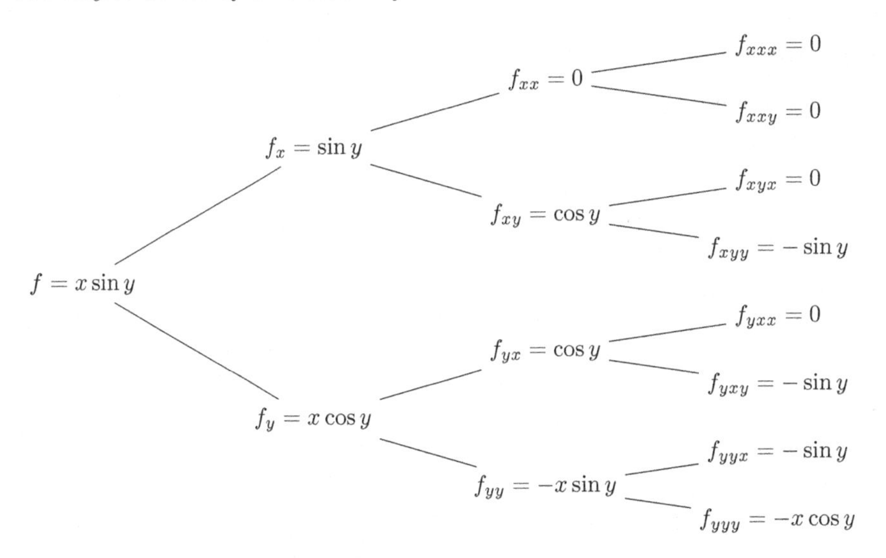

Solution. We have

Thus

$$T_3(0,0) = 0+0+0+0+\frac{1}{2}\cdot 1(x-0)(y-0)+\frac{1}{2}\cdot 1(x-0)(y-0)+0+\cdots+0 = xy \tag{9.18}$$

which gives us $x \sin y \approx xy$ near $(0,0)$. ■

Note how nicely Example 9.6.2 illustrates *Clairaut's Theorem* (Theorem 9.5.4).

Of course, Taylor Series and the related theorems all translate nicely to functions of more than two variables, though the notation becomes quite cumbersome. For more detail, see [59].

As stated in the single variable case, Taylor's Formula can be very useful in Optimization. Hence we will close with the multivariable version.

Theorem 9.6.3 (Multivariable Taylor's Formula). *Suppose $\boldsymbol{x} = (x_1, x_2, \ldots, x_n)$ and $\boldsymbol{x}^* = (x_1^*, x_2^*, \ldots, x_n^*)$ are points in $\mathbb{R}^n$. Suppose further that $f(\boldsymbol{x})$ is a function of n variables with continuous first and second partial derivatives on some open set containing the line segment joining $\boldsymbol{x}$ and $\boldsymbol{x}^*$. Then there exists some $\boldsymbol{c}$ on the line segment such that*

$$f(\boldsymbol{x}) = f(\boldsymbol{x}^*) + \nabla f(\boldsymbol{x}^*) \cdot (\boldsymbol{x} - \boldsymbol{x}^*) + \frac{1}{2}(\boldsymbol{x} - \boldsymbol{x}^*) \cdot Hf(\boldsymbol{c})(\boldsymbol{x} - \boldsymbol{x}^*).$$

9.7 Exercises

Exercise 9.1. *The Gamma Function*

$$\Gamma(x) := \int_0^\infty t^{x-1}e^{-t}\,dt$$

is commonly used in probability theory. The function holds for any complex number $a+bi$ where $a>0$ but we will only consider $x \in \mathbb{R}$. Use integration by parts to show that $\Gamma(x+1) = x\Gamma(x)$.

Exercise 9.2. *Consider $\Gamma(x)$ from Exercise 9.1.*

a) *Evaluate $\Gamma(1)$.*
b) *Using Exercise 9.1 and part a), calculate $\Gamma(2)$, $\Gamma(3)$, and $\Gamma(4)$.*
c) *Let n be a positive integer. State and prove a conjecture about $\Gamma(n)$ (as this is a result on the positive integers, a proof by induction [Section A.4] is in order).*

Exercise 9.3. *A function $f(x)$ is* convex *over $\mathbb{R}$ if for all $\lambda \in [0,1]$*

$$f(\lambda x_1 + (1-\lambda)x_2) \le \lambda f(x_1) + (1-\lambda)f(x_2) \tag{9.19}$$

(see Definition 18.1.1). Show that if a function is convex, then its second derivative $f''(x)$ must be greater than 0.

Exercise 9.4. a) *Use the Mean Value Theorem (Theorem 9.1.7) and Theorem 9.1.5 to prove that if $f'(x) = 0$ over an interval (a,b), then $f(x)$ is constant over that interval.*
b) *Use the result from the previous part of this exercise to prove that if $f'(x) = g'(x)$, then $f(x) = g(x) + c$ where c is some constant.*

Exercise 9.5. *Prove each of the following:*

a) *Rolle's Theorem (Theorem 9.1.6).*
b) *Theorem 9.1.5.*
c) *The Mean Value Theorem (Theorem 9.1.7).*
d) *The Extended Mean Value Theorem (Theorem 9.1.8).*
e) *The Mean Value Theorem for Integrals (Theorem 9.1.9).*
f) *The generalized Mean Value Theorem for Integrals (Theorem 9.1.10).*
g) *The Fundamental Theorem of Calculus (Theorem 9.1.11).*

Exercise 9.6. *Let $f(x)$ be a function that has as its power series centered at a*

$$f(x) = c_0 + c_1(x-a) + c_2(x-a)^2 + \cdots + c_n(x-a)^n + \cdots.$$

Using the power series representation find $f(a)$, $f'(a)$, $f''(a)$, $\ldots$, $f^{(n)}(a)$ thus verifying 9.5 in Definition 9.2.5.

Exercise 9.7. *Prove Taylor's Theorem (Theorem 9.2.3). [*HINT*: use the Fundamental Theorem of Calculus.]*

Exercise 9.8. *Prove Lagrange's Remainder Theorem (Theorem 9.2.4). [*HINT*: use the Taylor's Theorem and the Mean Value Theorem for Integrals.]*

Exercise 9.9. *a) Find the Taylor series expansion of* $f(x) = \cos x$ *at* $a = 0$.
b) State $T_4(x)$*, the fourth degree Taylor approximation, of* $f(x) = \cos x$ *at* $a = 0$.
c) Show $R_4(x)$*, the remainder between* $T_4(x)$ *and* $f(x) = \cos x$ *at* $a = 0$ *is bounded above by* $\frac{x^5}{5!}$.

Exercise 9.10. *a) Find the Taylor series expansion of* $f(x) = \sin x$ *at* $a = 0$.
b) State $T_2(x)$*, the fourth degree Taylor approximation, of* $f(x) = \sin x$ *at* $a = 0$.
c) Show $R_2(x)$*, the remainder between* $T_2(x)$ *and* $f(x) = \sin x$ *at* $a = 0$ *is bounded above by* $\frac{x^3}{3!}$.
d) What is the largest the error can be when using $T_2(x) = x$ *to approximate* $f(x) = sinx$ *over the interval* $(-0.5, 0.5)$*? (This shows that the approximation* $\sin x \approx x$ *near 0 for* simple harmonic motion *in Physics is a good approximation.)*

Exercise 9.11. *Use Taylor's Formula (Theorem 9.2.7) to show that* $x = 1$ *is a strict global minimizer of* $f(x) = x^4 - 4x^3 + 6x^2 - 4x + 1$ *over* $\mathbb{R}$.

Exercise 9.12. *Find an example where Newton's Method will not converge.*

Exercise 9.13. *Use Newton's Method to find the solution of the equation* $sin(x) = x$ *to to four decimal places (i.e. do enough iterations so that it is clear the first four digits do not change).*

Exercise 9.14. *Use Newton's Method to find the root of* $x^3 - x + 1$. *Show that this is the only root of this equation.*

Exercise 9.15. *Show that the derivative of the logistic function*

$$\sigma(x) = \frac{e^x}{1 + e^x} = \frac{1}{1 + e^{-x}}$$

satisfies $\sigma'(x) = \sigma(x)\sigma(-x)$. *Find* $\sigma''(x)$ *and write it in terms of* $\sigma(x)$ *and* $\sigma(-x)$. *Do the same for* $\sigma^{(n)}(x)$.

Exercise 9.16. *Let* $f(x) = \frac{1}{2}x^T Ax + \mathbf{b}^T x + \mathbf{c}$ *where* A *is a symmetric matrix. Calculate* ∇f *and the Hessian* Hf.

Exercise 9.17. *In Machine Learning, it is common to use the sum-of-squares function*

$$E(\mathbf{w}) = \frac{1}{2} \sum_{n=1}^{N} [\sum_{j=0}^{M} (w_j x^j) - y_n]^2$$

where x and y are regarded as constants (given data), and the w_js are variables (chosen parameters for regression). Take the partial derivative with respect to w_j, $j = 0, \ldots, M$; set the equation to 0 to find the minimum, and rewrite this formula as a system of linear equations.

Exercise 9.18. *In the previous problem, we assumed that the critical point we found was a minimum without doing a derivative test. What property about the function allows us to state that it must be a minimum?*

Exercise 9.19. *Find the third-order Taylor polynomial for $f(x, y) = xe^y$ at $(0, 0)$.*

Exercise 9.20. *Use the Multivariable Taylor's Formula (Theorem 9.6.3) to show that $f(x, y) = x^2 + y^2$ has $\boldsymbol{x} = (0, 0)^T$ as a global minimizer. [HINT: Follow the comments following the single variable statement (Theorem 9.2.7) and show $\nabla f((0, 0)^T) = \boldsymbol{0}$ while noting that for $\boldsymbol{x}^* \neq (0, 0)^T$, $\boldsymbol{x}^* \cdot \boldsymbol{x}^* > \boldsymbol{0}$.]*

10

A Calculus Approach to Nonlinear Programming

As not all problems and constraints are linear, we now consider the class of problems where any or all of the objective function and constraints are nonlinear. As nonlinear programming questions belong to a harder class of problems than linear programming questions, it is not surprising that many solution techniques exist and are required for these kinds of problems. As some nonlinear programming problems can be solved using familiar tools from Calculus, we start here.

10.1 Using Derivatives to Find Extrema of Functions of a Single Variable

Please see Chapter 9 for a review of the necessary Calculus technniques.

We begin by identifying the location of potential extreme values.

Definition 10.1.1 (Critical Point). *Let $x = c$ be a point in the domain of a function $f(x)$. c is called a* **critical point of** *f if c is in the domain of $f(x)$ and either $f'(c) = 0$ or does not exist.*

The following theorem explains why critical points are critical.

Theorem 10.1.2 (Fermat's Theorem for Local Extrema). *If $x = c$ is a local extrema and $f(x)$ is differentiable at c, then $f'(c) = 0$.*

Proof. Suppose $x = c$ is a location of a local maximum. Then by definition, there is some $\epsilon > 0$ such that for all x in $(c - \epsilon, c + \epsilon)$, $f(x) \leq f(c)$ and we thus have $f(x) - f(c) \leq 0$. For x where $c < x < c + \epsilon$, we have $x - c > 0$ and therefore $\frac{f(x)-f(c)}{x-c} \leq 0$. Thus

$$\lim_{x \to c^+} \frac{f(x) - f(c)}{x - c} \leq 0. \tag{10.1}$$

For x where $c - \epsilon < x < c$, we have then $x - c < 0$ and hence $\frac{f(x)-f(c)}{x-c} \geq 0$. Thus

$$\lim_{x \to c^-} \frac{f(x) - f(c)}{x - c} \geq 0. \tag{10.2}$$

DOI: 10.1201/9780367425517-10

But $f(x)$ is differentiable at c, so by 10.1

$$f'(c) = \lim_{x \to c^+} \frac{f(x) - f(c)}{x - c} \leq 0. \tag{10.3}$$

Likewise, by 10.2

$$f'(c) = \lim_{x \to c^-} \frac{f(x) - f(c)}{x - c} \geq 0. \tag{10.4}$$

As both $f'(c) \leq 0$ and $f'(c) \geq 0$, $f'(c) = 0$.

The case when $x = c$ is the location of a local minimum is similar and is Exercise 10.1. □

Hence *Fermat's Theorem for Local Extrema* gives us a method for identifying the location of possible extrema (namely, any extreme values of a function over an open interval must occur at critical points), we need a technique to help identify whether a given critical number is a local extrema or not and when it is, if it is a relative minimum or maximum.

First,

Theorem 10.1.3 (Increasing/Decreasing Test). *Suppose $f(x)$ is differentiable over an interval (a, b).*

i) If $f'(x) > 0$ over (a, b), then $f(x)$ is increasing over (a, b).
ii) If $f'(x) < 0$ over (a, b), then $f(x)$ is decreasing over (a, b).
iii) If $f'(x) = 0$ over (a, b), then $f(x)$ is constant over (a, b).

Proof. We prove part *i)* and leave the proofs of the remaining parts as an exercise.

Suppose $f'(x) > 0$ over (a, b). Let x_1, x_2 be points in the interval with $x_1 < x_2$. Since $f(x)$ is differentiable over (a, b), by Theorem 9.1.5, $f(x)$ is continuous over $[x_1, x_2]$. Of course, $f(x)$ is then also differentiable over (x_1, x_2). Thus, by the Mean Value Theorem (Theorem 9.1.7), there is a c in (x_1, x_2) such that $f'(c) = \frac{f(x_2) - f(x_1)}{x_2 - x_1}$. As $f'(c) > 0$ and $x_2 \neq x_1$, $f(x_2) - f(x_1) > 0$ giving $f(x_2) > f(x_1)$. As x_1 and x_2 are arbitrary points in (a, b), $f(x)$ is increasing over the interval. □

Of course, as the derivative gives the slope of the tangent line and, thus, the instantaneous rate of change of the function, the results of Theorem 10.1.3 are intuitively obvious (but we do not rely on intuition for truth in Mathematics). With this, we now have

Theorem 10.1.4 (The First Derivative Test). *Suppose $x = c$ is a critical point of $f(x)$ in some open interval I and I contains no other critical points of $f(x)$. Further suppose $f(x)$ is continuous over I and differentiable at every point in I except possibly at c. Then*

i) if $f'(x) > 0$ for $x \in I$ with $x < c$ but $f'(x) < 0$ for $x \in I$ with $x > c$, then $f(x)$ has a ***local maximum*** *at $x = c$;*

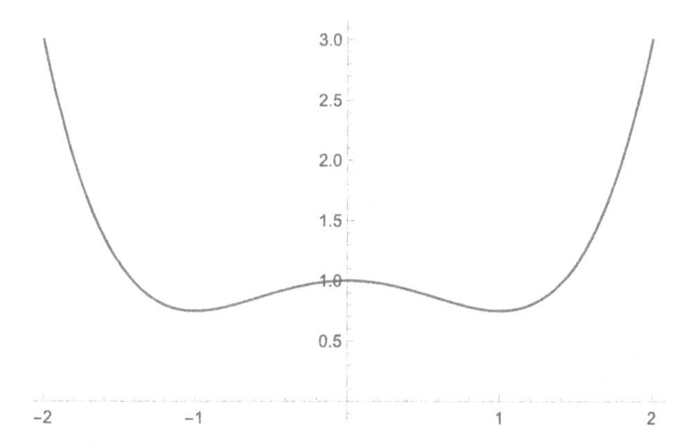

FIGURE 10.1
$f(x) = \frac{1}{4}x^4 - \frac{1}{2}x^2 + 1$ in Example 10.1.6.

*ii) if $f'(x) < 0$ for $x \in I$ with $x < c$ but $f'(x) > 0$ for $x \in I$ with $x > c$, then $f(x)$ has a **local minimum** at $x = c$; or*

iii) if $f'(x)$ does not change sign at $x = c$, then $f(x)$ does not have any local extrema at $x = c$.

Or, if you prefer to not evaluate a function multiple times and do not mind differentiating one more time

Theorem 10.1.5 (The Second Derivative Test)**.** *Let $x = c$ be a critical point of $f(x)$ with $f'(c) = 0$ and suppose $f(x)$ and $f'(x)$ are differentiable and $f''(x)$ over some interval I containing c. Then*

*i) if $f''(c) > 0$, then $f(x)$ has a **local minimum** at $x = c$;*

*ii) if $f''(c) < 0$, then $f(x$ has a **local maximum** at $x = c$; and*

iii) if $f''(c) = 0$, then the test is inconclusive.

Note that statement *iii*) of Theorem 10.1.5 is not a mistake (see Exercise 10.5).

Example 10.1.6. *Use Calculus to find the (local) extreme values of $f(x) = \frac{1}{4}x^4 - \frac{1}{2}x^2 + 1$.*

Solution. We have $f'(x) = x^3 - x = x(x^2 - 1)$, thus the critical numbers are $x = -1, 0, 1$. Using the First Derivative Test would require making a sign chart, thus we instead observe that $f''(-1) = f''(1) = 2 > 0$, thus $f(-1) = f(1) = \frac{3}{4}$ are local minimums whereas $f''(0) = -1 < 0$, hence $f(0) = 1$ is a local maximum. The graph of $f(x)$ is shown in Figure 10.1. ∎

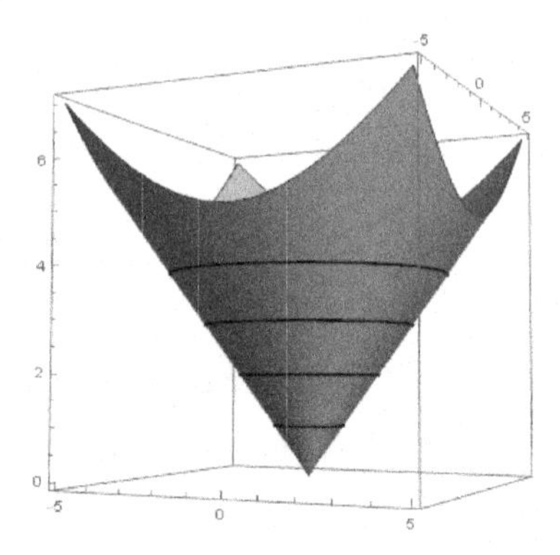

FIGURE 10.2
Level curves $z = 1, 2, 3, 4$ on the surface of $f(x, y) = \sqrt{x^2 + y^2}$.

10.2 Calculus and Extrema of Multivariable Functions

The analogous statements for optimization with multivariable functions are similar to their single-variable counterparts but often with some additional detail[1]. We begin with the multivariable[2] version of the first derivative

Definition 10.2.1 (The Gradient). *Let $z = f(x, y)$ be a function of two variables. Then the* gradient *of $f(x, y)$ is the vector*

$$\nabla f(x, y) := \left\langle \frac{\partial f}{\partial x}, \frac{\partial f}{\partial y} \right\rangle.$$

The two-variable definition extends naturally to any function of n variables where $n > 2$.

We now remind the reader of some important properties of the gradient. Note that if $z = f(x, y)$, that is, $f(x, y)$ is a function of two variables, any input into the function is a point in the $x - y$ plane and the corresponding output is a point in space. A curve where $f(x, y) = k$ for some constant k is called a *level curve*. The level curves for $z = f(x, y) = 1, 2, 3, 4$ on the surface of the cone $f(x, y) = \sqrt{x^2 + y^2}$ are shown in Figure 10.2.

Remark 10.2.2. *Let $f(x, y)$ be a differentiable function. Then the gradient of $f(x, y)$, $\nabla f(x, y) := \left\langle \frac{\partial f}{\partial x}, \frac{\partial f}{\partial y} \right\rangle$,*

---

[1] As mentioned in the previous chapter, there are some concerns with derivatives of multivariable functions existing and considering *directional derivatives* instead of just focusing on partials with respect to particular variables; but we wish to not be a Calculus text. There is an excellent treatment of this in [57].

[2] Our presentation will be for functions of two variables, but the results extend naturally to functions of more variables.

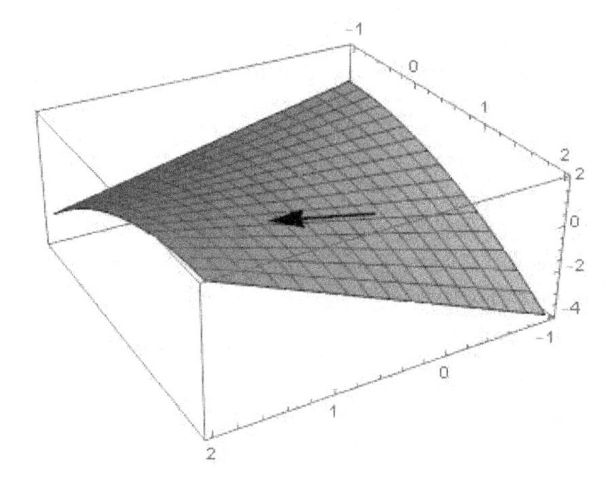

FIGURE 10.3
$\nabla f(1,1)$ pointing in the direction of the greatest increase out of $(1,1,\frac{1}{2})$ in Example 10.2.3. Here the y axis runs left-right and the x axis is coming out of the page.

1. *is a vector orthogonal to the level curves of* $f(x,y)$,
2. $\nabla f(a,b)$ *is a vector which always points in the direction of the maximal increase of* $f(x,y)$ *from the point* (a,b), *and*
3. $-\nabla f(a,b)$ *is a vector which always points in the direction of the maximal decrease of* $f(x,y)$ *from the point* (a,b),

Proofs of the gradient properties noted in Remark 10.2.2 can be found in most multivariable Calculus texts. A quality presentation of these can be found in [57].

Example 10.2.3. *Consider* $f(x,y) = -\frac{1}{2}y^2 + xy$. *Then* $\nabla f(x,y) = \langle y, -y+x\rangle$ *and* $\nabla f(1,1) = \langle 1,0\rangle$. *A glance at Figure 10.3 shows this vector pointing in the direction of the greatest increase in* $f(x,y)$ *when leaving* $(1,1,\frac{1}{2})$.

Equipped with the multivariable version of the derivative, we now have the tools to address optimizing multivariable functions.

Definition 10.2.4 (Critical Point – two variables). *A point* (a,b) *is a* critical point *of* $f(x,y)$ *if*

- *Both* $f_x(a,b) = 0$ *and* $f_y(a,b) = 0$ *(i.e.* $\nabla f(a,b) = \langle 0,0\rangle$*), or*
- $f_x(a,b)$ *is undefined, or*
- $f_y(a,b)$ *is undefined.*

As well, we have the following multivariable version of Fermat's Theorem for Local Extrema from single variable calculus:

Theorem 10.2.5. *[Two-variable Fermat's Theorem for Local Extrema] If $f(x, y)$ has a local minimum or a local maximum at (a, b), then (a, b) is a critical point.*

The proof of Theorem 10.2.5 is very similar to the proof of the single variable case (Theorem 10.1.2) and is left as Exercise 10.13.

Once a multivariable function's critical points are known, we can follow the same routine as in the single variable case to check for extrema. But a Multivariable First Derivative Test is too cumbersome, so we instead use

Theorem 10.2.6 (The Second Partial Derivative Test for a Function of Two Variables).
Let (a, b) be a critical point of $f(x, y)$ where $f(x, y)$ is a function with continuous 2nd-*order partial derivatives on some open disk containing (a, b). Put*

$$D = D(a,b) = \det(Hf(a,b)) = \begin{vmatrix} f_{xx} & f_{xy} \\ f_{yx} & f_{yy} \end{vmatrix} (a,b) = f_{xx}(a,b)f_{yy}(a,b) - [f_{xy}(a,b)]^2 \tag{10.5}$$

where Hf is the Hessian of f. If

i) if $D > 0$ and $f_{xx}(a, b) > 0$,then $f(a, b)$ is a local minimum;
ii) if $D > 0$ and $f_{xx}(a, b) < 0$, then $f(a, b)$ is a local maximum;
iii) if $D < 0$, then $f(a, b)$ is not a local extreme value (in this case, the point (a, b) is called a **saddle point***); or*
iv) if $D = 0$, the test is inconclusive.

Note that the D of the Second Partial Derivative Test is the determinant of the Hessian of f and by Clairaut's Theorem (9.5.4) we have that $f_{xy}(a, b) = f_{yx}(a, b)$. Recall that for a quadratic function $f(x) = ax^2 + bx + c$ has $D = b^2 - 4ac$ as its discriminant as, by the Quadratic Formula, this D determines ("discriminates") the nature of the roots of $f(x)$. Similarly, as the D of The Second Partial Derivative Test 10.2.6 discriminates how the critical points behave, this D is referred to as the *discriminant* of $f(x, y)$.

A proof of The Second Partial Derivative Test for a Function of Two Variables involves directional derivatives, so it will be excluded here. Again, a quality presentation of a proof can be found in [57].

We now present a classic example using these tools to find an extreme value.

Example 10.2.7. *Find the points on the cone $z^2 = x^2 + y^2$ closest to the point* $(4, 2, 0)$.

Solution. Let's let d represent the distance from an arbitrary point on the cone to the point $(4, 2, 0)$. Hence $d(x, y, z) = \sqrt{(x-4)^2 + (y-2)^2 + (z-0)^2}$. Since we are on the cone $z^2 = x^2 + y^2$, we use this constraint to eliminate a variable and obtain the unconstrained geometric programing problem

$$d(x, y, z) = d(x, y) = \sqrt{(x-4)^2 + (y-2)^2 + x^2 + y^2}. \tag{10.6}$$

Since $d(x, y)$ is nonnegative, minimizing d^2 will give us the point(s) that minimize d. Hence we seek to

$$\text{Minimize } f(x,y) = d^2(x,y) = (x-4)^2 + (y-2)^2 + x^2 + y^2. \tag{10.7}$$

To find critical points of d^2:

$$f_x(x,y) = d_x^2(x,y) = 2(x-4) + 2x = 4x - 8, \text{ and} \tag{10.8}$$

$$f_y(x,y) = d_y^2(x,y) = 2(y-2) + 2y = 4y - 4. \tag{10.9}$$

These partial derivatives are zero at $(2, 1)$ and are never undefined, therefore the only critical point is $(2, 1)$. Moreover,

$$f_{xx} = d_{xx}^2(x,y) = 4 \tag{10.10}$$

$$f_{yy} = d_{yy}^2(x,y) = 4 \text{ and} \tag{10.11}$$

$$f_{xy} = d_{xy}^2(x,y) = f_{yx} = d_{yx}^2(x,y) = 0. \tag{10.12}$$

Hence $D(2, 1) = 16$ and since $f_{xx}(2, 1) > 0$, by The Second Partial Derivative Test (Theorem 10.2.6), the point $(2, 1)$ minimizes d^2. Since $z^2 = x^2 + y^2$, when $x = 2$ and $y = 1$, $z = \pm\sqrt{5}$.

Thus there are two points on the cone $z^2 = x^2 + y^2$ that are a minimum distance from $(4, 2, 0)$, namely the points $(2, 1, \sqrt{5})$ and $(2, 1, -\sqrt{5})$. ■

The critical numbers in Example 10.2.7 were indeed places where extrema of the function were located. Let us try one with a not as satisfying outcome.

Example 10.2.8. *Find the extrema of the hyperbolic paraboloid*

$$z = f(x,y) = \frac{x^2}{2} - \frac{y^2}{3}. \tag{10.13}$$

Solution. We have

$$\nabla f(x,y) = \langle x, -\frac{2}{3}y \rangle \tag{10.14}$$

giving the only critical point as $(0, 0)$. The Hessian of $f(x, y)$ is

$$H = \begin{bmatrix} 1 & 0 \\ 0 & -\frac{2}{3} \end{bmatrix} \tag{10.15}$$

so $D(0, 0) = \det\left(H|_{x=0,y=0}\right) = -\frac{2}{3} < 0$ and by the Second Partial Derivative Test (Theorem 10.2.6) $(0, 0)$ is a saddle point and $f(x, y) = \frac{x^2}{2} - \frac{y^2}{3}$ has no extreme values. ■

Example 10.2.8 is illustrated in Figure 10.4.

For functions of three or more variables, the above technique's generalization involves *quadratic forms* and whether they are *positive definite*, *negative definite*, or *semidefinite*. We will consider these matters in Chapter 12.

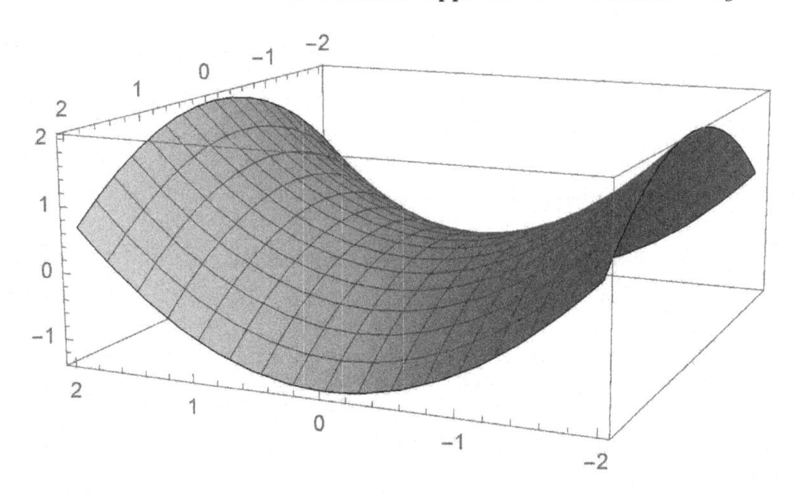

FIGURE 10.4
A saddle point from Example 10.2.8.

10.3 Exercises

Exercise 10.1. *Prove Theorem 10.1.2 for the case of a local minimum.*

Exercise 10.2. *Prove parts ii) and iii) of Theorem 10.1.3.*

Exercise 10.3. *Prove the single-variable Second Derivative Test (Theorem 10.1.4).*

Exercise 10.4. *Prove parts i) and ii) of the single-variable Second Derivative Test (Theorem 10.1.5).*

Exercise 10.5. *To show part iii) of the single-variable Second Derivative Test (Theorem 10.1.5):*

i) *Let $f(x) = x^4$. Use the First Derivative Test (Theorem 10.1.4) to show that $f(x)$ has a local minimum at $x = 0$. Note $f''(0) = 0$.*
ii) *Let $g(x) = -x^4$. Use the First Derivative Test (Theorem 10.1.4) to show that $g(x)$ has a local maximum at $x = 0$. Note $g''(0) = 0$.*
iii) *Let $h(x) = x^3$. Use the First Derivative Test (Theorem 10.1.4) to show that $h(x)$ does not have a local extreme value at $x = 0$. Note $h''(0) = 0$.*

Exercise 10.6. *In high school algebra, one learns that the that a quadratic function $f(x) = ax^2 + bx + c$ has exactly one extreme value (a maximum if $a < 0$ and a minimum if $a > 0$). We are taught that extreme value occurs at the vertex and are led to believe this by considering the graph of $f(x)$. Use calculus to properly prove this result from high school algebra.*

Exercise 10.7. *Find all local extremes of* $f(x) = 2x^3 - 6x^2 - 18x$ *over the domain of the function. Justify any claim that an extreme value is a maximum or minimum.*

Exercise 10.8. *Find all local extremes of* $g(x) = 3x^4 + 4x^3 - 36x^2$ *over the domain of the function. Justify any claim that an extreme value is a maximum or minimum.*

Exercise 10.9. *Find all local extremes of* $h(x) = 3x^4 - 4x^3 - 6x^2 + 12x$ *over the domain of the function. Justify any claim that an extreme value is a maximum or minimum.*

Exercise 10.10. *Find all local extremes of* $k(x) = e^x(x^3 - 12x)$ *over the domain of the function. Justify any claim that an extreme value is a maximum or minimum.*

Exercise 10.11. *Find any extreme values, if they exist – of the function* $f(x, y) = xy^2$ *subject to the constraint that* $x + y = 0$*. [Hint: solve the constraint for one variable and use this to eliminate a variable in the objective function, then proceed as normal.] Explain why the Second Derivative Test fails here.*

Exercise 10.12. *In Statistics, the* Normal Distribution *with mean* μ *and standard deviation* σ *has as its probability density function*

$$f(x) = \frac{1}{\sigma\sqrt{2\pi}} e^{-\frac{1}{2}\left(\frac{x-\mu}{\sigma}\right)^2}$$

(the graph of $f(x)$ *is the familiar bell-curve).*

a) Show $f'(x) = -\left(\frac{x-\mu}{\sigma}\right) f(x)$*.*
b) Show $x = \frac{\mu}{\sigma}$ *is the only critical number.*
c) Using the First Derivative Test (Theorem 10.1.4), show $f(x)$ *has a (local) maximum at* $x = \frac{\mu}{\sigma}$ *State the local maximum and provide a mathematical argument that this is actually a global max.*

Exercise 10.13. *Prove Theorem 10.2.5; namely, if* $f(x, y)$ *has an extreme value at* (a, b)*, then* (a, b) *is a critical point. [Hint: since* a *and* b *are fixed, consider the single-variable functions* $g(x) = f(x, b)$ *and* $h(y) = f(a, y)$ *and refer to the proof of the single variable case (Theorem 10.1.2)].*

Exercise 10.14. *Find all critical points for the stated function and determine if each critical point is a relative minimum, relative maximum, or a saddle point.*

a) $f(x, y) = e^{xy}$
b) $g(x, y) = x^2 - 4x + 2y^2 + 4y + 5$
c) $h(x, y) = x^2y^2$
d) $k(x, y) = x^3 - 12xy + y^3$
e) $t(x, y) = \frac{x^2+y^2}{2} + 2(y - x) - 3\ln(xy)$

11

Constrained Nonlinear Programming: Lagrange Multipliers and the KKT Conditions

11.1 Lagrange Multipliers

This section continuous the use of Calculus to answer nonlinear optimization questions; in particular, we will explore Lagrange multipliers. As this technique generalizes to the important *KKT conditions*, the material deserves its own separate treatment. Here we will seek extreme values of a function f not over its entire domain, but rather maximums and minimums where the function meets the constraint curve g. It can be proven (see [57]) that f has its extrema on g at points $\mathbf{x}_\star$ where the tangent vectors of these two functions are parallel; that is $\nabla f(\mathbf{x}_\star) = \lambda \nabla g(\mathbf{x}_\star)$ where λ is a non-zero scalar. This fact leads to the following optimization technique:

Method 11.1.1 (The Method of Lagrange Multipliers)**.** *If $f(x, y, z)$ has a relative extremum at (x_*, y_*, z_*) subject to the constraint $g(x, y, z) = 0$ where f and g both have continuous first-order partial derivatives, then $\nabla f(x_*, y_*, z_*)$ and $\nabla g(x_*, y_*, z_*)$ are parallel. Thus, if $\nabla g(x_*, y_*, z_*) \neq \mathbf{0}$, $\nabla f(x_*, y_*, z_*) = \lambda \nabla g(x_*, y_*, z_*)$ for some real number λ.*

Note that *the Method of Lagrange Multipliers* helps locate critical points. More work should be done to very that a maximum or minimum exist at any critical number. To simplify notation, we have stated a version of the method for a function of three variables, but the method works for functions of any number of variables.

To illustrate the method, let us return to Example 10.2.7.

Example 11.1.2. *Find the points on the cone $z^2 = x^2 + y^2$ closest to the point* $(4, 2, 0)$.

Solution. If we let (x, y, z) be as arbitrary point, then its distance from $(4, 2, 0)$ is given by $d(x, y, z) = \sqrt{(x-4)^2 + (y-2)^2 + z^2}$, but since $d \geq 0$, the minimum of d occurs at the same place as the minimum of d^2. As well we want

DOI: 10.1201/9780367425517-11

our points to be on the cone $z^2 = x^2 + y^2$, thus our problem is to

$$\text{Minimize: } f(x,y,z) = d^2 = (x-4)^2 + (y-2)^2 + z^2 \tag{11.1}$$
$$\text{Subject to: } g(x,y,z) = x^2 + y^2 - z^2 = 0. \tag{11.2}$$

So

$$\nabla f(x,y,z) = \langle 2x-8, 2y-4, 2z\rangle \text{ and} \tag{11.3}$$
$$\nabla g(x,y,z) = \langle 2x, 2y, -2z\rangle. \tag{11.4}$$

Thus we have the system of equations

$$2x - 8 = \lambda 2x \tag{11.5}$$
$$2y - 4 = \lambda 2y \tag{11.6}$$
$$2z = -2\lambda z \tag{11.7}$$
$$x^2 + y^2 - z^2 = 0 \tag{11.8}$$

Rearranging equation (11.7) gives us

$$0 = 2z + 2\lambda z = 2z(1+\lambda)$$

which implies that either $z = 0$ or $\lambda = -1$.

Case 1: $z = 0$

If $z = 0$, then by equation (11.8), $x^2 + y^2 = 0$. But this implies that $x = y = 0$ (since each term is squared, we are adding two nonnegative quantities and getting 0). Thus we obtain $(0,0,0)$, but realizing that $\nabla g(0,0,0) = \langle 0,0,0\rangle$ means this is not a solution (see the last line of Method 11.1.1).

Case 2: $\lambda = -1$

If $\lambda = -1$, then substituting into equation (11.5) gives us

$$2x - 8 = -2x \Leftrightarrow 4x = 8,$$

i.e. $x = 2$.

As well, substituting $\lambda = -1$ into equation (11.6) gives us

$$2y - 4 = -2y \Leftrightarrow 4y = 4,$$

i.e. $y = 1$.

These values substituted into equation (11.8) gives us $z = \pm\sqrt{5}$.

Since the cone extends forever up and down in the z direction, we realize these points $(2, 1, \pm\sqrt{5})$ are where the distance is minimized and that there is no maximum. (A mathematical argument: putting $y = 0$ and $x = z$ satisfy the constraint; letting $x, z \to \infty$ we see $d^2 \to \infty$.) ■

Note that, in the example, $\nabla f(2,1,\pm\sqrt{5}) = \langle -4, -2, \pm 2\sqrt{5}\rangle$ and $\nabla g(2,1,\pm\sqrt{5}) = \langle 4, 2, \mp 2\sqrt{5}\rangle$ thus illustrating the comment immediately preceding Method 11.1.1.

Lagrange multipliers may also be used when there are two constraints:

Method 11.1.3 (The Method of Lagrange Multipliers). *If $f(x, y, z)$ has a relative extremum at (x_*, y_*, z_*) subject to the constraints $g(x, y, z) = 0$ and $h(x, y, z) = 0$ where f, g, and h have continuous first-order partial derivatives, then if $\nabla g(x_*, y_*, z_*) \neq \mathbf{0}$ and $\nabla h(x_*, y_*, z_*) \neq \mathbf{0}$, $\nabla f(x_*, y_*, z_*) = \lambda \nabla g(x_*, y_*, z_*) + \mu \nabla h(x_*, y_*, z_*)$ for real numbers λ and μ.*

Lagrange's Method does extend beyond two constraints and its generalization is the topic of the next section.

11.2 The KKT Conditions

As we saw in the first two sections of Chapter 10, Calculus can be used to find optimal values of functions. Unfortunately, this is a multi-step process in that we first find critical points and then must do a little more work to determine if we have a (local!) maximum or minimum at any of these critical points (and even more work if we want to address global extreme values). One tool to check if a critical point is indeed where a function has an extreme value is to use the First Derivative Test (the single variable version is Theorem 10.1.4). If a nonlinear programming problem has appropriate constraints (specifically, satisfy some regularity conditions), the *KKT Conditions* are essentially a first derivative test to verify that a solution is indeed optimal. Moreover, as we will see shortly, the KKT conditions can be used to solve a nonlinear programming problem that has inequality constraints in a manner that generalizes Lagrange Multipliers.

The KKT Conditions were introduced in a 1951 paper by Harold W. Kuhn and Albert W. Tucker [40]. Later is was discovered that William Karush had developed the conditions in his 1939 Master's Thesis at the University of Chicago [38]. These conditions are commonly referred to as the Karush-Kuhn-Tucker Conditions though some older literature may merely call them the Kuhn-Tucker (KT) Conditions.

Theorem 11.2.1 (Karush-Kuhn-Tucker (KKT) Conditions). *Consider the following optimization problem:*

$$\begin{aligned} \textit{Maximize: } & f(x_1, \ldots, x_n) && (11.9)\\ \textit{subject to: } & g_i(x_1, \ldots, x_n) \leq 0 \textit{ for } i = 1, \ldots, m && (11.10) \end{aligned}$$

where $f, g_1, \ldots, g_m$ have continuous first derivatives. Define the Lagrangian

$$L(x_1, \ldots, x_n) := f(x_1, \ldots, x_n) - \sum_{i=1}^{m} \lambda_i g_i(x_1, \ldots, x_n)$$

where the λ_i are real numbers. Then if $(x_1^*, \ldots, x_n^*)$ is a maximizer of f, the following conditions are satisfied:

$$\frac{\partial L}{\partial x_j} = \frac{\partial f}{\partial x_j} - \sum_{i=1}^{m} \lambda_i \frac{\partial g_i}{\partial x_j} = 0 \qquad \text{for } j = 1, 2, \ldots, n; \tag{11.11}$$

$$\lambda_i \geq 0 \qquad \text{for } i = 1, 2, \ldots, m; \tag{11.12}$$

$$\lambda_i g_i = 0 \qquad \text{for } i = 1, 2, \ldots, m; \quad \text{and} \tag{11.13}$$

$$g_i(x_1^*, \ldots, x_n^*) \leq 0 \qquad \text{for } i = 1, 2, \ldots, m. \tag{11.14}$$

A few notes are in order. The inequalities in (11.14) are merely satisfying the constraints. (11.11), (11.12), and (11.13) are the KKT conditions. More importantly, the statement of the theorem is not quite correct; certain regularity conditions need to be met for the theorem to hold. A proof of Theorem 11.2.1 can be found in [26] as well as a discussion of the conditions on which the theorem holds. At the time of this writing, Wikipedia's entry lists many of the regularity conditions. Both sources also discuss sufficient conditions for Theorem 11.2.1.

Example 11.2.2. *We will use the KKT conditions to find the circle centered at $(3, 5)$ of smallest radius that intersects the region given by $y \leq 2$ and $x \leq 1$. Note that since the radius of a circle nonnegative, we may minimize r^2. Thus, we seek to solve*

$$\text{Maximize: } f(x, y) = -r^2 = -(x-3)^2 - (y-5)^2 \tag{11.15}$$

$$\text{subject to: } g_1(x, y) = y - 2 \leq 0 \tag{11.16}$$

$$g_2(x, y) = x - 1 \leq 0 \tag{11.17}$$

Solution. First we have

$$L(x, y) = -(x-3)^2 - (y-5)^2 - \lambda_1(y-2) - \lambda_2(x-1).$$

Assuming some regularity condition is meet, any maximizer will satisfy the KKT conditions

$$\frac{\partial L}{\partial x} = -2(x-3) - \lambda_2 = 0, \tag{11.18}$$

$$\frac{\partial L}{\partial y} = -2(y-5) - \lambda_1 = 0, \tag{11.19}$$

$$\lambda_1(y-2) = 0, \tag{11.20}$$

$$\lambda_2(x-1) = 0, \tag{11.21}$$

$$\lambda_1, \lambda_2 \geq 0.$$

Equation 11.18 gives $\lambda_2 = 6 - 2x$. Substituting into (11.21) leads to $x = 3$ or 1, but as $x = 3$ is not feasible, $x = 1$. Similar work with the other pair of equations yields $y = 5$ or 2 giving that $y = 2$. Note that these solutions are

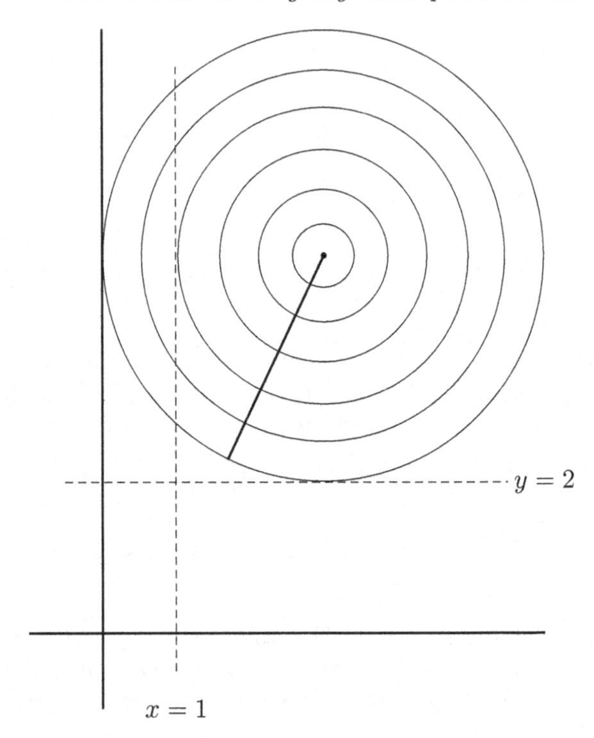

FIGURE 11.1
Minimizing the radius in Example 11.2.2.

independent, thus $x = 1$ or $y = 2$. The first circle to reach $x = 1$ has radius 2, which is not large enough to satisfy the constraint $y \leq 2$, thus the first circle to reach $y = 2$ is the circle with center $(3, 5)$ having the smallest radius and intersecting both half-planes (this circle has radius 3). This example is represented in Figure 11.1. ■

11.3 Exercises

Exercise 11.1. *Use the Method of Lagrange Multipliers (Method 11.1.1) to find the point(s) on the circle $x^2 + y^2 = 8$ whose coordinates have the largest sum as well as any point(s) that have the smallest sum (i.e. find the maximum and the minimum of $f(x, y) = x + y$ over the boundary of the circle).*

Exercise 11.2. *Explain why the Method of Lagrange Multipliers fails to find extreme values of $f(x, y) = xy^2$ subject to $x + y = 0$. Compare this with Exercise 10.11.*

Exercise 11.3. *Use the Method of Lagrange Multipliers to find the point on* $f(x, y, z) = xyz$ *that is closest to the origin.*

Exercise 11.4. *(From [57]) Use the Method of Lagrange Multipliers to find the point(s) in* $\mathbb{R}^3$ *closest to the origin that lie on both the cylinder* $x^2+y^2 = 4$ *and the plane* $x - 3y + 2z = 6$ *(thus the constraints are* $g(z, y, z) = x^2 + y^2 - 4$ *and* $h(z, y, z) = x - 3y + 2z - 6$*).*

Exercise 11.5. *(From [57]) Use the Method of Lagrange Multipliers to find the point(s) in* $\mathbb{R}^3$ *closest to the origin that lie on both the cone* $z^2 = x^2 + y^2$ *and the plane* $x + 2y = 6$ *(thus the constraints are* $g(z, y, z) = x^2 + y^2 - z^2$ *and* $h(z, y, z) = x + 2z - 6$*).*

Exercise 11.6. *Use the KKT conditions to find the circle centered at* $(3, 4)$ *of smallest radius that intersects the region given by* $x \leq 2$ *and* $y \leq 1$*. Note that since the radius of a circle nonnegative, it is sufficient to minimize* r^2*. You may assume any necessary regularity condition is met and you do not have to justify that your solution is a local or global min.*

Exercise 11.7. *(From [26]) Use the KKT conditions to solve the following Geometric Programming problem. You may assume any necessary regularity condition is met and you do not have to justify that your solution is a local or global max.*

$$\begin{aligned} \textit{Maximize: } & f(x_1, x_2, x_3, x_4) = -x_1^2 + 2x_2^2 + 4x_3^2 - 3x_4^2 \\ \textit{subject to: } & x_1 + 3x_2 + 4x_3 - 2x_4 \leq 0 \\ & x_1 + x_2 + x_3 + x_4 \leq 0 \\ & 4x_1 + 3x_2 + 2x_3 + x_4 \leq 0 \end{aligned}$$

[The solution is the point $(366, -168, -43, -155)$*.]*

Exercise 11.8. *(From [26]) Use the KKT conditions to solve the following Geometric Programming problem. You may assume any necessary regularity condition is met and you do not have to justify that your solution is a local or global max.*

$$\begin{aligned} \textit{Maximize: } & f(x_1, x_2, x_3) = 2x_1^2 - x_2^2 - 3x_3^2 \\ \textit{subject to: } & x_1 + 2x_2 + x_3 \leq 1 \\ & 4x_1 + 3x_2 + 2x_3 \leq 2 \end{aligned}$$

[The solution is the point $(0, 0, 0)$*.]*

12

Optimization Involving Quadratic Forms

12.1 Quadratic Forms

In this chapter we provide a tidy second-order condition for extreme values. The result is a nice application of Taylor's Formula (Theorem 9.6.3) involving quadratic forms.

Definition 12.1.1 (Quadratic Form). *A polynomial of the form*

$$a_{11}{x_1}^2+a_{22}{x_2}^2+\cdots+a_{nn}{x_n}^2+2a_{12}x_1x_2+2a_{13}x_1x_3+\cdots+2a_{n-1,n}x_{n-1}x_n$$
$$=\sum_{i=1}^{n} a_{ii}{x_i}^2+\sum_{i<j} 2a_{ij}x_ix_j \tag{12.1}$$

is called a quadratic form*; that is, it is a homogeneous polynomial where each term has total degree 2.*

It can be shown that every quadratic form has associated a matrix $A = \{a_{ij}\}$; in particular that

$$\mathbf{x}^T A\mathbf{x} = \mathbf{x}\cdot A\mathbf{x}\sum_{i=1}^{n} a_{ii}{x_i}^2+\sum_{i<j} 2a_{ij}x_ix_j \tag{12.2}$$

where $A = \{a_{ij}\}$ and $\mathbf{x} = [x_1,\ldots,x_n]^T$. For example, $A = \begin{bmatrix} 2 & 3 \\ 2 & 1 \end{bmatrix}$ is the matrix associated with the quadratic form $f(x_1, x_2) = 2{x_1}^2 + {x_2}^2 + 5x_1x_2$; that is

$$\mathbf{x}^T A\mathbf{x}=\begin{bmatrix} x_1 \\ x_2 \end{bmatrix}^T \begin{bmatrix} 2 & 3 \\ 2 & 1 \end{bmatrix} \begin{bmatrix} x_1 \\ x_2 \end{bmatrix} = \begin{bmatrix} 2x_1+2x_2 \\ 3x_1+x_2 \end{bmatrix}^T \begin{bmatrix} x_1 \\ x_2 \end{bmatrix} = 2x_1^2+5x_1x_2+x_2^2 = f(x_1, x_2). \tag{12.3}$$

From this example it is easy to see that a quadratic polynomial in two variables factors very nicely, namely

$$ax_1^2+2bx_1x_2+cx_2^2 = \begin{bmatrix} x_1 \\ x_2 \end{bmatrix}^T \begin{bmatrix} a & b_1 \\ b_2 & c \end{bmatrix} \begin{bmatrix} x_1 \\ x_2 \end{bmatrix} \tag{12.4}$$

where $b_1 + b_2 = 2b$. Much is known of factoring quadratic forms of more than two variables and matrix factorization is considered in Chapter 5.

DOI: 10.1201/9780367425517-12

There are many wonderful properties to explore regarding quadratic forms, but we focus our attention on those immediately relevant to Optimization. First, some necessary definitions:

Definition 12.1.2. *Consider a real, symmetric matrix A and its quadratic form $\mathbf{x}^T A\mathbf{x} = \mathbf{x} \cdot A\mathbf{x}$. Then A and its quadratic form are said to be*

- positive definite *if $\mathbf{x} \cdot A\mathbf{x} > 0$ for all $\mathbf{x} \neq \mathbf{0}$,*
- positive semidefinite *if $\mathbf{x} \cdot A\mathbf{x} \geq 0$ for all $\mathbf{x}$,*
- negative definite *if $\mathbf{x} \cdot A\mathbf{x} < 0$ for all $\mathbf{x} \neq \mathbf{0}$,*
- negative semidefinite *if $\mathbf{x} \cdot A\mathbf{x} \leq 0$ for all $\mathbf{x}$, and*
- indefinite *if $\mathbf{x} \cdot A\mathbf{x}$ takes on both positive and negative values.*

Note that these definitions specifically involve the quadratic form but are extended to the associated matrix A; that is we say A is positive semidefinite if its associated quadratic form $\mathbf{x} \cdot A\mathbf{x}$ is positive semidefinite, etc.

12.2 Definite and Semidefinite Matrices and Optimization

As for why the definitions from the previous section are important in Optimization, suppose $\mathbf{x} = (x_1, x_2, \ldots, x_n)$ and $\mathbf{x}^* = (x_1^*, x_2^*, \ldots, x_n^*)$ are distinct points in $\mathbb{R}^n$. Suppose further that $f(\mathbf{x})$ is a function of n variables with continuous first and second partial derivatives in $\mathbb{R}^n$. Then by the Multivariable Taylor's Formula (Theorem 9.6.3), there exists some $\mathbf{c}$ such that

$$f(\mathbf{x}) = f(\mathbf{x}^*) + \nabla f(\mathbf{x}^*) \cdot (\mathbf{x} - \mathbf{x}^*) + \frac{1}{2}(\mathbf{x} - \mathbf{x}^*) \cdot Hf(\mathbf{c})(\mathbf{x} - \mathbf{x}^*). \tag{12.5}$$

Now if x^* is a critical point of $f(\mathbf{x})$, (12.5) becomes

$$f(\mathbf{x}) = f(\mathbf{x}^*) + 0 + \frac{1}{2}(\mathbf{x} - \mathbf{x}^*) \cdot Hf(\mathbf{c})(\mathbf{x} - \mathbf{x}^*) \tag{12.6}$$

and if $Hf(c)$ is positive definite (note that the Hessian is always symmetric via Clairaut's Theorem), we have

$$f(\mathbf{x}) = f(\mathbf{x}^*) + \text{ something positive,} \tag{12.7}$$

thus

$$f(\mathbf{x}^*) < f(\mathbf{x})$$

for all $\mathbf{x} \neq \mathbf{x}^*$ establishing $\mathbf{x}^*$ as a strict global maximizer of $f(\mathbf{x})$. We have therefore established the first part of the following theorem (the other parts are proven by the same technique and are Exercise 12.4).

Theorem 12.2.1. *Suppose that $\boldsymbol{x}^*$ is a critical point of $f(\boldsymbol{x})$ having continuous first and second partial derivatives on $\mathbb{R}^n$ and that $Hf(\boldsymbol{x})$ is the Hessian of $f(\boldsymbol{x})$. Then $\boldsymbol{x}^*$ is a*

- *strict global minimizer for $f(\boldsymbol{x})$ if $Hf(\boldsymbol{x})$ is positive definite,*
- *global minimizer for $f(\boldsymbol{x})$ if $Hf(\boldsymbol{x})$ is positive semidefinite,*
- *strict global maximizer for $f(\boldsymbol{x})$ if $Hf(\boldsymbol{x})$ is negative definite,*
- *global maximizer for $f(\boldsymbol{x})$ if $Hf(\boldsymbol{x})$ is negative semidefinite, and*
- *saddle point of $f(\boldsymbol{x})$ if $Hf(\boldsymbol{x})$ is indefinite.*

Our earlier statements do not prove the final part of the previous theorem regarding when a critical point of $f(\mathbf{x})$ is a saddle point. This result is stated and proven as Theorem 1.3.7 in [43].

12.3 The Role of Eigenvalues in Optimization

Determining the definiteness of a symmetric matrix can be quite daunting (see the discussion on pages 13–19 in [43], for example). Fortunately, the magic of eigenvalues can save us from some heavy work (a review of eigenvalues is presented in Section 4.4.1).

Theorem 12.3.1. *For a symmetric real matrix A,*

- *A is* positive definite *if and only if all the eigenvalues of A are positive,*
- *A is* positive semidefinite *if and only if all the eigenvalues of A are nonnegative,*
- *A is* negative definite *if and only if all the eigenvalues of A are negative,*
- *A is* negative semidefinite *if and only if all the eigenvalues of A are nonpositive, and*
- *A is* indefinite *if and only if A has both positive and negative eigenvalues.*

A proof of this theorem can be found on pages 29–30 in [43].

We will not consider all the known ways to check the definiteness of a symmetric matrix, but will include the following easily established result:

Theorem 12.3.2. *The diagonal $n \times n$ matrix* $\begin{bmatrix} d_1 & 0 & \cdots & 0 \\ 0 & d_2 & \cdots & 0 \\ \vdots & \vdots & \ddots & \vdots \\ 0 & 0 & \ldots & d_n \end{bmatrix}$ *has its diagonal entries as its eigenvalues.*

By expansion by minors (Theorem 4.3.18),

Proof.

$$0 = \det\left(\begin{bmatrix} d_1-\lambda_1 & 0 & \cdots & 0 \\ 0 & d_2-\lambda_2 & \cdots & 0 \\ \vdots & \vdots & \ddots & \vdots \\ 0 & 0 & \dots & d_n-\lambda_n \end{bmatrix}\right) = (d_1-\lambda_1)(d_2-\lambda_2)\cdots(d_n-\lambda_n),$$

yielding the result. □

Example 12.3.3. *Find all extreme values and saddle points of*

$$f(x,y,z) = x^2 + y^2 + z^2 - 6yz.$$

Solution. We have

$$\frac{\partial f}{\partial x} = 2x \tag{12.8}$$

$$\frac{\partial f}{\partial y} = 2y - 6z \tag{12.9}$$

$$\frac{\partial f}{\partial z} = 2z - 6y \tag{12.10}$$

which leads to $(0,0,0)$ as the only critical point.

We have

$$Hf = \begin{bmatrix} 2 & 0 & 0 \\ 0 & 2 & -6 \\ 0 & -6 & 2 \end{bmatrix} \tag{12.11}$$

and obtain the its determinants by

$$0 = \det\left(\begin{bmatrix} 2-\lambda & 0 & 0 \\ 0 & 2-\lambda & -6 \\ 0 & -6 & 2-\lambda \end{bmatrix}\right) = (2-\lambda)\det\left(\begin{bmatrix} 2-\lambda & -6 \\ -6 & 2-\lambda \end{bmatrix}\right) \tag{12.12}$$

$$= (2-\lambda)\left[(2-\lambda)^2 - 36\right] \tag{12.13}$$

$$= (2-\lambda)(\lambda-8)(\lambda+4) \tag{12.14}$$

which gives eigenvalues $\lambda = 2$, -4, and 8.

Thus by Theorems 12.2.1 and 12.3.1, $(0,0,0)$ is a saddle point of $f(x,y,z) = x^2 + y^2 + z^2 - 6yz$. ■

Note that this technique works even if the objective function is not a quadratic form.

Example 12.3.4. *Find all extreme values and saddle points of*

$$f(x,y,z) = x^2 + y^4 + z^2 - 2x + 2z + 1.$$

Solution. We have

$$\frac{\partial f}{\partial x} = 2x - 2 \tag{12.15}$$
$$\frac{\partial f}{\partial y} = 4y^3 \tag{12.16}$$
$$\frac{\partial f}{\partial z} = 2z + 2 \tag{12.17}$$

which leads to $(1, 0, -1)$ as the only critical point.

We have

$$Hf = \begin{bmatrix} 2 & 0 & 0 \\ 0 & 12y^2 & 0 \\ 0 & 0 & 2 \end{bmatrix} \tag{12.18}$$

which, by Theorem 12.3.2, has all positive eigenvalues. Thus by Theorems 12.2.1 and 12.3.1, $(1, 0, -1)$ is a strict global minimizer for $f(x, y, z) = x^2 + y^4 + z^2 - 2x + 2z + 2$. ■

We will close this chapter by stating a beautiful result regarding optimizing a quadratic form over inputs constrained to be unit vectors.

Theorem 12.3.5. *Suppose the matrix A associated with the quadratic form $\mathbf{x}^T A\mathbf{x}$ has eigenvalues $\lambda_1 \geq \lambda_2 \geq \cdots \geq \lambda_n$. Then for all $\mathbf{x}$ satisfying $||\mathbf{x}|| = 1$,*

$$\lambda_1 \geq f(\mathbf{x}) \geq \lambda_n$$

where

- *$\lambda_1 = f(\mathbf{x}_1)$ with $\mathbf{x}_1$ the unit eigenvector corresponding to λ_1,*
- *$\lambda_n = f(\mathbf{x}_n)$ where $\mathbf{x}_n$ is the unit eigenvector corresponding to λ_n,*
- *and $||\cdot||$ is the Euclidean norm.*

Note that Theorem 12.3.5 tells us that if we wish to find the maximum and minimum values of a quadratic the form $f(\mathbf{x}) = \mathbf{x}^T A\mathbf{x}$ subject to $||\mathbf{x}|| = 1$, then the maximum of $f(\mathbf{x})$ is the largest eigenvalue λ_1 of A and it occurs when $\mathbf{x}$ is a unit eigenvector corresponding to λ_1 and the minimum is the smallest eigenvalue λ_n of A also occurring when $\mathbf{x}$ is an eigenvector corresponding to λ_n.

Example 12.3.6. *Find the largest rectangle that can be inscribed in the circle $x^2 + y^2 = 1$.*

Solution. We wish to maximize $f(x, y) = xy$ subject to $x^2 + y^2 = 1$. From (12.4),

$$xy = \begin{bmatrix} x \\ y \end{bmatrix}^T \begin{bmatrix} 0 & 1/2 \\ 1/2 & 0 \end{bmatrix} \begin{bmatrix} x \\ y \end{bmatrix}. \tag{12.19}$$

From Section 4.4.1, to find the eigenvalues of $A = \begin{bmatrix} 0 & 1/2 \\ 1/2 & 0 \end{bmatrix}$,

$$\det(A - \lambda I) = \det\left(\begin{bmatrix} -\lambda & 1/2 \\ 1/2 & -\lambda \end{bmatrix}\right) = \lambda^2 - 1/4 = 0, \tag{12.20}$$

thus A has eigenvalues $\lambda = \pm 1/2$. Hence by Theorem 12.3.5, max $f(x, y) = 1/2$.

Without showing the details, we mention that the eigenvector associated with the eigenvalue $\lambda = 1/2$ is $[\sqrt{2}/2, \sqrt{2}/2]^T$ which gives us that, subject to the constraint $x^2 + y^2 = 1$, max $f(x, y) = xy$ is $1/2$ which occurs when $x = \sqrt{2}/2$ and $x = \sqrt{2}/2$. ■

12.4 Keywords

quadratic form, positive or negative definite matrix, positive or negative semidefinite matrix.

12.5 Exercises

Exercise 12.1. *Classify each of the following matrices according to whether they are positive or negative definite or semidefinite or indefinite.*

a. $\begin{bmatrix} 1 & 0 & 0 \\ 0 & 2 & 0 \\ 0 & 0 & 3 \end{bmatrix}$

b. $\begin{bmatrix} 1 & 0 & 0 \\ 0 & -2 & 0 \\ 0 & 0 & 3 \end{bmatrix}$

c. $\begin{bmatrix} -1 & 0 & 0 \\ 0 & -2 & 0 \\ 0 & 0 & -3 \end{bmatrix}$

d. $\begin{bmatrix} 1 & 0 & 0 \\ 0 & 3 & 2 \\ 0 & 2 & 4 \end{bmatrix}$

e. $\begin{bmatrix} 1 & 0 & 2 \\ 0 & 3 & 0 \\ 2 & 0 & 4 \end{bmatrix}$

f. $\begin{bmatrix} 2 & 0 & 2 \\ 0 & -5 & 0 \\ 2 & 0 & -2 \end{bmatrix}$

g. $\begin{bmatrix} 3 & 1 & 2 \\ 1 & 5 & 3 \\ 2 & 3 & 7 \end{bmatrix}$

Exercise 12.2. *Write the quadratic form associated with each matrix in Exercise 12.1.*

Exercise 12.3. *Using the techniques from this chapter, find the extreme values and saddle points of the following functions:*

a. $f(x, y, z) = x^2 + y^2 + z^2 + 4xz$
b. $f(x, y) = x^8 + 16xy + y^4$
c. $f(x, y, z) = 4x^2 + 2y^2 + 2z^2 - 4xy - 2yz - 2z + 1$
d. $f(x, y, z) = x^2 + y^2e^x + z^4$

Exercise 12.4. *Prove the last four parts of Theorem 12.2.1.*

Exercise 12.5. *Given a collection of n data points* $\{(x_1, y_1), \ldots, (x_n, y_n)\}$*, the goal of* linear regression *is to find values m and b such that the line* $y = mx+b$ *best fits the data. One way to do determine the best fit of a line is to minimize how far the points are from a potential line by minimizing the square of the distance each data point is from the line; that is:*

$$\text{Minimize } S(m, b) = \sum_{i=1}^{n} (y_i - mx_i - b)^2. \tag{12.21}$$

Use the techniques of this chapter to show that the values of m and b that minimize $S(m, b)$ *are*

$$m = \frac{n\overline{x}\,\overline{y} - \sum\limits_{i=1}^{n} x_i y_i}{n(\overline{x})^2 - \sum\limits_{i=1}^{n} x_i^2}$$

and

$$b = \overline{y} - m\overline{x}$$

where $\overline{x} = \frac{x_1+\cdots+x_n}{n}$ *and* $\overline{y} = \frac{y_1+\cdots+y_n}{n}$.

13

Iterative Methods

13.1 Newton's Method for Optimization

We saw in Section 9.3 an iterative process that helped us approximate the zeros of a function. We can also mimic this idea to obtain approximations of extreme points of a function.

13.1.1 Single-Variable Newton's Method for Optimization

In the case of finding zeros of a function $f(x)$, Newton's method approximates $f(x)$ by a linear function (the tangent line at a nearby point) then finds the zero of this linear function. This time, the idea of Newton's Method for function optimization is to approximate a function $f(x)$ by its 2nd order Taylor polynomial $T_2(x)$ [see Section 9.6.1], then find a critical point of $T_2(x)$ and iterate with the hope of finding an approximation of a critical point of $f(x)$. That is, if $f(x)$ is twice differentiable, then near $x_0 = a$

$$f(x) \approx T_2(x) = f(a) + f'(a)(x-a) + \frac{f''(a)}{2}(x-a)^2 \tag{13.1}$$

and

$$0 = f'(x) \approx T_2{}'(x) = f'(a) + f''(a)(x-a) \tag{13.2}$$

Therefore

$$x \approx a - \frac{f'(a)}{f''(a)}. \tag{13.3}$$

So we have

Method 13.1.1 (Single-Variable Newton's Method for Optimization). *Let $f(x)$ be a twice differentiable function and x_0 a starting point. Then an approximation of a critical number of $f(x)$ is obtained by the iteration*

$$x_{n+1} = x_n - \frac{f'(x_n)}{f''(x_n)}.$$

Example 13.1.2. *Use* Newton's Method for Optimization *to find critical points of $f(x) = \frac{1}{7}x^7 - 3x^5 + 22x^3 - 80x$ with starting points $x_0 = 1, 2, 3$.*

DOI: 10.1201/9780367425517-13

TABLE 13.1
Obtaining Critical Numbers via Newton's Method in Example 13.1.2 (Truncated)

n	x_n	x_n	x_n
0	1	2	3
1	1.358974358974	2.333333333333	2.880341880341
2	1.411478349258	2.233620026571	2.835093055278
3	1.414205695493	2.236069384087	2.828557057505
4	1.414213562307	2.236067977500	2.828427175454
5	1.414213562373	2.236067977499	2.828427124746
6	1.414213562373	2.236067977499	2.828427124746
7	1.414213562373	2.236067977499	2.828427124746
$\vdots$	$\vdots$	$\vdots$	$\vdots$
	$\sqrt{2}$	$\sqrt{5}$	$\sqrt{8}$

Solution. Here we have

$$f'(x) = x^6 - 15x^4 + 66x^2 - 80 \tag{13.4}$$

and

$$f''(x) = 6x^5 - 60x^3 + 132x. \tag{13.5}$$

The iterative values for the starting points are given in Table 13.1. Some calculus and algebra will show that $f(x)$ does have as its positive critical points $x = \sqrt{2}, \sqrt{5}, \sqrt{8}$. The graph of the function in Example 13.1.2 appears in figure 13.1. ■

As in Newton's Method for finding zeros, this technique will fail if $f''(a) = 0$ or is too close to 0.

13.1.2 Multivariable Newton's Method for Optimization

The multivariable version of Newton's Method for Optimization is as follows:

Method 13.1.3 (Multivariable Newton's Method for Optimization)**.** *Let $f : \mathbb{R}^n \to \mathbb{R}$ be a twice differentiable function and $\mathbf{x}_0$ a starting point. Then an approximation of a critical number of $f(\mathbf{x}) = f(\langle x_1, \ldots, x_n \rangle)$ is obtained by the iteration*

$$\mathbf{x}_{k+1} = \mathbf{x}_k - \frac{\nabla f(\mathbf{x}_n)}{Hf(\mathbf{x}_n)} = \mathbf{x}_k - \nabla f(\mathbf{x}_n)[Hf(\mathbf{x}_n)]^{-1} \tag{13.6}$$

where ∇f is the gradient of f, $[Hf]^{-1}$ is the inverse[1] *of the Hessian of f, and the product is matrix multiplication.*

[1]Of course, we are abusing the notation by writing division involving a matrix; but this

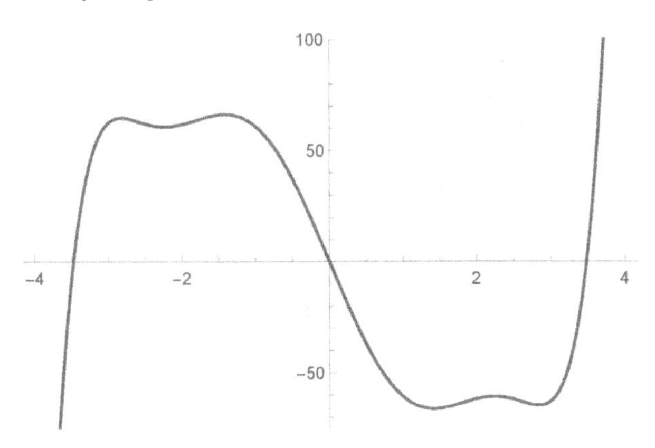

FIGURE 13.1
$f(x) = \frac{1}{7}x^7 - 3x^5 + 22x^3 - 80x$ in Example 13.1.2.

We illustrate the technique with a two-variable example.

Example 13.1.4. *Use Newton's Method to find a critical number of* $f(x, y) = x^2 + y^2 - 2x - 4y + 4$ *using* $\langle 0, 0 \rangle$ *as a starting point. A graph of* $f(x, y)$ *is given in figure 13.2.*

Solution. We have

$$\nabla f(x, y) = \langle 2x - 2, 2y - 4 \rangle \tag{13.7}$$

and

$$Hf(x, y) = \begin{bmatrix} 2 & 0 \\ 0 & 2 \end{bmatrix} \tag{13.8}$$

giving

$$[Hf(x, y)]^{-1} = \frac{1}{4} \begin{bmatrix} 2 & 0 \\ 0 & 2 \end{bmatrix} = \begin{bmatrix} \frac{1}{2} & 0 \\ 0 & \frac{1}{2} \end{bmatrix}. \tag{13.9}$$

Thus to iterate we use

$$\begin{aligned} \langle x_{n+1}, y_{n+1} \rangle &= \langle x_n, y_n \rangle - \langle 2x - 2, 2y - 4 \rangle \begin{bmatrix} \frac{1}{2} & 0 \\ 0 & \frac{1}{2} \end{bmatrix} && (13.10) \\ &= \langle x_n, y_n \rangle - \langle x_n - 1, y_n - 2 \rangle && (13.11) \\ &= \langle 1, 2 \rangle. \end{aligned}$$

So we obtain a solution without using values for x_n and y_n. And as $f(x, y) = x^2 + y^2 - 2x - 4y + 4 = (x - 1)^2 + (y - 2)^2$, it is easy to see from Calculus that $\langle 1, 2 \rangle$ is the location of the minimum value. ■

statement looks like the single variable version of Newton's Method for finding critical points making it easier for students to believe and remember.

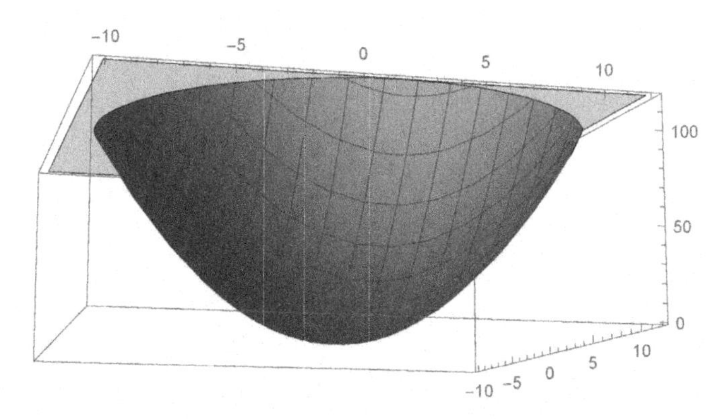

FIGURE 13.2
$f(x, y) = x^2 + y^2 - 2x - 4y + 4$ in Example 13.1.4.

Something interesting has happened in Example 13.1.4: the starting point was irrelevant and we have converged to the solution in one iteration. This illustrates

Theorem 13.1.5. *If $f(\mathbf{x})$ is linear or quadratic and the Hessian is invertible, then Newton's Method for Optimization will converge to a critical number after one iteration regardless of the starting point.*

Proof. If $f(\mathbf{x})$ is quadratic, then it can be written in the form

$$f(\mathbf{x}) = \frac{1}{2}\mathbf{x} \cdot A\mathbf{x} + \mathbf{b} \cdot \mathbf{x} + \mathbf{c} = \frac{1}{2}\mathbf{x}^T A\mathbf{x} + \mathbf{b} \cdot \mathbf{x} + \mathbf{c}. \tag{13.12}$$

Thus $\nabla f(\mathbf{x}) = A\mathbf{x} + \mathbf{b}$ and $Hf(\mathbf{x}) = A$. Suppose an extrema of $f(\mathbf{x})$ occurs at $\mathbf{x}_\star$, then $\mathbf{x}_\star$ is a critical point of $f(\mathbf{x})$ and by the higher dimensional version of Theorem 10.2.5,

$$\mathbf{0} = \nabla f(\mathbf{x}_\star) = A\mathbf{x}_\star + \mathbf{b} \tag{13.13}$$

and $A\mathbf{x}_\star = -\mathbf{b}$. Then for an arbitrary starting point $\mathbf{x}_0$ Newton's Method for Optimization gives

$$\begin{aligned}
\mathbf{x}_1 &= \mathbf{x}_0 - \nabla f(\mathbf{x}_0)[Hf(\mathbf{x}_0)]^{-1} && (13.14)\\
&= \mathbf{x}_0 - (A\mathbf{x}_0 + \mathbf{b})A^{-1} && (13.15)\\
&= \mathbf{x}_0 - (A\mathbf{x}_0 + -A\mathbf{x}_\star)A^{-1} && (13.16)\\
&= \mathbf{x}_0 - (\mathbf{x}_0 + -\mathbf{x}_\star) && (13.17)\\
&= \mathbf{x}_\star. && (13.18)
\end{aligned}$$

We obtain the proof for the case when $f(\mathbf{x})$ is linear by letting A be the zero matrix. □

We now illustrate Newton's Method for Optimization with a little more difficult example.

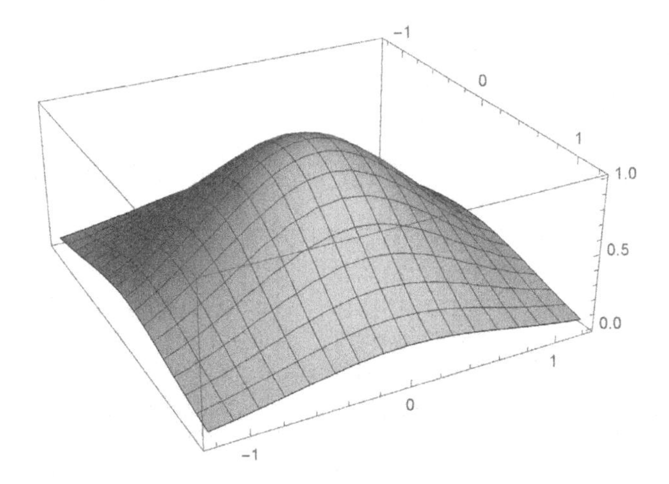

FIGURE 13.3
$f(x,y) = e^{-(x^2+y^2)}$ in Example 13.1.6.

Example 13.1.6. *Find a critical point of* $f(x,y) = e^{-(x^2+y^2)}$ *using Newton's Method for Optimization with* $(1,1)$ *as a starting point.*

Solution. Here we have

$$\nabla f(x,y) = \langle -2xe^{-(x^2+y^2)}, -2ye^{-(x^2+y^2)} \rangle \tag{13.19}$$

and

$$Hf(x,y) = \begin{bmatrix} (4x^2-2)e^{-(x^2+y^2)} & 4xye^{-(x^2+y^2)} \\ 4xye^{-(x^2+y^2)} & (4y^2-2)e^{-(x^2+y^2)} \end{bmatrix}. \tag{13.20}$$

So

$$\nabla f(1,1) = \langle -2e^{-2}, -2e^{-2} \rangle \tag{13.21}$$

and

$$[Hf(1,1)]^{-1} = \begin{bmatrix} 2e^{-2} & 4e^{-2} \\ 4e^{-2} & 2e^{-2} \end{bmatrix}^{-1} = \frac{-1}{12e^{-4}} \begin{bmatrix} 2e^{-2} & -4e^{-2} \\ -4e^{-2} & 2e^{-2} \end{bmatrix} = \begin{bmatrix} -\frac{e^2}{6} & \frac{e^2}{3} \\ \frac{e^2}{3} & -\frac{e^2}{6} \end{bmatrix}. \tag{13.22}$$

So

$$\mathbf{x}_1 = \langle 1,1 \rangle - \left\langle \frac{-2}{e^2}, \frac{-2}{e^2} \right\rangle \cdot \begin{bmatrix} \frac{-e^2}{6} & \frac{e^2}{3} \\ \frac{e^2}{3} & \frac{-e^2}{6} \end{bmatrix} = \langle 1,1 \rangle - \left\langle -\frac{1}{3}, -\frac{1}{3} \right\rangle = \left\langle \frac{4}{3}, \frac{4}{3} \right\rangle. \tag{13.23}$$

We have worked too hard here, as some algebra gives

$$(x_{k+1}, y_{k+1}) = (x_k, y_k) + \frac{1}{2x_k^2 + 2y_k^2 - 1}(x_k, y_k). \tag{13.24}$$

TABLE 13.2
Obtaining Critical Numbers via Newton's Method in Example 13.1.6 (Truncated)

Iteration n	Starting Point (x_0, y_0) $x_n = y_n$	Starting Point (x_0, y_0) $x_n = y_n$
0	1.000000000000	0.3333333333333
1	1.333333333333	−0.2666666666666
2	1.551515151515	0.1060041407867
3	1.731321808277	−0.0049888591355
4	1.888859336357	0.0000004967146
5	2.031187480116	0.0000000000000
6	2.162207402112	0.0000000000000
7	2.284362124161	0.0000000000000
8	2.399308754889	0.0000000000000
9	2.508235897244	0.0000000000000
10	2.612032173325	0.0000000000000
11	2.711383557644	0.0000000000000
12	2.806833289165	0.0000000000000
13	2.898820641006	0.0000000000000
14	2.987707043542	0.0000000000000
15	3.073794283751	0.0000000000000
16	3.157337545150	0.0000000000000
17	3.238554977370	0.0000000000000
18	3.317634865661	0.0000000000000
19	3.394741100051	0.0000000000000
20	3.470017414222	0.0000000000000
⋮	⋮	⋮

The iterations of this example with starting point $(1, 1)$ are given in the second column of Table 13.2. Note that it appears the values are diverging, so $(1, 1)$ was an unfortunate choice for a starting point as the curve is too flat at this point. The third column uses $(\frac{1}{3}, \frac{1}{3})$ as a starting point and with this we do converge to the actual critical point. ■

As seen in this example and as before with the original Newton's Method, we do have concerns with good and bad starting points, convergence, and rates of convergence. This are addressed in any good Numerical Analysis textbook.

Example 13.1.7. *Use Newton's Method for Optimization to find the minimum of $f(x, y) = x^2 + y^2 + 2xy$.*

Solution. By Theorem 13.1.5, we can hopes for this approach in that we may get our solution in one iteration regardless of the starting point. We have

$\nabla f(x, y) = \langle 2x + 2y, 2y + 2x \rangle$ and

$$Hf(x, y) = \begin{bmatrix} 2 & 2 \\ 2 & 2 \end{bmatrix}.$$

Unfortunately $\det(Hf(x, y)) = 0$, thus by the Fundamental Theorem of Invertible Matrices (Theorem 4.3.54), $Hf(x, y)$ is not invertible. This does not mean that there is no solution; what it means is that Newton's Method does not work and we have to try something else. ■

13.2 Steepest Descent (or Gradient Descent)

Another technique for minimizing a nonlinear function is the method of *Steepest Descent* which is also known as *Gradient Descent* (be careful: there is a technique for approximating integrals which is also called "Steepest Descent"). This method works for any function $f(\mathbf{x})$ that has continuous first partial derivatives in $\mathbb{R}^n$.

Algorithm 13.2.1 Gradient Descent for Nonlinear Programming.

Input: A nonlinear programming problem.

1: To minimize a nonlinear function $f(\mathbf{x})$, choose a starting point $\mathbf{x}_0$ from the domain of $f(\mathbf{x})$.
2: Form a sequence $\{\mathbf{x}_0, \mathbf{x}_1, \mathbf{x}_2, \mathbf{x}_3, \ldots\}$ using the recurrence
3: **loop**
4: $\quad \mathbf{x}_{n+1} = \mathbf{x}_n - \lambda_n \nabla f(\mathbf{x_n})$
5: $\quad$ where λ_n is the value of λ_n that minimizes $\phi_n(\lambda_n) = f(\mathbf{x_n} - \lambda_n \nabla f(\mathbf{x_n}))$ with $\lambda_n \geq 0$.
6: **end loop**

Output: Under the right conditions, the $\{\mathbf{x}_0, \mathbf{x}_1, \mathbf{x}_2, \mathbf{x}_3, \ldots\}$ converge to a point $\mathbf{x}_\star$ that is a minimizer of $f(\mathbf{x})$.

Note the λ_n is the **step size** for each iteration. The method has us walk downhill in the direction of the steepest descent until that direction takes us downhill no more (mathematically, we are minimizing $f(\mathbf{x})$ in the direction of the steepest descent). A nice way to consider the idea of steepest decent is to consider going for a hike in the foothills of Scotland. Let us say we park our car at the lowest point in the foothills and then begin our walk. After some time on our hike the fog rolls in and visibility is zero; that is, we cannot see anything at all. A good strategy would be to feel at our feet, find the biggest drop, then walk as far as we can in that direction. We then feel at our feet for the next biggest drop and repeat the process.

Example 13.2.1. *Use Steepest Descent to minimize the function in Example 13.1.4 using* $(0, 0)$ *as a starting point.*

Solution. $\mathbf{x}_0 = (0,0)$ and $\nabla f((x,y)) = (2x-2, 2y-4)$. Hence $\nabla f((0,0)) = (-2,-4)$ and we must minimize

$$\phi_0(\lambda_0) = f(2\lambda_0, 4\lambda_0) \tag{13.25}$$
$$= 4\lambda_0^2 + 16\lambda_0^2 - 4\lambda_0 - 16\lambda_0 + 4 \tag{13.26}$$
$$= 4 - 20\lambda_0 + 20\lambda_0^2. \tag{13.27}$$

To minimize $\phi_0(\lambda_0)$, note $\phi_0'(\lambda_0) = -20 + 40\lambda_0$ hence $\phi_0(\lambda_0)$ has $\lambda_0 = \frac{1}{2}$ as a critical point. The second derivative test verifies $\phi_0(\lambda_0)$ has a minimum at $\lambda_0 = \frac{1}{2}$ as $\phi_0''(\frac{1}{2}) = 40 > 0$. Thus

$$\mathbf{x}_1 = \mathbf{x}_0 - \frac{1}{2}\nabla f(\mathbf{x}_0) \tag{13.28}$$
$$= (0,0) - \frac{1}{2}(-2,-4) \tag{13.29}$$
$$= (1,2). \tag{13.30}$$

For the second iteration, $\mathbf{x}_1 = (1,2)$ hence $\nabla f((1,2)) = (0,0)$ and regardless of the step size, the process will not move from $(1,2)$ thus we have reached the minimizer in one iteration. ■

This technique is not without its concerns. We will consider them after the next example.

Example 13.2.2. *Compute the first three terms of the Steepest Descent sequence (i.e. perform three iterations) for*

$$f(x,y) = 4x^2 - 4xy + 2y^2$$

with initial point $(2,3)$.

Solution:

$\mathbf{x}_0 = (2,3)$ and $\nabla f((x,y)) = (8x-4y, -4x+4y)$. Hence $\nabla f((2,3)) = (4,4)$ and we must minimize

$$\phi_0(\lambda_0) = f(2-4\lambda_0, 3-4\lambda_0) \tag{13.31}$$
$$= (2-4\lambda_0)^2 - 4(2-4\lambda_0)(3-4\lambda_0) + 2(3-4\lambda_0)^2 \tag{13.32}$$
$$= 10 - 32\lambda_0 + 32\lambda_0^2. \tag{13.33}$$

To minimize $\phi_0(\lambda_0)$, note $\phi_0'(\lambda_0) = -32 + 64\lambda_0$ hence $\phi_0(\lambda_0)$ has $\lambda_0 = \frac{1}{2}$ as a critical point. The second derivative test verifies $\phi_0(\lambda_0)$ has a minimum at $\lambda_0 = \frac{1}{2}$ as $\phi_0''(\frac{1}{2}) = 64 > 0$. Thus

$$\mathbf{x}_1 = \mathbf{x}_0 - \frac{1}{2}\nabla f(\mathbf{x}_0) \tag{13.34}$$
$$= (2,3) - \frac{1}{2}(4,4) \tag{13.35}$$
$$= (0,1). \tag{13.36}$$

For the second iteration, $\mathbf{x_1} = (0,1)$ hence $\nabla f((0,1)) = (-4,4)$ and we must minimize

$$\begin{aligned}\phi_1(\lambda_1) &= f(4\lambda_1, 1-4\lambda_1) && (13.37)\\ &= 4(4\lambda_1)^2 - 4(4\lambda_1)(1-4\lambda_1) + 2(1-4\lambda_1)^2 && (13.38)\\ &= 2 - 32\lambda_1 + 160\lambda_1^2. && (13.39)\end{aligned}$$

To minimize $\phi_1(\lambda_1)$, note $\phi_1'(\lambda_1) = -32 + 320\lambda_1$ hence $\phi_1(\lambda_1)$ has $\lambda_1 = \frac{1}{10}$ as a critical point. The second derivative test verifies $\phi_1(\lambda_1)$ has a minimum at $\lambda_1 = \frac{1}{10}$ as $\phi_1''(\frac{1}{10}) = 320 > 0$. Thus

$$\begin{aligned}\mathbf{x_2} &= \mathbf{x_1} - \frac{1}{10}\nabla f(\mathbf{x_1}) && (13.40)\\ &= (0,1) - \frac{1}{10}(-4,4) && (13.41)\\ &= \left(\frac{2}{5}, \frac{3}{5}\right). && (13.42)\end{aligned}$$

For the third iteration, $\mathbf{x_2} = (\frac{2}{5}, \frac{3}{5})$ hence $\nabla f((\frac{2}{5}, \frac{3}{5})) = (\frac{4}{5}, \frac{4}{5})$ and we must minimize

$$\begin{aligned}\phi_2(\lambda_2) &= f\left(\frac{2}{5} - \frac{4}{5}\lambda_2, \frac{3}{5} - \frac{4}{5}\lambda_2\right) && (13.43)\\ &= 4\left(\frac{2}{5} - \frac{4}{5}\lambda_2\right)^2 - 4\left(\frac{2}{5} - \frac{4}{5}\lambda_2\right)\left(\frac{3}{5} - \frac{4}{5}\lambda_2\right) + 2\left(\frac{3}{5} - \frac{4}{5}\lambda_2\right)^2 && (13.44)\\ &= \frac{1}{25}(10 - 32\lambda_2 + 32\lambda_2^2). && (13.45)\end{aligned}$$

To minimize $\phi_2(\lambda_2)$, note $\phi_2'(\lambda_2) = -\frac{32}{25}\lambda_2 + \frac{64}{25}\lambda_2$ hence $\phi_2(\lambda_2)$ has $\lambda_2 = \frac{1}{2}$ as a critical point. The second derivative test verifies $\phi_2(\lambda_2)$ has a minimum at $\lambda_2 = \frac{1}{2}$ as $\phi_2''(\frac{1}{2}) = \frac{64}{25} > 0$. Thus

$$\begin{aligned}\mathbf{x_3} &= \mathbf{x_2} - \frac{1}{2}\nabla f(\mathbf{x_2}) && (13.46)\\ &= \left(\frac{2}{5}, \frac{3}{5}\right) - \frac{1}{2}\left(\frac{2}{5}, \frac{2}{5}\right) && (13.47)\\ &= \left(0, \frac{1}{5}\right). && (13.48)\end{aligned}$$

For the next iteration,

$$\phi_3(\lambda_3) = \frac{1}{25}(160\lambda_3 - 32\lambda_3 + 2\lambda_3^2) \qquad (13.49)$$

which has a minimum when $\lambda_3 = \frac{1}{10}$ giving $x_4 = \left(\frac{2}{25}, -\frac{3}{5}\right)$.

TABLE 13.3
Obtaining Critical Numbers via Steepest Descent in Example 13.2.2 (Truncated)

n	x_n	y_n	λ_n
0	2.000000	3.000000	0.500000
1	0.000000	1.000000	0.100000
2	0.400000	0.600000	0.500000
3	0.000000	0.200000	0.100000
4	0.080000	−0.600000	0.445205
⋮	⋮	⋮	

The iterations in Example 13.2.2 are given in Table 13.3.
NOTES:

1. It is not difficult to see that the minimum value of $f(x, y)$ occurs at $(0, 0)$ [we could verify this quickly using methods from Calc 3].
2. Using Solver in Microsoft's Excel gives a solution of $\min f = 2.1027 \times 10^{-13}$ which occurs at $(-0.0000002, -0.00000046)$ [If you use Solver to answer this, remember to use "GRG Nonlinear" and uncheck the "Make Unconstrained Variables Non-Negative" box]. The message included with Solver's solution is "Solver has converged to the current solution. All Constraints are satisfied. Solver has performed 5 iterations for which the objective did not move significantly. Try a smaller convergence setting, or a different starting point". Solver has performed a modified version of what we did above for 5 iterations, noticed that f was not changing much, then quit. We can have Solver do more iterations from this starting point by changing the convergence tolerance: in Solver, go to OPTIONS then GRG NONLINEAR then change CONVERGENCE to something smaller than the default 0.0001.
3. The previous note points out a disadvantage of *Steepest Descent*: if the function starts to level off, the method will converge very slowly.
4. It would be great fun to discuss the mathematics of what is going on here (why this works, that at each iteration we move orthogonally from the previous, convergence, etc.). If interested, please referene a good advanced Numerical Analysis text.

13.2.1 Generalized Reduced Gradient

Computationally, finding derivatives is expensive, yet alone that we need also find a zero. To make matters worse, we are minimizing a second function to find the minimum of the objective function. This is great in theory, but computationally insane, so we instead do something much more reasonable and fix the step size. This (and a few more tricks) are what is done in the *Generalized Reduced Gradient* algorithm used in most software including *Solver.*

13.3 Additional Geometric Programming Techniques

Additional techniques which we will not consider in this text include:

- Least Squares Optimization and Minimum Norms,
- Secant Methods, and
- penalty methods.

13.4 Exercises

Exercise 13.1. *Use Newton's Method for Optimization to find the extreme values of* $x^3 - x + 1$.

Exercise 13.2. *Do the first two iterations of Steepest Descent to minimize* $f(x, y) = 2x^2 + y^2 - 2xy$ *with a starting point of* $(2, 3)$.

Exercise 13.3. *Do the first two iterations of Steepest Descent to minimize* $g(x, y) = x^2 + y^2 - 2x - 2y - xy$ *with a starting point of* $(0, 0)$. *Comment on what happens here.*

Exercise 13.4. *Give an example for which the proof of Theorem 9.1.5 would not apply. (Hint: is* $[Hf(x_o)]^{-1}$ *always invertible?)*

Exercise 13.5. *Newton's Method for Optimization requires finding the inverse of the Hessian. Consider a function of two variables* $f(x, y)$. *In terms of the second partial derivatives (*f_{xx}, f_{yy}, *and* f_{xy} *and/or* f_{yx}*), under what conditions is the Hessian not invertible?*

Exercise 13.6. *Consider an arbitrary, simple graph* G *on* n *vertices with* m *edges (as we will see in Chapter 21, each edge is determined by two vertices and no more than one edge can exist between a given pair of vertices). Consider assigning each vertex the color* red *or* blue *and then once the vertices are colored, color the existing edges by the following scheme:*

- *if both of the endpoints of an edge are red, color the edge* red,
- *if both of the endpoints of an edge are blue, color the edge* blue, *and*
- *if one endpoint is red and the other blue, color the edge* purple.

QUESTION: *For an arbitrary graph, what vertex-coloring scheme will lead to the most purple edges? (If you can answer this, engineers, physicists, computer scientists, and discrete mathematicians will be very interested. This is the open* "Maximum Cut" *problem and is known to be* NP-Hard *and has many applications.)*

One approach to solving this is to introduce a decision variable x_i for each vertex v_i and put

$$x_i = \begin{cases} 0 & \textit{if } v_i \textit{ is to be colored red or} \\ 1 & \textit{if } v_i \textit{ is to be colored blue.} \end{cases} \tag{13.50}$$

We then answer the question by maximizing the m-term function

$$\textit{Maximize } f(x_1, \ldots, x_n) = \sum_{v_i v_j \in E(G)} (x_i - x_j)^2. \tag{13.51}$$

As this is quadratic, we should run to Newton's Method – which would converge to the critical point in one iteration regardless of the starting point.

Explain why Newton's Method fails for this important open problem.

14

Derivative-Free Methods

14.1 The Arithmetic Mean-Geometric Mean Inequality (AGM)

Our approach to solving some nonlinear optimization problems in this section will be through what is known as the *Arithmetic Mean-Geometric Mean Inequality.*

We can derive the inequality in the two variable case as follows:

$$0 \leq (\sqrt{x_1} - \sqrt{x_2})^2 = x_1 - 2\sqrt{x_1}\sqrt{x_2} + x_2 \tag{14.1}$$

and note that since we are taking square roots we necessarily have $x_1, x_2 \geq 0$ (as we will see shortly, the generalized AGM requires the variables to be positive). A little bit of algebra yields

$$\sqrt{x_1 x_2} \leq \frac{x_1 + x_2}{2}. \tag{14.2}$$

The left hand side of (14.2) is the ***geometric mean*** of x_1 and x_2 whereas the right hand side of (14.2) is the ***arithmetic mean*** of x_1 and x_2. A close inspection of (14.1) reveals that we have equality in the AGM exactly when $x_1 = x_2$ (this is the only way to get the right side to be 0 in this equation). It will turn out this fact will be incredibly useful when we use the AGM to solve optimization problems.

In general,

Theorem 14.1.1 (The Arithmetic Mean-Geometric Mean Inequality (AGM)). *If $x_1, \ldots, x_n$ are positive real numbers and if $\delta_1, \ldots, \delta_n$ are also positive real numbers such that $\delta_1 + \cdots + \delta_n = 1$, then*

$$\prod_{i=1}^{n} {x_i}^{\delta_i} \leq \sum_{i=1}^{n} \delta_i x_i$$

with equality[1] *if and only if $x_1 = x_2 = \cdots = x_n = \sum_{i=1}^{n} \delta_i x_i$.*

[1] As we will see, this is where the magic happens with using the AGM to solve optimization questions; we get equality, i.e. reach our bound, exactly when the things that we are averaging are the same. I tell my students that if they wanot to sound smart during a math talk, whenever there is an inequality they should ask: "Is the bound tight? If so, does anything magical happen when there is equality?"

DOI: 10.1201/9780367425517-14

This version of the AGM is not hard to prove but involves the idea of convexity and will be proven in Section 18.5.

We illustrate the method by considering the following problem:

Example 14.1.2. *A customer has requested an order of boxes from the Springfield Box Company. The order asks for boxes to have a volume of 27 cubic feet but has no specific requirement on the dimensions of the box. Vincenzo Natali is responsible for fulfilling this request, but would like some expert analysis on designing the boxes. In particular, Mr. Natali would like to know what dimensions his company should make the boxes so that they satisfy the customer's request for the desired volume but minimize the amount of cardboard used.*

We will assume the cardboard has a given thickness and we will put $x =$ the length, $y =$ the width, and $z =$ the height of the manufactured box. The problem is then modeled as

$$\text{Minimize Surface Area: } S(x, yz) = 2xy + 2xz + 2yz \text{ subject to } V = xyz = 27 \tag{14.3}$$

where clearly $x, y, z > 0$.

Solution 1: (Using partial derivatives.)
We first solve the constraint function for z and obtain

$$z = \frac{27}{xy}. \tag{14.4}$$

Substituting this into the objective function gives us the problem:

$$\text{Minimize } S(x, y) = 2xy + 2x \cdot \frac{27}{xy} + 2y \cdot \frac{27}{xy} = 2xy + 54y^{-1} + 54x^{-1}. \tag{14.5}$$

So

$$\frac{\partial S}{\partial x} = 2y - \frac{54}{x^2} \quad \text{and} \tag{14.6}$$

$$\frac{\partial S}{\partial y} = 2x - \frac{54}{y^2}. \tag{14.7}$$

Thus we have critical points when

$$2yx^2 - 54 = 0 \quad \text{and} \tag{14.8}$$

$$2xy^2 - 54 = 0. \tag{14.9}$$

Therefore the left-hand sides are the same which gives us $2yx^2 = 2xy^2$ and some algebra yields $x = y$. Substituting into equation (14.8)

$$2y^3 - 54 = 0 \tag{14.10}$$

showing

$$y = 3. \tag{14.11}$$

Hence $x = 3$ and from $V = 27$ we also see that $z = 3$. Thus the Springfield Box Company can satisfy the customer's request using the least amount of cardboard by making a cube with each side 3 feet.

Solution 2: (Using Lagrange multipliers.)
We have $\nabla S = \lambda \nabla V$,

$$\langle 2y + 2z, 2x + 2z, 2x + 2y \rangle = \lambda \langle yz, xz, xy \rangle, \tag{14.12}$$

thus

$$2y + 2z = \lambda yz \tag{14.13}$$
$$2x + 2z = \lambda xz \tag{14.14}$$
$$2x + 2y = \lambda xy \tag{14.15}$$
$$xyz = 27 \tag{14.16}$$

Thus

$$\lambda = \frac{2y + 2z}{yz} = \frac{2x + 2z}{xz} = \frac{2x + 2y}{xy} \tag{14.17}$$

so, by symmetry, $x = y = z$. Since $xyz = 27$, we obtain $x = y = z = 3$.

The previous two solution techniques illustrate an important concept when solving optimization questions. The method of LaGrange Multipliers is built to include the constraint in its approach to answering the question. As calculus techniques are not engineered this way, we must incorporate the constraint(s) in some way.

Highlight 14.1.3. *If a geometric programming solution technique does not consider any stated constraints (as in using derivative tests or Newton's Method), the constraints are to be incorporated into the problem by using them to eliminate a variable in the objective function before the technique is applied.*

Solution 3: (Using the AGM inequality.)

$$
\begin{aligned}
SA = S(x, y, z) &= 2xy + 2xz + 2yz && (14.18)\\
&= 2 \cdot 3 \left(\frac{xy + xz + yz}{3} \right) && (14.19)\\
&= 6 \left(\frac{xy}{3} + \frac{xz}{3} + \frac{yz}{3} \right) && (14.20)\\
\text{(by the AGM)} \quad &\geq 6(xy)^{\frac{1}{3}} \cdot (xz)^{\frac{1}{3}} \cdot (yz)^{\frac{1}{3}} && (14.21)\\
&= 6 \cdot (x^2 y^2 z^2)^{\frac{1}{3}} && (14.22)\\
&= 6 \cdot V^{\frac{2}{3}} = 54. && (14.23)
\end{aligned}
$$

So the surface area will be the smallest when we have equality in (14.21). Also by the AGM, this minimum occurs when the things we are averaging are the same, namely

$$xy = xz = yz.$$

This gives us the system of equations

$$xy = xz \tag{14.24}$$
$$xy = yz \tag{14.25}$$
$$xz = yz \tag{14.26}$$

which easily yields $x = y = z$ and using the volume constraint we see that we minimize the surface area when $x = y = z = 3$ feet.

The AGM solution to the situation faced by the Springfield Box Company is quite beautiful but the beauty does not end there. As is often the case in Optimization, one minimum is another's maximum and the problem has a dual:

Example 14.1.4 (The Dual of Example 14.1.2). *Maximize* $V(x, y, z) = xyz$ *subject to* $SA = 2xy + 2xz + 2yz = 54$ *where* $x, y, z > 0$.

Solution.

$$\begin{aligned} V = xyz &= (x^2)^{\frac{1}{2}}(y^2)^{\frac{1}{2}}(z^2)^{\frac{1}{2}} && (14.27)\\ &= (xy \cdot xz \cdot yz)^{\frac{1}{2}} && (14.28)\\ &= [(xy)^{\frac{1}{3}} \cdot (xz)^{\frac{1}{3}} \cdot (yz)^{\frac{1}{3}}]^{\frac{3}{2}} && (14.29)\\ \text{(AGM)} &\leq \left[\frac{xy}{3} + \frac{xz}{3} + \frac{yz}{3}\right]^{\frac{3}{2}} && (14.30)\\ &= \left[\frac{1}{6}(2xy + 2xz + 2yz)\right]^{\frac{3}{2}} && (14.31)\\ &= \left[\frac{1}{6}(54)\right]^{\frac{3}{2}} = 27. && (14.32) \end{aligned}$$

So the volume is no bigger than 27 and (again) by the AGM we reach equality exactly when the things we are averaging are the same; that is $xy = xz = yz$ giving $x = y = z = 3$. ■

Example 14.1.2 was not so bad, but we had to be a little clever to get the AGM to work for us in its dual (Example 14.1.4). Let us do another example where the weights δ_i to use the AGM require a bit of thought.

Example 14.1.5. *Maximize* $f(x, y, z) = xy^2z$ *subject to* $x + y + z^2 = 7$ *with* $x, y, z > 0$.

Solution. Since there are three terms, let us try averaging those and see what happens:

$$
\begin{aligned}
7 &= x + y + z^2 && (14.33)\\
&= 3\left(\frac{x}{3} + \frac{y}{3} + \frac{z^2}{3}\right) && (14.34)\\
&\geq 3(x^{\frac{1}{3}}y^{\frac{1}{3}}z^{\frac{2}{3}}) \text{ (by the AGM)} && (14.35)\\
&= 3(xyz^2)^{\frac{1}{3}}. && (14.36)
\end{aligned}
$$

We do not have a constant times $f(xyz)$ to a power, so this approach has not helped us. Note that we have one too many factors of z and not enough factors of y, so we will try the following:

$$
\begin{aligned}
7 &= x + y + z^2 && (14.37)\\
&= \frac{x}{2} + \frac{x}{2} + \frac{y}{4} + \frac{y}{4} + \frac{y}{4} + \frac{y}{4} + z^2 && (14.38)\\
&= 7\left(\frac{1}{7}\cdot\frac{x}{2} + \frac{1}{7}\cdot\frac{x}{2} + \frac{1}{7}\cdot\frac{y}{4} + \frac{1}{7}\cdot\frac{y}{4} + \frac{1}{7}\cdot\frac{y}{4} + \frac{1}{7}\cdot\frac{y}{4} + \frac{1}{7}\cdot z^2\right) && (14.39)\\
&\geq 7\left(\frac{x}{2}\right)^{\frac{1}{7}}\left(\frac{x}{2}\right)^{\frac{1}{7}}\left(\frac{y}{4}\right)^{\frac{1}{7}}\left(\frac{y}{4}\right)^{\frac{1}{7}}\left(\frac{y}{4}\right)^{\frac{1}{7}}\left(\frac{y}{4}\right)^{\frac{1}{7}}(z^2)^{\frac{1}{7}} \text{ (by the AGM)} && (14.40)\\
&= 7\left(\frac{x^2y^4z^2}{2^{10}}\right)^{\frac{1}{7}} = 7\left(xy^2z\right)^{\frac{2}{7}}\left(2^{-\frac{10}{7}}\right). && (14.41)
\end{aligned}
$$

Thus

$$2^{\frac{10}{7}} \geq [f(x, y, z)]^{\frac{2}{7}} \tag{14.42}$$

giving

$$f(x, y, z) \leq 2^5 = 32. \tag{14.43}$$

Thus $Max\ f(x, y, z) = 32$ which occurs when

$$\frac{x}{2} = \frac{y}{4} = z^2 \tag{14.44}$$

giving $x = 2z^2$ and $y = 4z^2$. Substituting these into the constraint tells us this max occurs when $z = 1$, $x = 2$, and $y = 4$. ■

Example 14.1.5 illustrates that using the AGM is not always straight forward. However, the following algorithm determines the weights for us.

14.2 Weight Finding Algorithm for the AGM

We may be confronted with trying to optimize a multivariable function of the form

$$\begin{aligned} \text{Minimize } g(\mathbf{x}) &= c_1 x_1^{\alpha_{11}} x_2^{\alpha_{12}} \cdots x_m^{\alpha_{1m}} + \cdots + c_n x_1^{\alpha_{n1}} x_2^{\alpha_{n2}} \cdots x_m^{\alpha_{nm}} \\ &= \sum_{i=1}^{n} \left(c_i \prod_{j=1}^{m} x_j^{\alpha_{ij}} \right) \end{aligned} \tag{14.45}$$

where $c_i > 0$, α_{ij} are real numbers, and $x_1, \ldots, x_m > 0$. Such a $g(\mathbf{x})$ is not necessarily a polynomial as the exponents can be noninteger, but this function resembles a polynomial and is called a *posynomial*. We refer to this problem as a ***Geometric Programming Problem*** **(GP)**, and it is a perfect candidate for the AGM. As we have seen in Example 14.1.5, figuring the weights δ_i to use the AGM can be tricky, so let us develop a technique which will determine each δ_i.

Let us introduce weights δ_i into 14.45 which we can rewrite as

$$\text{Minimize } g(\mathbf{x}) = \sum_{i=1}^{n} \delta_i \left(\frac{c_i \prod_{j=1}^{m} x_j^{\alpha_{ij}}}{\delta_i} \right) \tag{14.46}$$

$$\text{where } \delta_i > 0 \qquad \text{(the \textbf{Positivity Constraints})} \tag{14.47}$$

$$\text{and } \sum_{i=1}^{n} \delta_i = 1. \quad \text{(the \textbf{Normality Condition})} \tag{14.48}$$

This is a perfect situation to use the AGM inequality, so

$$g(x) = \sum_{i=1}^{n} \delta_i \left(\frac{c_i \prod_{j=1}^{m} x_j^{\alpha_{ij}}}{\delta_i} \right) \tag{14.49}$$

$$\text{(AGM)} \geq \prod_{i=1}^{n} \left(\frac{c_i \prod_{j=1}^{m} x_j^{\alpha_{ij}}}{\delta_i} \right)^{\delta_i} \tag{14.50}$$

$$= \prod_{i=1}^{n} \left(\frac{c_i}{\delta_i} \right)^{\delta_i} \cdot \prod_{i=1}^{n} \left(\prod_{j=1}^{m} x_j^{\alpha_{ij}\delta_i} \right) \tag{14.51}$$

$$= \prod_{i=1}^{n} \left(\frac{c_i}{\delta_i} \right)^{\delta_i} \cdot \prod_{j=1}^{m} x_j^{\sum_{i=1}^{n} \alpha_{ij}\delta_i} \tag{14.52}$$

$$= \prod_{i=1}^{n} \left(\frac{c_i}{\delta_i} \right)^{\delta_i} \tag{14.53}$$

where we have equality in line (14.53) for each fixed j if we put

$$\sum_{i=1}^{n} \alpha_{ij}\delta_i = 0 \text{ for } j = 1, 2, \ldots, m. \quad (\text{the } \textbf{Orthogonality Condition}) \tag{14.54}$$

As well, the product in line (14.53) is so important that we name it

$$\prod_{i=1}^{n} \left(\frac{c_i}{\delta_i}\right)^{\delta_i} = v(\boldsymbol{\delta}).$$

Since we are using the AGM the posynomial $g(\mathbf{x})$ is at a minimum when the ***Dual Geometric Programming Problem*** **(DGP)** is at a maximum, that is

$$\text{Maximize: } v(\boldsymbol{\delta}) = \prod_{i=1}^{n} \left(\frac{c_i}{\delta_i}\right)^{\delta_i} \tag{14.55}$$

$$\text{Subject to: } \delta_i > 0, \qquad (\text{the } \textbf{Positivity Constraint}), \tag{14.56}$$

$$\sum_{i=1}^{n} \delta_i = 1, \quad (\text{the } \textbf{Normality Condition}), \text{ and} \tag{14.57}$$

$$\sum_{i=1}^{n} \alpha_{ij}\delta_i = 0 \quad (\text{the } \textbf{Orthogonality Condition}). \tag{14.58}$$

where $j = 1, \ldots, m$. Note that, in words, α_{ij} is the exponent of variable x_j in the ith term of the objective posynomial.

Thus we have developed the **W***eight Finding Algorithm for the* **AG***M (WAG)*[2]:

Method 14.2.1 (WAG). *If we wish to minimize the posynomial*

$$\textit{Minimize } g(x) = c_1 x_1^{\alpha_{11}} x_2^{\alpha_{12}} \cdots x_m^{\alpha_{1m}} + \cdots + c_n x_1^{\alpha_{n1}} x_2^{\alpha_{n2}} \cdots x_m^{\alpha_{nm}}$$
$$= \sum_{i=1}^{n} \left(c_i \prod_{j=1}^{m} x_j^{\alpha_{ij}} \right) \tag{14.59}$$

where $c_i > 0$, α_{ij} are real numbers, and $x_1, \ldots, x_m > 0$, we proceed as follows:
Step 0:
Form the ***Dual Geometric Programming Problem (DGP)****, namely*

$$\max_{\delta} \{v(\boldsymbol{\delta})\} = \max_{\delta} \{v(\delta_1, \ldots, \delta_2)\} = \max_{\delta} \left\{ \prod_{i=1}^{n} \left(\frac{c_i}{\delta_i}\right)^{\delta_i} \right\}$$

subject to the constraints stated in the next step.

---

[2]In [43], this is called the *Geometric Programming Procedure*. As there are many kinds of geometric programming problems and this technique only applies to a certain class (namely, ones suitable for applying the AGM), we have opted for a different name.

Step 1:
Compute the set F of feasible vectors $\boldsymbol{\delta}$ in $\mathbb{R}^n$ for $v(\boldsymbol{\delta})$ such that

$$\delta_i > 0, \qquad \textit{(the \textbf{Positivity Constraint})}, \tag{14.60}$$

$$\sum_{i=1}^{n} \delta_i = 1, \quad \textit{(the \textbf{Normality Condition}), and} \tag{14.61}$$

$$\textit{for each fixed } j \ \sum_{i=1}^{n} \alpha_{ij}\delta_i = 0 \quad \textit{(the \textbf{Orthogonality Condition})}. \tag{14.62}$$

where $j = 1, \ldots, m$. Note that in this situation ***the j's range over the number of variables*** *(i.e. over $1, 2, \ldots, m$) and* ***the i's range over the number of terms in the posynomial*** *(i.e. over $1, 2, \ldots, n$)*

We do this by first solving the linear equations in the **Normality** *and* **Orthogonality** *conditions then impose the* **Positivity** *condition on the resulting solution.*

Step 2:
If the set F of feasible vectors from step 1

1. *is empty, then STOP. The Geometric Programming problem (GP) has no solution.*
2. *consists of a single vector, call it $\boldsymbol{\delta}^*$, then $\boldsymbol{\delta}^*$ is a solution to the Dual Geometric Programming problem (DGP) and we proceed to step 4.*
3. *has more than one vector, go to step 3.*

Step 3:
Of all the vectors in F, find the one that maximizes the dual function; that is,

$$\max_{\delta} \{v(\boldsymbol{\delta})\} = \max_{\delta} \{v(\delta_1, \ldots, \delta_n)\} = \max_{\delta} \left\{ \prod_{i=1}^{n} \left(\frac{c_i}{\delta_i} \right)^{\delta_i} \right\}. \tag{14.63}$$

Now go to step 4.

Step 4:
Given a solution $\boldsymbol{\delta}^*$ of the DGP, by the "if and only if" of the AGM, the solution $\mathbf{x}^* = \langle x_1^*, x_2^*, \ldots, x_m^* \rangle$ of the initial GP problem is found by solving

$$\begin{aligned}
\textit{(The first term in the posynomial)} \quad & u_1(\mathbf{x}^*) = c_1 x_1^{\alpha_{11}} \cdots x_m^{\alpha_{1m}} = \delta_1 \cdot v(\boldsymbol{\delta}) \\
\textit{(The second term in the posynomial)} \quad & u_2(\mathbf{x}^*) = c_2 x_1^{\alpha_{21}} \cdots x_m^{\alpha_{2m}} = \delta_2 \cdot v(\boldsymbol{\delta}) \\
& \vdots \\
\textit{(The } n^{th} \textit{ term in the posynomial)} \quad & u_n(\mathbf{x}^*) = c_n x_1^{\alpha_{n1}} \cdots x_m^{\alpha_{nm}} = \delta_n \cdot v(\boldsymbol{\delta})
\end{aligned}$$

This system of equations can be made into a system of linear equations by taking logarithms.

14.3 The AGM, Newton's Method, and Reduced Gradient Compared

Using Newton's Method for Optimization involves derivatives and (in the multivariable case) the inverse of a matrix. Though not difficult for humans, this can be computationally expensive and, as such, Newton's Method is not a preferred choice for a go-to technique to have a computer solve a geometric programming question. As such, the (Generalized) Reduced Gradient technique is preferred, but it is not without its concerns.

We illustrate this with an example.

Example 14.3.1. *Minimize* $f(x, y, z) = x^2 + y^2 + z$ *subject to* $\frac{z}{xy} = 2$ *and* $x, y, z > 0$.

Solution. (via Newton's Method) As the function is quadratic, by Theorem 13.1.5 we can be excited that this may be the best approach since (if the method works) we will have convergence to the extreme value in one iteration regardless of our starting point. Using the constraint to eliminate a variable, the problem becomes

$$\text{Minimize } f(x, y) = x^2 + y^2 + 2xy \text{ subject to } x, y > 0. \tag{14.64}$$

Thus

$$\nabla f(x, y) = \langle 2x + 2y, 2y + 2x \rangle \tag{14.65}$$

and

$$Hf(x, y) = \begin{bmatrix} 2 & 2 \\ 2 & 2 \end{bmatrix}. \tag{14.66}$$

Since $\det([Hf(x, y)]) = 0$, the Hessian is not invertible and Newton's Method fails. ■

Let $f(x, y) = 3x^2 + y^2 - 2xy - 8x + 3y + 20$ and consider finding its minimum.

14.4 Exercises

Exercise 14.1. *Use the AGM to minimize* $f(x) = x^2 + \frac{1}{x^2}$ *where* $x > 0$.

Exercise 14.2. *A farmer intends to build a 60,000 square foot rectangular livestock pen. One side of the pen is over terrain that needs cleared which has a cost of $10 per linear foot. The cost for the remaining three sides of fence is $5 per linear foot. Use the AGM to find the dimensions of the pen that give the desired area at a minimum cost.*

Exercise 14.3. *For $a, b, c > 0$, use the AGM to prove $(a+b)(b+c)(a+c) \geq 8abc$.*

Exercise 14.4. *a) The* harmonic mean *of two positive numbers a and b is defined to be $\frac{2ab}{a+b}$. Show that the harmonic mean of two positive numbers a and b is less than or equal to the geometric mean with equality if and only if $a = b$.*

b) The root mean squared *of two positive numbers a, b is defined to be $\sqrt{\frac{a^2+b^2}{2}}$ (this is also referred to as the* quadratic mean *of a and b). Show that the root mean squared of two positive numbers a and b is greater than or equal to the arithmetic mean with equality if and only if $a = b$.*

c) Write a summary statement with necessary conditions involving all four of the means mentioned in this problem.

Exercise 14.5. *Consider minimizing $P = \frac{2}{xy} + xy + x + y$ where x and y are positive real numbers.*

i) Solve this on a computer using Steepest Descent (Generalized Reduced Gradient) starting at $(10, 10)$.

ii) Repeat part i) using various starting points. Compare your different answers.

iii) Solve this using Evolutionary Programming by first bounding x and y above by 10^8.

iv) Repeat solving this by Evolutionary Programming but now bound the variables above by 10^6, then 10^4, and then 10^2. Compare the answers and runtime when using the different upper bounds.

v) Find $\min P$ and the values of the decision variables that give this P by using the AGM.

Exercise 14.6. *Consider minimizing $P = 5x^2 + \frac{x}{y^2} + \frac{2y^3}{x^2}$ where x and y are positive real numbers. Attempt to solve this via Newton's Method and Steepest Descent using $(1, 1)$ and other values as starting points. Also attempt to solve this using the AGM. Explain why all three techniques fail.*

15

Search Algorithms

As mentioned in Section 2.4, not all optimization problems can be solved with a deterministic algorithm guaranteed to find the solution. For example, we may not be able to use the gradient of the objective function to lead us to the optimal value because the gradient is too expensive to calculate or because it is not defined. All is not lost, however; we can use a heuristic to guide our search in a way that makes finding the optimal value more likely than just trying points at random, while offering no absolute guarantees of success.

15.1 Evolutionary Algorithms

Suppose we are hoping to maximize $f(\mathbf{x}) = x_1^3 + x_2x_3 - x_2x_4^2 + x_3x_5 + \sqrt{x_4x_5}$ where for each i, $0 \leq x_i \leq 10$. We can approach this problem using techniques from the previous chapter, but let us attempt a different approach meant to mimic evolution (for this and other biologically inspired algorithms, see [6]). At each step in the process, we will consider a *population* of possible solutions and the *fitness* of each, in this case the value of f. We will then create a new population by adding *variation* to the solutions and applying *selection*.

We will begin with a population of 10 randomly selected data points in the feasible region and consider the value of f (the fitness) at each point. In evolutionary terms, each point represents an individual with a single *chromosome* encoding five *genes*, one per component of $\mathbf{x}$.

Data generated using a 10-sided die.

To introduce variation, we will randomly *crossover* 5 pairs of chromosomes by swapping genes between them (one from each column; displayed in boxes) and *mutate* another randomly chosen 7 genes (in light gray *italics*) by replacing them with new random values.

If we imagine applying these operations repeatedly, we can see that repeated crossovers alone would ultimately create an exhaustive (but not systematic) search over the space defined by the original set of genes. Repeated mutations of this sort will generate new points, but those points are no more likely to be the optimal value than our original population. And if we need find the optimal value or a point close to it, it is no more likely to persist to subsequent generations than any other point. Thus, just introduction

DOI: 10.1201/9780367425517-15

TABLE 15.1
Initial Population to Begin Evolutionary Programming

Chromosome	x_1	x_2	x_3	x_4	x_5	Fitness
1	7.726	5.090	0.965	1.115	2.160	463.4
2	9.763	3.676	0.829	4.306	5.867	875.4
3	2.502	1.601	6.251	6.600	5.757	−1.918
4	8.720	2.567	1.637	0.961	7.274	679.4
5	0.416	6.550	3.427	0.146	9.466	55.99
6	8.540	6.620	9.397	0.711	3.514	716.3
7	5.596	9.254	2.785	2.230	9.051	184.7
8	0.682	7.355	2.418	6.491	2.021	−283.3
9	0.194	4.775	6.506	9.317	2.334	−363.6
10	6.863	8.177	1.189	6.474	1.313	−5.268

TABLE 15.2
Crossover and Mutation in the Process of Evolutionary Programming

Chromosome	x_1	x_2	x_3	x_4	x_5	Fitness
1	7.726	5.090	0.965	1.115	2.160	463.4
2	9.763	3.676	6.251	4.306	5.867	927.1
3	2.502	2.567	0.829	6.600	*2.164*	−88.45
4	8.720	1.601	1.637	*0.866*	9.466	682.8
5	*0.237*	*10.00*	3.427	0.711	7.274	56.43
6	8.540	6.620	9.397	0.146	3.514	718.6
7	5.596	9.254	2.785	2.230	9.051	184.7
8	0.194	7.355	2.418	6.491	*0.698*	−288.3
9	0.682	4.775	6.506	9.317	2.334	−363.3
10	*9.505*	*0.155*	1.189	6.474	1.313	856.9

variation alone is no different than a random search; to do better than random search we also need to apply selection.

To apply selection, we will create a new population by choosing pairs of individuals from the existing population in Table 15.3 and applying crossover and mutation to generate two individuals for the new population. We will make these choices with replacement, so that a given individual can be chosen more than once. And we will make weighted choices so that the individuals with higher fitness have a higher chance of being selected. This weighting adds bias to our random search so that we tend to look in the neighborhood of the higher fitness values we have already considered. We'll define the probability p that the individual i with rank R_i is chosen to be $p_i = (\max_i R_i - R_i + 1)/\sum_i R_i$.

TABLE 15.3
Population with Rank and Selection Probability in Evolutionary Programming

Chromosome	x_1	x_2	x_3	x_4	x_5	Fitness	Rank	p
1	7.726	5.090	0.965	1.115	2.160	463.4	5	0.109
2	9.763	3.676	6.251	4.306	5.867	927.1	1	0.182
3	2.502	2.567	0.829	6.600	2.164	−88.45	8	0.055
4	8.720	1.601	1.637	0.866	9.466	682.8	4	0.127
5	0.237	10.00	3.427	0.711	7.274	56.43	7	0.073
6	8.540	6.620	9.397	0.146	3.514	718.6	3	0.145
7	5.596	9.254	2.785	2.230	9.051	184.7	6	0.091
8	0.194	7.355	2.418	6.491	0.698	−288.3	9	0.364
9	0.682	4.775	6.506	9.317	2.334	−363.3	10	0.018
10	9.505	0.155	1.189	6.474	1.313	856.9	2	0.164

TABLE 15.4
Population After One Round of Selection in Evolutionary Programming

Chromosome	x_1	x_2	x_3	x_4	x_5	Fitness	Rank	p
1	5.596	9.254	2.785	0.866	9.051	222.1	8	0.055
2	0.237	1.601	3.427	0.711	7.274	31.89	10	0.018
3	9.505	0.155	1.189	6.474	1.313	856.9	4.5	0.118
4	9.763	3.676	6.251	4.306	5.867	927.1	1.5	0.173
5	9.763	3.676	6.251	4.306	5.867	927.1	1.5	0.173
6	8.540	6.620	1.189	0.146	3.514	635.5	6	0.091
7	5.596	1.722	2.785	2.230	9.051	201.2	9	0.036
8	9.505	0.155	1.189	6.474	3.514	861.4	3	0.145
9	7.726	6.620	0.965	1.115	2.160	463.0	7	0.073
10	9.505	0.155	1.189	6.474	1.313	856.9	4.5	0.118

The mean fitness of the population in Table 15.3 is 315, while the mean fitness of the population in Table 15.4 is 598, so we've already achieved a substantial improvement. If we continue to repeat this process, over time the population will tend to converge around the optimal value – with no guarantees about how long that will take, however.

Variation and selection are generic concepts, and we have multiple options for how to implement them. For example, mutation does not have to involve replacement with another random draw from the domain. We could instead choose a δ and randomly add or subtract that amount to the gene chosen for mutation. This keeps the search more local to the current points, which can be

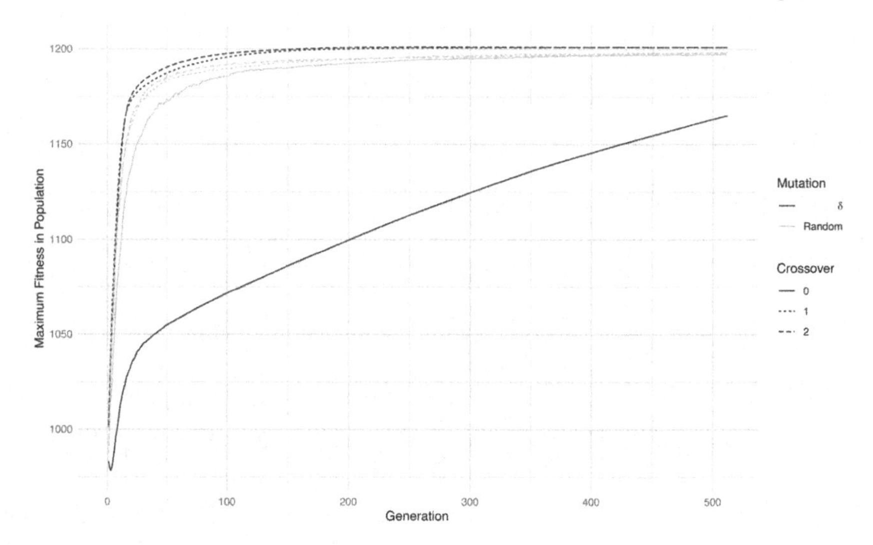

FIGURE 15.1
Maximum fitness over time by crossover and mutation strategy.

advantageous for points close to the optimal value but can slow down search if the population is far from it. δ also represents a step size, which can limit the precision of the final result; shrinking δ over time can allow for increased precision without unnecessarily slowing down the initial search.

The crossover operation can also be modified. One can perform an evolutionary search without it, one can use a single crossover point as illustrated above or one can use multiple crossover points. On this particular task, the best combination in terms of fastest convergence is to use δ mutations with a single crossover. δ mutation alone performs slowest because it can only take small steps; adding crossover introduces a means for taking big steps as well. There is little difference between one or multiple crossover points. Random mutations with one crossover seems to perform slightly worse and cannot quite reach the absolute maximum, possibly because it can only take big steps. Figure 15.1 shows results averaged over 64 iterations.

The rates of all these processes can be important as well. If little variation is introduced with each generation, resources can be wasted checking the same points repeatedly. Introducing a lot of variation in each generation pushes this approach closer to an equivalence with random search, since any signal selected from the previous generation will likely get replaced with a new variation. Some experimentation may be needed to identify appropriate rates for mutation and crossover.

Different selection strategies are possible as well. We assigned p based on fitness rank. We could also have chosen to define p proportional to the fitness (with appropriate shifting so that there are no negative probabilities). Given

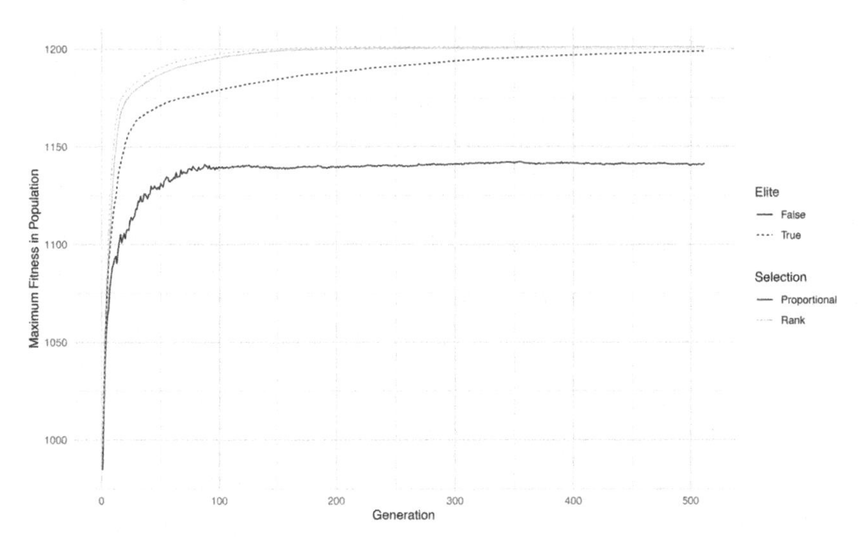

FIGURE 15.2
Maximum fitness over time by selection strategy.

the wide range of starting fitness values, this would strongly bias the next generation around the highest fitness individual, which may or may not be advantageous depending on how close it is to the optimal value. In general, one wants to maintain more population diversity early on, which in this case is better achieved with rank selection.

Another consideration in selection is whether to completely replace the previous population with each generation or to maintain one or more elite individuals with the highest fitness from the previous generation. Preserving elite individuals prevents the population from drifting away from a good solution even if it is optimal, at the cost of possibly biasing the population towards a local optimum that was found first. Figure 15.2 compares results for different selection strategies averaged over 64 iterations, all using the δ mutation and one crossover approach.

Now, for this particular problem where the objective function is differentiable and possesses a single maximum on the feasible region, a gradient ascent method will provide a guaranteed result for much less computation than an evolutionary algorithm, regardless of the variation or selection strategies employed. Indeed, the observant student may note that the function is monotonic increasing in x_1, x_3, and x_5 so the optimal $\mathbf{x}$ will have the value of 10 for each of those components, greatly simplifying the problem and narrowing the search space before the evolutionary algorithm has crawled out of the primordial soup.

But consider the reverse problem: Given a set of N pairs of the form $(\mathbf{x}, y)$, find the polynomial g that minimizes the mean squared error

$\frac{1}{N}\sum_j [y_j - g(\mathbf{x_j})]^2$. Can we define a gradient over all g and readily compute it? If there is no obvious answer, a search like an evolutionary algorithm may be more efficient in terms of overall time spent than working out the appropriate gradient and then applying gradient descent.

We need to choose a representation for possible solutions that is amenable to the variation and selection of an evolutionary algorithm. There are multiple options. For example, we could just treat each polynomial as a character string like `x1^3+x2x3` and define a mutation as a swap of one character for another. But then we have the potential to generate invalid expressions like `xx9++` or `^^^^^^`. We can potentially waste a lot of time searching in unfruitful regions of the space. Another possible representation is to use a tree structure; this can make for a space with more valid candidate solutions but the implementation can get complex. For simplicity, let's use a basic grammatical approach ([49]) that imposes some structure on the candidate solutions such that they are always valid while keeping the details straightforward.

We'll say that a polynomial is made up of terms, with each term having three parts: an operator, a vector component, and an exponent. Examples of terms include $/x_2^3$, $-x_4^{1/3}$, and $*x_1^1$. We'll further specify that the first term always starts with the $+$ operator. For the initial problem, we restricted the x_i components within the range $[0, 10]$ because the search over all $\mathbb{R}$ would never terminate. Similarly, to keep our search more bounded we'll use a finite set of possible exponents: 1/3, 1/2, 1, 2, and 3. For operators, we'll choose $+$, $-$, $*$, $/$, and %.

Our chromosomes then will be a sequence of the form vector component–exponent–operator–vector component–exponent–... . We can represent these as symbols directly or assign them numbers that index into the relevant set. A mutation will replace an element with one randomly chosen from the appropriate set; an operator will always replace an operator and an exponent will always replace a an exponent. Crossover will the same as before, swapping elements of the same type from the same position. To evaluate fitness, we'll convert each sequence in to an expression, evaluate it on each $\mathbf{x_j}$, and compute the mean squared error. Table 15.5 shows a possible starting population along with fitness (mean squared error) an a sample set of (x, y) pairs.

Given the orders of magnitude difference in fitness values and the possibility for undefined results from exceeding the maximum value a given computer may allow, rank selection seems appropriate here. And since the mutation operation is a random one, we'll preserve a few elite individuals so that if the correct polynomial is found the population doesn't drift away from it.

An observant student may note that a term like x_1^3 might make a larger contribution to the error function than a lower-order term. Thus the $\sqrt{x_4 x_5}$ term may be harder to find as part of the search. We can make allowances for behavior like this by permitting the expressions to grow over time. This increases the likelihood that the higher-order terms are found first in a smaller space and then the neighborhood of expressions containing those terms is

TABLE 15.5
Population of Polynomials

Chromosome	Polynomial	Fitness
1	$x_4^2 * x_2^{1/3} * x_1^2 + x_3^2 \,\%\, x_5^3 \,\%\, x_2^2 \,\%\, x_4^{1/2} - x_5^{1/2}/x_2^2$	$4.07E8$
2	$x_3^3 * x_4^2 + x_5^1 * x_4^3 + x_3^{1/3}/x_3^{1/2} + x_1^{1/2} \,\%\, x_4^{1/2} - x_2^{1/2}$	$3.62E8$
3	$x_3^{1/2} + x_2^1 * x_3^1 \,\%\, x_5^2 + x_3^2 * x_5^1 \,\%\, x_4^{1/2} + x_2^3 * x2^{1/3}$	$7.58E5$
4	$x_4^{1/3} - x_3^3 + x_1^1 \,\%\, x_2^3 - x_4^{1/2} + x_5^3 - x_4^{1/2}/x_2^2 * x_3^1$	$4.16E10$
5	$x_2^3/x_1^{1/3} \,\%\, x_1^{1/3} \,\%\, x_3^1/x_1^3 - x_4^2 - x_4^1/x_1^{1/3} * x2^{1/3}$	∞[1]
6	$x_3^3 * x_3^1/x_1^1 - x_1^1 - x_2^{1/3} \,\%\, x_4^{1/2} + x_2^2 - x_2^3 * x_2^{1/2}$	$8.30E7$
7	$x_5^{1/3} * x_1^{1/3} + x_4^{1/3} + x_3^{1/2} + x_2^2 - x_3^3 * x_4^1 + x_4^1 \,\%\, x_4^2$	$5.09E6$
8	$x_3^3 * x_2^3 - x_5^2 * x_4^1 \,\%\, x_2^1 * x_1^{1/2} - x_2^{1/2}/x_2^{1/2}/x_4^{1/3}$	$1.82E10$
9	$x_2^{1/2} - x_4^2 * x_3^2 + x_5^2/x_1^{1/3}/x_5^3 \,\%\, x_2^1 - x_1^3 + x_5^{1/3}$	$1.31E9$
10	$x_1^{1/3} + x_3^{1/3} + x_4^{1/3} + x_2^2 - x_4^{1/3} * x_4^{1/2} * x_1^{1/3} * x_1^1/x_4^2$	$1.61E5$

[1]Actually just larger than the largest floating point value the software could represent

searched for the lower-order terms, rather than trying to search the large space of all possible expressions at once.

To do this, we'll introduce new variation operations, insertion and deletion. Insertion will add a term (the full combination of operator, vector component, and exponent) and deletion will remove a term. For simplicity, we can add or remove from the end of the chromosome. Mutations will stay the same. Crossover will need to be modified to handle chromosomes of unequal length. The easiest way to handle this is simply to not perform crossover if the chosen crossover point is beyond the end of one of the chromosomes, but we could also restrict the crossover point to be within the shorter chromosome or use modular arithmetic to keep the crossover point within the length of each chromosome.

Table 15.6 shows the best fitting polynomials from multiple iterations of an evolutionary algorithm using this representation and these variation and selection strategies. Note that each expression is unique, even though the output of each will be the same. This is a useful feature for the search process because it doesn't have to find the only solution in a very large, high dimensional space; it can find one of many equivalent solutions which are spread out in different regions of that space.

The applications of evolutionary algorithms needn't be restricted to mathematical problems. For example, we could use a different grammar and set of symbols to create computer programs instead of algebraic expressions and then evaluate their performance on a task of interest. That task needn't have a quantitative output for selection, either. There are selection strategies that use relative fitness rather than absolute fitness, such as tournament selection

TABLE 15.6
Sample of Evolved Polynomials

Chromosome	Polynomial
1	$x_1^3 - x_4^2 * x_2^1 + x_2^{1/2} * x_3^1 * x_2^{1/2} + x_3^3 * x_5^1 / x_3^2 + x_4^{1/2} * x_5^{1/2}$
2	$x_1^3 - x_4^2 * x_2^1 - x_4^{1/2} + x_4^{1/2} + x_5^1 * x_3^1 + x_3^1 * x_2^{1/2} * x_2^{1/2} + x_4^{1/3} \% x_4^{1/3} \% x_4^2 + x_4^{1/2} * x_5^{1/2}$
3	$x_1^3 - x_4^2 * x_2^1 + x_5^2 \% x_5^2 \% x_5^{1/2} + x_2^1 * x_3^1 + x_5^1 * x_3^1 + x_4^{1/2} * x_5^{1/2}$
4	$x_1^3 - x_4^2 * x_2^1 + x_3^1 * x_2^{1/2} * x_2^{1/2} + x_3^1 * x_5^{1/2} * x_5^{1/2} + x_5^{1/2} * x_5^{1/3} / x_5^{1/3} * x_4^{1/2}$
5	$x_1^3 - x_4^2 * x_2^1 + x_3^1 * x_2^1 + x_3^3 / x_3^2 + x_4^{1/2} * x_5^{1/2} + x_5^1 * x_3^1 - x_3^1$
6	$x_1^1 - x_4^2 * x_2^1 + x_1^3 + x_3^1 * x_4^3 / x_4^3 * x_2^1 + x_5^1 * x_3^1 - x_1^1 + x_5^{1/2} * x_4^{1/2}$
7	$x_1^3 - x_4^2 * x_2^1 + x_4^{1/2} * x_5^{1/2} + x_2^1 * x_3^1 + x_5^1 * x_3^1 - x_5^1 + x_5^1$
8	$x_1^3 - x_4^2 * x_2^1 + x_3^{1/2} * x_2^1 * x_3^{1/2} + x_3^3 / x_3^2 * x_5^1 + x_5^3 \% x_5^3 + x_4^{1/2} * x5^1 / x5^{1/2}$
9	$x_1^3 - x_2^1 * x_4^2 + x_3^1 * x_5^1 + x_5^{1/2} * x_4^{1/2} + x_4^{1/3} + x_3^1 * x_3^{1/3} / x_3^{1/3} * x_2^1 - x_4^{1/3}$
10	$x_1^3 - x_4^2 * x_2^1 + x_3^1 * x_5^1 + x_5^{1/2} * x_4^{1/2} + x_3^1 * x_2^1$

where two individuals at a time are randomly chosen and compared head-to-head, with the winner being allowed to reproduce. Using this approach, we can evolve music or visual art, where the objective function is the aesthetic preferences of one or more human evaluators.

15.2 Ant Foraging Optimization

One way to differentiate optimization methods, particularly heuristics, is by what they remember and how they forget. In the simplest gradient descent scenario, a single point is remembered at a time, and it is forgotten as soon as the gradient is followed to the next point. Evolutionary algorithms remember a population of candidate solutions, and the ones that are not selected for reproduction are forgotten. Ant foraging optimization (first introduced by Dorigo *et al* [19]; see [6] and [7] for an overview of further developments), another family of heuristics inspired by biology, likewise remembers a set of candidate solutions, but forgetting is based on which candidates are "visited" least often.

When ants are foraging, they deposit chemical signals known as *pheromones* on the ground wherever they travel. When they find food, they follow that pheromone trail back to their nest, while other ants follow it to the food. Over time, the pheromones evaporate; if a trail is frequently used, more pheromone will be deposited to offset evaporation, while infrequently used trails will fade away. As a consequence, the trail to the nearest food source will have the strongest signal because the round trip takes less time and gets replenished more quickly.

Modeling pheromone deposits can be used to find shortest paths in physical space for robots or drones, or to find the shortest or most efficient routes for software agents traversing a computer network of possibly unknown and changing topology. Individual agents can deposit pheromone as they go and make the choice of their next move based on the current pheromone concentrations in their area around them. A physical or simulated evaporation process can them remove some pheromone.

This model of remembering and forgetting through pheromone strength can also be adapted for maximization and minimization tasks. For a finite combinatorial search, the search space can be modeled as a graph. Each possible element of a candidate solution is a node in the graph, and edges connect adjacent nodes. In the network topology scenario or problems like the traveling salesman, nodes are locations and edges are available routes from that location.

For tasks like the symbolic regression problem from the previous section, nodes represent values that could fill positions in the expression string. To build the graph, begin with a start position node. This node has edges to nodes for each of the five vector components that could occupy the first position of the string. Each of those five nodes has edges to nodes for each of the five exponent values that could occupy the second position. Each of those five nodes has edges to nodes for each of the five operator nodes, plus a node representing string termination since it would be valid to end the expression here. Those operator nodes (but not the string termination node) have edges to nodes for the five vector components (different nodes than before), and so on up to some maximum expression length, at which point the exponent nodes only have a single edge each to the string termination node.

Once the graph is defined, the ant optimization algorithm loops through two steps. First, a population of ants each travels a path through the graph to construct a candidate solution. The ant starts at the starting node and chooses an edge to travel along based on the pheromone level associated with each edge. If the pheromone from node i to node j is τ_{ij}, then an ant at node i chooses to travel to node j with probability P_{ij}:

$$P_{ij} = \frac{\tau_{ij}}{\sum_{k=1}^{K} \tau_{ik}}$$

where nodes 1 through K are the neighbors of node i. Once each ant completes its path, that path can be evaluated by the objective function.

The second step in the algorithm is to update the pheromone for every edge. The update has an evaporation component ρ and a deposit component δ_{ij}.

$$\tau_{ij}(t+1) = \tau_{ij}(t)(1-\rho) + \delta_{ij}$$

The evaporation parameter ρ is a free parameter which impacts how quickly the algorithm converges; the higher ρ is, the more the ants will base their decision on the most recent paths chosen.

The pheromone deposit parameter δ_{ij} depends on whether edge ij was part of any ant's path and how that path was evaluated by the objective function. If there are m ants, and ant k at time t constructed a candidate solution $S^k(t)$ as a set of edges, then δ_{ij} can be defined as:

$$\delta_{ij} = \sum_{k=1}^{m} \Delta\tau_{ij}^{k}(t)$$

with $\Delta\tau_{ij}^{k}(t) = Q/F(S^k(t))$ if the edge ij is in $S^k(t)$ and 0 otherwise. $F(S^k)$ is the objective function value for candidate solution $S^k(t)$ and Q is another constant parameter. If the optimization requires maximization instead of minimization, one can use $\Delta\tau_{ij}^{k}(t) = QF(S^k(t))$ instead.

For the symbolic regression problem, recall that certain terms in the expression contribute more to the mean squared error than others. But the pheromone update amount is the same for every edge in candidate solution. As a result, the algorithm may converge too quickly on paths that contain the correct higher order terms and whichever lower order terms happened to be chosen along with them. Multiple variations on the basic ant algorithm exist to prevent this premature convergence, using different schemes to ensure that all of the edges have some chance of being chosen. The MAX-MIN variation ([56]) sets a minimum and maximum value of τ and does not let the pheromone update step exceed these bounds. The ant colony variation ([20]) constructs paths with a modified process; at each node, if a uniform random number q drawn from $U(0,1)$ is less than a parameter q_0 then the edge with the highest pheromone value is chosen, otherwise one of the other edges from the current node is chosen with equal probability. The ant colony variation also uses a different pheromone update process; there is an evaporation step after each ant choose an edge so that subsequent ants are more likely to choose different edges and only the best-so-far candidate solution contributes to the pheromone deposit step at the end of the iteration.

Ant foraging optimization can be extended to searches of continuous spaces, such as finding the vector that maximizes the function in the previous section. One approach is known as $\text{ACO}_{\mathbb{R}}$ [55]. The memory of the algorithm is now represented by an archive of solutions with a fixed size; when better solutions are found, the lowest quality solutions are removed from the archive and thus forgotten. Pheromone concentration is modeled implicitly via probability density functions which have higher densities around the points represented by the solution archive in proportion to the quality of each solution. The pheromone update process is then simply a matter of adding the latest round of solutions to the archive and removing the worst ones such that the archive size stays the same.

Part IV

Convexity and the Fundamental Theorem of Linear Programming

16

Important Sets for Optimization

This chapter gives us the tools to discuss an essential element of any good study of Optimization: convexity. In addition to being its own branch of nonlinear programming, we will later use the tools from this chapter to prove the Fundamental Theorem of Linear Programming.

16.1 Special Linear Combinations

Let

$$\mathbf{x}_1 = \begin{bmatrix} x_{11} \\ x_{12} \\ \vdots \\ x_{1m} \end{bmatrix}, \mathbf{x}_2 = \begin{bmatrix} x_{21} \\ x_{22} \\ \vdots \\ x_{2m} \end{bmatrix}, \cdots, \mathbf{x}_k = \begin{bmatrix} x_{k1} \\ x_{k2} \\ \vdots \\ x_{km} \end{bmatrix} \tag{16.1}$$

be $n \times 1$ matrices (i.e. column vectors) with real entries. A *linear combination* of $\mathbf{x}_1, \mathbf{x}_2, \ldots, \mathbf{x}_k$ is any expression of the form

$$\sum_{i=1}^{k} \alpha_i \mathbf{x}_i \tag{16.2}$$

where each α_i is a real number. When we have $\alpha_1 + \alpha_2 + \cdots + \alpha_k = 1$ in 16.2, we refer to the linear combination as an *affine combination* . When each $\alpha_i \geq 0$, $\sum_{i=1}^{k} \alpha_i \mathbf{x}_i$ is called a *conical* or *nonnegative combination*. Lastly, a linear combination that is both an affine and conical combination (i.e. both $\sum_{i=1}^{k} \alpha_1 = 1$ and $\alpha_i \geq 0$ are satisfied) is called a *convex combination.*

Example 16.1.1. *Consider the standard basis vectors of* $\mathbb{R}^3$*:* $\mathbf{i} = \langle 1, 0, 0 \rangle$, $\mathbf{j} = \langle 0, 1, 0 \rangle$*, and* $\mathbf{k} = \langle 0, 0, 1 \rangle$*. Then the linear combination of the standard basis vectors*

$$\frac{1}{2}\mathbf{i} - \frac{1}{3}\mathbf{j} + \frac{5}{6}\mathbf{k} = \left\langle \frac{1}{2}, -\frac{1}{3}, \frac{5}{6} \right\rangle \tag{16.3}$$

is an ***affine combination*** *of the vectors since* $\frac{1}{2} - \frac{1}{3} + \frac{5}{6} = 1$*, but* ***not conical combination*** *of these vectors since one coefficient is negative.*

$$\frac{1}{3}\mathbf{i} + \frac{1}{4}\mathbf{j} + \frac{1}{5}\mathbf{k} = \left\langle \frac{1}{3}, \frac{1}{4}, \frac{1}{5} \right\rangle \tag{16.4}$$

DOI: 10.1201/9780367425517-16

is a ***conical combination*** *but* ***not affine combination*** *of the vectors; and*

$$\frac{1}{2}\mathbf{i} + \frac{1}{3}\mathbf{j} + \frac{1}{6}\mathbf{k} = \left\langle \frac{1}{2}, \frac{1}{3}, \frac{1}{6} \right\rangle \tag{16.5}$$

is a ***convex combination*** *(and equivalently both* ***affine*** *and* ***conical****).*

16.2 Special Sets

Equipped with the terminology of the previous section, we now introduce some important sets.

Definition 16.2.1. *Let S be a nonempty collection of $n \times 1$ matrices (i.e. column vectors) with real entries.*

1. *The* linear hull *or the* span *of S is*

$$L(S) := \left\{ \sum_{i=1}^{k} \alpha_i \mathbf{x}_i \mid \mathbf{x}_i \in S \text{ and } \alpha_i \in \mathbb{R} \right\}$$

   where the sum is taken over all positive integers; that is, k is arbitrary. Hence $L(S)$ is the collection of all finite linear combinations of the vectors of S.
2. *The* affine hull *of S is*

$$aff(S) := \left\{ \sum_{i=1}^{k} \alpha_i \mathbf{x}_i \mid \mathbf{x}_i \in S, \alpha_i \in \mathbb{R}, \text{ and } \sum_{i=1}^{k} \alpha_i = 1 \right\}$$

   where k is again arbitrary. In other words, $aff(S)$ is the collection of all finite affine combinations of the vectors of S.
3. *The* conical hull *of S is*

$$coni(S) := \left\{ \sum_{i=1}^{k} \alpha_i \mathbf{x}_i \mid \mathbf{x}_i \in S, \alpha_i \in \mathbb{R}, \text{ and } \alpha_i \geq 0 \right\}$$

   where k is arbitrary. Thus $coni(S)$ is the collection of all finite conical (nonnegative) combinations of the vectors of S. And
4. *The* convex hull *of S is*

$$conv(S) := \left\{ \sum_{i=1}^{k} \alpha_i \mathbf{x}_i \mid \mathbf{x}_i \in S, \alpha_i \in \mathbb{R}, \alpha_i \geq 0, \text{ and } \sum_{i=1}^{k} \alpha_i = 1 \right\}$$

   where k is an arbitrary positive integer giving $conv(S)$ to be the collection of all finite convex combinations of the vectors of S.

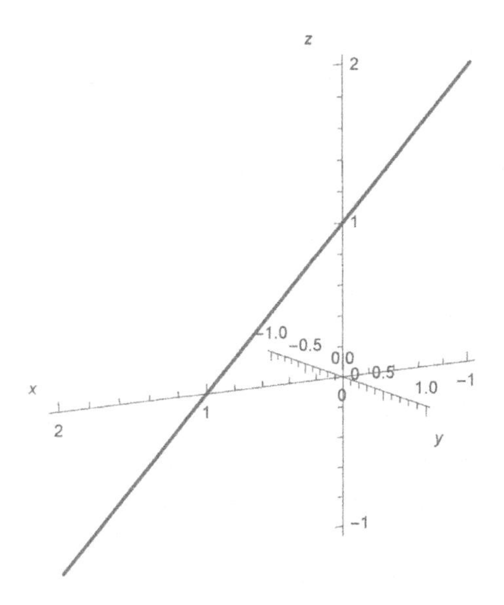

FIGURE 16.1
The affine hull of S in Example 16.2.2.

The significance of hulls will be discussed shortly. First, an example illustrating the definitions.

Example 16.2.2. *Let* $\mathbf{x}_1 = [1, 0, 0]^T$, $\mathbf{x}_2 = [0, 0, 1]^T$ *and put* $S = \{\mathbf{x}_1, \mathbf{x}_2\}$. *Then*

$$\begin{aligned} L(S) &:= \{\alpha_1\mathbf{x}_1 + \alpha_2\mathbf{x}_2 \mid \alpha_1, \alpha_2 \in \mathbb{R}\} \\ &= \{\alpha_1[1, 0, 0]^T + \alpha_2[0, 0, 1]^T \mid \alpha_1, \alpha_2 \in \mathbb{R}\} \qquad (16.6) \\ &= \{[\alpha_1, 0, \alpha_2]^T \mid \alpha_1, \alpha_2 \in \mathbb{R}\} \qquad (16.7) \end{aligned}$$

which is all points in $\mathbb{R}^3$ *with second coordinate 0; i.e. the* $x - z$ *plane.*

$$\begin{aligned} aff(S) &:= \{\alpha_1\mathbf{x}_1 + \alpha_2\mathbf{x}_2 \mid \alpha_1, \alpha_2 \in \mathbb{R} \text{ with } \alpha_1 + \alpha_2 = 1\} \\ &= \{\alpha_1[1, 0, 0]^T + \alpha_2[0, 0, 1]^T \mid \alpha_1, \alpha_2 \in \mathbb{R} \text{ with } \alpha_1 + \alpha_2 = 1\} \qquad (16.8) \\ &= \{[\alpha_1, 0, 1 - \alpha_1]^T \mid \alpha_1, \alpha_2 \in \mathbb{R}\} \qquad (16.9) \end{aligned}$$

i.e. the line through $(1, 0, 0)$ *and* $(0, 0, 1)$*.* $Aff(S)$ *is illustrated in Figure 16.1.*

$$\begin{aligned} coni(S) &:= \{\alpha_1\mathbf{x}_1 + \alpha_2\mathbf{x}_2 \mid \alpha_1, \alpha_2 \in \mathbb{R} \text{ and } \alpha_1, \alpha_2 \geq 0\} \qquad (16.10) \\ &= \{[\alpha_1, 0, \alpha_2]^T \mid \alpha_1, \alpha_2 \in \mathbb{R} \text{ and } \alpha_1, \alpha_2 \geq 0\} \qquad (16.11) \end{aligned}$$

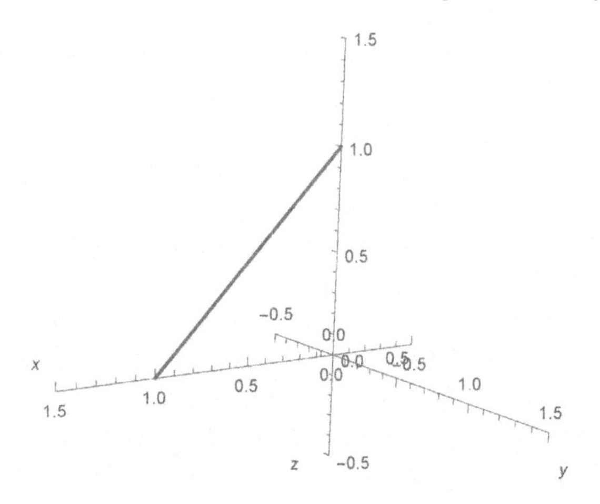

FIGURE 16.2
The convex hull of S in Example 16.2.2.

that is, the positive quadrant of the $x - z$ plane.

$$conv(S) := \{\alpha_1 \mathbf{x}_1 + \alpha_2 \mathbf{x}_2 \mid \alpha_1, \alpha_2 \in \mathbb{R}; \alpha_1, \alpha_2 \geq 0; \text{ and } \alpha_1 + \alpha_2 = 1\} \tag{16.12}$$

$$= \{\alpha_1 [1,0,0]^T + \alpha_2 [0,0,1]^T \mid \alpha_1, \alpha_2 \in \mathbb{R}_{\geq 0}; \text{ and } \alpha_1 + \alpha_2 = 1\} \tag{16.13}$$

$$= \{[\alpha_1, 0, 1 - \alpha_1]^T \mid \alpha_1 \in \mathbb{R} \text{ and } 0 \leq \alpha_1 \leq 1\} \tag{16.14}$$

$$= aff(S) \cap coni(S).$$

which is the line segment from $(0,0,1)$ to $(1,0,0)$. $Conv(S)$ is illustrated in Figure 16.2.

A better illustration of convex hull is to consider a finite collection S of points in $\mathbb{R}^2$. Then the convex hull of those points, $conv(S)$, is the collection of all line segments between the points, the collection of all line segments between the points on those segments, etc. This ends up giving a region that would also be found by putting a rubber band around the figure until all of S is in the rubber band. This is illustrated with the following diagram.

Example 16.2.3. *(Convex hull of a set of points in $\mathbb{R}^2$.)*

Note that in Example 16.2.3, $conv(S)$ is the entire region enclosed by the boundary line segments.

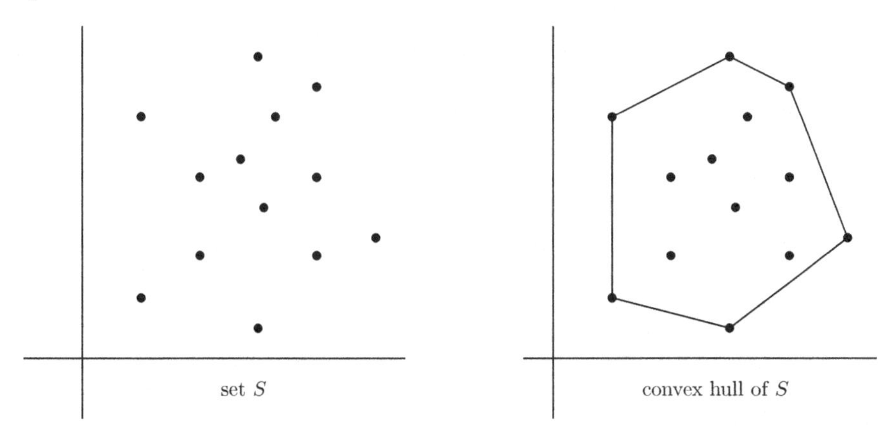

Examples 16.2.2 and 16.2.3 will help us form geometric interpretations of the special sets of which we are interested. Remembering these geometric interpretations will be quite helpful. In particular,

Highlight 16.2.4. *(Geometric interpretations of affine, conical, and convex.)*

- ***affine combination*** *means line;*
- ***convex combination*** *means line segment;*
- ***conical hull*** *is a collection of rays emitting from the origin and all the points between the rays; and*
- ***convex hull*** *means putting a rubber band around the outside of all the points.*

Relations among the special sets are

Observation 16.2.5.

$$S \subseteq conv(S) \subseteq aff(S) \subseteq L(S)$$
$$S \subseteq conv(S) \subseteq coni(S) \subseteq L(S)$$

We now consider an interesting result that is hopefully not too surprising.

Proposition 16.2.6. *S is a subspace of $\mathbb{R}^n$ if and only if $S = L(S)$.*

Proof. ($\Leftarrow$) By Exercise 16.1 $L(S)$ is a subspace of $\mathbb{R}^n$. Hence, since $S = L(S)$, S is a subspace.
($\Rightarrow$) Suppose S is a subspace of $\mathbb{R}^n$. We show $S = L(S)$ by showing that each set is a subset of the other. ($\subseteq$) If $\mathbf{s} \in S$, then by the definition of linear hull, $1\mathbf{s} \in L(S)$ and thus $S \subseteq L(S)$. ($\supseteq$) As we need to now prove $L(S) \subseteq S$, we proceed by induction on the number of terms in the linear combination from $L(S)$: Let $\alpha\mathbf{s} \in L(S)$. Since S is a subspace, it is closed under scalar multiplication and thus $\alpha\mathbf{s} \in S$. Now suppose that any linear combination from $L(S)$ with k terms is in S; that is $\alpha_1 s_1 + \cdots \alpha_k s_k \in S$ for any collection

$s_1, \ldots, s_k \in S$ and $\alpha_1, \ldots, \alpha_k \in \mathbb{R}$. Consider an arbitrary linear combination with $k+1$ terms from $L(S)$. Then

$$\sum_{i=1}^{k+1} \alpha_i s_i = \sum_{i=1}^{k} \alpha_i s_i + \alpha_{k+1} s_{k+1}$$

where the summation on the right-hand side is in S by the induction hypothesis and $\alpha_{k+1} s_{k+1}$ is in S by the base step. Since S is a subspace, this sum is then also in S. Hence, by induction, $L(S) \subseteq S$ giving $S = L(S)$. □

16.3 Special Properties of Sets

In the previous section, we introduced special sets. We now consider special properties of sets.

Definition 16.3.1. *Let S be a set, $x, y \in S$, and α, β real numbers.*

- *If $\alpha x + \beta y \in S$ whenever $\alpha + \beta = 1$, then S is said to be an* affine set.
- *If $\alpha x + \beta y \in S$ whenever $\alpha, \beta \geq 0$, then S is said to be a* convex cone with vertex at the origin *or, more simply, a* conical set.
- *If $\alpha x + \beta y \in S$ whenever $\alpha + \beta = 1$ and $\alpha, \beta \geq 0$, then S is said to be an* convex set.

The hulls in Example 16.2.2 will serve well as examples of sets with these special properties. Note also that the geometric interpretations of the hulls in Highlight 16.2.4 also hold for sets with these special properties.

It is important to observe that various hulls discussed in Section 16.2 are in some sense optimal. Let S be a set. Then the convex hull of $conv(S)$ is the smallest convex set that contains S. By "smallest" here we mean that any other convex set that contains S also contains the convex hull of S, $conv(S)$. The same statement can be made regarding the other two properties. It is case that these matters are formally addressed in Proposition 16.3.7 and Exercise 16.8.

To summarize:

Remark 16.3.2.

- *$aff(S)$ is the smallest affine set that contains S. Hence $S = aff(S)$ is a quick way to state that the set S is an affine set.*
- *$coni(S)$ is the smallest conical set that contains S. Hence $S = coni(S)$ is a quick way to state that the set S is a conical set.*
- *$conv(S)$ is the smallest convex set that contains S. Hence $S = conv(S)$ is a quick way to state that the set S is a convex set.*

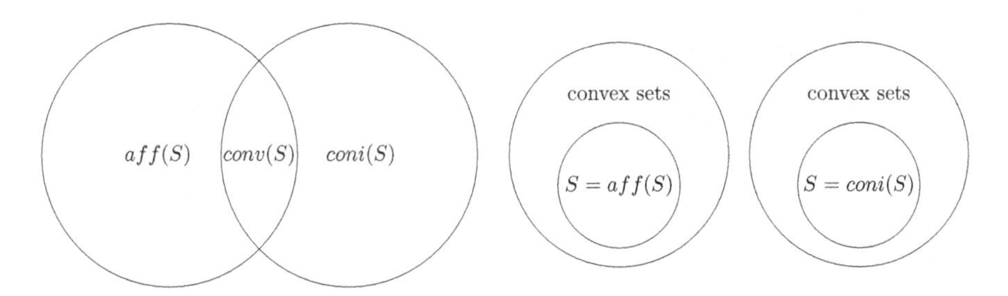

FIGURE 16.3
Relationship among affine, conical, and convex hulls and sets.

Let us now consider the relationships that exist between these properties.

Proposition 16.3.3. *Let S be a set. If S is affine, then it is also convex. Similarly, if S is conical, then it is also convex.*

Proof. Let S be an affine set. Then for any collection $s_1, s_2, \ldots s_k$ of elements from S and for any collection of real numbers $\alpha_1, \alpha_2, \ldots, \alpha_k$ where $\sum_{i=1}^{k} \alpha_i = 1$, we have $\sum_{i=1}^{k} \alpha_i s_i \in S$. Hence it follows that if we restrict the α_i to be nonnegative, the linear combination is still in S. Thus S is also convex.

The proof that if S is conical, then it is also convex is similar and is Exercise 16.3. □

It is important to notice the difference between the two results in these two recent sections: Proposition 16.3.3 is a result about *sets* whereas Observation 16.2.5 is a result about *hulls*. These are illustrated in Figure 16.3.

Example 16.3.4. *Determine whether or not $S = \{[x, y]^T \mid y \geq x; x, y \in \mathbb{R}\}$ is affine, conical, or convex.*

Solution. As $y_i \geq x_i$ for both points, $[-1, 1]^T$ and $[1, 1]^T$ are in S. But the affine combination $-2[-1, 1]^T + 1[1, 1]^T = [-1, 0]^T$ is not in S, hence S is not affine.

Let $[x_1, y_1]^T$ and $[x_2, y_2]^T$ be in S and $\alpha_1, \alpha_2 \geq 0$. Then an arbitrary conical combination of elements of S is of the form

$$\alpha_1 \begin{bmatrix} x_1 \\ y_1 \end{bmatrix} + \alpha_2 \begin{bmatrix} x_2 \\ y_2 \end{bmatrix} = \begin{bmatrix} \alpha_1 x_1 + \alpha_2 x_2 \\ \alpha_1 y_1 + \alpha_2 y_2 \end{bmatrix}. \tag{16.15}$$

But

$$\alpha_1 y_1 + \alpha_2 y_2 \geq \alpha_1 x_1 + \alpha_2 x_2 \text{ since } y_1 \geq x_1, y_2 \geq x_2, \tag{16.16}$$

establishing that the arbitrary conical combination meets the membership qualifications of S thus showing that S is conical.

Lastly, since S is conical we have by Proposition 16.3.3 that S is also convex. ■

Example 16.3.5. *Determine whether or not* $T = \{[x,y]^T \mid x+y \leq 5; x,y \geq 0\}$ *is affine, conical, or convex.*

Solution. Note that $[0,0]^T$ and $[5,0]^T$ are in T but that the affine combination $-1[0,0]^T + 2[5,0]^T = [10,0]^T$ is not in T, hence T is not affine.

For $[0,0]^T$ and $[5,0]^T$ that the conical combination $1[0,0]^T + 2[5,0]^T = [10,0]^T$ is not in T, hence T is not conical.

Let $[x_1,y_1]^T$ and $[x_2,y_2]^T$ be in T and $\alpha_1+\alpha_2 = 1$ with $\alpha_1, \alpha_2 \geq 0$. Then an arbitrary convex combination of elements of T is of the form

$$\alpha_1 \begin{bmatrix} x_1 \\ y_1 \end{bmatrix} + \alpha_2 \begin{bmatrix} x_2 \\ y_2 \end{bmatrix} = \begin{bmatrix} \alpha_1 x_1 + \alpha_2 x_2 \\ \alpha_1 y_1 + \alpha_2 y_2 \end{bmatrix}. \tag{16.17}$$

But

$$\begin{aligned} \alpha_1 x_1 + \alpha_2 x_2 &\geq \alpha_1 \cdot 0 + \alpha_2 \cdot 0 = 0 \ \text{ since } x_1, x_2 \geq 0, \\ \alpha_1 y_1 + \alpha_2 y_2 &\geq \alpha_1 \cdot 0 + \alpha_2 \cdot 0 = 0 \ \text{ since } y_1, y_2 \geq 0, \end{aligned} \tag{16.18}$$

and

$$\begin{aligned} \alpha_1 x_1 + \alpha_2 x_2 + \alpha_1 y_1 + \alpha_2 y_2 &\leq (\alpha_1 x_1 + \alpha_1 y_1) + (\alpha_2 x_2 + \alpha_2 y_2) && (16.19) \\ &\leq \alpha_1 5 + \alpha_2 5 \ \text{ since } x+1+y_1, x_2+y_2 \leq 5 && (16.20) \\ &= 5 \ \text{ since } \alpha_1 + \alpha_2 = 1 && (16.21) \end{aligned}$$

establishing that the arbitrary convex combination meets the membership qualifications of S showing that S is convex. ■

The next example considers the convexity of a familar object.

Example 16.3.6 (Convexity of a Circle). *We show that any circle is convex by showing that* $C = \{[x_1,x_2]^T \mid x_1^2 + x_2^2 \leq r^2\}$ *is convex. Note that since a rigid transformation of a circle does not change its convexity, we may without loss of generality suppose* C *is centered at the origin.*

Proof. An arbitrary convex combination of elements from C is of the form $[\alpha x_1 + \beta y_1, \alpha x_2 + \beta y_2]^T$ where $[x_1,x_2]^T, [y_1,y_2]^T \in C$ and $0 \leq \alpha, \beta \leq 1$ with $\alpha + \beta = 1$. Hence we establish the convexity of C if we can show

$$(\alpha x_1 + \beta y_1)^2 + (\alpha x_2 + \beta y_2)^2 \leq r^2. \tag{16.22}$$

But $(\alpha x_1 + \beta y_1)^2 + (\alpha x_2 + \beta y_2)^2$

$$\begin{aligned} &= \alpha^2(x_1^2 + x_2^2) + \beta^2(y_1^2 + y_2^2) + 2\alpha\beta(x_1 y_1 + x_2 y_2) && (16.23) \\ &\leq (\alpha^2 + \beta^2) r^2 + 2\alpha\beta(x_1 y_1 + x_2 y_2) && (16.24) \\ &= (\alpha^2 + \beta^2) r^2 + 2\alpha\beta [x_1,x_2]^T \cdot [y_1,y_2]^T && (16.25) \\ &= (\alpha^2 + \beta^2) r^2 + 2\alpha\beta \left|\left|[x_1,x_2]^T\right|\right| \cdot \left|\left|[y_1,y_2]^T\right|\right| \cos\theta && (16.26) \\ &\leq (\alpha^2 + \beta^2) r^2 + 2\alpha\beta \cdot r \cdot r \cdot 1 && (16.27) \\ &= (\alpha^2 + 2\alpha\beta + \beta^2) r^2 = r^2. && (16.28) \end{aligned}$$

□

Now that we understand the important properties of sets, let us tie these ideas together with the important sets.

Proposition 16.3.7. *Let $S \in \mathbb{R}^n$. Then*

i) S is an affine set if and only if $S = aff(S)$,
ii) S is an conical set if and only if $S = coni(S)$, and
iii) S is an convex set if and only if $S = conv(S)$.

We prove part $i)$ and leave the rest as Exercise 16.5.

Proof. ($\Leftarrow$) Suppose $S = aff(S)$; i.e. S is closed under all finite affine combinations. Hence for any $x, y \in S$ and $a, b \in \mathbb{R}$ with $a + b = 1$, $ax + by \in S$, establishing that S is affine.
($\Rightarrow$) Now suppose S is an affine set. In the same manner as Proposition 16.2.6, we will show $S = aff(S)$ by showing set containment both ways. ($\subseteq$) $S \subseteq aff(S)$ is clear by the definition of affine hull. ($\supseteq$) We again proceed by induction on the number of terms in the affine linear combinations from $aff(S)$. When $k = 1$, $\alpha s \in aff(S)$ is also clearly in S since α must be 1. Suppose for some positive k that any k-term affine combination from $aff(S)$ is also in S; that is $\sum_{i=1}^{k} \alpha_i s_s$ where $\sum_{i=1}^{k} \alpha_i = 1$ in $aff(S)$ is also in S. Consider $\sum_{i=1}^{k+1} \alpha_i s_i$ in $aff(S)$. Put $\alpha = \alpha_1 + \cdots + \alpha_k$. Then

$$\sum_{i=1}^{k+1} \alpha_i s_i = \alpha_1 s_1 + \cdots + \alpha_k s_k + \alpha_{k+1} s_{k+1} \tag{16.29}$$

$$= \alpha \left(\frac{\alpha_1}{\alpha} s_1 + \cdots + \frac{\alpha_k}{\alpha} s_k \right) + \alpha_{k+1} s_{k+1}. \tag{16.30}$$

Since $\frac{\alpha_1}{\alpha} + \cdots + \frac{\alpha_k}{\alpha} = \frac{\alpha}{\alpha} = 1$, the sum in the parentheses is in S by the induction hypothesis and since $\alpha + \alpha_{k+1} = 1$, the resulting two-term sum is in S since S is affine. Thus, by induction, $aff(S) \subseteq S$ giving $S = aff(S)$. □

We now establish why affine sets are so important.

Proposition 16.3.8. *$S \subseteq \mathbb{R}^n$ is affine if and only if $S - \mathbf{s}_0$ is a subspace of $\mathbb{R}^n$ for all $\mathbf{s}_0 \in S$.*

Proof. ($\Rightarrow$) Suppose $\mathbf{s}_0 \in S$ with S an affine set. We show that $S - \mathbf{s}_0$ is a subspace by showing it is closed under linear combinations. To do this, suppose $\mathbf{x}, \mathbf{y} \in S - \mathbf{s}_0$ and α, β are any real numbers. But $\mathbf{x}, \mathbf{y} \in S - \mathbf{s}_0$ imply that $\mathbf{x} + \mathbf{s}_0, \mathbf{y} + \mathbf{s}_0 \in S$. We also have $\alpha + \beta + (1 - \alpha - \beta) = 1$ and since S is affine

$$\alpha(\mathbf{x} + \mathbf{s}_0) + \beta(\mathbf{y} + \mathbf{s}_0) + (1 - \alpha - \beta)\mathbf{s}_0 \in S \tag{16.31}$$

$$\Leftrightarrow \alpha\mathbf{x} + \beta\mathbf{y} + (\alpha + \beta)\mathbf{s}_0 - (\alpha + \beta)\mathbf{s}_0 + \mathbf{s}_0 \in S \tag{16.32}$$

$$\Leftrightarrow \alpha\mathbf{x} + \beta\mathbf{y} + \mathbf{s}_0 \in S \tag{16.33}$$

$$\Leftrightarrow \alpha\mathbf{x} + \beta\mathbf{y} \in S - \mathbf{s}_0 \tag{16.34}$$

and thus $S - \mathbf{s}_0$ is a subspace for every $\mathbf{s}_0 \in S$.
($\Leftarrow$) For the necessary condition, suppose $S - \mathbf{s}_0$ is a subspace for every $\mathbf{s}_0 \in S$, that $\mathbf{x}, \mathbf{y} \in S$, that $\alpha, \beta \in \mathbb{R}$ with $\alpha + \beta = 1$. We then have $\mathbf{x} - \mathbf{s}_0, \mathbf{y} - \mathbf{s}_0 \in S$ and since $S - \mathbf{s}_0$ is a subspace

$$\alpha(\mathbf{x} - \mathbf{s}_0) + \beta(\mathbf{y} - \mathbf{s}_0) \in S - \mathbf{s}_0. \tag{16.35}$$

But

$$\alpha(\mathbf{x} - \mathbf{s}_0) + \beta(\mathbf{y} - \mathbf{s}_0) = \alpha\mathbf{x} + \beta\mathbf{y} - (\alpha + \beta)\mathbf{s}_0 = \alpha\mathbf{x} + \beta\mathbf{y} - \mathbf{s}_0 \tag{16.36}$$

so $\alpha\mathbf{x} + \beta\mathbf{y} - \mathbf{s}_0 \in S - \mathbf{s}_0$ giving $\alpha\mathbf{x} + \beta\mathbf{y} - \mathbf{s}_0 \in S - \mathbf{s}_0$ which establishes that S is affine. □

The significance of what we have just shown is

Highlight 16.3.9. *An affine set is just a translate of some subspace.*

Hence we can think of an affine set as a subspace that is lacking $\mathbf{0}$. We add that the subspace of which a given affine set is a translate is unique. The proof is left as an exercise (Exercise 16.9).

Since an affine set is almost a vector space

Definition 16.3.10 (Dimension of a Set)**.** *The* dimension of an affine set *is the dimension of the subspace of which it is a translate. The* dimension of a set *is the dimension of the set's affine hull.*

16.4 Special Objects

We have explored important sets and important properties that are in Optimization. Let us now consider the geometric objects essential to the field.

We all know that $ax + by = c$ is an equation of a line in $\mathbb{R}^2$. Hence we may think of its graph as $\{(x, y) \mid ax + by = c\}$ which is the same as $\{[x, y]^T \mid [a, d][x, y]^T = c\}$. A plane plays the same role in $\mathbb{R}^3$ as a line does in $\mathbb{R}^2$. A plane has as it equation $ax + by + cz = d$ which means we may think of its graph as $\{[x, y, z]^T \mid [a, b, c][x, y, z]^T = d\}$. These notions generalize to

Definition 16.4.1 (Hyperplane)**.** *Let* $\mathbf{c} = [c_1, c_2, \ldots c_n]^T$ *be a nonzero vector in* $\mathbb{R}^n$ *and* b *a real number. We define a* hyperplane *in* $\mathbb{R}^n$ *to be*

$$H(\mathbf{c}, b) := \{[x_1, x_2, \ldots x_n]^T \mid \mathbf{c}^T\mathbf{x} = b\}.$$

Note that a hyperplane in $\mathbb{R}^n$ is an $n - 1$ dimensional affine subset of $\mathbb{R}^n$ (see Exercise 16.10).

Note that a line divides $\mathbb{R}^2$ in half and a plane does the same for $\mathbb{R}^3$. We capture this notion in the following definitions.

Definition 16.4.2 (Half-space). *Let* $\mathbf{c} = [c_1, c_2, \ldots c_n]^T$ *be a nonzero vector in* $\mathbb{R}^n$ *and* b *a real number. For the hyperplane* $H(\mathbf{c}, b)$ *we define the* upper half space *to be*

$$H^+(\mathbf{c}, b) := \{\mathbf{x} \mid \mathbf{c}^T\mathbf{x} \geq b\}$$

and the lower half space *to be*

$$H^-(\mathbf{c}, b) := \{\mathbf{x} \mid \mathbf{c}^T\mathbf{x} \leq b\}$$

Definition 16.4.3 (Polyhedron, Polytope). *An intersection of finitely many half-spaces is called a* polyhedron *(in Greek* poly = *"many" and* hedron = *"face of a geometric solid"). The plural of polyhedron is* polyhedra, *though* polyhedrons *has become acceptable. If the intersection is bounded, the resulting object is referred to as a* polytope *(thus a polytope is a bounded polyhedron).*

Example 16.4.4. *All of the feasible regions in Chapter 6 are examples of polyhedra.*

16.5 Exercises

Exercise 16.1. *Let* $S \subseteq \mathbb{R}^n$. *Use induction and Exercise 4.29 to show that* $L(S)$ *is a subspace of* $\mathbb{R}^n$.

Exercise 16.2. *Prove the set relations in Observation 16.2.5.*

Exercise 16.3. *Complete the proof of Proposition 16.3.3; that is, if a set* S *is conical, then* S *is also convex.*

Exercise 16.4. *Consider* $S = \{(-3, 1), (2, 2), (1, 0)\} \subset \mathbb{R}^2$. *Draw* $aff(S)$, $coni(S)$, *and* $conv(S)$. *Notice how this illustrates the first part of Figure 16.3 (this was proposed by my student Edison Hauptman immediately after he sat through a lecture on the material).*

Exercise 16.5. *a. Prove part ii) of Proposition 16.3.7*
b. Prove part iii) of Proposition 16.3.7

Exercise 16.6. *Prove the following:*

a. Any intersection of affine sets is an affine set.
b. Any intersection of conical sets is a conical set.
c. Any intersection of convex sets is a convex set.

Exercise 16.7. *Example 16.3.6 shows that any circle is a convex set. A key step in the proof is (16.24) showing that* $x_1y_1 + x_2y + 2 \leq r^2$ *which was established using that the vector's magnitude is less than the radius and a property of the dot product. Establish this inequality by*

1. *using cases; that is, going through the four possibilities:*
   i) $x_1 \leq y_1, x_2 \leq y_2$,
   ii) $x_1 \leq y_1, x_2 > y_2$,
   iii) $x_1 > y_1, x_2 \leq y_2$, *and*
   iv) $x_1 > y_1, x_2 > y_2$;
2. *using Calculus to show that* $\max(x_1y_1 + x_2y_2) \leq r^2$*; or*
3. *using the AGM.*

Exercise 16.8. *Prove the following:*

a. *Let* S *be a set and put* $\mathcal{F} = \{A_\alpha \mid S \subseteq A_\alpha, A_\alpha \text{ is affine}\}$*; i.e.* $\mathcal{F}$ *is the family of all affine sets that contain* S *as a subset. Let* I *be the collection of indices of the sets in* $\mathcal{F}$*. Prove* $aff(S) = \bigcap_{\alpha \in I} A_\alpha$*. (This shows that* $aff(S)$ *is the smallest affine set containing* S*.)*
b. *Let* S *be a set and put* $\mathcal{G} = \{A_\alpha \mid S \subseteq A_\alpha, A_\alpha \text{ is conical}\}$*; i.e.* $\mathcal{G}$ *is the family of all conical sets that contain* S *as a subset. Let* J *be the collection of indices of the sets in* $\mathcal{G}$*. Prove* $aff(S) = \bigcap_{\alpha \in J} A_\alpha$*. (This shows that* $coni(S)$ *is the smallest conical set containing* S*.)*
c. *Let* S *be a set and put* $\mathcal{H} = \{A_\alpha \mid S \subseteq A_\alpha, A_\alpha \text{ is convex}\}$*; i.e.* $\mathcal{H}$ *is the family of all conical sets that contain* S *as a subset. Let* K *be the collection of indices of the sets in* $\mathcal{H}$*. Prove* $aff(S) = \bigcap_{\alpha \in K} A_\alpha$*. (This shows that* $conv(S)$ *is the smallest convex set containing* S*.)*

Exercise 16.9. *Proposition 16.3.8 establishes that an affine set is a translate of some subspace. Prove this subspace is unique.*

Exercise 16.10. *Let* $H(\mathbf{c}, b)$ *be a hyperplane in* $\mathbb{R}^n$*. Show that* $H(\mathbf{c}, b)$ *is an affine set and that its dimension is* $n - 1$*.*

17

The Fundamental Theorem of Linear Programming

In Section 6.1, we approached solving a manufacturing problem for Lincoln Outdoors by reasoning geometrically. From the starting point of that geometric reasoning, we formed the Fundamental Theorem of Linear Programming. This chapter will provide the mathematical justification for our geometric reasoning and, in particular, we will prove that very important theorem.

17.1 The Finite Basis Theorem

With the machinery from the previous chapter, we may now formally state the vague understanding we have had of why the Simplex Method for answering Linear Programming problems works. We begin with an important structural fact about the feasible regions of LP problems. But first, some notation:

Definition 17.1.1 (Sumset). *Let A and B be sets. Then*

$$A + B := \{a + b \mid a \in A \text{ and } b \in B\},$$

where $A + B$ is called a sumet *or the* Minkowski sum *of sets A and B.*

In other words, the sumset $A + B$ is the collection of all possible sums of an element from A with an element from B.

Example 17.1.2. *Let $A = \{-1, 2, 4\}$ and $B = \{0, 3\}$. Then*

$$\begin{aligned} A + B &= \{-1 + 0, -1 + 3, 2 + 0, 2 + 3, 4 + 0, 4 + 3\} \\ &= \{-1, 2, 2, 5, 4, 7\} \\ &= \{-1, 2, 4, 5, 7\}. \end{aligned}$$

Though it has little to do with how we will use sumsets, it is interesting to note that in the example $|A + B| \neq |A||B|$ where $|A|$ denotes the number of elements in the set A. Certainly we have $|A + B| \leq |A||B|$ but, more interestingly, from Combinatorics, we have $|A + B| \geq |A| + |B| - 1$ with equality if and only if A and B are arithmetic progressions with the same common difference. This result is known as the *Cauchy-Davenport Theorem.*

DOI: 10.1201/9780367425517-17

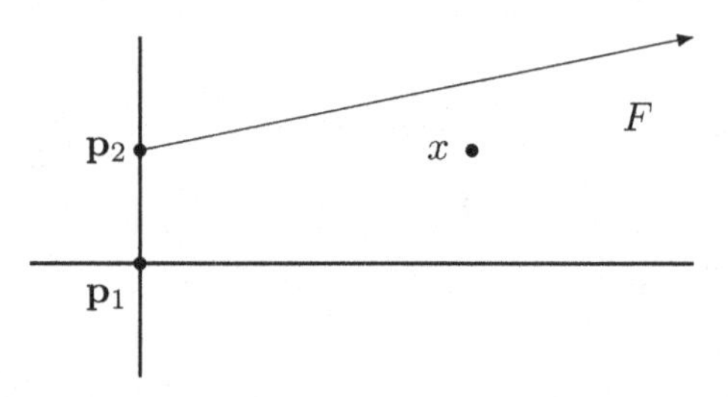

FIGURE 17.1
The region F in Example 17.1.4.

We now state a very useful theorem whose proof is beyond the scope of this text.

Theorem 17.1.3 (The Finite Basis Theorem). *Let F be a convex polyhedron with extreme points $P = \{\mathbf{p}_1, \mathbf{p}_2, \ldots, \mathbf{p}_n\}$. Then there exist* directional vectors $D = \{\mathbf{d}_1, \mathbf{d}_2, \ldots, \mathbf{d}_k\}$ *such that*

$$F = conv\{\mathbf{p}_i\} + coni\{\mathbf{d}_j\}$$

where $i = 1, 2, \ldots, n$ and $j = 1, 2, \ldots, k$ or, in the language of sumsets

$$F = conv(P) + coni(D).$$

That is, F is the convex combination of its extreme points and a conical combination of its direction vectors.

For a proof of the theorem, see [50]. As well, we note that Theorem 17.1.3 is also known as *The Minkowski-Weyl Theorem.*

We illustrate the Finite Basis Theorem with the following example:

Example 17.1.4. *Let $F = \{(x, y)|x, y \in \mathbb{R}; y \leq 0.2x + 1; x, y \geq 0\}$. Note that F has $P = \{\mathbf{p}_1 = \langle 0, 0\rangle, \mathbf{p}_2 = \langle 0, 1\rangle\}$ and for direction vectors we can use $D = \{\mathbf{d}_1 = \langle 1, 0\rangle, \mathbf{d_2} = \langle 1, 5\rangle\}$. Note that a choice of direction vectors is not unique.*

For $\mathbf{x} = \langle 3, 1\rangle \in F$ we have by Theorem 17.1.3 that

$$(1 - A)\langle 0, 0\rangle + A\langle 0, 1\rangle + B\langle 1, 0\rangle + C\langle 1, 5\rangle = \langle B + C, A + 5C\rangle = \langle 3, 1\rangle \tag{17.1}$$

where $0 \leq A \leq 1$ and $B, C \geq 0$. One way to write $\mathbf{x}$ as a convex combination of the extreme points of F and a conical combination of direction vectors would thus be

$$\langle 3, 1\rangle = 1\langle 0, 0\rangle + 0\langle 0, 1\rangle + \frac{14}{5}\langle 1, 0\rangle + \frac{1}{5}\langle 1, 5\rangle. \tag{17.2}$$

17.2 The Fundamental Theorem of Linear Programming

Given The Finite Basis Theorem, we are now equipped to establish the main result of this chapter. We will do all of the heavy lifting in the following lemma.

Lemma 17.2.1. *Suppose $f(\mathbf{x})$ has a maximum over the convex polyhedron F and F has $P = \{\mathbf{p_1}, \mathbf{p_2}, \ldots, \mathbf{p}_n\}$ as its extreme points and $D = \{\mathbf{d_1}, \ldots, \mathbf{d_k}\}$ as its direction vectors. Then for each $j = 1, 2, \ldots, k$, we have $f(\mathbf{d_j}) \leq 0$.*

Proof. Since $f(\mathbf{x})$ has a maximum over F, there exist M in $\mathbb{R}$ such that $f(\mathbf{x}) \leq M$ for all $\mathbf{x}$ in F. Suppose for contradiction that $f(d_t) > 0$ for some t where $1 \leq t \leq k$. By the Finite Basis Theorem

$$F = conv(P) + coni(D)$$

which means that the range of the function $f(\mathbf{x})$ over the set F is

$$
\begin{aligned}
f(F) &= f\left(conv(P) + coni(D)\right) && (17.3)\\
&= f\left(\sum_{i=1}^{n} \alpha_i \mathbf{p_i} + \sum_{j=1}^{k} \beta_j \mathbf{d_j}\right) \text{ where } \alpha_i, \beta_j \geq 0 \text{ and } \sum_{i=1}^{n} \alpha_i = 1 && (17.4)\\
&= \sum_{i=1}^{n} \alpha_i f(\mathbf{p_i}) + \sum_{j=1}^{k} \beta_j f(\mathbf{d_j}) \text{ since } f \text{ is linear} && (17.5)\\
&\leq \sum_{i=1}^{n} \alpha_i M + \sum_{j=1}^{k} \beta_j f(\mathbf{d_j}) && (17.6)\\
&= M + \sum_{j=1}^{k} \beta_j f(\mathbf{d_j}). && (17.7)
\end{aligned}
$$

But as $\mathbf{d_t} > 0$, since $\beta_t > 0$ we a value of f exceeding M contradicting that it is a maximum. Thus, for all $j = 1, \ldots, k$, we must have that $f(\mathbf{d_j}) \leq 0$. □

We may now prove the chapter's main attraction.

Theorem 17.2.2 (The Fundamental Theorem of Linear Programming). *If the linear function $f(\mathbf{x}) = c_1x_1 + c_2x_2 + \cdots + c_nx_n$ has a maximum over the convex polyhedron $F = \{\mathbf{y} \mid A\mathbf{y} \leq b\}$, then $f(\mathbf{x})$ attains its maximum at one of the extreme points of F.*

Proof. By assumption, there exists $\mathbf{x}^* \in F$ such that $f(\mathbf{x}^*) = M \geq f(\mathbf{x})$ for all $\mathbf{x} \in F$. The Finite Basis Theorem (Theorem 17.1.3) gives us that there exist extreme points $P = \{\mathbf{p}_1, \mathbf{p}_2, \ldots, \mathbf{p}_n\}$ and direction vectors $D = \{\mathbf{d}_1, \mathbf{d}_1, \ldots, \mathbf{d}_k\}$ such that $F = conv(P) + coni(D)$. Since there are only finitely many extreme points $\mathbf{p_i}$, there exists $N \in \{1, 2, \ldots, n\}$ such that

$$f(\mathbf{p}_N) \geq f(\mathbf{p}_i) \text{ for } i = 1, 2, \ldots, n. \tag{17.8}$$

Thus we have that there are nonnegative real numbers $\alpha_1, \ldots, \alpha_n$ and $\beta_1, \ldots, \beta_k$ with $\alpha_1 + \cdots + \alpha_n = 1$ such that

$$
\begin{aligned}
M = f(\mathbf{x}^*) &= f\left(\sum_{i=1}^{n} \alpha_i \mathbf{p}_i + \sum_{j=1}^{k} \beta_j \mathbf{d}_j\right) && && (17.9)\\
&= \sum_{i=1}^{n} \alpha_i f(\mathbf{p}_i) + \sum_{j=1}^{k} \beta_j f(\mathbf{d}_j) && \text{since } f \text{ is linear} && (17.10)\\
&\geq \sum_{i=1} i^n \alpha_i f(\mathbf{p}_i) && \text{by Lemma 17.2.1} && (17.11)\\
&\geq \sum_{i=1}^{n} \alpha_i f(\mathbf{p}_N) && \text{by 17.8} && (17.12)\\
&= f(\mathbf{p}_N) \sum_{i=1}^{n} \alpha_i && && (17.13)\\
&= f(\mathbf{p}_N)
\end{aligned}
$$

Thus f achieves its max at an extreme point of its feasible region.

The remainder of the proof is left as Exercise 17.5. □

Note that Lemma 17.2.1 does not say that $f(\mathbf{d}_j)$ is always nonnegative. It says that if f has a maximum over F, then the value of f along the direction vectors cannot be positive (else f would grow without bound and therefore not have a max, which is exactly how the proof proceeds). The same is true of the corresponding version of the lemma for when f has a minimum over F. This is Exercise 17.4.

17.3 For Further Study

A very thorough treatment of the material in this chapter can be found in Alexander Schrijver's *Theory of Linear and Integer Programming* [50].

17.4 Exercises

Exercise 17.1. *Let* $A = \{1, 3, 5, 7, 9\}$ *and* $B = \{2, 3, 5, 7\}$*. Find the sumset* $A + B$*.*

Exercise 17.2. *Write* $\mathbf{x} = \langle 1, 10\rangle$ *as a convex combination of* P *and a conical combination of* D *from Example 17.1.4.*

Exercise 17.3. *Referencing The Finte Basis Theorem (Theorem 17.1.3), identify P and select an appropriate D for the feasible region F in Example 6.1.1. Pick a point in the interior of F and write it the form $conv(P) + coni(D)$.*

Exercise 17.4. *Prove the proper version of Lemma 17.2.1 for when f has a minimum over its feasible region F.*

Exercise 17.5. *Prove the remaining part of The Fundamental Theorem of Linear Programming for when f has a minimum over F. This will require Exercise 17.4.*

18

Convex Functions

18.1 Convex Functions of a Single Variable

We turn our view now from sets to functions and consider the class of convex functions. Much like linear functions, convex functions are very appealing. The graphs of linear functions are easily understood and their properties are quite convenient. Each of these facts makes linear programming rather straightforward. Likewise, convex functions have graphs with convenient characteristics. Convex functions are useful in many areas of mathematics, but play an especially important role in probability and optimization due to their special properties. Calculus students will recognize two of the properties of this section by the terms *concave up* and *concave down*, but we now call these properties by their proper names.

Definition 18.1.1 (Single Variable Convex Function). *Let C be a convex subset of a real vector space and $f : C \to \mathbb{R}$ a function of one variable. f is said to be* convex *if for all $x_1, x_2 \in C$ and for all $\lambda \in [0, 1]$,*

$$f(\lambda x_1 + (1 - \lambda)x_2) \leq \lambda f(x_1) + (1 - \lambda)f(x_2). \tag{18.1}$$

If this inequality is strict when $x_1 \neq x_2$, then the function is said to be strictly convex.

Note that since $C \subseteq \mathbb{R}$ is a convex set, C is necessarily an interval (see Exercise 18.2).

Example 18.1.2. *Show $f(x) = x^2$ is a convex function.*

Solution. Note that the domain of the function is $\mathbb{R}$ which is a convex set. Let $\lambda \in [0, 1]$ and $x \leq y$ be real numbers. If $\lambda = 1$, we have

$$f(\lambda x + (1 - \lambda)y) = f(x) = \lambda f(x) + (1 - \lambda)f(y). \tag{18.2}$$

DOI: 10.1201/9780367425517-18

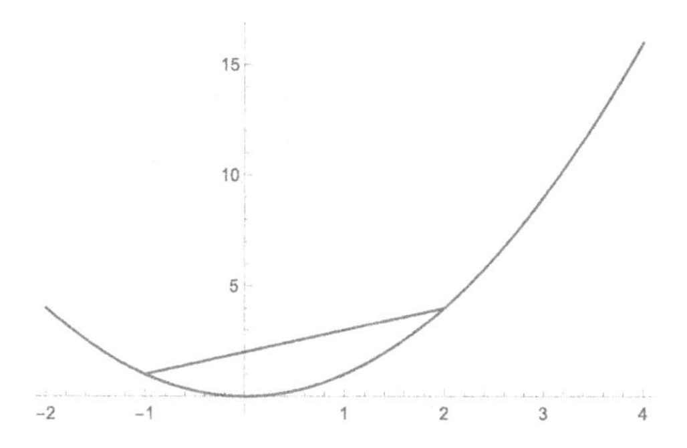

FIGURE 18.1
$f(x) = x^2$ with line segment from $x = -1$ to $x = 2$ in Example 18.1.2.

When $\lambda \neq 1$

$$\begin{aligned} 0 &\leq (x-y)^2 = x^2 - 2xy + y^2 && (18.3)\\ &\iff 2xy \leq x^2 + y^2 && (18.4)\\ &\iff (1-\lambda)\lambda 2xy \leq (1-\lambda)\lambda(x^2+y^2) \quad \text{since } 1-\lambda, \lambda \geq 0 && (18.5)\\ &\iff 2\lambda xy - 2\lambda^2 xy \leq \lambda x^2 - \lambda^2 x^2 + \lambda y^2 - \lambda^2 y^2 && (18.6)\\ &\iff \lambda^2 x^2 + 2\lambda(1-\lambda)xy - \lambda y^2 \leq \lambda x^2 - \lambda^2 y^2 && (18.7)\\ &\iff \lambda^2 x^2 + 2\lambda(1-\lambda)xy + y^2 - \lambda y^2 \leq \lambda x^2 + y^2 - \lambda^2 y^2 && (18.8)\\ &\iff \lambda^2 x^2 + 2\lambda(1-\lambda)xy + (1-\lambda)y^2 \leq \lambda x^2 + (1-\lambda^2)y^2 && (18.9)\\ &\iff [\lambda x + (1-\lambda)y]^2 \leq \lambda x^2 + (1-\lambda^2)y^2 && (18.10) \end{aligned}$$

thus establishing for $f(x) = x^2$ that

$$f(\lambda x + (1-\lambda)y) \leq \lambda f(x_1) + (1-\lambda)f(x_2).$$

■

Example 18.1.2 is illustrated in Figure 18.1 with $x = -1$ and $y = 2$. These correspond to the points $(-1, 1)$ and $(2, 4)$ and notice that the line segment joining the points lies above the graph of $f(x) = x^2$. This is not an isolated event as this holds for any line segment connecting points on the graph and is an important property of convex functions.

Highlight 18.1.3 (Geometric Meaning of Convex Function). *If f is a convex function, then any line segment connecting points on the graph of f will lie above the graph of f.*

Note that We have not necessarily drawn this conclusion from one example. If we look at 18.1 long enough, we see that the left-hand side considers functional values of any input from $[x_1, x_2]$ and that they are never bigger than any point on the line segment from $(x_1, f(x_1))$ to $(x_2, f(x_2))$. Note there is a second geometric interpretation of convex functions – one involving tangent lines; and we explore this after the following more elementary result.

Theorem 18.1.4. *If f is a convex function defined on an open interval (a,b), then f is continuous over (a,b).*

The proof of Theorem 18.1.4 is Exercise 18.5.

Now let us suppose $f(x)$ is differentiable over some interval I and fix $a \in I$. By 9.4, the equation of the line tangent to $f(x)$ at $x = a$ is

$$y = f'(a)[x-a] + f(a).$$

If this tangent line were to always lie below the graph of $f(x)$, that would mean for any $b \in I$ that

$$f'(a)[b-a] + f(a) \leq f(b) \tag{18.11}$$

where the left-hand side is the value of the input b on the tangent line and the right-hand side is the value of b in the function (this is illustrated in Figure 9.4 where $a = 1$ and b can be any value). All of this was for a fixed value $x = a$, but this a can be arbitrary. Thus we have the following observation

Observation 18.1.5. *Let f be a differentiable function over an interval I. Then every tangent line to f lies below the graph of f if and only if for any $a, b \in I$, $f'(a)[b-a] + f(a) \leq f(b)$.*

Theorem 18.1.6 (First-Order Condition of Convexity)**.** *If f is differentiable over an interval I, then f is convex over I if and only if for all $a, b \in I$*

$$f'(a)[b-a] + f(a) \leq f(b). \tag{18.12}$$

Proof. ($\Longrightarrow$) Suppose $f(x)$ is convex, $\lambda \in (0,1]$, and $a, b \in I$ with $a \neq b$. Then

$$f([1-\lambda]a + \lambda b) = f(a + \lambda[b-a]) \leq (1-\lambda)f(a) + \lambda f(b) \tag{18.13}$$

$$\Longleftrightarrow f(a + \lambda[b-a]) - f(a) \leq \lambda[f(b) - f(a)] \tag{18.14}$$

$$\Longleftrightarrow \frac{f(a + \lambda[b-a]) - f(a)}{\lambda} \leq f(b) - f(a) \tag{18.15}$$

$$\Longleftrightarrow \frac{f(a + \lambda[b-a]) - f(a)}{\lambda[b-a]}[b-a] \leq f(b) - f(a) \tag{18.16}$$

and letting $\lambda \to 0$ gives (here the $\lambda[b-a]$ palys the role of $h = \Delta\, x$ in the defintion of the derivative)

$$f'(a)[b-a] \leq f(b) - f(a)$$

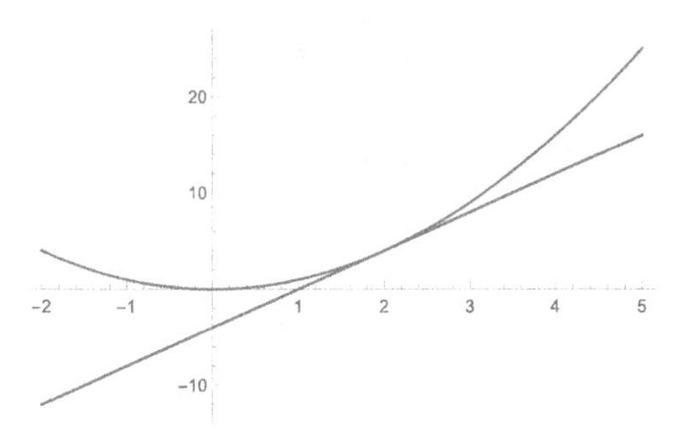

FIGURE 18.2
$f(x) = x^2$ with tangent line $4x - 4$ at the point $(2, 4)$.

. ($\Longleftarrow$) Now suppose for all $a, b \in I$ that $f'(a)[b - a] + f(a) \leq f(b)$. Let $c := \lambda a + (1 - \lambda)b$. Then by assumption

$$f(a) \geq f(c) + f'(c)(a - c) \tag{18.17}$$
$$\text{and } f(b) \geq f(c) + f'(c)(b - c). \tag{18.18}$$

Multiplying the first equation by λ and the second by $1 - \lambda$ then adding gives

$$\begin{aligned} \lambda f(a) + (1 - \lambda)f(b) &\geq [\lambda + 1 - \lambda]f(c) + f'(c)[\lambda a + (1 - \lambda)b] \\ &\quad - [\lambda + 1 - \lambda]f'(c) \cdot c & (18.19) \\ &= f(c) + f'(c) \cdot c - f'(c) \cdot c & (18.20) \\ &= f(c) := f(\lambda a + (1 - \lambda)b). & (18.21) \end{aligned}$$

□

Note that by Observation 18.1.5 the first-order condition of convexity means that $f(x)$ is convex if and only if any line tangent to $f(x)$ over interval I lies below the graph of $f(x)$ This is illustrated in Figure 18.2. Also note that it is clear from the proof of Theorem 18.1.6 that

Remark 18.1.7. *$f(x)$ is strictly convex if and only if the inequality in 18.12 is strict.*

Note that Remark 18.1.7 says also says the following:

Highlight 18.1.8. *The first-order condition of convexity preserves strictness.*

It is important to note that the first order condition of convexity depends on f being differentiable. This is an important observation as it is possible for a function to be convex, but not necessarily differentiable over its entire domain. See Exercise 18.7 for an example.

Let us now return to Example 18.1.2 but this time use the first-order condition of convexity rather than the definition.

Example 18.1.9. *Use the Theorem 18.1.6 to show that $f(x) = x^2$ is a convex function over $\mathbb{R}$.*

Solution. Let a, b be elements in the trivially convex set $\mathbb{R}$. Since

$$0 \leq (a-b)^2 = a^2 - 2ab + b^2 \tag{18.22}$$

we have

$$2a(b-a) + a^2 = 2ab - a^2 \leq b^2 \tag{18.23}$$

establishing 18.12. ■

Going back to 18.11, we can carry our thinking a little further than we have by noting that for a fixed b, if $a < b$, then 18.11 is equivalent to

$$f'(a) \leq \frac{f(b) - f(a)}{b - a} \tag{18.24}$$

but if $a > b$, then 18.11 is equivalent to

$$f'(a) \geq \frac{f(b) - f(a)}{b - a}. \tag{18.25}$$

This means

Observation 18.1.10. *Let f be a differentiable function over an interval I. Then*

i. the secant lines to the right of a have a larger slope than the tangent line at $x = a$ and
ii. the secant lines to the left of a have a smaller slope than the tangent line at $x = a$.

In particular, the slope of the chord formed by two points on the curve is increasing.

We formalize the final statement in the observation by

Proposition 18.1.11. *f is convex over (a, b) if and only if $\frac{f(u)-f(s)}{u-s} \leq \frac{f(t)-f(u)}{t-u}$ whenever $a < s < u < t < b$.*

Proof. ($\Longrightarrow$) Suppose f is convex and $a < s < u < t < b$. Let $h(x) = \frac{f(t)-f(s)}{t-s}$ be the equation of the secant line through $(s, f(s))$ and $(t, f(t))$. Then

$$\frac{f(u) - f(s)}{u - s} \leq \frac{h(u) - h(s)}{u - s} \quad \text{by convexity of } f(x) \text{ and that } f(s) = h(s) \tag{18.26}$$

$$= \frac{h(t) - h(u)}{t - u} \quad \text{since the slope of } h(x) \text{ is constant} \tag{18.27}$$

$$\leq \frac{f(t) - f(u)}{t - u} \quad \text{by convexity of } f(x) \text{ and that } f(t) = h(t) \tag{18.28}$$

and thus the cords' slopes are increasing.

($\Longleftarrow$) We prove the contrapositive of the sufficient condition, namely that if $f(x)$ is not convex, then the slope of the chords is decreasing:

$$\frac{f(u)-f(s)}{u-s} > \frac{h(u)-h(s)}{u-s} \quad \text{by non-convexity of } f(x) \text{ and that } f(s)=h(s) \tag{18.29}$$

$$= \frac{h(t)-h(u)}{t-u} \quad \text{since the slope of } h(x) \text{ is constant} \tag{18.30}$$

$$> \frac{f(t)-f(u)}{t-u} \quad \text{by non-convexity of } f(x) \text{ and that } f(t)=h(t), \tag{18.31}$$

i.e. the slope of the chords is decreasing. □

As there is a relationship between convexity and first derivatives, there is also a relationship between convexity and second derivatives.

Theorem 18.1.12 (Second-Order Condition of Convexity). *If f is twice differentiable on I then f is convex over I if and only if $f''(x) \geq 0$ for all $x \in I$.*

Proof. ($\Longrightarrow$) Suppose f is convex and $x, y \in I$ with $x < y$. By Theorem 18.1.6,

$$f(y) \geq f(x) + f'(x)(y-x) \tag{18.32}$$

$$\text{and } f(x) \geq f(y) + f'(y)(x-y) \tag{18.33}$$

thus

$$f(y) - f(x) \geq f'(x)(y-x) \tag{18.34}$$

$$\text{and } f(y) - f(x) \leq -f'(y)(x-y) = f'(y)(y-x) \tag{18.35}$$

which gives

$$f'(x)(y-x) \leq f(y) - f(x) \leq f'(y)(y-x). \tag{18.36}$$

Dividing the terms by (nonzero) $y-x$ yields

$$f'(x) \leq \frac{f(y)-f(x)}{y-x} \leq f'(y). \tag{18.37}$$

In particular, we have

$$0 \leq f'(y) - f'(x) \tag{18.38}$$

in which we can again divide both sides by $y-x$ giving

$$0 \leq \frac{f'(y)-f'(x)}{y-x}. \tag{18.39}$$

By letting $y \to x$, we obtain

$$0 \leq \lim_{y \to x} \frac{f'(y)-f'(x)}{y-x} =: f''(x). \tag{18.40}$$

($\Longleftarrow$) Now suppose $f''(x) \geq 0$ for all $x \in I$. Let $x, y \in I$ with $x < y$. By the Extended Mean Value Theorem (Theorem 9.1.8), there exists a in $[x, y]$ such that

$$f(y) = f(x) + f'(x)(y-x) + \frac{1}{2}f''(a)(y-x)^2 \geq f(x) + f'(x)(y-x) \quad (18.41)$$

since $f''(z) \geq 0$. Thus by the First-Order Condition of Convexity, $f(x)$ is convex. □

Taking the limit in the proof of Theorem 18.1.12 has an interesting consequence. Namely

Remark 18.1.13. *If $f''(x) > 0$ for every x in I, then f is strictly convex over I. The converse is not necessarily true.*

Proof. The first part of the remark is clear from replacing each $\leq$ in ($\Longrightarrow$) in the proof of the theorem with $<$. For a counterexample regarding the converse, consider $f(x) = x^4$ and let a,b be in $\mathbb{R}$ with $a < b$. Since

$$0 < 3a^4 + b^4 = 3a^4 - 4b^4 + b^4 \quad (18.42)$$

$$< 3a^4 - 4a^3b + b^4 \quad (\text{since } -a > -b) \quad (18.43)$$

which is equivalent to

$$b^4 > 4a^3b - 3a^4 = 4a^3(b-a) + a^4 \quad (18.44)$$

establishing that f satisfies the first-order condition for convexity and thus $f(x) = x^4$ is strictly convex. But $f''(x) = 12x^2$ and, unfortunately, $f''(0) = 0$. □

Highlight 18.1.14. *Note that unlike the first-order condition of convexity, Remark 18.1.13 points out that strictness is not preserved in the second-order condition of convexity.*

Each of the properties we have developed are important characteristics of convex functions. As we are concerned with optimization, the apex of our pursuit is the following theorem:

Theorem 18.1.15. *Let C be a convex subset of $\mathbb{R}$ and $f : C \to \mathbb{R}$ a convex function. Then any local minimum of f over C is a global minimum of f over C. Moreover, if f is strictly convex, then there is a unique x^* in C such that $f(x^*)$ is the global minimum of $f(x)$.*

Proof. Suppose $f(c)$ is a local minimum of $f(x)$ over C for some c in C. This means that there is a positive number ϵ such that $f(x) \geq f(c)$ whenever $c - \epsilon \leq x \leq c + \epsilon$. Let y be an arbitrary element in C. Since $f(x)$ is convex, for any $0 \leq \lambda \leq 1$ we have

$$f(\lambda y + [1-\lambda]c) \leq \lambda f(y) + [1-\lambda]f(c) \quad (18.45)$$

and rearranging terms gives

$$f(c + \lambda[y - c]) \leq \lambda[f(y) - f(c)] + f(c). \tag{18.46}$$

The inequality 18.46 holds for all λ in $[0, 1]$, so we choose a nonzero λ small enough such that $c - \epsilon \leq c + \lambda(y - c) \leq c + \epsilon$ and thus, since $f(c)$ is a local minimum, 18.46 becomes

$$f(c) \leq f(c + \lambda[y - c]) \leq \lambda[f(y) - f(c)] + f(c). \tag{18.47}$$

Subtracting $f(c)$ from both sides then dividing by nonzero λ gives the desired

$$f(c) \leq f(y)$$

for arbitrary y in C, establishing that $f(c)$ is a global minimum of $f(x)$ over C. If $f(x)$ is strictly convex, then all of the above inequalities are strict and c is the unique global minimizer. □

Theorem 18.1.15 is wonderful, but we can actually show something stronger and establish the most useful result in this chapter.

Theorem 18.1.16. *If f is a differentiable convex function over an interval I, then if c is a critical of point of f in I, then $f(c)$ is the global minimum of f over I.*

Proof. Suppose c is a critical point of f over I and that x is any other value in I. By Theorem 18.1.6 we have

$$f(x) \geq f'(c)[x - c] + f(c) = f(c) \tag{18.48}$$

establishing $f(c)$ as the global minimum. □

Other examples of convex functions are included in the exercises.

18.2 Concave Functions

Definition 18.2.1 (Single Variable Concave Function).
Let C be a convex subset of a real vector space and $h : C \to \mathbb{R}$ a function of one variable. h is said to be concave *if for all $x_1, x_2 \in C$ and for all $\lambda \in [0, 1]$,*

$$h(\lambda x_1 + (1 - \lambda)x_2) \geq \lambda h(x_1) + (1 - \lambda)h(x_2). \tag{18.49}$$

If this inequality is strict when $x_1 \neq x_2$, then the function is said to be strictly concave.

Example 18.2.2. *Show $h(x) = -x^2$ is a concave function.*

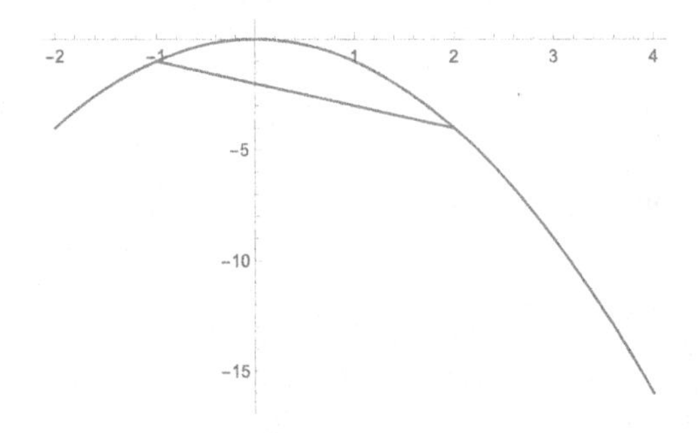

FIGURE 18.3
$f(x) = -x^2$ with line segment from $x = -1$ to $x = 2$ in Example 18.2.2.

Solution. Negating each line of the argument in Example 18.1.2 and noting that $h(x) = -f(x)$ establishes the result. ∎

What occurred in Example 18.2.2 is no accident:

Theorem 18.2.3. *$h(x)$ is concave if and only if $-h(x)$ is convex,*

Example 18.2.2 is illustrated in Figure 18.3 with $x = -1$ and $y = 2$. These correspond to the points $(-1, -1)$ and $(2, -4)$ and notice that the line segment joining the points lies below the graph of $f(x) = -x^2$. As with convexity, this is not an isolated event and holds for any line segment connecting points on the graph and is an important property of concave functions.

Highlight 18.2.4 (Geometric Meaning of Concave Function). *If $f(x)$ is a concave function, then any line segment connecting points on the graph of $f(x)$ will lie below the graph of $f(x)$.*

As with convexity this property is formally established via the *first-order condition of concavity.* Note that each of the results for convex functions translates naturally to concave functions, Their statements and proofs are left as exercises (see Exercise 18.11).

18.3 Graphs of Convex and Concave Functions

With the tools from Chapter 16, we have an alternate way to define both convex and concave functions by focusing on their graphs. First, convex functions.

As one of the meanings of the Greek prefix epi is "above", the *epigraph* of a function is the collection of points that lie above the function. That is,

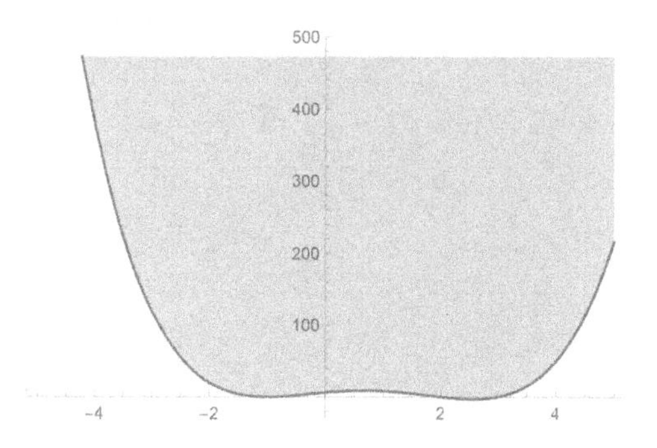

FIGURE 18.4
The epigraph of $f(x) = x^4 - 3x^3 - 3x^2 + 7x + 6$.

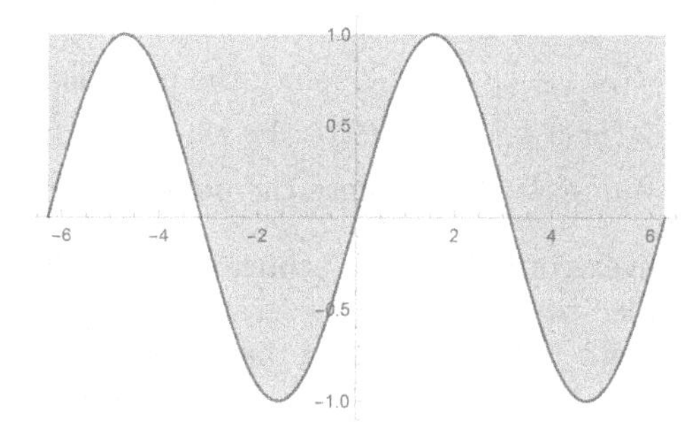

FIGURE 18.5
The epigraph of $f(x) = \sin x$.

Definition 18.3.1 (Epigraph).

$$epi(f) := \{(x, y) \mid x \text{ in the domain of } f \text{ and } y \geq f(x)\}.$$

Examples of the epigraph of familiar functions are given in Figures 18.4 and 18.5.

The following theorem will explain how convex functions got their name:

Theorem 18.3.2. *Let $f : C \to \mathbb{R}$ where C is a convex set. Then f is a convex function if and only if $epi(f)$ is a convex set.*

Proof. Let E be the epigraph of f.
($\Longleftarrow$) Suppose $(x_1, y_1), (x_2, y_2) \in E$. Since these points are in E note that we have $y_1 \geq f(x_1)$ and $y_2 \geq f(x_2)$. Further suppose f is a convex function and $0 \leq \lambda \leq 1$. Then $\lambda x_1 + [1 - \lambda]x_2$ is in C since x_1, x_2 are in C and C is convex.

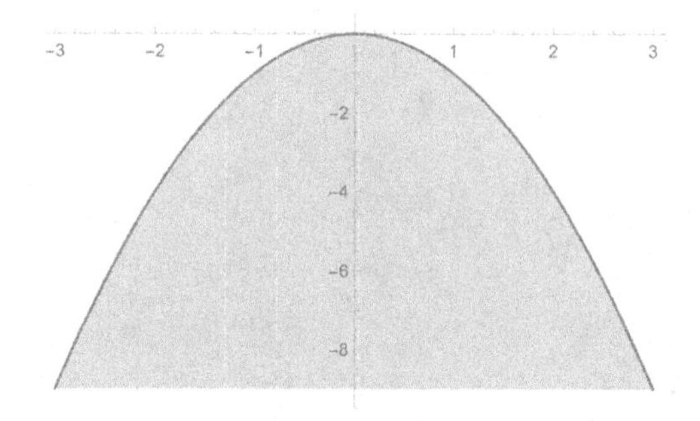

FIGURE 18.6
The hypograph of $f(x) = -x^2$.

Thus

$$\begin{aligned}\lambda x_1 + [1-\lambda]x_2 &\leq f(\lambda x_1 + [1-\lambda]x_2) \quad \text{since the point is in } E && (18.50)\\ &\leq \lambda f(x_1) + [1-\lambda]f(x_2) \quad \text{by the convexity of } f(x) && (18.51)\\ &\leq \lambda y_1 + [1-\lambda]y_2 \quad \text{since the points are in } E, && (18.52)\end{aligned}$$

showing that the convex combination of arbitrary points in E is again in E. ($\Longrightarrow$) We now consider two points on the graph of f, (x_1, y_1) and (x_2, y_2), thus $f(x_1) = y_1$ and $f(x_2) = y_2$. Think of the points as vectors and since E is convex,

$$\lambda[x_1, y_1]^T + (1-\lambda)[x_2, y_2]^T = [\lambda x_1 + (1-\lambda)x_2, \lambda y_1 + (1-\lambda)y_2]^T \in E \quad (18.53)$$

thus

$$\begin{aligned}f(\lambda x_1 + (1-\lambda)x_2) &\leq \lambda y_1 + (1-\lambda)y_2 && (18.54)\\ &= \lambda f(x_1) + [1-\lambda]f(x_2) && (18.55)\end{aligned}$$

establishing the convexity of f. □

Concave functions once again enjoy properties analogous to convex functions. First, a definition.

Definition 18.3.3 (Hypograph).

$$hypo(f) := \{(x, y) \mid x \text{ in the domain of } f \text{ and } y \leq f(x)\}.$$

An example of the hypograph of a function is given in Figure 18.6.

The result of the following theorem will not be a surprise. Its proof is left as Exercise 18.12.

Theorem 18.3.4. *Let $f : C \to \mathbb{R}$ where C is a convex set. Then f is a concave function if and only if $hypo(f)$ is a convex set.*

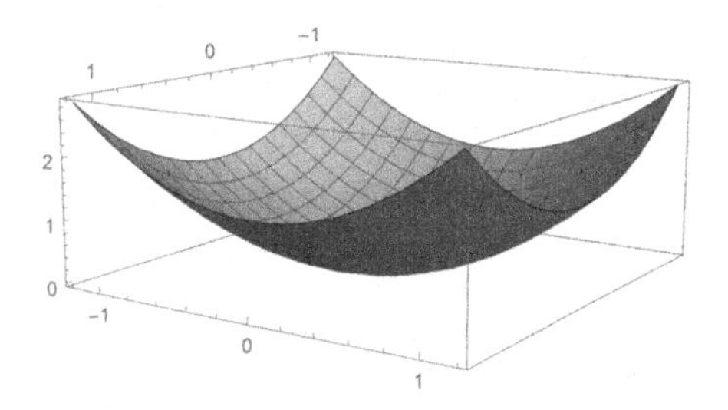

FIGURE 18.7
$f(\mathbf{x}) = {x_1}^2 + {x_2}^2$.

18.4 Multivariable Convex Functions

All of the results for convex functions of a single variable carry over to multivariable functions, including the geometric interpretations.

Definition 18.4.1 (Multivariable Convex Function). *Suppose $f : C \to \mathbb{R}$ where $C \subseteq \mathbb{R}^n$ and C is a convex set. Then f is a* convex function *on C if and only if for every $\mathbf{x}, \mathbf{y}$ in C and any $0 \leq \lambda \leq 1$,*

$$f(\lambda\mathbf{x} + [1-\lambda]\mathbf{y}) \leq \lambda f(\mathbf{x}) + [1-\lambda] f(\mathbf{y}). \tag{18.56}$$

f is strictly convex *if the inequality in 18.56 is strict.*

Example 18.4.2. *For $\mathbf{x} = [x_1.x_2]^T \in \mathbb{R}^2$, define $f(\mathbf{x}) = {x_1}^2 + {x_2}^2$. We show that f is a convex function.*

Solution. Let $\mathbf{a}, \mathbf{b} \in \mathbb{R}^2$ and $0 \leq \lambda \leq 1$. Then

$$f(\lambda\mathbf{a} + [1-\lambda]\mathbf{b}) = (\lambda a_1 + [1-\lambda]b_1)^2 + (\lambda a_2 + [1-\lambda]b_2)^2 \tag{18.57}$$

$$\leq \lambda {a_1}^2 + [1-\lambda]{b_1}^2 + \lambda {a_2}^2 + [1-\lambda]{b_2}^2 \text{ by Example 18.1.2} \tag{18.58}$$

$$= \lambda f(\mathbf{a}) + [1-\lambda] f(\mathbf{b}) \tag{18.59}$$

establishing that f satisfies 18.56. ∎

Example 18.4.2 is shown in Figure 18.7. Note that we can also talk about epigraphs of multivariable functions and the results are the same as in the single variable case and we can also observe that geometrically, as in the single variable case, the definition of a multivariable convex function means that any line segment between two points on the graph lies above the graph; specifically in its epigraph.

Definition 18.4.3. *As with the single variable case, f is* (strictly) concave *over a convex set C if and only if $-f$ is (strictly) convex.*

As all of the results for convexity translate easily to results to concavity, we will state only convexity results and leave the analogous concavity results to the reader.

Note that Definition 18.4.1 involves only two variables. We may inductively extend equation 18.56 to any number of variables.

Theorem 18.4.4. *Suppose $C \subseteq \mathbb{R}^n$ is a convex set and $f : C \to \mathbb{R}$. If $\lambda_1, \lambda_2, \ldots, \lambda_k$ are nonnegative numbers and $\lambda_1 + \lambda_2 + \cdots + \lambda_k = 1$, then f is convex if and only if*

$$(\lambda_1 \mathbf{x}_1 + \lambda_2 \mathbf{x}_2 + \cdots + \lambda_k \mathbf{x}_k) \leq \lambda_1 f(\mathbf{x}_1) + \lambda_2 f(\mathbf{x}_2) + \cdots + \lambda_k f(\mathbf{x}_k). \quad (18.60)$$

Moreover, if f is strictly convex on C and each $\lambda_i \neq 0$, then equality holds in 18.60 if and only if $\mathbf{x}_1 = \mathbf{x}_2 = \cdots = \mathbf{x}_k$.

Proof. ($\Longleftarrow$) Suppose 18.60 holds for any positive integer k. When $k = 2$, this is the definition of convexity.
($\Longrightarrow$) Now suppose f is convex. When $k = 1$, $\lambda = 1$ and 18.60 holds trivially. When $k = 2$ we again have the definition of a convex function.

For the inductive step, suppose 18.60 holds for some $k \geq 1$. Suppose $\lambda_1, \ldots, \lambda_k, \lambda_{k+1} \geq 0$ with $\lambda_1 + \cdots + \lambda_k + \lambda_{k+1} = 1$ and that $\mathbf{x}_1, \ldots, \mathbf{x}_k, \mathbf{x}_{k+1}$ are in C. If $\lambda_{k+1} = 0$, then we have the result by the induction hypothesis. Suppose $\lambda_{k+1} \neq 0$ and put

$$\Lambda = \lambda_2 + \cdots + \lambda_k + \lambda_{k+1} \text{ and} \quad (18.61)$$

$$\mathbf{X} = \frac{\lambda_2}{\Lambda}\mathbf{x}_2 + \cdots + \frac{\lambda_k}{\Lambda}\mathbf{x}_k + \frac{\lambda_{k+1}}{\Lambda}\mathbf{x}_{k+1}. \quad (18.62)$$

Since

$$\frac{\lambda_2}{\Lambda} + \cdots + \frac{\lambda_k}{\Lambda} + \frac{\lambda_{k+1}}{\Lambda} = \frac{\lambda_2 + \cdots + \lambda_{k+1}}{\lambda_2 + \cdots + \lambda_{k+1}} = 1, \quad (18.63)$$

18.62 satisfies the induction hypothesis. Thus

$$\begin{aligned}
&f(\lambda_1 \mathbf{x}_1 + \cdots + \lambda_{k+1}\mathbf{x}_{k+1}) \\
&\quad = f(\lambda_1 \mathbf{x}_1 + [\lambda_2 \mathbf{x}_2 + \cdots + \lambda_{k+1}\mathbf{x}_{k+1}]) && (18.64)\\
&\quad = f\left(\lambda_1 \mathbf{x}_1 + \Lambda\left[\frac{\lambda_2}{\Lambda}\mathbf{x}_2 + \cdots + \frac{\lambda_{k+1}}{\Lambda}\mathbf{x}_{k+1}\right]\right) && (18.65)\\
&\quad = f(\lambda_1 \mathbf{x}_1 + \Lambda \mathbf{X}) && (18.66)\\
&\quad \leq \lambda_1 f(\mathbf{x}_1) + \Lambda f(\mathbf{X}) \text{ by the convexity of } f(\mathbf{x}) && (18.67)\\
&\quad = \lambda_1 f(\mathbf{x}_1) + \Lambda f\left(\frac{\lambda_2}{\Lambda}\mathbf{x}_2 + \cdots + \frac{\lambda_{k+1}}{\Lambda}\mathbf{x}_{k+1}\right) && (18.68)\\
&\quad \leq \lambda_1 f(\mathbf{x}_1) + \Lambda\left[\frac{\lambda_2}{\Lambda}f(\mathbf{x}_2) + \cdots + \frac{\lambda_{k+1}}{\Lambda}f(\mathbf{x}_{k+1})\right] \text{ by the IH} && (18.69)\\
&\quad = \lambda_1 f(\mathbf{x}_1) + \lambda_2 f(\mathbf{x}_2) + \cdots + \lambda_{k+1} f(\mathbf{x}_{k+1}). && (18.70)
\end{aligned}$$

We also establish the second statement in the theorem by induction. It is clear that if $\mathbf{x}_1 = \mathbf{x}_2 = \cdots = \mathbf{x}_k$, then equality holds in 18.60. For the other direction, when $k = 1$ the result is trivial. Suppose $k = 2$, thus $\mathbf{X} = \mathbf{x}_2$. If f is strictly convex over C, then the inequality in 18.67 is a strict inequality (since f is strictly convex) unless

$$\mathbf{x}_1 = \mathbf{X} = \mathbf{x}_2. \tag{18.71}$$

Suppose the statement holds for some $k \geq 1$. Again, the inequality in 18.67 is a strict inequality unless

$$\mathbf{x}_1 = \mathbf{X}. \tag{18.72}$$

By the induction hypothesis, 18.69 is an equality unless

$$\mathbf{x}_2 = \cdots = \mathbf{x}_{k+1}. \tag{18.73}$$

Thus in 18.72

$$\mathbf{x}_1 = \mathbf{X} := \frac{\lambda_2}{\Lambda}\mathbf{x}_2 + \cdots + \frac{\lambda_k}{\Lambda}\mathbf{x}_k + \frac{\lambda_{k+1}}{\Lambda}\mathbf{x}_{k+1} \tag{18.74}$$

$$= \frac{\lambda_2}{\Lambda}\mathbf{x}_2 + \cdots + \frac{\lambda_k}{\Lambda}\mathbf{x}_2 + \frac{\lambda_{k+1}}{\Lambda}\mathbf{x}_2 \tag{18.75}$$

$$= \frac{\Lambda}{\Lambda}\mathbf{x}_2 = \mathbf{x}_2 \tag{18.76}$$

giving $\mathbf{x}_1 = \mathbf{x}_2 = \cdots = \mathbf{x}_{k+1}$. □

As with single variable functions, there are convenient tools for establishing convexity of multivariable functions; namely the first and second-order conditions of convexity. Recall that for single-variable functions, the first-order condition of convexity states that lines tangent to convex functions lie on or below the graph of the convex function. In dimensions higher than 2, the role of lines are played by hyperplanes. Thus the multivariable version of the first-order of convexity will give us that tangent hyperplanes to convex functions lie on or below the graph of the convex function. Note that If the reader needs a refresher on the operators that play the role of the first and second derivative for multivariable functions, please see Section 9.5.

Example 18.4.5. *Find the hyperplane tangent to* $f(\mathbf{x}) = {x_1}^2 + {x_2}^2$ *at the point* $(2, 1)$.

Solution. We showed in Example 18.4.2 that $f(\mathbf{x}) = {x_1}^2 + {x_2}^2$ is convex. The multivariable version of 9.4, that is, the hyperplane tangent to $f(\mathbf{x})$ at $\mathbf{x} = \mathbf{a}$, is

$$y = f(\mathbf{a}) + \nabla f(\mathbf{a})[\mathbf{x} - \mathbf{a}]. \tag{18.77}$$

Note that $\nabla f(\mathbf{x}) = [2x_1 2x_2]$, thus the tangent hyperplane is

$$\mathbf{y} = f([21]^T) + \nabla f([21]^T)[\mathbf{x} - [21]^T] = 4x_2 + 2x_1 - 5. \tag{18.78}$$

The graph of $f(\mathbf{x}) = {x_1}^2 + {x_2}^2$ as well as its tangent hyperplane at the point $(2, 1)$ are shown in Figure 18.8. ■

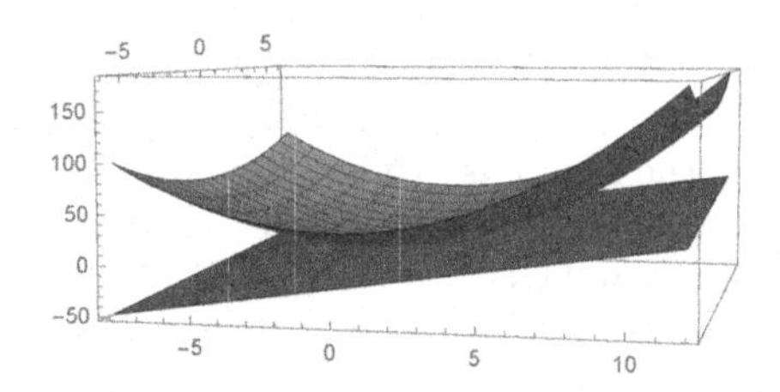

FIGURE 18.8
$f(\mathbf{x}) = {x_1}^2 + {x_2}^2$ and tangent hyperplane at $(2, 1)$.

Theorem 18.4.6 (Multivariable First-Order Condition of Convexity). *If f has continuous first partial derivatives over a convex set C, then f is convex over C if and only if for all $\mathbf{a}, \mathbf{b} \in C$*

$$\nabla f(\mathbf{a}) \cdot (\mathbf{b} - \mathbf{a}) + f(\mathbf{a}) \leq f(\mathbf{b}). \tag{18.79}$$

Moreover, f is strictly convex *if and only if the inequality in 18.79 is strict.*

Proof. ($\Longrightarrow$) Suppose f is convex and that $\mathbf{a}, \mathbf{b} \in C$ with $0 < \lambda \leq 1$. Then

$$f(\mathbf{a} + \lambda[\mathbf{b} - \mathbf{a}]) = f([1 - \lambda]\mathbf{a} + \lambda\mathbf{b}) \leq \lambda f(\mathbf{b}) + [1 - \lambda]f(\mathbf{a}) \tag{18.80}$$

giving

$$\left(\frac{f(\mathbf{a} + \lambda[\mathbf{b} - \mathbf{a}]) - f(\mathbf{a})}{\lambda}\right) \leq f(\mathbf{b}) - f(\mathbf{a}) \tag{18.81}$$

hence

$$\left(\frac{f(\mathbf{a} + \lambda[\mathbf{b} - \mathbf{a}]) - f(\mathbf{a})}{\lambda[\mathbf{b} - \mathbf{a}]}\right) [\mathbf{b} - \mathbf{a}] \leq f(\mathbf{b}) - f(\mathbf{a}). \tag{18.82}$$

As $\lambda \to 0$, $\lambda[\mathbf{b} - \mathbf{a}] \to 0$, yielding

$$\nabla f(\mathbf{a}) \cdot (\mathbf{b} - \mathbf{a}) \leq f(\mathbf{b}) - f(\mathbf{a}) \tag{18.83}$$

which gives the desired result

$$\nabla f(\mathbf{a}) \cdot (\mathbf{b} - \mathbf{a}) + f(\mathbf{a}) \leq f(\mathbf{b}). \tag{18.84}$$

($\Longleftarrow$) Now suppose that for all $\mathbf{a}, \mathbf{b} \in C$ that

$$f(\mathbf{a}) + \nabla f(\mathbf{a}) \cdot (\mathbf{b} - \mathbf{a}) \leq \mathbf{b}. \tag{18.85}$$

Let $\mathbf{c}$ and $\mathbf{d}$ be arbitrary points in C. Note that since C is convex,

$$\mathbf{h} := \lambda\mathbf{c} + (1 - \lambda)\mathbf{d} \in C. \tag{18.86}$$

Some algebra gives

$$\mathbf{d} = \frac{\mathbf{h} - \lambda\mathbf{c}}{1 - \lambda} = \frac{(1 - \lambda)\mathbf{h} - \lambda\mathbf{c} + \lambda\mathbf{h}}{1 - \lambda} = \mathbf{h} - \frac{\lambda}{1 - \lambda}(\mathbf{c} - \mathbf{h}). \tag{18.87}$$

Applying 18.85 to $\mathbf{h}$ and $\mathbf{c}$ as well as to $\mathbf{h}$ and $\mathbf{d}$ yields

$$f(\mathbf{h}) + \nabla f(\mathbf{h}) \cdot (\mathbf{c} - \mathbf{h}) \leq \mathbf{c} \quad \text{and} \tag{18.88}$$

$$f(\mathbf{h}) + \nabla f(\mathbf{h}) \cdot (\mathbf{c} - \mathbf{h}) \frac{-\lambda}{1-\lambda} \leq \mathbf{d}. \tag{18.89}$$

Multiplying 18.88 by λ and 18.89 by $1 - \lambda$ then adding gives

$$f(\mathbf{h}) := f(\lambda \mathbf{c} + [1-\lambda]\mathbf{d}) \leq \lambda \mathbf{c} + [1-\lambda]\mathbf{d} \tag{18.90}$$

thus establishing that f is convex over the set C.

Regarding the second statement, ($\Longrightarrow$) for $\lambda \neq 1$ with $\mathbf{a} \neq \mathbf{b}$, f strictly convex means the inequality in 18.80 is strict, and the result in this direction follows. ($\Longleftarrow$) For the other direction, if 18.85 is strict, then so are the inequalities in 18.88 and 18.89, and the result follows. □

Theorem 18.4.7 (Multivariable Second-Order Condition of Convexity)**.** *Suppose f has continuous second partial derivatives over some open convex set C. If the Hessian, Hf of f is positive semidefinite over C then f is convex over C. If Hf is positive definite over C, then f is strictly convex over C.*

Proof. Suppose $\mathbf{x}$ and $\mathbf{y}$ are arbitrary points in C. Since C is open, by Taylor's Theorem together with Lagrange's Remainder Theorem, there is an $\mathbf{x}^*$ on the line segment joining $\mathbf{x}$ and $\mathbf{y}$ such that

$$f(\mathbf{y}) = f(\mathbf{x}) + \nabla f(\mathbf{x}) \cdot (\mathbf{y} - \mathbf{x}) + \frac{1}{2}(\mathbf{y} - \mathbf{x}) \cdot Hf(\mathbf{x}^*)(\mathbf{y} - \mathbf{x}). \tag{18.91}$$

As $(\mathbf{y} - \mathbf{x}) \cdot Hf(\mathbf{x}^*)(\mathbf{y} - \mathbf{x})$ is a quadratic form (see Chapter 12), if Hf is positive semidefinite over C, then the quadratic form is nonnegative and

$$f(\mathbf{y}) \leq f(\mathbf{x}) + \nabla f(\mathbf{x}) \cdot (\mathbf{y} - \mathbf{x}). \tag{18.92}$$

The result then follows from Theorem 18.4.6. Likewise, if Hf is positive definite over C, then the quadratic form is positive and the desired results follow from the same argument. □

Also as with single variable functions, multivariable convex functions have extremely convenient properties regarding optimization:

Theorem 18.4.8. *Let C be a convex subset of $\mathbb{R}^n$ and $f : C \to \mathbb{R}$ a convex function. Then any local minimum of f over C is a global minimum of f over C. If f is strictly convex, then any local minimizer of f is the unique global minimizer.*

Proof. Suppose $f(\mathbf{x}^*)$ is a local minimum of $f(\mathbf{x})$ over C. Then there exists an $\epsilon > 0$ such that for all $\mathbf{x}$ with $||\mathbf{x}^*|| - \epsilon \leq ||\mathbf{x}|| \leq ||\mathbf{x}^*|| + \epsilon$ we have $f(\mathbf{x}) \geq f(\mathbf{x}^*)$ (here

$$||\cdot||$$

is the Euclidean norm). Let $\mathbf{y}$ be any arbitrary point in C and we have the result if we can show $f(\mathbf{x}^*) \leq f(\mathbf{y})$. Choose $0 < \lambda < 1$ small enough that

$$||\lambda(\mathbf{y} - \mathbf{x}^*)|| < \epsilon \tag{18.93}$$

and that

$$\mathbf{x}^* + \lambda(\mathbf{y} - \mathbf{x}^*) \in C. \tag{18.94}$$

Thus

$$||\mathbf{x}^* + \lambda(\mathbf{y} - \mathbf{x}^*) - \mathbf{x}^*|| < \epsilon \tag{18.95}$$

and since $f(x^*)$ is a local minimum

$$f(\mathbf{x}^*) \leq f(\mathbf{x}^* + \lambda(\mathbf{y} - \mathbf{x}^*)) \tag{18.96}$$
$$= f(\lambda \mathbf{y} + [1 - \lambda]\mathbf{x}^*) \tag{18.97}$$
$$\leq \lambda f(\mathbf{y}) + [1 - \lambda] f(\mathbf{x}^*) \tag{18.98}$$

with the last inequality by the convexity of $f(x)$. Rearranging and dividing by λ gives

$$f(x^*) \leq f(y)$$

establishing x^* as a minimizer. If $f(\mathbf{x})$ is strictly convex, then the above inequalities are strict and it follows that $\mathbf{x}^*$ is the unique global minimizer. □

Also, as in the single-variable case, the very useful

Theorem 18.4.9. *If f is a convex function over C with continuous first partial derivatives, then any critical of point $\mathbf{c}$ of f in C, is a global minimizer of f over C.*

Proof. Suppose $\mathbf{c}$ is a critical point of f over C and that $\mathbf{x}$ is any other value in C. By Theorem 18.4.6

$$f(\mathbf{x}) \geq \nabla f(\mathbf{c}) \cdot (\mathbf{x} - \mathbf{c}) + f(\mathbf{c}) = f(\mathbf{c}) \tag{18.99}$$

as $\nabla f(\mathbf{c}) = 0$, establishing $f(c)$ as the global minimum. □

Establishing that a function is convex using the definition or either the first or second-order conditions of convexity can sometimes be challenging. The following theorem can be a useful aid in establishing convexity.

Theorem 18.4.10 (The Algebra of Convex Functions).

i) If $f_1, f_2, \ldots, f_k$ are convex functions on a set $C \subseteq \mathbb{R}^n$, then

$$F(\mathbf{x}) := f_1(\mathbf{x}) + f_2(\mathbf{x}) + \cdots + f_k(\mathbf{x})$$

is a convex function over C. Moreover, if any of the f_i is strictly convex, then F is strictly convex.

ii) If f is a (strictly) convex function over C and c is a positive number, then cf is also a (strictly) convex function.

iii) If f is a (strictly) convex function over C and g is a (strictly) increasing convex function defined on the range of f in $\mathbb{R}$, then $g \circ f$ is a (strictly) convex function over C.

Proof. A proper proof of *i*) is by induction on the number of terms k in the finite sum forming $F(\mathbf{x})$. $k = 1$ is trivial, so suppose $k = 2$, that is f_1 and f_2 are both convex over C and that $0 \leq \lambda \leq 1$. Then for any $\mathbf{a}$ and $\mathbf{b}$ in C,

$$(f_1 + f_2)(\lambda\mathbf{a} + [1-\lambda]\mathbf{b}) := f_1(\lambda\mathbf{a} + [1-\lambda]\mathbf{b}) + f_2(\lambda\mathbf{a} + [1-\lambda]\mathbf{b}) \quad (18.100)$$

$$\leq \lambda f_1(\mathbf{a}) + [1-\lambda]f_1(\mathbf{b}) + f_2(\mathbf{a}) + [1-\lambda]f_2(\mathbf{b}) \quad (18.101)$$

$$=: \lambda(f_1 + f_2)(\mathbf{a}) + [1-\lambda](f_1 + f_2)(\mathbf{b}) \quad (18.102)$$

establishing the base case for $k = 2$. The rest of the induction proof is left to the reader (see Exercise 18.13) and note that if either f_1 or f_2 is strictly convex, then the above inequality is strict establishing the strict convexity of F.

The proof of *ii*) is left as an Exercise 18.13.

Regarding *iii*), suppose $f : C \to \mathbb{R}$ is convex and put R_f = the range of f. Further suppose $g : R_f \to \mathbb{R}$ is an increasing convex function and that $\mathbf{a}, \mathbf{b} \in C$ with $0 \leq \lambda \leq 1$. Since f is convex

$$f(\lambda\mathbf{a} + [1-\lambda]\mathbf{b}) \leq \lambda f(\mathbf{a}) + [1-\lambda]f(\mathbf{b}) \quad (18.103)$$

and therefore since g is an increasing convex function on the range of f

$$g(f(\lambda\mathbf{a} + [1-\lambda]\mathbf{b})) \leq g(\lambda f(\mathbf{a}) + [1-\lambda]f(\mathbf{b})) \quad (18.104)$$

$$\leq \lambda g(f(\mathbf{a})) + [1-\lambda]g(f(\mathbf{b})) \quad (18.105)$$

establishing that $g \circ f$ is convex. Moreover, if f is strictly convex and g is strictly increasing, the inequality in 18.104 is strict giving the composition to be strictly convex. □

We illustrate the utility of Theorem 18.4.10 with the following examples:

Example 18.4.11. *Determine the convexity of a) $f(x_1, x_2, x_3) = {x_1}^2 + {x_2}^2 + {x_3}^2$, b) $g(x_1, x_2, x_3) = e^{{x_1}^2 + {x_2}^2 + {x_3}^2}$, and c) $h(x_1, x_2) = 2{x_1}^2 + {x_2}^2 - \ln x_1 x_2$.*

Solution. a) By Example 18.1.2, $f_1(x_1, x_2, x_3) = {x_1}^2$, $f_2(x_1, x_2, x_3) = {x_2}^2$, and $f_3(x_1, x_2, x_3) = {x_3}^3$ are each convex, therefore by Theorem 18.4.10, part *i*), $f(x_1, x_2, x_3) = {x_1}^2 + {x_2}^2 + {x_3}^2$ is convex.

b) By part *a*) of this exercise and part *ii*) of Theorem 18.4.10, $g(x_1, x_2, x_3) = e^{{x_1}^2 + {x_2}^2 + {x_3}^2}$ is convex.

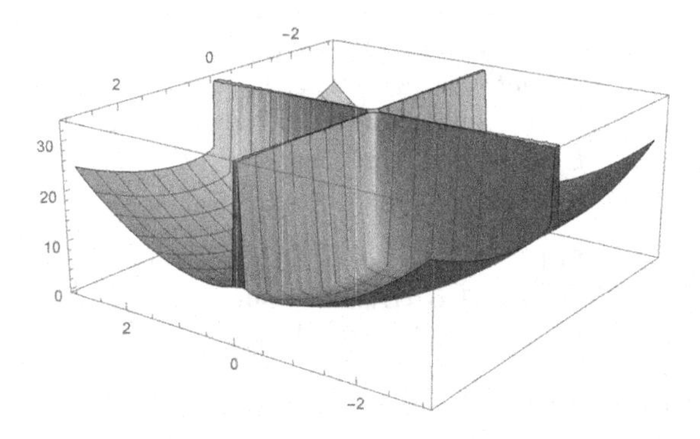

FIGURE 18.9
Strictly convex $h(x_1, x_2) = 2{x_1}^2 + {x_2}^2 - \ln x_1 x_2$ from Example 18.4.11.

c) By Example 18.1.2 together with part *ii)* of Theorem 18.4.10, $h_1(x_1, x_2) = 2{x_1}^2$ and $h_2(x_1, x_2) = {x_2}^2$ are both convex. $h_3(x_1, x_2) = -\ln x_1 x_2$ has as its Hessian $\begin{bmatrix} 1/{x_1}^2 & 0 \\ 0 & 1/{x_2}^2 \end{bmatrix}$ with the associated quadratic form

$$[x_1, x_2] \begin{bmatrix} 1/{x_1}^2 & 0 \\ 0 & 1/{x_2}^2 \end{bmatrix} \begin{bmatrix} x_1 \\ x_2 \end{bmatrix} = [1/x_1, 1/x_2] \begin{bmatrix} x_1 \\ x_2 \end{bmatrix} = 2 > 0, \tag{18.106}$$

thus the Hessian is positive definite and by Theorem 18.4.7 $h_3(x_1, x_2) = -\ln x_1 x_2$ is strictly convex. Thus it follows from Theorem 18.4.10 that $h(x_1, x_2) = 2{x_1}^2 + {x_2}^2 - \ln x_1 x_2$ is strictly convex. $h(x_1, x_2)$ is pictured in Figure 18.9.

■

As with the single variable case, analogous results to multivariable convex functions exist for multivariable concave functions. See Exercise 18.16.

18.5 Mathematical Results from Convexity

The Generalized Arithmetic Mean-Geometric Mean Theorem was introduced without proof in Section 14.1. We now have the equipment to establish the result.

Theorem 18.5.1 (The Arithmetic Mean-Geometric Mean Inequality (AGM)). *If $x_1, \ldots, x_n$ are positive real numbers and if $\delta_1, \ldots, \delta_n$ are also positive real numbers such that $\delta_1 + \cdots + \delta_n = 1$, then*

$$\prod_{i=1}^{n} {x_i}^{\delta_i} \leq \sum_{i=1}^{n} \delta_i x_i \tag{18.107}$$

with equality if and only if

$$x_1 = x_2 = \cdots = x_n = \sum_{i=1}^{n} \delta_i x_i. \tag{18.108}$$

Proof. Let $f(x) = -\ln x$. Since $f''(x) = \frac{1}{x^2} > 0$, we have that $f(x) = -\ln x$ is strictly convex by Theorem 18.1.12. Since f is strictly convex, by Theorem 18.4.4

$$-\ln\left(\sum_{i=1}^{n} \delta_i x_i\right) = f(\sum_{i=1}^{n} \delta_i x_i) \leq \sum_{i=1}^{n} \delta_i f(x_i) = -\sum_{i=1}^{n} \delta_i \ln(x_i) \tag{18.109}$$

with equality if and only if $x_1 = x_2 = \cdots = x_n$. Thus

$$\ln\left(\sum_{i=1}^{n} \delta_i x_i\right) \geq \sum_{i=1}^{n} \delta_i \ln(x_i) = \sum_{i=1}^{n} \ln({x_i}^{\delta_i}) = \ln(\prod_{i=1}^{n} {x_i}^{\delta_i}) \tag{18.110}$$

with equality if and only if $x_1 = x_2 = \cdots = x_n$. But since $\ln x$ is strictly increasing,

$$\sum_{i=1}^{n} \delta_i x_i \geq \prod_{i=1}^{n} {x_i}^{\delta_i} \tag{18.111}$$

with equality if and only if $x_1 = x_2 = \cdots = x_n$.

Now we put $X = x_1 = x_2 = \cdots = x_n$. Then

$$\prod_{i=1}^{n} {x_i}^{\delta_i} = \prod_{i=1}^{n} X^{\delta_i} = X^{(\sum_{i=1}^{n} \delta_i)} = X^1 = X, \tag{18.112}$$

establishing 18.108. □

Note that Theorem 18.5.1 is one form of Jensen's Inequality and is an acceptable alternate name of the theorem.

Another useful named inequality is

Theorem 18.5.2 (Young's Inequality). *Suppose p and q are real numbers both greater than 1 satisfying $p^{-1} + q^{-1} = 1$. If x and y are positive real numbers, then*

$$xy \leq \frac{x^p}{p} + \frac{y^q}{q}$$

with equality precisely when $x^p = y^q$.

Proof. Since $\frac{1}{p}, \frac{1}{q} > 0$ and $\frac{1}{p} + \frac{1}{q} = 1$, by the AGM

$$xy = (x^p)^{\frac{1}{p}}(y^q)^{\frac{1}{q}} \leq \frac{x^p}{p} + \frac{y^q}{q} \tag{18.113}$$

with equality if and only if $x^p = y^q$. □

What follows is a useful extension of Young's Inequality to vectors. First, a definition.

Recall that the magnitude or Euclidean norm of a vector is $\mathbf{x} = [x_1, x_2, \ldots, x_n]$ is

$$||\mathbf{x}|| := \sqrt{x_1{}^2 + x_2{}^2 + \cdots + x_n{}^2}.$$

We generalize this to

Definition 18.5.3 (p norm). *Let p be a real number with $p \geq 1$. For a vector $\mathbf{x}$, define its p **norm** to be*

$$||\mathbf{x}||_p := \sqrt[p]{x_1{}^p + x_2{}^p + \cdots + x_n{}^p}.$$

With this, we may now state and prove

Corollary 18.5.4 (Hölder's Inequality). *Suppose p and q are real numbers with $p, q \geq 1$ which also satisfy $\frac{1}{p} + \frac{1}{q} = 1$ and that $\mathbf{x} = [x_1, x_2, \ldots, x_n]$ and $\mathbf{y} = [y_1, y_2, \ldots, y_n]$ are vectors in $\mathbb{R}^n$. Then*

$$\sum_{i=1}^{n} |x_i y_i| \leq \left(\sum_{i=1}^{n} |x_i|^p\right)^{\frac{1}{p}} \left(\sum_{i=1}^{n} |y_i|^q\right)^{\frac{1}{q}} = ||\mathbf{x}||_p ||\mathbf{y}||_q. \tag{18.114}$$

Proof. If either $\mathbf{x} = \mathbf{0}$ or $\mathbf{y} = \mathbf{0}$, then both sides of 18.114 are 0 and the inequality is satisfied. Suppose that neither $\mathbf{x}$ or $\mathbf{y}$ is $\mathbf{0}$, then for each $1 \leq i \leq n$, Young's Inequality (Theorem 18.5.2) gives

$$\frac{|x_i y_i|}{||\mathbf{x}||_p ||\mathbf{y}||_q} \leq \frac{1}{p} \frac{|x_i|^p}{||\mathbf{x}||_p^p} + \frac{1}{q} \frac{|y_i|^q}{||\mathbf{y}||_q^q}. \tag{18.115}$$

Summing over $1 \leq i \leq n$ and noting that $||\mathbf{x}||_p^p = \sum_{i=1}^{n} |x_i|^p$ and $||\mathbf{y}||_q^q = \sum_{i=1}^{n} |y_i|^q$,

$$\frac{1}{||\mathbf{x}||_p ||\mathbf{y}||_q} \sum_{i=1}^{n} |x_i y_i| \leq \frac{1}{p} \frac{1}{||\mathbf{x}||_p^p} \sum_{i=1}^{n} |x_i|^p + \frac{1}{q} \frac{1}{||\mathbf{y}||_q^q} \sum_{i=1}^{n} |y_i|^q = \frac{1}{p} + \frac{1}{q} = 1. \tag{18.116}$$

Multiplying both sides of 18.116 by $||\mathbf{x}||_p ||\mathbf{y}||_q$ gives

$$\sum_{i=1}^{n} |x_i y_i| \leq ||\mathbf{x}||_p ||\mathbf{y}||_q = \left(\sum_{i=1}^{n} |x_i|^p\right)^{\frac{1}{p}} \left(\sum_{i=1}^{n} |y_i|^q\right)^{\frac{1}{q}}. \tag{18.117}$$

□

A corollary to the corollary is

Corollary 18.5.5 (The Cauchy-Schwarz Inequlity). *Let $\mathbf{x} = [x_1, x_2, \ldots, x_n]$ and $\mathbf{y} = [y_1, y_2, \ldots, y_n]$ be vectors in $\mathbb{R}^n$. Then*

$$|\mathbf{x} \cdot \mathbf{y}| \leq ||\mathbf{x}||\ ||\mathbf{y}||.$$

The proof of the Cauchy-Schwarz Inequality is left as an exercise (Exercise 18.17).

Our "named inequalities from convexity" mini-section will now be complete with

Theorem 18.5.6 (Minkowski's Inequality)**.** *If* $\mathbf{x} = [x_1, x_2, \ldots, x_n]$ *and* $\mathbf{y} = [y_1, y_2, \ldots, y_n]$ *are vectors in* $\mathbb{R}^n$ *and if* $p \geq 1$*, then*

$$||\mathbf{x}+\mathbf{y}||_p = \left(\sum_{i=1}^{n} |x_i + y_i|^p\right)^{\frac{1}{p}} \leq \left(\sum_{i=1}^{n} |x_i|^p\right)^{\frac{1}{p}} + \left(\sum_{i=1}^{n} |y_i|^p\right)^{\frac{1}{p}} = ||\mathbf{x}||_p + ||\mathbf{y}||_p. \tag{18.118}$$

Proof. If $\mathbf{x} = \mathbf{0}$ or $\mathbf{y} = \mathbf{0}$, then 18.118 is 0 is an equality. Hence suppose $\mathbf{x} \neq \mathbf{0}$ and $\mathbf{y} \neq \mathbf{0}$. If $p = 1$, then for each $1 \leq i \leq n$ $|x_i + y_i| \leq |x_i| + |y_i|$ holds by the Triangle Inequality (Theorem B.2.3), and we get 18.118 by summing over $1 \leq i \leq n$. Thus we now also suppose that $p > 1$.

For this case, for $t > 0$, define $f(t) = t^p$. As t is positive and $p > 1$,

$$f''(t) = p(p-1)t^{p-2} > 0 \tag{18.119}$$

establishing that f is strictly convex by the second-order condition of convexity (Theorem 18.1.12). Since

$$\frac{||x||_p}{||x||_p + ||y||_p} + \frac{||y||_p}{||x||_p + ||y||_p} = 1, \tag{18.120}$$

we have for each i by the convexity of $f(t) = t^p$ that

$$\begin{aligned} &\left(\frac{||\mathbf{x}||_p}{||\mathbf{x}||_p + ||\mathbf{y}||_p} \frac{|x_i|}{||\mathbf{x}||_p} + \frac{||\mathbf{y}||_p}{||\mathbf{x}||_p + ||\mathbf{y}||_p} \frac{|y_i|}{||\mathbf{y}||_p}\right)^p \\ &\quad < \frac{||x||_p}{||x||_p + ||y||_p} \left(\frac{|x_i|}{||\mathbf{x}||_p}\right)^p + \frac{||y||_p}{||x||_p + ||y||_p} \left(\frac{|y_i|}{||\mathbf{y}||_p}\right)^p. \end{aligned} \tag{18.121}$$

Summing over i,

$$\begin{aligned} &\sum_{i=1}^{n} \left(\frac{|x_i + y_i|}{||\mathbf{x}||_p + ||\mathbf{y}||_p}\right)^p \\ &\quad \leq \sum_{i=1}^{n} \left(\frac{|x_i| + |y_i|}{||\mathbf{x}||_p + ||\mathbf{y}||_p}\right)^p \quad \text{by the triangle inequality} && (18.122) \\ &\quad = \sum_{i=1}^{n} \left(\frac{||\mathbf{x}||_p}{||\mathbf{x}||_p + ||\mathbf{y}||_p} \frac{|x_i|}{||\mathbf{x}||_p} + \frac{||\mathbf{y}||_p}{||\mathbf{x}||_p + ||\mathbf{y}||_p} \frac{|y_i|}{||\mathbf{y}||_p}\right)^p && (18.123) \\ &\quad \leq \frac{||\mathbf{x}||_p}{||\mathbf{x}||_p + ||\mathbf{y}||_p} \sum_{i=1}^{n} \left(\frac{|x_i|}{||\mathbf{x}||_p}\right)^p + \frac{||\mathbf{y}||_p}{||\mathbf{x}||_p + ||\mathbf{y}||_p} \sum_{i=1}^{n} \left(\frac{|y_i|}{||\mathbf{y}||_p}\right)^p \quad \text{by 18.121} && (18.124) \\ &\quad = \frac{||\mathbf{x}||_p}{||\mathbf{x}||_p + ||\mathbf{y}||_p} \frac{||\mathbf{x}||_p^p}{||\mathbf{x}||_p^p} + \frac{||\mathbf{y}||_p}{||\mathbf{x}||_p + ||\mathbf{y}||_p} \frac{||\mathbf{y}||_p^p}{||\mathbf{y}||_p^p} = 1. && (18.125) \end{aligned}$$

Thus a little multiplication gives

$$\sum_{i=1}^{n} |x_i + y_i|^p \leq (||\mathbf{x}||_p + ||\mathbf{y}||_p)^p \tag{18.126}$$

which gives the result once we take p^{th} roots of both sides. □

18.6 Exercises

Exercise 18.1. *In Example 18.1.2, we showed that $f(x) = x^2$ is a convex function over $\mathbb{R}$ by using the definition of a convex function. As an alternative approach, establish convexity using i) the first order condition of convexity (Theorem 18.1.6) and ii) the second-order condition of convexity (Theorem 18.1.12).*

Exercise 18.2. *Let $C = [0, 1] \cup (2, 3) \subset \mathbb{R}$. Show that C is not a convex set.*

Exercise 18.3. *Show that any linear function $f(x) = ax + b$ is both convex and concave.*

Exercise 18.4. *Suppose f and g are convex functions over an open interval I and that $c \geq 0$. Show that fg, c, and $f \pm c$ are each convex over I.*

Exercise 18.5. *Prove Theorem 18.1.4 [HINT: use the Squeeze Theorem].*

Exercise 18.6. *Consider the non-continuous function*

$$f(x) := \begin{cases} 0 \text{ when } 0 \leq x < 1 \\ \pi \text{ when } x = 1. \end{cases}$$

Show that f is convex over $[0, 1]$ (thus showing that the interval in Theorem 18.1.4 must be open).

Exercise 18.7. *Show that $f(x) = |x|$ is not differentiable at $x = 0$ but is convex over $\mathbb{R}$.*

Exercise 18.8. *Show that*

a) $f(x) = e^{ax}$ is a convex function over $\mathbb{R}$ where $a \in \mathbb{R}$;
b) $g(x) = x^a$ is a convex function over $\mathbb{R}_{++} := (0, \infty)$ for $a \leq 0$ or $a \geq 1$ [Note that $\mathbb{R}_{+} := [0, \infty)$];
c) $h(x) = -x^a$ is a convex function over $\mathbb{R}_{++}$ for $0 \leq a \leq 1$;
d) $k(x) = |x|^a$ for $a \geq 1$ is a convex function over $\mathbb{R}$;
e) $l(x) = -\ln x$ is a convex function over $\mathbb{R}_{++}$;

f) $m(x) = x \ln x$ is a convex function over $\mathbb{R}_{++}$; and
g) $n(x) = \frac{1}{x}$ is convex over $(0, \infty)$ but concave over $(-\infty, 0)$.

Exercise 18.9. *A function f is said to be* strongly convex *if for all a, b in its domain*

$$f(\lambda a + (1-\lambda)b) \leq \lambda f(a) + (1-\lambda)f(b) - \lambda(1-\lambda)\frac{m}{2}\left(||a-b||_k\right)^2 \quad (18.127)$$

where $m \geq 0$ and $||\cdot||_k$ is any norm. Note 18.127 is equivalent to saying $f(x) - \frac{m}{2}\left(||x||_k\right)^2$ is convex; i.e. f is at least as convex as a quadratic function. Also, 18.127 reduces to the definition of a convex function when $m = 0$. Show that f being strongly convex implies f is strictly convex. Provide a counterexample showing that the converse is false.

Exercise 18.10. *Let $f : C \to \mathbb{R}$ where $C \subseteq \mathbb{R}$ and C is a convex set and f is a concave function. Let $F = -f$. Restate Definition 18.1.1 in terms of F.*

Exercise 18.11. *State and prove concave versions of the following convex results:*

a) Theorem 18.1.4,
b) Observation 18.1.5,
c) Theorem 18.1.6,
d) Remark 18.1.7/Highlight 18.1.8,
e) Theorem 18.1.12,
f) Remark 18.1.13/Highlight 18.1.14 (this will involve a proof and a counterexample), and
g) Theorem 18.1.15.

Exercise 18.12. *Prove Theorem 18.3.4.*

Exercise 18.13. *a) Complete the induction proof in part i) of Theorem 18.4.10.*
b) Prove part ii) of Theorem 18.4.10.

Exercise 18.14. *Let I be a real interval and define $M(x) = \max\{x_\alpha \mid x_\alpha \in I\}$. Show that $M(x)$ is a convex function.*

Exercise 18.15. *Determine the convexity over the reals of*

a) $f(x_1, x_2) = {x_1}^2 - 2x_1x_2 + 4{x_2}^2$
b) $g(x_1, x_2, x_3) = 3{x_1}^2 + 3{x_2}^2 + 3{x_3}^2 - 2x_1x_2 - 2x_1x_3 - 2x_2x_3$ and
c) $h(x_1, x_2, x_3) = {x_1}^2 + {x_2}^2 + {x_3}^2 - \ln x_1x_2x_3$.

Exercise 18.16. *a) Let $f : C \to \mathbb{R}$ where $C \subseteq \mathbb{R}^n$ and C is a convex set and f is a concave multivariable function. Let $F = -f$. Restate Definition 18.4.1 in terms of F.*
b) State and prove the multivariable concave version of Theorem 18.4.4.
c) State and prove the multivariable concave version of Theorem 18.4.6.

d) State and prove the multivariable concave version of Theorem 18.4.7.
e) State and prove the multivariable concave version of Theorem 18.4.8.
f) State and prove the multivariable concave version of Theorem 18.4.9.
g) State and prove the multivariable concave version of Theorem 18.4.10.

Exercise 18.17. *Prove the Cauchy-Schwarz Inequality (Corollary 18.5.5).*

19

Convex Optimization

Convex optimization involves optimizing a convex function over a convex set. A large number of problems, arising from a variety of application areas, can be formulated as convex optimization problems. These formulations are particularly attractive because convex optimization problems can be solved efficiently in theory, namely in polynomial time, and most of the time in practice. Perhaps this should not come as a surprise in light of Theorem 18.1.15, which establishes that all locally optimal points of a convex function are globally optimal.

This chapter is about a few important concepts in convex optimization and is organized as follows. Subsection 19.1 introduces the mathematical definition of a convex optimization problem along with a few examples. Like linear programs, convex optimization problems also have dual problems. We provide a procedure for deriving the dual a convex optimization problem in Subsection 19.2. In Subsection 19.3, we introduce and analyze an algorithm for solving unconstrained convex optimization problems. We also discuss other algorithms for solving convex optimization problems.

19.1 Convex Optimization and Applications

A convex optimization problem is a problem of the form

$$\begin{aligned} \min_{x \in \mathbb{R}^n} \quad & f(x) \\ \text{s.t.} \quad & g_i(x) \leq 0 \quad i = 1, \ldots, m \end{aligned} \tag{19.1}$$

where $f : \mathbb{R}^n \to \mathbb{R}$ and $g_1, \ldots, g_m : \mathbb{R}^n \to \mathbb{R}$ are convex functions. It can be shown that the feasible region of (19.1) is a convex set; we leave this as Exercise 19.1. Accordingly, convex optimization involves minimizing a convex function over a convex set.

Let us consider a few examples, some of which we are already familiar with from previous chapters.

DOI: 10.1201/9780367425517-19

Example 19.1.1. *Consider the linear program*

$$\begin{aligned} \min_{x\in\mathbb{R}^n} \quad & c^\top x \\ s.t. \quad & Ax \geq b \\ & x \geq 0, \end{aligned}$$

where $(A, b, c) \in \mathbb{R}^{(m-n)\times n} \times \mathbb{R}^{(m-n)} \times \mathbb{R}^n$. *Let* a_i *denote the* i*-th row of* A. *The linear program has* n *variables,* $m - n$ *linear constraints of the form* $b_i - a_i^\top x \leq 0$, *and* n *linear constraints of the form* $-x_i \leq 0$. *It follows that the linear program is an instance of* (19.1) *with* $f(x) := c^\top x$ *and*

$$g_i(x) := \begin{cases} b_i - a_i^\top x & i = 1, \ldots, m-n \\ -x_i & i = m-n+1, \ldots, m. \end{cases}$$

Next we consider a convex quadratic optimization problem in Example (19.1.2) below. Before considering the example, it is worthwhile to recall that not all quadratic functions are convex. Indeed, the one-dimensional function $f(x) = x^2$ is convex, while the one-dimensional function $f(x) = -x^2$ is not.

Example 19.1.2. *A fundamental class of convex optimization problems are quadratic optimization problems of the form*

$$\begin{aligned} \min_{x\in\mathbb{R}^n} \quad & \frac{1}{2}x^\top Qx + c^\top x \\ s.t. \quad & Ax \geq b, \end{aligned} \tag{19.2}$$

where $(A, b, c) \in \mathbb{R}^{m\times n} \times \mathbb{R}^m \times \mathbb{R}^n$ *and* $Q \in \mathbb{R}^{n\times n}$ *is symmetric positive definite. Because* Q *is symmetric positive definite, the objective* $\frac{1}{2}x^\top Qx + c^\top x$ *is strictly convex (see Exercise 19.3). It follows that* (19.2) *is a convex optimization problem as the feasible region is described by a system of linear inequalities.*

Next we consider two special instances of (19.2) that arise in machine learning and data science.

Example 19.1.3. Support vector machines *(SVMs) are a machine learning model used for binary classification. We provide a brief and informal description of SVMs here, but we discuss the model with more care in Chapter 29.*

In this model, we have labeled data $(a_1, y_1), \ldots, (a_m, y_m) \in \mathbb{R}^n \times \{-1, 1\}$, *and we would like to find a hyperplane that separates the data points in* $\mathbb{R}^n$ *(by their labels* -1 *and* 1*) as much as possible, in some sense. We show in Chapter 29 that we can formulate this task as the quadratic optimization problem*

$$\begin{aligned} \min_{x\in\mathbb{R}^n, b\in\mathbb{R}} \quad & x^\top x \\ s.t. \quad & y_i(a_i^\top x - b) \geq 1 \quad i = 1, \ldots, m \\ & x \geq 0. \end{aligned}$$

Note that this optimization problem is indeed an instance of (19.2).

Example 19.1.4. *Another special instance of* (19.2) *is linear regression*

$$\min_{x\in\mathbb{R}^n} \|Ax-b\|^2 = x^\top A^\top Ax - 2b^\top Ax + b^\top b,$$

where $(A,b) \in \mathbb{R}^{k\times n} \times \mathbb{R}^k$. *Linear regression is a fundamental problem in machine learning and data science, and we discuss it in more detail in Chapter 27.*

In contrast to the examples discussed above, linear regression is an unconstrained optimization problem. In other words, it is an instance of (19.1) *with* $m = 0$.

A related unconstrained optimization problem (also discussed in Chapter 27) is Lasso

$$\min_{x\in\mathbb{R}^n} \|Ax-b\|^2 + \lambda\|x\|_1,$$

where $\lambda \in \mathbb{R}_+$ *is a model parameter. In further contrast to all of the examples discussed so far (including linear regression), Lasso does not have a differentiable objective function due to the term* $\lambda\|x\|_1$. *Indeed, in one dimension, this term is just* $\lambda|x|$, *a scaled absolute value function.*

So far we have seen that linear and certain quadratic optimization problems are convex, but there are much more sophisticated convex optimization problems (that are still of practical interest). Let us take a look at one of these problems that arises in data science.

Example 19.1.5. *Suppose we are given data points* $a_1, \ldots, a_m \in \mathbb{R}^n$ *and asked to compute an* ellipsoid *that (i) contains these points and (ii) is of minimum* volume. *What is an ellipsoid and what is its volume? Let us take some time to unpack this!*

An ellipsoid *is a set* $E(Q,c) \subseteq \mathbb{R}^n$ *of the form*

$$E(Q,c) = \{x \in \mathbb{R}^m : (x-c)^\top \Sigma (x-c) \leq 1\}$$

that is parameterized in terms of $c \in \mathbb{R}^n$ *and a symmetric positive definite matrix* $\Sigma \in \mathbb{R}^{n\times n}$. *The point* c *is called the* center *of the ellipsoid, and* Σ *is called the* shape *matrix.*

How should we think about this object? Some intuition is as follows. In two dimensions, an ellipsoid is just an ellipse with center c, *and* Σ *captures it shape. And more generally, we can obtain any ellipsoid by simply translating and stretching the unit ball centered at the origin; see Exercise 19.4.*

The volume *of the ellipsoid* $E(Q,c)$ *is*

$$\frac{\pi^{n/2}}{\Gamma(n/2+1)} \frac{1}{\sqrt{\det Q}},$$

where $\Gamma(\cdot)$ *is the gamma function. Note that the volume of an ellipsoid does not depend on the location of its center, which makes sense.*

The minimum volume ellipsoid that contains $a_1, \ldots, a_m$ is of interest in data science because it tells us about the "shape" of the data. In particular, it informs us of the directions in which the data varies the most and least.

Let us return to the problem at hand. In order to find the minimum volume ellipsoid containing $a_1, \ldots, a_m$, we claim it is sufficient to solve the optimization problem

$$\begin{aligned} \min_{(Q,c)\in\mathbb{R}^{n\times n}\times\mathbb{R}^n} \quad & \det Q^{-1/2} \\ s.t. \quad & (a_i - c)^\top Q (a_i - c) \le 1 \quad i = 1, \ldots, m \\ & Q \succ 0. \end{aligned} \tag{19.3}$$

First, notice that the variables of (19.3) *are a matrix Q and a vector c. Observe that the first m constraints ensure that the ellipsoid contains $a_1, \ldots, a_m$, respectively. And the last constraint $Q \succ 0$ is shorthand for writing Q should be positive definite. We leave it to the reader to show that the objective is correct; see Exercise 19.5.*

Unfortunately, (19.3) *is not a convex optimization problem. However, we can resolve this in two steps. First, we perform the change of variables*

$$M = Q^{1/2} \text{ and } z = Q^{1/2} c$$

to obtain the optimization problem

$$\begin{aligned} \min_{M,z} \quad & \det M^{-1} \\ s.t. \quad & (Ma_i - z)^\top (Ma_i - z) \le 1 \quad i = 1, \ldots, m \\ & M \succ 0. \end{aligned} \tag{19.4}$$

Next, we apply a logarithmic transform to the objective (which does not change the optimal solution as $\log(\cdot)$ *is a monotone function) to obtain*

$$\begin{aligned} \min_{M,z} \quad & -\log\det M \\ s.t. \quad & (Ma_i - z)^\top (Ma_i - z) \le 1 \quad i = 1, \ldots, m \\ & M \succ 0. \end{aligned} \tag{19.5}$$

Problem (19.4) *is a convex optimization problem; see Exercise 19.6.*

19.2 Duality

Recall from Chapter 6 that every linear program has a dual linear program. Here we develop a duality theory for convex optimization. First we introduce a general three step procedure for constructing dual problems. Next we establish some properties of the dual problems, and then we demonstrate how to use the three step procedure to construct the dual of the linear programming problem in Example 19.1.1 and the quadratic optimization problem in Example 19.1.2.

We consider nonlinear optimization problems of the form

$$\begin{aligned} \min_{x \in \mathbb{R}^n} \quad & f(x) \\ \text{s.t.} \quad & g_i(x) \leq 0 \quad i = 1, \ldots, m \\ & x \in X, \end{aligned} \tag{19.6}$$

where $f : \mathbb{R}^n \to \mathbb{R}$ and $g_1, \ldots, g_m : \mathbb{R}^n \to \mathbb{R}$ are functions, and $X \subseteq \mathbb{R}^n$. So the feasible region of (19.6) is

$$\{x \in \mathbb{R}^n : g_i(x) \leq 0, i = 1, \ldots, m\} \cap X.$$

Note that we have not assumed that $f : \mathbb{R}^n \to \mathbb{R}$ and $g_1, \ldots, g_m : \mathbb{R}^n \to \mathbb{R}$ are convex functions.

Below we provide the three step procedure for constructing the dual problem of (19.6).

1. First, we construct the *Lagrangian function* $L : \mathbb{R}^n \times \mathbb{R}^m_+ \to \mathbb{R}$ defined by

$$L(x, u) := f(x) + u^\top g(x).$$

Note that u is a nonnegative vector. Intuitively, we can use the Lagrangian function to convert a constrained optimization problem into an unconstrained optimization problem as follows. Consider minimizing the Lagrangian with respect to x over X for a fixed value of u. Then, the term $u^\top g(x)$ penalizes violating the constraints $g_i(x) \leq 0, i = 1, \ldots, m$.

2. Next, we construct the *dual function* $D : \mathbb{R}^m \to \mathbb{R}$ defined by

$$D(u) := \min_{x \in X} L(x, u). \tag{19.7}$$

We will see that the in the context of linear programming and quadratic optimization, the optimization problem in (19.7) admits a closed form solution (in terms of u). We will substitute this closed form solution into (19.7) to obtain an analytic description of $D(u)$.

3. Finally, we construct the *dual problem* defined by

$$\begin{aligned} \max_{u \in \mathbb{R}^m} \quad & D(u) \\ \text{s.t.} \quad & u \geq 0. \end{aligned} \tag{19.8}$$

It follows that whenever we have an analytic description for $D(u)$, problem (19.8) will have an analytic description like (19.6).

Before considering some examples, we establish some properties of the dual problem. Proposition 19.2.1 below shows that the dual problem satisfies a weak duality relationship with (19.6). Interestingly, the proof does not require that f and $g_1, \ldots, g_m$ are convex. In other words, weak duality holds for general nonlinear optimization problems.

Proposition 19.2.1. *Suppose that the primal problem* (19.6) *and dual problem* (19.8) *are feasible. Then the optimal values* p^* *and* d^* *of the primal and dual are finite, respectively, and*

$$p^* \geq d^*.$$

Proof. Suppose $\hat{x}$ is feasible for (19.6) and $\hat{u}$ is feasible for (19.8). Then

$$\begin{aligned} f(\hat{x}) &\geq f(\hat{x}) + \hat{u}^\top g(\hat{x}) \\ &\geq \min_{x \in X} f(x) + \hat{u}^\top g(x) \\ &= D(\hat{u}). \end{aligned}$$

It follows that the objective values p^* and d^* of the primal and dual, respectively, are finite, and $p^* \geq d^*$. □

Next we show that whenever f and $g_1, \ldots, g_m$ are convex, the dual function (19.7) is concave on $\{u \in \mathbb{R}^m : u \geq 0\}$, implying we can write the dual problem (19.8) as a convex minimization problem (by simply minimizing $-D(u)$ over the convex region $\{u \in \mathbb{R}^m : u \geq 0\}$).

Proposition 19.2.2. *Suppose that* f *and* $g_1, \ldots, g_m$ *are convex. Then the dual function* (19.7) *is concave on* $\{u \in \mathbb{R}^m : u \geq 0\}$.

Proof. Let $u_1, u_2 \in \mathbb{R}^m_+$ and $u = \lambda u_1 + (1 - \lambda)u_2$, where $\lambda \in [0, 1]$. Then

$$\begin{aligned} D(u) &= \min_{x \in X} f(x) + u^\top g(x) \\ &= \min_{x \in X} \lambda(f(x) + u_1^\top g(x)) + (1 - \lambda)(f(x) + u_2^\top g(x)) \\ &\geq \lambda(\min_{x \in X} f(x) + u_1^\top g(x)) + (1 - \lambda)(\min_{x \in X} f(x) + u_2^\top g(x)) \\ &= \lambda D(u_1) + (1 - \lambda)D(u_2), \end{aligned}$$

where the inequality follows from the fact that f and $g_1, \ldots, g_m$ are convex. Thus, $D(u)$ is concave on $\mathbb{R}^m_+$. □

Unlike linear programming, strong duality does not always hold in the context of (19.6), even if (19.6) is a convex optimization problem. However, there are a few mild sufficient conditions that ensure strong duality holds. For example, if the feasible region contains a *Slater's point*, namely a point $x \in \mathbb{R}^n$ such that $g_i(x) < 0$ for all $i = 1, \ldots, m$ and $x \in \text{int}(X)$, then it can be shown that strong duality holds. The proof of this is outside of the scope of this chapter, but we state the result in Theorem 19.2.3 below.

Theorem 19.2.3. *Suppose f and $g_1, \ldots, g_m$ are convex and the feasible region of (19.6) contains a Slater's point. Let p^* be the optimal value of (19.6) and d^* be the optimal value of (19.8). Then strong duality holds, namely*

$$p^* = d^*.$$

Let us apply the three step procedure introduced above to some examples. First we consider the linear programming problem from Example 19.1.1.

Example 19.2.4. *Consider the linear program*

$$\begin{aligned} \min_{x \in \mathbb{R}^n} \quad & c^\top x \\ s.t. \quad & Ax \geq b \\ & x \in X, \end{aligned}$$

where $X = \{x \in \mathbb{R}^n : x \geq 0\}$. The Lagrangian function is defined by

$$L(x, u) = c^\top x + u^\top (b - Ax) = u^\top b + (c - A^\top u)^\top x.$$

And hence the dual function is given by

$$\begin{aligned} D(u) = u^\top b + \min_{x \in \mathbb{R}^n} \quad & (c - A^\top u)^\top x \\ s.t. \quad & x \geq 0. \end{aligned} \tag{19.9}$$

Consider the minimization problem in (19.9) for a fixed value of $u \in \mathbb{R}^m$. If $A^\top u \leq c$, then the optimal solution is $x = 0$. Otherwise, there is no optimal solution because we can choose x in a way that makes the objective arbitrarily small. From this discussion, it follows that

$$D(u) = \begin{cases} u^\top b & A^\top u \leq c \\ -\infty & \textit{otherwise.} \end{cases} \tag{19.10}$$

Recall that the dual problem is defined by

$$\begin{aligned} \max_{u \in \mathbb{R}^m} \quad & D(u) \\ s.t. \quad & u \geq 0. \end{aligned} \tag{19.11}$$

And hence it follows from (19.10) that the dual of the linear program is

$$\begin{aligned} \max_{u \in \mathbb{R}^m} \quad & b^\top u \\ s.t. \quad & A^\top u \leq c \\ & u \geq 0, \end{aligned}$$

which is also a linear program. This of course should come as no surprise in light of Chapter 6.

Example 19.2.5. *Recall the quadratic optimization problem from Exercise 19.1.2:*

$$\begin{aligned} \min_{x\in\mathbb{R}^n} \quad & \frac{1}{2}x^\top Qx + c^\top x \\ s.t. \quad & Ax \geq b, \end{aligned}$$

where Q is symmetric positive semidefinite. In this context, we will take $X = \mathbb{R}^n$.

The Lagrangian function is then defined by

$$L(x,u) = \frac{1}{2}x^\top Qx + c^\top x + u^\top(b - Ax) = u^\top b + \frac{1}{2}x^\top Qx + (c - A^\top u)^\top x,$$

and hence the dual function is defined by

$$D(u) = \min_{x\in\mathbb{R}^n} L(x,u) = u^\top b + \min_{x\in\mathbb{R}^n} (c - A^\top u)^\top x + \frac{1}{2}x^\top Qx \tag{19.12}$$

For a fixed value of u, the problem

$$\min_{x\in\mathbb{R}^n} (c - A^\top u)^\top x + \frac{1}{2}x^\top Qx$$

is a convex quadratic problem, and so by Theorem 18.1.15, each of its optimal solutions x must satisfy

$$0 = \nabla_x L(x,u) = c - A^\top u + Qx,$$

or equivalently (because Q is positive definite),

$$x = -Q^{-1}(c - A^\top u).$$

Substituting this into (19.12) gives

$$D(u) = u^\top b - \frac{1}{2}(c - A^\top u)^\top Q^{-1}(c - A^\top u).$$

Thus, the dual problem is

$$\begin{aligned} \max_{u\in\mathbb{R}^m} \quad & u^\top b - \frac{1}{2}(c - A^\top u)^\top Q^{-1}(c - A^\top u) \\ s.t. \quad & u \geq 0. \end{aligned}$$

19.3 Subgradient Descent

We present and analyze an algorithm called *subgradient descent* that is used to solve unconstrained optimization problems of the form

$$\min_{x\in\mathbb{R}^n} f(x), \tag{19.13}$$

where $f : \mathbb{R}^n \to \mathbb{R}$ is convex.

Let us begin by considering the definition of a subgradient.

Definition 19.3.1. *We say that $g \in \mathbb{R}^n$ is a* subgradient *of f at $x \in \mathbb{R}^n$ if*

$$f(y) \geq f(x) + g^\top (y - x) \text{ for all } y \in \mathbb{R}^n.$$

The notion of a subgradient generalizes the notion of a gradient to the context of general (not necessarily differentiable) convex functions. To see this, consider the first order characterization of convexity presented in Chapter 18.

It can be shown that the notion of a subgradient is well-defined in the sense that there exists a subgradient of f at each point $x \in \mathbb{R}^n$. We will denote the nonempty set of subgradients of f at x by

$$\partial f(x) := \{g \in \mathbb{R}^n : f(y) \geq f(x) + g^\top (y - x) \text{ for all } y \in \mathbb{R}^n\}.$$

In order to analyze subgradient descent, we will make some assumptions about f. We will assume that (19.13) has an optimal solution $x^* \in \mathbb{R}^n$, and we will let $f^* := f(x^*)$ denote its objective value. We will also assume that f is Lipschitz continuous, namely

$$|f(x) - f(y)| \leq L\|x - y\| \text{ for all } x, y \in \mathbb{R}^n.$$

See Algorithm 19.3.1 for a description of subgradient descent. We see that at iteration i of the algorithm, we update the current iterate x_i by taking a step of size t_i in the opposite direction of the subgradient. It follows that subgradient descent is a natural generalization of gradient descent (see Chapter 13).

Algorithm 19.3.1 Subgradient Descent for (19.13)

Input: Initial point $x^0 \in \mathbb{R}^n$ and step-sizes t_i, $i = 0, 1, \ldots$.

1: **for** i = 0,1,... **do**
2: $\quad x^{i+1} \leftarrow x^i - t_i g_i$, where $g_i \in \partial f(x^i)$
3: **end for**

Theorem 19.3.2. *After running Algorithm 19.3.1 for k iterations, it holds that*

$$\min_{0 \leq i \leq k} f(x^i) - f^* \leq \frac{L^2 \sum_{i=0}^k t_i^2 + \|x - x^0\|^2}{2 \sum_{i=0}^k t_i}.$$

We prove Theorem 19.3.2 below, but first we note that the theorem allows us to answer the following practical question. For a given tolderance $\epsilon > 0$, how long do we need to run subgradient descent in order to compute a point x such that $f(x) - f^* \leq \epsilon$?

Corollary 19.3.3. *Let $\epsilon > 0$. If we run Algorithm 19.3.1 for*

$$k = \left\lceil \frac{\|x^0 - x^*\|^2 L^2}{\epsilon^2} \right\rceil - 1$$

iterations using the step-size $t_i = \frac{\epsilon}{L^2}$*, then*

$$\min_{0 \le i \le k} f(x^i) - f^* < \epsilon.$$

Proof. We leave the proof as Exercise 19.8. □

Now we prove Theorem 19.3.2.

Proof of Theorem 19.3.2. Let $x \in \mathbb{R}^n$. We first show that if $g \in \partial f(x)$, then

$$\|g\| \le L. \tag{19.14}$$

Clearly (19.14) holds for $g = 0$, so suppose $g \ne 0$. From the subgradient inequality,

$$f(x+g) \ge f(x) + g^\top g = f(x) + \|g\|^2,$$

and hence

$$\|g\|^2 \le f(x+g) - f(x) \le |f(x+g) - f(x)| \le L\|g\|,$$

where the last inequality follows from Lipschitz continuity. Dividing through by $\|g\|$ yields (19.14).

Now we are prepared to establish the theorem. For $i = 0, \ldots, k$, it follows from Step 2 in Algorithm 19.3.1 that

$$\begin{aligned} \|x^{i+1} - x^*\|^2 &= \|x^i - t_i g_i - x^*\|^2 \\ &= \|x^i - x^*\|^2 + +t_i^2\|g_i\|^2 + 2t_i g_i^\top (x^* - x^i) \\ &\le \|x^i - x^*\|^2 + t_i^2\|g_i\|^2 + 2t_i(f^* - f(x^i)) \\ &\le \|x^i - x^*\|^2 + t_i^2 L^2 + 2t_i(f^* - f(x^i)), \end{aligned} \tag{19.15}$$

where the first inequality follows from the subgradient inequality and the second inequality from (19.14). Rearranging (19.15), we obtain

$$2t_i f(x^i) \le 2t_i f^* + t_i^2 L^2 + \|x^i - x^*\|^2 - \|x^{i+1} - x^*\|^2. \tag{19.16}$$

Observe that

$$\begin{aligned} 2\sum_{i=0}^{k} t_i \min_{0 \le i \le k} f(x^i) &\le 2\sum_{i=0}^{k} t_i f(x^i) \\ &\le 2\sum_{i=0}^{k} t_i f^* + L^2 \sum_{i=0}^{k} t_i^2 + \|x^0 - x^*\|^2 - \|x^{k+1} - x^*\|^2 \\ &\le 2\sum_{i=0}^{k} t_i f^* + L^2 \sum_{i=0}^{k} t_i^2 + \|x^0 - x^*\|^2, \end{aligned} \tag{19.17}$$

where the second inequality follows from summing (19.16) over $i = 0, \ldots, k-1$. Rearranging (19.17) gives

$$\min_{0 \le i \le k} f(x^i) - f^* \le \frac{\sum_{i=0}^{k} t_i^2 L^2 + \|x^0 - x\|^2}{2\sum_{i=0}^{k} t_i},$$

which is the desired result. □

Let us see how to apply subgradient descent to Lasso (which recall does not have a differentiable objective function) from Example 19.1.4.

Example 19.3.4. *We are interested in applying subgradient descent to*

$$\min_{x \in \mathbb{R}^n} \|Ax - b\|^2 + \lambda \|x\|_1.$$

In order to apply Algorithm 19.3.1, we need to derive a formula for the subgradient of the objective function at an arbtrary point $x \in \mathbb{R}^n$. The first term $\|Ax - b\|^2$ in the objective is differentiable, and its gradient is

$$\nabla \|Ax - b\|^2 = \nabla \left(x^\top A^\top Ax - 2b^\top Ax + b^\top b \right) = 2A^\top Ax - 2A^\top b.$$

A subgradient of the second term $\lambda \|\cdot\|_1$ at x is

$$\lambda \mathrm{sgn}(x),$$

where $\mathrm{sgn} : \mathbb{R}^n \to \{-1, 0, 1\}^n$ *is the function defined by*

$$[\mathrm{sgn}(x)]_i := \begin{cases} -1 & x_i < 0 \\ 0 & x_i = 0 \\ 1 & x_i > 0 \end{cases}$$

for $i = 1, \ldots, n$. Thus, a subgradient at x is

$$2A^\top Ax - 2A^\top b + \lambda \mathrm{sgn}(x).$$

We are now prepared to apply subgradient descent to Lasso; see Algorithm 19.3.2.

Algorithm 19.3.2 Subgradient Descent for Lasso.

Input: Initial point $x^0 \in \mathbb{R}^n$ and step-sizes t_i, $i = 0, 1, \ldots$.

1: **for** i = 0,1,... **do**

2: $\quad x^{i+1} \leftarrow x^i - t_i(2A^\top Ax^i - 2A^\top b + \lambda \mathrm{sgn}(x^i))$

3: **end for**

We have not addressed one detail. How do we choose the step-sizes? It is not immediately clear from Corollary 19.3.3 because the objective function is not Lipshitz continuous. Addressing this is outside the scope of this book; however, we note that it is possible to establish a result similar to Corollary 19.3.3 for subgradient descent applied to Lasso.

Algorithms like gradient descent and subgradient descent are *first order methods* because they at most use first order derivatives (as opposed to using second or higher order derivatives). If we use higher order derivatives, then we can develop algorithms (like Newton's method) that converge faster. However, there is a disadvantage to using these algorithms; they have a higher per iteration operation cost. This is one reason why first order methods are often the algorithm of choice in modern large-scale machine learning and data science applications that involve millions of variables.

While subgradient descent is an algorithm for unconstrained problems, it can also be applied to certain constrained problems. There are a variety of methods for solving constrained optimization problems, like the ellipsoid algorithm and interior-point methods, but they are outside the scope of this book.

19.4 Exercises

Exercise 19.1. *Consider the convex optimization problem* (19.1). *Prove that the feasible region is a convex set.*

Exercise 19.2. *Prove that the linear program discussed in Example 19.1.1 is a special case of the quadratic optimization problem discussed in Example 19.1.2.*

Exercise 19.3. *Consider the quadratic function* $f : \mathbb{R}^n \to \mathbb{R}$ *defined by*

$$f(x) = \frac{1}{2}x^\top Qx + c^\top x,$$

where Q *is symmetric positive definite. Show that* f *is strictly convex.*

Exercise 19.4. *Consider the ellipsoid*

$$E(Q, c) = \{x \in \mathbb{R}^n : (x - c)^\top Q(x - c) \leq 1\},$$

where $c \in \mathbb{R}^n$ *and* $Q \in \mathbb{R}^{n\times n}$ *is symmetric positive definite. Show that there is a matrix* $M \in \mathbb{R}^{n\times n}$ *and a vector* $z \in \mathbb{R}^n$ *such that*

$$E(Q, c) = \{M(x + z) : \|x\| \leq 1\}.$$

Exercise 19.5. *Show that the objective of* (19.3) *is correct. Hint: Use the following two facts.*

1. *For invertible matrices* A *and* B*, it holds that* $\det(AB) = \det(A)\det(B)$.
2. *It holds that* $\det(A^{-1}) = 1/\det(A)$ *for an invertible matrix* A.

Exercise 19.6. *Show that the set of symmetric positive semidefinite matrices*

$$\{M \in \mathbb{R}^{n \times n} : x^\top M x \geq 0 \text{ for all } x \in \mathbb{R}^n\}.$$

is a convex cone.

Exercise 19.7. *Derive the dual problem of the convex optimization problem*

$$\begin{aligned} \min_{x \in \mathbb{R}^n} \quad & x^\top x \\ s.t. \quad & Ax = b. \end{aligned}$$

Exercise 19.8. *Prove Corollary 19.3.3.*

Part V

Combinatorial Optimization

20

An Introduction to Combinatorics

20.1 Introduction

The simplest classification of different areas of math is to split math into analytic and discrete subdivisions. Analysis would be mathematics over $\mathbb{R}$ where continuity is assumed and discrete mathematics would be mathematics over $\mathbb{Z}$ where continuity does not apply[1]. "Analysis is the line; discrete math is the dots". Calculus belongs to the analysis side of mathematics as well as areas like Topology and Differential Equations where as the discrete side is home to Combinatorics, Graph theory and Abstract Algebra. Of course, there can be overlap and a class in Number Theory could be taught from any of an Analytic, Algebraic, or Combinatorial approach.

Almost all of our study to this point has involved mathematics from analysis but there was one exception: Integer Linear Programming. It is worthwhile to remember that Integer Linear Programming problems (discrete optimization) were infinitely more difficult than their continuous counterparts (Linear Programming over the Reals); in particular, Dakin's Branch and Bound as well as Gomory Cuts are not guaranteed to converge to a globally optimal solution, but the Simplex Method always gives the globally optimal solution in a finite number of steps. With continuity not in play, tools from Calculus cannot be used and new approaches must be considered. Hence this chapter.

At its most basic level, Combinatorics is the science of counting. If Jeff has 3 different sports coats, 7 different shirts, and 4 different slacks, then (being a mathematician and not worrying about fashion) he has $3 \cdot 7 \cdot 4 = 84$ different combinations of outfits. As much fun as it would be to turn this into a Combinatorics text (we do have the prettiest proofs, as you shall see...), we instead introduce the basics that will be used in our study and recommend Miklós Bóna's excellent text [5] for those wishing a deeper exploration of the subject.

[1] Technically, discrete math is done over any finite or countably infinite set, but we do not need to go into that here.

DOI: 10.1201/9780367425517-20

20.2 The Basic Tools of Counting

Counting questions can always be stated in terms of sets and sequences. We thus present the tools in this context; but first, some notation.

Notation (Factorial Symbol). *Let n be a positive integer. Then $n! := n(n-1)(n-2)\cdots 3\cdot 2\cdot 1$.*

If we have nothing to add, then nothing should change; hence we regard the *empty sum* to be zero (the additive identity). Likewise, multiplying by nothing should not change anything, thus we regard the *empty product* as 1 (the multiplicative identity). Notice that the n in $n!$ tells us how many factors we have; thus, $0!$ has no factors and is the empty product. For this reason (and because it makes all the theorems work), we define $0! := 1$.

We should note that one thing that makes combinatorial problems difficult is how quickly the factorial function grows. Figure 3.1 does not go beyond $x = 5$ and $x!$ is already towering over other familiar functions. Table 25.1 in Chapter 25 also shows how quickly the values of $n!$ can get incredibly large.

Now on to our counting machinery.

20.2.1 When We Add, Subtract, Multiply, or Divide

Our study of counting begins with the situations in which the counting can be done via a straight-forward arithmetic operation.

Theorem 20.2.1. *When we multiply: The Multiplication Principle. Let A and B be finite sets. Then the number of pairs (a, b) where $a \in A$ and $b \in B$ is $|A| \cdot |B|$.*

A less formal way to state this principle is "Suppose a task consists of a sequence of two processes and there are n_1 many ways to do the first process and n_2 many ways to do the second. Then there exist $n_1 n_2$ many ways to do the task".

Our earlier example involving Jeff's outfits in the introduction to this chapter is an application of *The Multiplication Principle.* We present one more:

Example 20.2.2. *Suppose a committee is to be formed from 15 women and 12 men where the committee is to have one woman and one man. By* The Multiplication Principle, *this can be done in* $15 \cdot 12 = 180$ *different ways.*

Theorem 20.2.3. *When we add: The Addition Principle. Let A and B be finite, disjoint sets (i.e. $A \cap B = \emptyset$). Then the number of elements c where $c \in A$ or $c \in B$ is $|A| + |B|$.*

A less formal way to state this principle is "Suppose a task consists of a sequence of doing one of two mutually exclusive processes; that is, the task is completed if we perform either a process N_1 or a different process N_2 where there is no overlap in the processes. If there are n_1 many ways to do the first process and n_2 many ways to do the second, then there exist $n_1 n_2$ many ways to do the task".

Example 20.2.4. *Suppose a personal identification number (PIN) for a bank account can consist of four or five digits. By* The Multiplication Principle *there are* 10^4 *ways to form a four digit password and* 10^5 *ways to choose a five digit password. Thus, by* The Addition Principle, *there are* $10^4 + 10^5 = 110{,}000$ *different ways to select such a PIN.*

Notice that there is a pair of important words that will alert us to when to multiply and when to add. When counting, *"and"* usually triggers multiplication where *"or"* is a good sign to use addition. Also, we may use Induction to extend both *The Multiplication Principle* and *The Addition Principle* to an arbitrary number of sets.

Theorem 20.2.5. *When we Subtract: The Inclusion-Exclusion Principle Let A and B be finite sets. Then the number of elements c where $c \in A$ or $c \in B$ is $|A| + |B| - |A \cap B|$.*

The intuitive reason for subtracting the elements in *both* A and B is because they have been double counted, as we see in the next example.

Example 20.2.6. *An Optimization class consisting only of Math and Computer Science majors has 17 Math majors, 12 Computer Science majors, and 5 students that are majoring in both subjects. By* The Inclusion-Exclusion Principle, *there are* $17 + 12 - 5 = 24$ *students in the course.*

We may also extend *The Inclusion-Exclusion Principle* to any number of sets. The case for three sets looks like this

$$|A \cup B \cup C| = |A| + |B| + |C| - |A \cap B| - |A \cap C| - |B \cap C| + |A \cap B \cap C|. \tag{20.1}$$

Of course, we must remove the elements counted multiple times; that is, the elements in the pairwise intersections. The addition at the end is to add back the elements that have been thrown out of the count twice.

The case for three sets was given due to the fact that the statement of *Inclusion-Exclusion* for an arbitrary number of sets is a bit cumbersome.

Theorem 20.2.7 (Generalized Inclusion-Exclusion.). *Let $A_1, A_2, \ldots, A_n$ be finite sets. Then*

$$\begin{aligned} |A_1 \cup \cdots \cup A_n| = \sum_{i=1}^{n} |A_i| - \sum_{1 \le i < j \le n} |A_i \cap A_j| \\ + \sum_{1 \le i < j < k \le n} |A_i \cap A_j \cap A_k| - \cdots + (-1)^{n-1} |A_1 \cap \cdots \cap A_n|. \end{aligned} \tag{20.2}$$

Theorem 20.2.8. *When we Divide: The Division Principle Let A and B be finite sets such where $f : A \to B$ is a $d - to - 1$ function (i.e. every element in B has d elements of A mapped to it). Then $|B| = |A|/d$.*

The standard example for *The Division Principle* involves seating arrangements at a circular table.

Example 20.2.9. *Suppose six people will sit together at a circular table and are curious how many possible ways they may arrange their choice of seats. As the table is circular, they have decided that what matters is not which seats people are in, but rather the order of the people at the table.*

Solution. Let us label the six people as A, B, C, D, E, and F and identify a chair as *Chair 1* and label the remaining chairs in order as we move clockwise around the table. By *the Multiplication Principle*, there are then $6 \cdot 5 \cdot 4 \cdot 3 \cdot 2 \cdot 1 = 6!$ ways to fill the seats with $ABCDEF$ one possible assignment. But $BCDEFA$, $CDEFAB$, $DEFABC$, $EFABCD$, and $FABCDE$ are five other arrangements that giving the same order, meaning each ordering as six possible particular seatings. This is true in general, that is, that every possible arrangement has 6 ways to occur, thus, by *the Division Principle*, there are $6!/6 = 5! = 120$ different seating arrangements for the 6 at the table. ■

It is worthwhile to make clear that each of the above counting principles is, in fact, a theorem which can be proven. The approach would begin with a mathematical understanding of counting: form a bijection from the collection of objects being counted into a subset of $\mathbb{N}$. As most of us are comfortable with trusting counting, we will forgo the proofs and refer curious readers to most discrete math, introduction to theoretical math, or basic Combinatorics textbooks.

20.2.2 Permutations and Combinations

Equipped with these basic counting tools, we may now build more useful equipment.

Definition 20.2.10 (Permutation). *Let S be a finite set with n elements. An* r-permutation *of S is a sequence (ordered list) of r elements from S.*

Theorem 20.2.11 (Number of r-Permutations without Repetition). *Let S be a finite set with n elements. Then for $r \leq n$, the number of r-permutations of S when there is no repetition of the elements from S is*

$$P(n,r) = n \cdot (n-1) \cdot \dots \cdot (n-r+1) = \frac{n!}{r!}.$$

Proof. By *the Multiplication Principle*, there are n choices for the first element, $n-1$ choices for the second element, ..., $n-(r-1) = n-r+1$ for the rth element. □

Example 20.2.12. *Suppose the senior class at a high school has 52 students. How many ways are there to select a president, vice-president, and a secretary/treasurer?*

Solution. We can think of choosing the officers as forming an ordered list of three of the classmates (a 3-permutation); hence the number of ways to pick the officers is $P(52,3) = 132{,}600$. ■

It is important to drive home that permutations count sequences; namely, they are used when order matters. When order does not matter, combinations are used.

Definition 20.2.13 (Combination). *Let S be a finite set with n elements. An* r-combination *of S is an unordered collection of r elements from S.*

The unordered collection of elements is not necessarily a set as we be interested in collecting elements with repetition. This collection would then be a *multiset.*

Theorem 20.2.14 (Number of r-Combinations without Repetition). *Let S be a finite set with n elements. Then for $r \leq n$, the number of r-combinations*[2] *of S is*

$$C(n,r) = \binom{n}{r} = \frac{n!}{r!(n-r)!}.$$

Note that the alternate notation $\binom{n}{r}$ for r-combinations is called a *binomial coefficient* and is read "n choose r". We will explore these and the reason for this name in the next section.

Proof. By Theorem 20.2.11, there are $P(n,r)$ r-permutations of the r elements. As order does not matter (set versus sequence) and there are $r!$ different orderings of a given collection of r elements, by the *Division Principle*, $C(n,r) = \frac{P(n,r)}{r!} = \frac{n\cdot(n-1)\cdot\dots\cdot(n-r+1)}{r!} = \frac{n\cdot(n-1)\cdot\dots\cdot(n-r+1)}{r!} \cdot \frac{(n-r)!}{(n-r)!} = \frac{n!}{r!(n-r)!}$. □

Example 20.2.15. *Suppose the senior class at a high school has 52 students. How many ways are there to form a three person committee from this senior class?*

Solution. As the order of the committee members does not matter, we can think of choosing the committee as forming an unordered collection (since the same person cannot be on the committee more than once, we are actually forming a subset). Thus we are counting the number of possible 3-combinations which is $C(52,3) = 22{,}100$. ■

[2]It is possible to extend the meaning of this symbol beyond nonnegative integers and to cases where $r > n$, but this is not relevant to our work.

Before continuing, we note that the binomial coefficients enjoy a special property.

Theorem 20.2.16 (Symmetric Property of Binomial Coefficients). *Let n and r be nonnegative integers with $r \leq n$. Then*

$$\binom{n}{r} = \binom{n}{n-r}.$$

Proof. We can easily prove this property by applying the formula for the symbol, but let us take this opportunity to use one of the more beautiful aspects of Combinatorics: proofs via a counting argument[3]. This is done by showing that the two sides of the equality count the same thing but in different ways; namely $\binom{n}{r}$ counts the number of ways to form an r person committee from a group of n people by choosing the r people to be on the committee. The right-hand side counts the same thing by instead choosing the $n - r$ people that are not on the committee[4]. □

Our excursion into permutations and combinations was in the context that objects were not permitted to be repeated (this is implied by the definition of set but not part of the definition of a sequence). Let us now consider the situation where objects can be repeated.

Example 20.2.17. *A lottery draws from a urn of ping-pong balls labeled 1 through 20. Four ping-pong balls are drawn to select the winning number and after a ball is selected it is put back into the urn before choosing the next number. The lottery rules state that the winning numbers must appear in the same order the numbers were drawn. How many ways are there to create a winning number in this lottery?*

Solution. Here we are counting the number of 4-permutations from a collection of 20 objects with repetition of objects permitted. Thus, by *the Multiplication Principle*, there are $20^4 = 160{,}000$ possible winning numbers. ■

From the example, we see that an easy application of *the Multiplication Principle* gives

Theorem 20.2.18 (r-Permutations with Repetition of Objects). *The number of r-permutations from a collection of n objects with repetition of objects permitted is n^r.*

Now the number of r-combinations with repetition of objects is a bit tricky, but a proper way of looking at the problem makes counting these quite easy.

---

[3]As claimed, Combinatorics does have the prettiest proofs.

[4]If your initial reaction to a counting argument is that it is not formal enough to qualify as a proof, you are not alone. Do realize, though, that the process of counting can be done formally by introducing a bijection, which we certainly can do here, but in most cases none of us distrust counting enough to go to this extreme level of formality.

Example 20.2.19. *An ice cream shop offers homemade ice cream available in 4 flavors. In how many ways can a customer take home 6 pints of ice cream? (Here we are regarding pints of the same flavor as the same; i.e. a flavor is permitted to be repeated.)*

Solution. As the order of the ice cream is irrelevant, this is a question where we count combinations but with the added difficulty of possible repetitions. A solution can be obtained by using the Addition Principle and considering all possible combinations of repeated flavors; but that is much too complicated. Instead, let us represent each pint taken home with an $*$ and we will separate flavor groupings by using a $|$. Thus $* | * * | | * * *$ represents a customer order of one pint of the first flavor, two pints of the second flavor, no pints of the third flavor, and three pints of the final flavor. This representation by "stars and bars" helps us see that the solution is that there are $C(6+4-1, 4-1) = \binom{9}{3} = \binom{9}{6} = 84$ possible combinations of 6 pints of the 4 flavors. ■

We thus have

Theorem 20.2.20 (r-Combinations with Repetition of Objects). *The number of r-combinations from a set of n objects with repetition possible is $C(n+r-1, n-1) = C(n+r-1, r)$.*

A summary of counting permutations and combinations is given in Table 20.1 at the end of Section 20.4.

20.3 The Binomial Theorem and Binomial Coefficients

20.3.1 Pascal's Triangle

Recall that Pascal's Triangle has the form:

$$
\begin{array}{ccccccccc}
 & & & & 1 & & & & \\
 & & & 1 & & 1 & & & \\
 & & 1 & & 2 & & 1 & & \\
 & 1 & & 3 & & 3 & & 1 & \\
1 & & 4 & & 6 & & 4 & & 1 \\
 & \vdots & & & \vdots & & & \vdots &
\end{array}
$$

where the outer diagonals are all ones and the inner entries are formed by summing the two entries above. This pattern was known by many before Pascal, but his name is attached as he dedicated a treatise to the triangle which included many previously undiscovered uses.

Pascal's Triangle is very useful for expanding binomials as its entries are the coefficients in the expansion. For example:

Example 20.3.1.

$$\begin{aligned}(2x-3y)^4 &= \mathit{1}(2x)^4(-3y)^0 + \mathit{4}(2x)^3(-3y)^1 + \mathit{6}(2x)^2(-3y)^2 + \mathit{4}(2x)^1(-3y)^3 \\ &\quad + \mathit{1}(2x)^0(-3y)^4 \\ &= 16x^4 - 96x^3y + 216x^2y^2 - 216xy^3 + 81y^4.\end{aligned}$$

But why does this work?

20.3.2 Binomial Coefficients

From the previous section, we have $C(n,r) = \binom{n}{r} = \frac{n!}{r!(n-r)!}$ where $n! := n(n-1)(n-2)\cdots 3\cdot 2\cdot 1$ and $0! := 1$. These are known as the *binomial coefficients* and are used to count the number of ways to select r objects (without replacement) from a group of n objects.

Notice that $\binom{1}{1} = 1$, $\binom{2}{0} = 1$, $\binom{2}{1} = 2$, $\binom{2}{2} = 1$, $\binom{3}{0} = 1$, $\binom{3}{1} = 3$, ... Do these numbers look familiar?

Our observations make in reasonable to conjecture that Pascal's Triangle is really

$$\begin{array}{ccccccccc} & & & & \binom{0}{0} & & & & \\ & & & \binom{1}{0} & & \binom{1}{1} & & & \\ & & \binom{2}{0} & & \binom{2}{1} & & \binom{2}{2} & & \\ & \binom{3}{0} & & \binom{3}{1} & & \binom{3}{2} & & \binom{3}{3} & \\ \binom{4}{0} & & \binom{4}{1} & & \binom{4}{2} & & \binom{4}{3} & & \binom{4}{4} \\ & \vdots & & & \vdots & & & \vdots & \end{array}$$

Our conjecture can be proven and is, in fact, known as the ***Binomial Theorem***:

20.3.3 The Binomial Theorem

Theorem 20.3.2 (The Binomial Theorem). *For and nonnegative integer n,*

$$(a+b)^n = \sum_{k=0}^{n} \binom{n}{k} a^{n-k} b^k.$$

We offer two proofs of the *Binomial Theorem*; a standard proof using induction[5] and then a beautiful counting argument.

The first proof by Induction on n involves using

Lemma 20.3.3 (Pascal's Identity). *For a nonnegative integer k and a positive integer j,*

$$\binom{k+1}{j} = \binom{k}{j-1} + \binom{k}{j}. \tag{20.3}$$

[5]The author would like to point out that he finds induction proofs soulless and boring; induction is certainly a valid proof technique, but it offers no insight into why the result is what it is.

Proof (of Pascal's Identity[6]*).*
From the formula for calculating binomial coefficients,

$$\begin{aligned}\binom{k}{j-1}+\binom{k}{j} &= \frac{k!}{(k-[j-1])!(j-1)!}+\frac{k!}{(k-j)!j!}\\ &= \frac{k!j}{(k-j+1)!j(j-1)!}+\frac{k!(k-j+1)}{(k-j+1)(k-j)!j!}\\ &= \frac{k!j+k!(k-j+1)}{(k-j+1)!j!}\\ &= \frac{k![j+k-j+1]}{(k-j+1)!j!}\\ &= \frac{k![k+1]}{(k-j+1)!j!}\\ &= \frac{(k+1)!}{(k+1-j)!j!}\\ &= \binom{k+1}{j}.\end{aligned}$$

This completes the proof of Pascal's Identity (which is called Pascal's Identity because (20.3) exactly states how to form Pascal's Triangle: the jth entry in the $(k+1)^{st}$ row is found by adding the $(j-1)^{st}$ and j^{th} entries in the kth row.)

It is worthwhile to note why this result is called Pascal's Identity. Recall that the entries of Pascal's Triangle are, in fact, the binomial coefficients. The method one uses to generate the entries in the triangle is exactly given by this identity (to get the jth entry in the $(k+1)^{st}$ row, add the $(j-1)^{st}$ and the jth entry in the kth row).

Now that we are equipped with this identity, we may prove the Binomial Theorem. Our proof is by induction on n.

Proof (of the Binomial Theorem).
(Base Step) Let $n = 0$. Then

$$1 = (a+b)^0 = \sum_{k=0}^{0}\binom{n}{k}a^{n-k}b^k = \binom{0}{0}a^0b^0 = 1.$$

Hence, the statement we are trying to prove is true for the base case $n = 0$.

(Inductive Step) Assume for $k \geq 0$ that

$$(a+b)^k = \sum_{m=0}^{k}\binom{k}{m}a^{k-m}b^m \quad \text{(this is the } \textit{induction hypothesis}\text{)}.$$

[6]Do you see why it is called Pascal's Identity? The answer will come after the proof.

Then

$$\begin{aligned}
(a+b)^{k+1} &= (a+b)(a+b)^k \\
&= (a+b)\sum_{m=0}^{k}\binom{k}{m}a^{k-m}b^m \text{ (by the induction hypothesis)} \\
&= a\left(\sum_{m=0}^{k}\binom{k}{m}a^{k-m}b^m\right) + b\left(\sum_{m=0}^{k}\binom{k}{m}a^{k-m}b^m\right) \\
&= \sum_{m=0}^{k}\binom{k}{m}a^{k-m+1}b^m + \sum_{m=0}^{k}\binom{k}{m}a^{k-m}b^{m+1} \\
&= a^{k+1} + \binom{k}{1}a^k b^1 + \cdots + \binom{k}{k}ab^k + \binom{k}{0}a^k b^1 \\
&\quad + \cdots + \binom{k}{k-1}ab^k + b^{k+1}
\end{aligned}$$

by combining like terms and using *Pascal's Identity*

$$\begin{aligned}
&= \binom{k+1}{0}a^{k+1} + \binom{k+1}{1}a^k b^1 + \cdots + \binom{k+1}{k}ab^k + \binom{k+1}{k+1}b^{k+1} \\
&= \sum_{m=0}^{k+1}\binom{k+1}{m}a^{k+1-m}b^m.
\end{aligned}$$

Since we have shown that the statement being true for the integer k implies the statement is true for $k+1$, the Principle of Mathematical Induction establishes the Binomial Theorem.

This completes the proof.

A Combinatorial Proof of the Binomial Theorem:

Proof. This proof involves counting and is absolutely soul-inspiringly beautiful (unlike proofs by Induction, which, we remind the reader, are soulless and boring).

Consider multiplying $(a+b)^n = \overbrace{(a+b)(a+b)\cdots(a+b)}^{n \text{ factors}}$. The only way to get an a^n term when we multiply is to choose an a from each of the n factors which is the same as choosing no $b's$. This can be done in $\binom{n}{0}$ many ways. Likewise, to get an $a^{n-1}b$ term, we would choose exactly one b when from then n binomials when multiplying and the remaining $n-1$ choices are $a's$. This can be done in $\binom{n}{1}$ ways. And so on. Thus

$$(a+b)^n = \sum_{k=0}^{n}\binom{n}{k}a^{n-k}b^k.$$

This completes the proof. □

20.3.4 Another Counting Argument

Since that experience was so wonderful, we offer another Combinatorial proof. Counting[7] can also be used to establish the following:

Theorem 20.3.4 (The Chairperson Identity). *For a nonnegative integer n,*

$$\sum_{k=1}^{n} k\binom{n}{k} = n2^{n-1}.$$

Proof. The left hand side counts the total number of ways to choose a committee of size 1 through n with chair by counting the number of ways to first form a committee of size k then select a chair from the k committee members.

The right hand side counts the same thing by first choosing a chair and then filling the arbitrary-sized committee. □

Note that not only are counting arguments elegant, they offer insight into why the result is what it is; and knowing the counting argument makes it very easy to remember the result.

20.3.5 The Multinomial Theorem

Since we have presented *the Binomial Theorem*, we also offer its extension, *the Multinomial Theorem*

The Binomial Theorem is, as the name suggests, a result regarding binomials. If having two terms is fun, then having three or more can only be more fun.

First, some notation:

Notation (Multinomial Coefficient). *Let $n, n_1, n_2 \dots, n_m$ be nonnegative integers where $n_1 + n_2 + \cdots + n_m = n$. Then we define the following symbol*

$$\binom{n}{n_1, n_2, \dots, n_m} := \frac{n!}{n_1!n_2!\cdots n_m!}.$$

(Of course this looks a lot like a binomial coefficient, but we will where it gets its name in the next theorem.)

Note that a multinomial coefficient with two terms reduces to a binomial coefficient; that is

$$\binom{n}{r, n-r} := \frac{n!}{r!(n-r)!} = \binom{n}{r} = \binom{n}{n-r}. \tag{20.4}$$

Equipped with the notation we may state

[7] We restate that if one feels that counting arguments are not formal enough, then please feel free to make the argument by introducing some bijections.

Theorem 20.3.5 (The Multinomial Theorem). *For and nonnegative integer n and any positive integer m,*

$$(x_1 + x_2 + \cdots + x_m)^n = \sum_{n_1+n_2+\cdots+n_m=n} \binom{n}{n_1, n_2, \ldots, n_m} {x_1}^{n_1} {x_2}^{n_2} \cdots {x_m}^{n_m}.$$

As with *the Binomial Theorem*, *the Multinomial Theorem* may be proven using either induction or a counting argument. Moreover, *the Multinomial Theorem* may be used to quickly expand multinomials just as *the Binomial Theorem* may be used to easily expand binomials.

Example 20.3.6. *By* the Multinomial Theorem,

$$\begin{aligned}(x+y+z)^2 &= \binom{2}{2,0,0}x^2 + \binom{2}{1,1,0}xy + \binom{2}{1,0,1}xz + \binom{2}{0,2,0}y^2 \\ &\quad + \binom{2}{0,1,1}yz + \binom{2}{0,0,2}z^2 \\ &= x^2 + 2xy + 2xz + 2y^2 + 2yz + z^2.\end{aligned}$$

Note that in this example determining the number of possible terms in the expansion of the multinomial is a nice application of counting 2-combinations with repetition (Theorem 20.2.20) from a set of 3 objects (the set $\{x, y, z\}$). Thus the final number of terms is $\binom{3+2-1}{2} = \binom{4}{2} = 6$.

20.4 Counting When Objects Are Indistinguishable

We now consider the case when we have some objects that are the same.

20.4.1 Permutations with Indistinguishable Objects

Example 20.4.1. *In how many ways can the letters of* ENGINEER *be rearranged?*

Solution. Note that there are 5 distinct letters E,N,G,I,R with the 3 E's being indistinguishable from each other as are the 2 N's. We have 8 spots to fill and will proceed in the order the letters appear (our choice of the order in which we place the letters does not matter... you are encouraged to check this for yourself). Starting with the E's, we have $\binom{8}{3}$ places we may place them. Thus there remain 5 positions where we can place the 2 N's, which can be done in $\binom{5}{2}$ ways. It follows that the remaining letters can respectively be placed in $\binom{3}{1}$, $\binom{2}{1}$, and $\binom{1}{1}$ many ways. Thus, by *the Multiplication Principle*, the possible number of arrangements of ENGINEER is

$$\binom{8}{3}\cdot\binom{5}{2}\cdot\binom{3}{1}\cdot\binom{2}{1}\cdot\binom{1}{1}=\frac{8!}{3!5!}\cdot\frac{5!}{2!3!}\cdot\frac{3!}{1!2!}\cdot\frac{2!}{1!1!}\cdot\frac{1!}{1!0!}=\frac{8!}{3!2!1!1!1!}$$
$$=\binom{8}{3,2,1,1,1}.$$

■

Generalizing the argument used the example establishes the following theorem:

Theorem 20.4.2 (Permutations with Indistinguishable Objects). *Given a collection of n objects of k indistinguishable types, the number of r-permutations of the n objects where there are n_1 indistinguishable objects of type 1, n_2 indistinguishable objects of type 2, ..., and n_k indistinguishable objects of type k is $\binom{n}{n_1,n_2,\cdots n_k}$.*

You can now see why we waited until after introducing *the Multinomial Theorem* as multinomial coefficients are used to count permutations with indistinguishable objects. The classical example of this problem is, as we saw above,

Example 20.4.3. *How many distinct arrangements are there of the word* TENNESSEE*?*

Solution. Here we have $n = 9$ and $k = 4$; that is, we have 9 total letters and 4 sets of indistinguishable letters. Given there are 4 E's, 2 each of N and S, and 1 T, by Theorem 20.4.2[8], the number of distinct rearrangements of the letters is

$$\left(\frac{9!}{4!2!2!1!}\right) = 3{,}780.$$

■

The previous example was the same as the first, thus shedding little light on the different ways in which this brand of questions can be asked. Let us then consider a slightly different arrangement of the problem.

Example 20.4.4. *A strand of Christmas lights has 4 red bulbs, 3 green bulbs, 5 yellow bulbs, and 2 blue bulbs. How many different arrangements of lights are possible for this strand?*

Solution. As what matters is the color of the bulbs, lights of the same color are regarded as indistinguishable. Thus there are

$$\binom{14}{4,3,5,2} = 2{,}522{,}520$$

possible arrangements of the lights. ■

[8]It is common to see Theorem 20.4.2 referred to as *the Multinomial Theorem*. It is possible to make a formal connection between the two, though an adequate amount of prayer and meditation will also reveal the relationship.

TABLE 20.1
Summary of Permutations and Combinations

Order?	Type	Repetitions?	Objects Indistinguishable?	Formula	Example
Yes	r-permutation	No	No	$\frac{n!}{(n-r)!}$	20.2.12
No	r-combination	No	No	$\binom{n}{r} = \binom{n}{n-r}$	20.2.15
Yes	r-permutation	Yes	No	n^r	20.2.17
No	r-combination	Yes	No	$\binom{n+r-1}{r} = \binom{n+r-1}{n-1}$	20.2.19
Yes	permutation	No	Yes	$\binom{n}{n_1, n_2, \cdots n_r}$	20.4.1
					20.4.3

20.4.2 Summary of Basic Counting Techniques

A summary of basic counting techniques is offered in Table 20.1 with a decision tree provided in the following diagram. When considering which technique to use to count, we must first ask if the order in which we select the objects matters or not. After this, we must then decide if objects can be repeated or if there are any objects that are indistinguishable. On this matter, it is important to understand the difference between what is meant by *repetition of objects* and as compared to when we have *indistinguishable objects.* In Example 20.2.17, a selected ping-pong ball is returned to the urn and can be chosen again but the individual ping-pong balls are very much distinguishable. In Example 20.4.3, though, we do not return a selected "s" to possibly be chosen again, but, rather, there is another "s" available that is indistinguishable from the first "s".

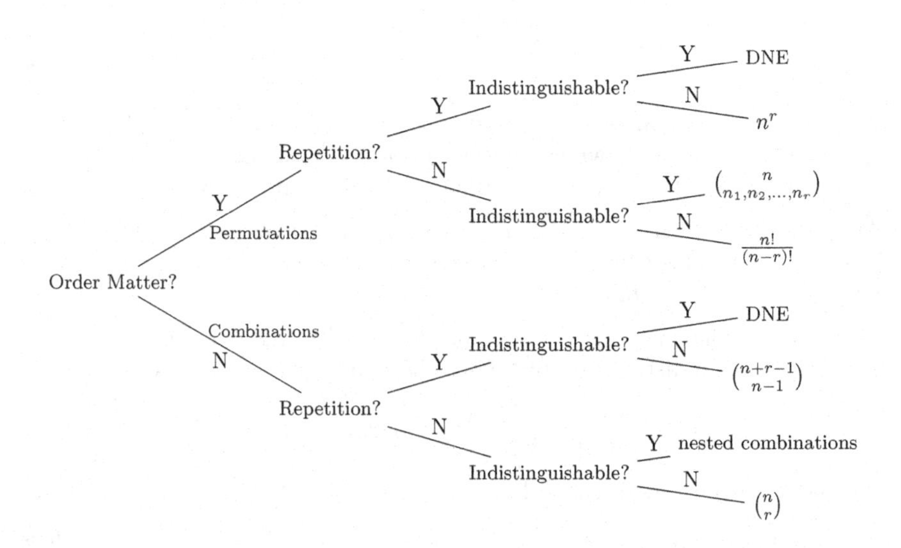

20.5 The Pigeonhole Principle

Though not necessary for our pursuits, no introduction to Combinatorics would be complete without introducing the following beautifully simple yet elegant principle. Though the Pigeonhole Principle is may not appear very sophisticated, it is quite useful and its applications are sometimes surprising.

Theorem 20.5.1 (The Pigeonhole Principle). *Let k be a positive integer. If $k+1$ or more objects are placed into k boxes, then at least one box has two or more of the objects.*

The Pigeonhole Principle gets its name from the days of messages being delivered by carrier pigeons with the pigeons playing the role of the objects and their cells the boxes. It sometimes is also called the *Dirichlet Drawer Principle* after the 19th century German mathematician G. Lejeune Dirichlet (Dirichlet was born in Belgium) who was fond of using this principle in his work. Though Dirichlet's name is attached, the principle was used in France in the 17th century [48] (see Exercise 20.18). The proof of Theorem 20.5.1 will follow directly from the proof of the Generalized Pigeonhole Principle (Theorem 20.5.6) with $N = k + 1$.

Example 20.5.2. *In a class of 13 students, at least two people share a birthday in the same month.*

Example 20.5.3. *If Lincoln strikes out 7 batters in 6 innings pitched, there must be at least one inning in which he struck out 2 or more batters.*

Example 20.5.4. *Let (A, B), (C, D), (E, F), (G, H), and (I, J) be distinct points in the standard Cartesian $x - y$ plane with integer coordinates. Show that at least one line segment joining pairs of points has a midpoint that has integer coordinates.*

Solution. The midpoint will have integer coordinates if the parities (even/odd) of the x coordinates and the y coordinates are respectively the same. By the Multiplication Principle (Theorem 20.2.1), there are 4 possible combinations of parities and as there are 5 points, by the Pigeonhole Principle there must be at least one pair of points whose x and y coordinates have parities that agree, thus the midpoint of the line segment joining them must have integer coordinates. ■

Example 20.5.5. *Let n be a positive integer. Show that n has a multiple whose digits are all 0 or 1.*

Solution. Consider the $n+1$ integers 1, 11, 111, ..., $\overbrace{111\cdots1}^{n+1 \text{ digits}}$. Note that in dividing each of these by n, there are only n possible remainders. But as there are $n+1$ numbers, by the Pigeonhole Principle there are two numbers that have the same remainder when doing this division. Subtracting the larger of these two numbers by the smaller gives a number that has only 0's and 1's as its digits and its remainder when divided by n must be 0 (note that not only is the number just made up of 0's and 1's, we have really shown it is a string of 1's followed by a string of 0's). ■

The Pigeonhole Principle can be generalized, but requires some notation.

Notation (Ceiling Function). *Let r be a real number. Then the* ceiling *of r is*

$$\lceil r \rceil = \text{ the smallest integer } k \geq r.$$

We have $\lceil \pi \rceil = 4$ and $\lceil -\frac{5}{2} \rceil = -2$.

Theorem 20.5.6 (Generalized Pigeonhole Principle). *Let N and k be positive integers. Then if N objects are placed into k boxes, there is at least one box with $\lceil N/k \rceil$ objects in it.*

Proof. Suppose for contradiction that no box contains more than $\lceil N/k \rceil$. Then the total number of objects is

$$N = k(\lceil N/k \rceil - 1) < k([N/k + 1] - 1) = N$$

which is contradiction since $N \not< N$. □

Note that the proof used the inequality $\lceil N/k \rceil < N/k + 1$ (see Exercise 20.16).

Example 20.5.7. *Since $\lceil 54/12 \rceil = 5$, by the Generalized Pigeonhole Principle, any class of* 54 *students must have at least* 5 *students that have a birthday in the same month.*

Example 20.5.8. *A standard deck of cards contains 52 cards distributed equally among four suits: hearts, diamonds, clubs, and spades. a) How many cards must an individual draw to guarantee that their hand has* 3 *cards from the same suit? b) How may cards must be drawn to guarantee the hand has* 3 *spades?*

Solution. a) By the Generalized Pigeonhole Principle, we need an integer N such that $\lceil N/4 \rceil > 3$ and the smallest such N is 9. *b)* As the we have selected a specific box (suit), the Generalized Pigeonhole Principle does not apply. A worst-case scenario application is needed, and it is possible for the player to draw all 13 each of the hearts, diamonds, and clubs before getting any spades, thus the answer is $13 + 13 + 13 + 3 = 42$. ■

20.6 Exercises

Exercise 20.1. *A password for an account on a particular website must be 8–16 characters long and contain at least one capital letter, at least one lowercase letter, and one of the following four symbols: @, !, \$, *. Furthermore, as long as these conditions are met the password can contain any combination of letters (lower or uppercase), numbers, or the above special symbols. How many possible passwords are there for this account?*

Exercise 20.2. *How many password combinations from the previous question are removed if we know the user of the account in question only uses a capital letter as the first character and only uses a single special character?*

Exercise 20.3. *How many ways can a function f map a domain D to a range R if D has 10 elements, R has 20 elements, and f is injective?*

Exercise 20.4. *How many ways can a function f map a domain D to a range R if D has 10 elements, R has 20 elements, and no element in R is mapped to by more than 2 elements of D?*

Exercise 20.5. *How many ways can a function f map a domain D to a range R if D has 10 elements, R has 5 elements, and no element in R is mapped to by the same number of elements in D?*

Exercise 20.6. *Partition the set $\{A, B, C, D, E\}$ into 4 subsets in a way such that for all subsets s_1, s_2, $|s_1| \neq |s_2|$, or prove you cannot.*

Exercise 20.7. *How many ways can a function f map a domain D to a range R if D has 9 elements, R has 5 elements, and no element in R is mapped to by the same number of elements in D?*

Exercise 20.8. *Show that the number of submatrices (including the empty submatrix) of an $m \times n$ matrix is $2^m 2^n$.*

Exercise 20.9. *Show that the number of square submatrices (including the empty submatrix) of an $m \times n$ matrix is*

$$\binom{M}{k} \sum_{i=0}^{N} \binom{N}{i}^2$$

where $M = \max\{m, n\}$, $N = \min\{m, n\}$, and $k = |m - n|$.

Exercise 20.10. *Use a counting argument to prove*

$$\sum_{k=0}^{n} \binom{n}{k}^2 = \binom{2n}{n}$$

where $\binom{2n}{n}$ are the central binomial coefficients *from Pascal's Triangle.*

Exercise 20.11. *If $m = n$ in Exercise 20.9 (i.e. the matrix is square), then by Exercise 20.10 there are $\binom{2n}{n}$ square submatrices of an $n \times n$ matrix. Calculate the first 10 terms of the sequence formed by these values: $\binom{0}{0}, \binom{2}{1}, \binom{4}{2}, \ldots$. Compare your answer with the appropriate sequence in the* Online Encyclopedia of Integer Sequences *[53].*

Exercise 20.12. *In how many different ways can the letters of* OPTIMIZATION *be arranged?*

Exercise 20.13. *In how many different ways can the letters of* MATHEMATICS *be arranged?*

Exercise 20.14. *Write a proper proof for Theorem 20.4.2.*

Exercise 20.15. *Prove the Pigeonhole Principle (Theorem 20.5.1) without using the Generalized Pigeonhole Principle (Theorem 20.5.6). [Hint: mimic the proof of Theorem 20.5.6].*

Exercise 20.16. *Prove that for real numbers m and n with $n \neq 0$ that $\lceil \frac{m}{n} \rceil < \frac{m}{n} + 1$. [Hint: there are two cases: $\frac{m}{n}$ is an integer and $\frac{m}{n}$ is not an integer.]*

Exercise 20.17. *Prove that no algorithm can list all of the subsets of a given set S in polynomial time.*

Exercise 20.18. *(From [48]) French writer Pierre Nicole wrote of* 17th *century Paris having over* 800,000 *residence and (as believed at the time) no one had more than* 200,000 *hairs on their head. Assuming that these numbers are correct and that no Parisian at this time was completely bald, argue that there had to be at least two Parisians with the same number of hairs on their head. Then use the Generalized Pigeonhole Principle (Theorem 20.5.6) to determine the minimum number of Parisians having the same number of hairs on their head.*

21

An Introduction to Graph Theory

Though technically a subset of Combinatorics, Graph Theory is deep enough to be regarded as its own subject. It also has an incredible propensity to be useful... especially in today's world with the internet (in particular, page rankings in web searches is one application), shipping of products (which has exploded with online purchasing), and even Machine Learning. It is used in Chemistry, Biology, Physics, Economics, telecommunication, social behavioral studies, ... anywhere there is a network.

21.1 Basic Definitions

Though we have seen many graphs in our study of mathematics, the graphs in Graph Theory have a particular meaning. As is almost always the case in this subject, a picture does more than words.

Example 21.1.1. *Let G and D be the following graphs:*

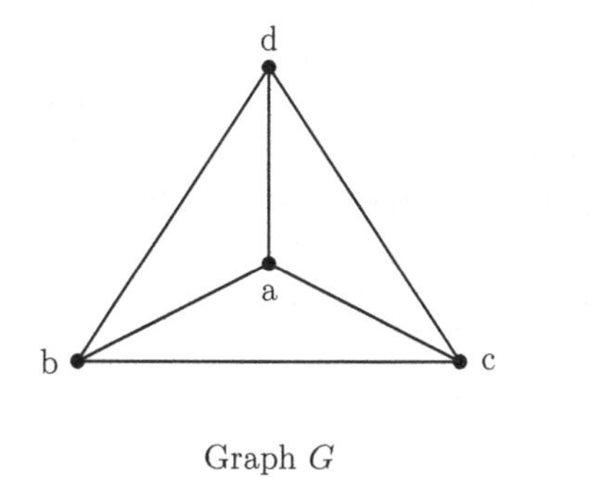

Graph G

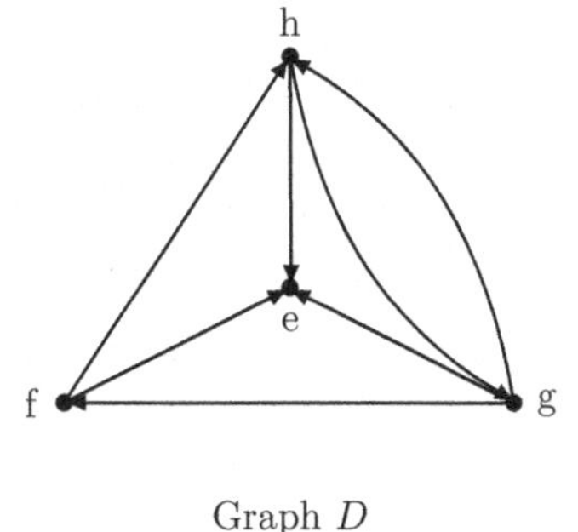

Graph D

What makes G a graph in Example 21.1.1 is that it consists of *vertices* and *edges*; that is, the *nodes* and *arcs* connecting the nodes. Thus

Definition 21.1.2 (Graph). *A* graph $G = (V, E)$ *is an order pair of sets where V is nonempty and finite and is the set of vertices and $E \subset V \times V$ is the set of edges.*

DOI: 10.1201/9780367425517-21

This is one of the best definitions in all of mathematics in that it has many important subtleties built in it. It may appear very simple, yet an entire area of mathematics is built on this definition (and note that the Handbook of Graph Theory is 1630 pages long![1]).

In exploring the buried elements of Definition 21.1.2, first note that the vertex set is always nonempty; that is, any graph will always have at least one vertex. The vertex set for the graph G in Example 21.1.1 is $V(G) = \{a, b, c, d\}$. As the edge set E is a subset of the Cartesian product of V with itself, it is allowed to be empty, thus it is possible for a graph to have no edges (such a graph is called an *empty graph*). Moreover, by this definition we identify an edge by its vertices. For example, ab is an edge in G. Note that technically we should write this edge as (a, b) since it is stated to be an ordered pair, but it is standard to avoid such cumbersome notation. You may also notice that we are cheating a bit here; (a, b) is an *ordered* pair, but we are going to ignore the order for our graph G in the example. Thus in G, the edge set is $E(G) = \{ab, ac, ad, bc, bd, cd\}$. There are times we care about a direction on the edges as in the graph D in Example 21.1.1. When, as in D, the orientation (direction) of the edge matters, the graph is called a *directed graph* or *digraph* for short. In digraphs, the initial vertex of the directed edge is listed first and the terminal vertex second. Thus D has vertex set $V(D) = \{e, f, g, h\}$ and edge set $E(D) = \{fe, fh, gf, ge, gh, he, hg\}$.

Note also that since the edge set is a set, repetition of elements is not permitted. Thus, an undirected graph can only at most one edge between vertices and a directed graph can have at most one directed edge between vertices (with both directions permitted). Graphs that have *multiedges*, that is multiple edges between vertices, are called *multigraphs* and are not considered here[2]. Observe that in D, directed edges gh and hg are not multiedges since they have different orientations. The number of vertices of a graph is called the *order* of the graph and is denoted by n where the number of edges of a graph is called the *size* and is denoted by m. In Example 21.1.1, G is a graph of order 4 and size 6 where D is a digraph of order 4 and size 7.

From Definition 21.1.2 we also get the very important fact that where we draw the vertices and how we draw the edges does not matter[3].

[1] We are being a little disingenuous here. *The Handbook of Graph Theory* does include the additional topics of weighted graphs, directed graphs, multigraphs, hypergraphs, and graph colorings, but these all build on Definition 21.1.2. Nonetheless, it is absolutely impressive that such a wealth of Mathematics comes from such a simple definition.

[2] Graph Theory is a relatively young area of Mathematics and, as such, definitions and notation are not yet completely standard and may vary by author. Most Graph Theorists mean by "graph" the definition that we have stated (with conveniently ignoring the "order" in "ordered pair"). As it is acceptable to consider a multigraph as a graph (thus requiring the edge set to be a multiset), some authors refer to what we have called a graph as a *simple graph*.

[3] There are, of course, a few exceptions, such as the important notion of planar graphs as well as the Geometric Heuristic for the Traveling Salesperson Problem.

Highlight 21.1.3. *All that matters in a graph is that there are vertices and that there either is an edge between a pair of vertices or there is not.*

Some terminology will be useful. Graphs that have only a few edges are commonly called *sparse graphs* whereas a graph with many edges is referred to as a *dense graph*. Of course, these are relative terms how many of the total possible edges we have, but they do get used often. Vertices that are joined by an edge are said to be *adjacent*. In Example 21.1.1, every pair of vertices in graph G is adjacent. When an edge exists, it is said to be *incident* with each of the vertices that determine it. Again, in G, edge bd is incident with vertex b as well as vertex d. The relationship works both ways and we may also say that vertex b is incident with edge bd. Adjacent vertices are said to be *neighbors* and the collection of all neighbors of a given vertex v is called the *neighborhood* of v and is denoted $N_G(v)$ or just $N(v)$ when the graph G is understood. In G, $N(a) = \{b, c, d\}$. The number of neighbors of a vertex is called the *degree* of the vertex, thus $\deg_G(a) = \deg(a) = 3$. The sum of the degrees of all the vertices in a graph is called the *total degree* of the graph; in other words, if G is a graph with vertex set V,

$$\sum_{v \in V(G)} \deg(v) = \text{ the total degree of } G$$

In digraphs, the notion of the degree of a vertex is replaced by the *in-degree* and the *out-degree*. Let u be a vertex in a digraph D. The *in-neighborhood* of u is

$$N^-(u) := \{v \in V(D) | vu \in E(D)\};$$

that is, the collection of vertices v in D such that edge vu is a directed edge into u. It will be no surprise that the *out-neighborhood* of u is[4]

$$N^+(u) := \{v \in V(D) | uv \in E(D)\};$$

that is, the collection of vertices v in D such that edge uv is a directed edge out of u. We then define the *in-degree* of a vertex u to be $\deg^-(u) = |N^-(u)|$ and the *out-degree* of a vertex u to be $\deg^+(u) = |N^+(u)|$. We also have $\deg(u) = \deg^-(u) + \deg^+(u)$; that is the degree of a vertex in a digraph is its total degree: the sum of its in-degree and out-degree.

A graph H is said to be a *subgraph* of a graph G if both $V(H) \subseteq V(G)$ and $E(H) \subseteq E(G)$. If either of the subset relations is proper, that is if $V(H) \subsetneq V(G)$ or $E(H) \subsetneq E(G)$, then H is said to be a *proper subgraph* of G. Further, H is a *spanning subgraph* of G if it is a subgraph of G and satisfies $V(H) = V(G)$. Lastly, suppose H is a subgraph of G with the property that for all $u, v \in V(H)$, u and v are adjacent in H if and only if u and v are adjacent in

[4]The superscripts "+" and "−" for the directed neighborhoods are borrowed from *network flows* (see Chapter 22). If one considers a flow in a network, the "+" edges are found by gazing forward in the network whereas the "−" edges are found by looking backward in the network.

G. When meeting these conditions, H is an *induced subgraph of* G. In other words, an induced subgraph inherits all the appropriate edges from its parent graph.

Example 21.1.4. *(Graph G and different subgraphs.)*

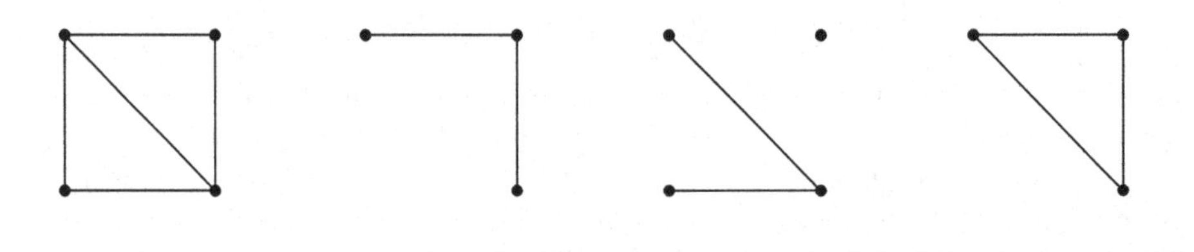

The last notion we consider in our introduction is the idea of sameness. As emphasized in Highlight 21.1.3, where or how we draw the vertices and edges does not matter. This can be formally addressed with the mathematical concept of an isomorphism, but (as previously stated regarding Graph Theory) a picture is better than a definition. In the Example 21.1.5, graphs G_1 and G_2 are *equal* as $V(G_1) = V(G_2)$ and $E(G_1) = E(G_2)$; that is, *equal graphs* have the same vertex and edge sets. Graph G_3 is not equal to either of the other two graphs as it has different vertex and edge sets (vertex f is not in either of $V(G_1)$ or $V(G_2)$), but if we relabel vertex f in G_3 as vertex a in graph G_3^*, then we would have $V(G_1) = V(G_2) = V(G_3^*)$ and $E(G_1) = E(G_2) = E(G_3^*)$. Since G_3 is just a relabelling of a graph that is equal to G_1 and G_2, we say G_3 is *isomorphic* to both G_1 and G_2.

Example 21.1.5. *(Three isomorphic graphs.)*

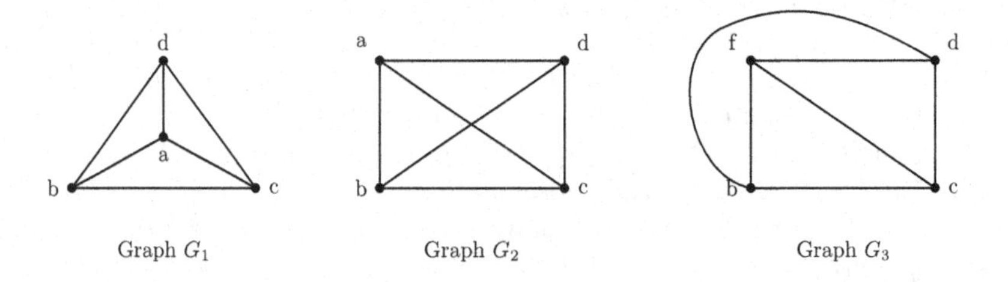

21.2 Special Graphs

21.2.1 Empty Graphs and the Trivial Graph

By definition, a graph must have at least one vertex but it does not have to have any edges. A graph on n vertices with an empty edge set is called an *empty graph of order* n. The simple graph consisting of only a single vertex is referred to as the *trivial graph*.

21.2.2 Walks, Trails, Paths, and Cycles

There are different types of journeys from one vertex to another. A *walk* in a graph is a sequence of adjacent vertices $v_0, v_1, v_2, \ldots, v_n$ and each edge $v_{i-1}v_i$. A *trail* is a walk that does not repeat an edge and a *path* is a walk that does not repeat a vertex (thus each path is necessarily a trail). The number of edges in a walk, trail, or path is said to be the *length* of the walk, trail, or path. Again using G from Example 21.1.1, *bdababc* is a walk of length 6, *bdabc* is a trail of length 4, and *bdc* is a path of length 2.

A path on n vertices is denoted P_n. Example 21.2.1 illustrates the first 5 paths.

Example 21.2.1. *(Paths of order* $n = 1, 2, 3, 4, 5$.*)*

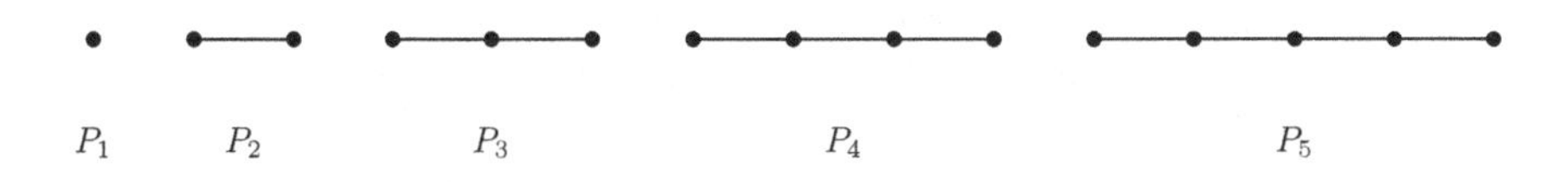

Adding an edge from the last vertex of a path to the first vertex creates a *cycle*. As multiedges are not permitted, there can be no cycle of order 2, thus cycles are only defined for 3 or more vertices. A cycle of order n is denoted C_n and the first 5 cycles are shown in Example 21.2.2 with C_6 and C_7 intentionally drawn a little differently. Cycles are involved in the very important Traveling Salesperson Problem (Chapter 25).

Example 21.2.2. *(Cylces of order* $n = 3, 4, 5, 6, 7$.*)*

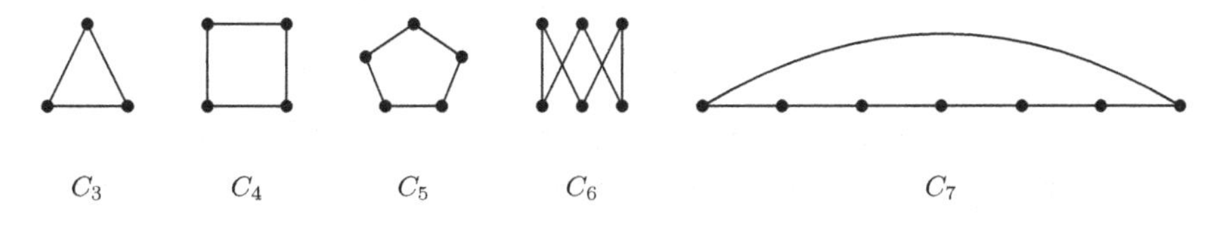

The cycle C_3 is often very useful in mathematics and is usually referred to as a *triangle*.

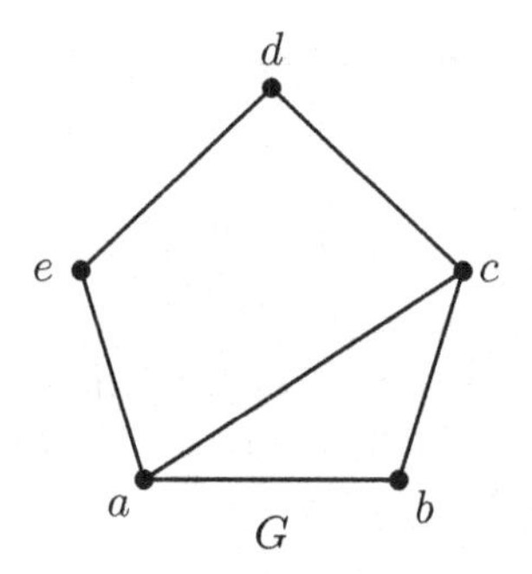

FIGURE 21.1
A cycle and Hamiltonian cycle.

Some special paths and cycles will be important in later work.

Definition 21.2.3 (Hamiltonian Path, Hamiltonian Cycle, Hamiltonian Graph). *A path the visits every vertex of a graph G is called a* Hamiltonian path. *Likewise, a cycle that passes through every vertex of a graph is called a* Hamiltonian cycle *or a* tour.

A graph that contains a Hamiltonian cycle is said to be a Hamiltonian graph.

We will note that *tour* is most commonly used but that Graph Theorist tend to favor the term *Hamiltonian cycle.* Optimal tours will be the focus of our study when answering the *Traveling Salesperson Problem* (Chapter 25).

In Figure 21.1, a, b, c, a is a cycle in G where a, b, c, d, e, a is a Hamiltonian cycle.

21.2.3 Trees

The previous section introduced the important notion of cycles. Cycles are incredibly relevant in Optimization as are graphs that have no cycles.

Definition 21.2.4 (tree, forest, leaf). *A connected*[5]*, acyclic graph is called a* tree *and a collection of trees is called a* forest. *A vertex of degree 1 in a tree is called a* leaf.

[5] *Connected* means exactly what it should mean. Formally, a *connected graph* is one where there always exits some path between any pair of vertices. *Connectivity* is addressed in Section 21.3.

Some examples of trees are shown in Example 21.2.5.

Example 21.2.5. *[Various trees.]*

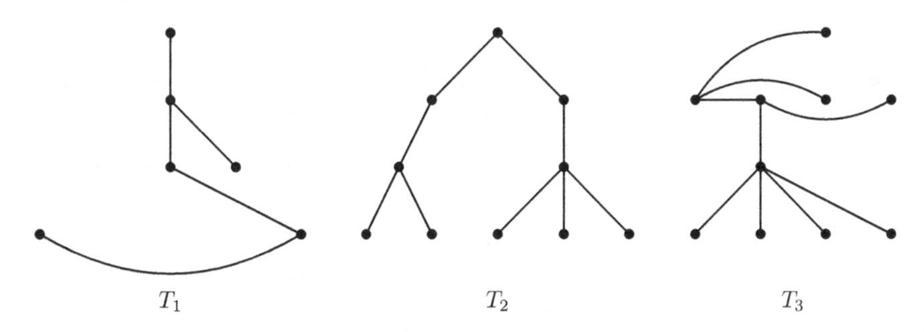

Notice that trees are, in a sense, optimal. In a tree, there are as many edges as we can have between existing vertices without having a cycle.

Before moving on, we will include a determining characteristic of a tree.

Theorem 21.2.6. *A graph T is a tree if and only if every pair of vertices in T are connected by a unique path.*

We will not write the details of the proof, but both directions rely on the fact that trees are acyclic. A proof can be found in [9].

Though it is bad form, we will also note that future Theorem 21.3.5 has a helpful corollary.

Corollary 21.2.7. *Every nontrivial tree has at least two leaves.*

We will prove the following useful result, though.

Theorem 21.2.8. *Let T_n be a tree of order n. Then T_n has size $n - 1$.*

Proof. The proof will be on the order of an arbitrary tree. As T_1 has size 0, the base case is easily established. For the induction hypothesis, suppose any tree of order n has size $n-1$ and consider a tree T of order $n+1$. By Corollary 21.2.7, T has at least 2 leaves. Let v be on of the leaves and consider $T - v$ (the graph T less vertex v and the edge incident with v). Since $T - v$ is a tree of order n, by the induction hypothesis, $T - v$ has size $n - 1$. Adding back leaf v and its only incident edge gives T which then has $n - 1 + 1 = n$ edges. □

21.2.4 Complete and Bipartite Graphs

We have mentioned that a graph will always have at least one vertex but may be void of edges. Such a graph is called an *empty graph* and empty graphs exist of every order. The complement of an empty graph – one in which every pair of distinct vertices is adjacent – is called a *complete graph*. A complete graph of order n is denoted K_n and the first few complete graphs are shown in Example 21.2.9.

Example 21.2.9. *(Complete graphs K_n of order $n = 1, 2, 3, 4, 5, 6$.)*

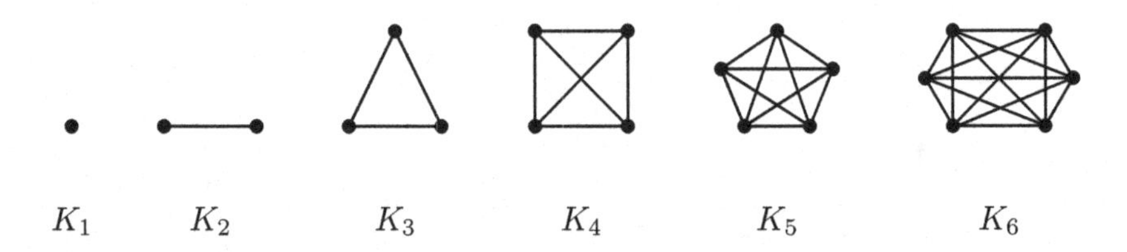

We have seen some of these graphs before. Notice that $K_1 = P_1$, $K_2 = P_2$, and that $K_3 = C_3$. An easy counting argument gives that the size of K_n is $\binom{n}{2}$ since every edge is determined by its vertices.

Definition 21.2.10 (Bipartite Graph). *A graph G is said to be bipartite if the vertex set of G can be partitioned into two disjoint sets U and V (called* partite sets*) such that no two vertices belonging to the same set are adjacent.*

To show a graph is bipartite, merely rewrite it making the partite sets clear as in the following example:

Example 21.2.11. *(Showing a graph to be bipartite.)*

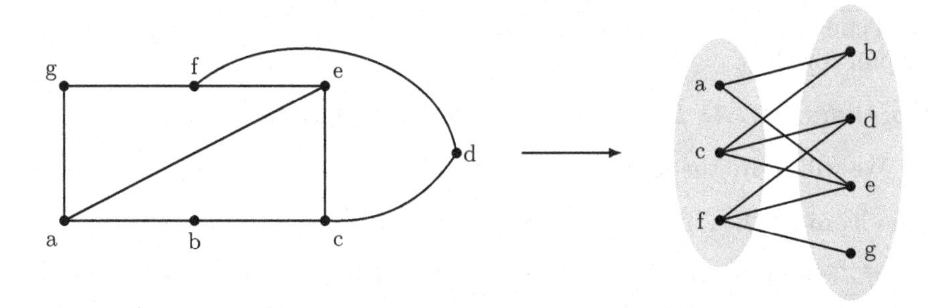

It is not hard to see that any tree is bipartite. Also, C_4 and C_6 as well as K_2 are bipartite whereas C_3, C_5, and C_7 are not bipartite. In fact, we have

$$C_n \text{ is } \begin{cases} \text{bipartite} & \text{if } n \text{ is even and} \\ \text{not bipartite} & \text{if } n \text{ is odd.} \end{cases}$$

Moreover, it is the case a graph G is bipartite if and only if it does not contain an odd cycle (a proof of this can be found in any Graph Theory text; see [9], for example).

Our last class of special graphs combines the previous two. A *complete bipartite graph* is a graph G whose vertex set can be partitioned into two disjoint sets U and V such that uv is an edge in G if and only if $u \in U$ and $v \in V$; that is, every edge in G has one of its vertices in U and the other in V. The notation for a complete bipartite graph is $K_{s,t}$.

Example 21.2.12. *(Complete bipartite graph $K_{3,4}$.)*

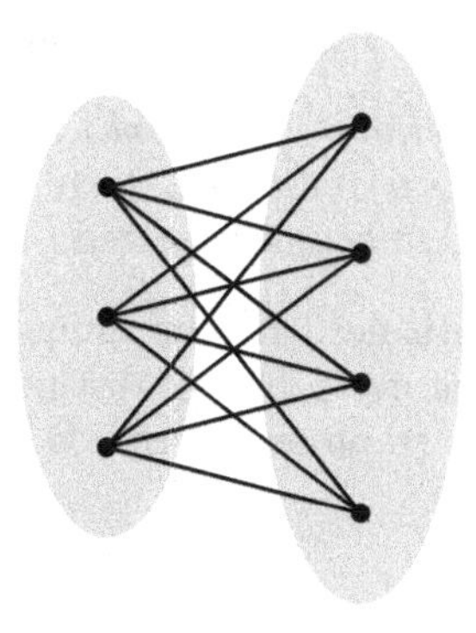

Since two partite sets is fun, more partite sets must be more fun. The notion of a bipartite graph may be extended to a *multipartite graph* where the definition is extended to G is a k-partite graph if its vertices can be partitioned into k disjoint nonempty sets such that no two vertices in the same partite set are adjacent. A *complete multipartite graph* is exactly what it should be and uses the same notation as a complete bipartite graph.

Example 21.2.13. *(Complete tripartite graph $K_{1,2,3}$.)*

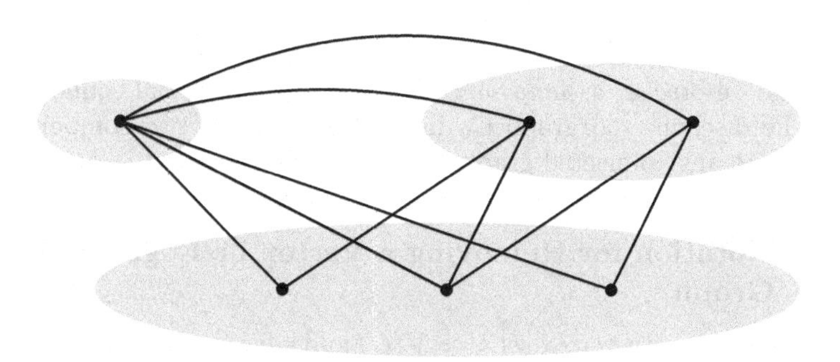

21.3 Vertex and Edge Cuts

Now we consider the useful notion of removing vertices and edges from a graph. First, the notion of connectivity is introduced.

21.3.1 Graph Connectivity

Though previously alluded to in our discussion of trees, we now offer a proper definition of what it means for a graph to be connected.

Definition 21.3.1. *(Connectivity in a Graph) A graph G is* connected *if for every pair of distinct vertices u and v in G there exists a $u - v$ path. A graph that is not connected is said to be* disconnected.

Note that the $u - v$ path need not be unique; many may exist, but as long as there is one between every pair of distinct vertices, the definition is satisfied. As well, the trivial graph on 1 vertex is regarded as being vacuously connected.

Example 21.3.2.

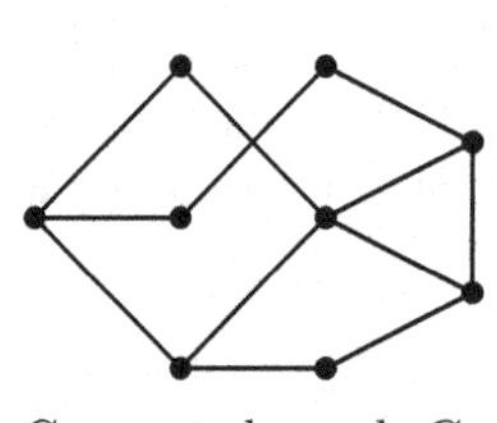

Connected graph G_1

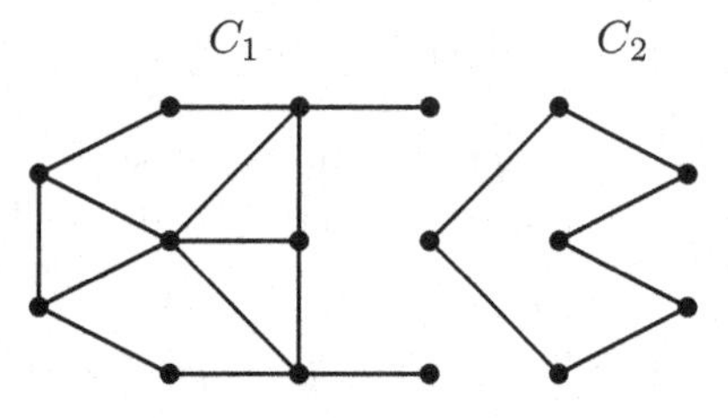

Disconnected graph G_2

The connected parts of the disconnected graph G_2 are referred to as *components*. To be precise, a *component* of a graph is a maximal connected subgraph. The disconnected graph G_2 in the example has two components C_1 and C_2 where any connected graph has only one component.

21.3.2 Notation for Removing a Vertex or Edge from a Graph

Let G be a graph with vertex set $V = V(G)$ and edge set $E = E(G)$. If $e \in E$, then by $G - e$ we mean the graph that results from removing edge e from G. Similarly, if $u \in V$, then by $G - u$ we mean the graph resulting from removing vertex u from G as well as any edges of G that are incident to u (the latter must occur as, by definition, every edge needs two vertices). As well, we may remove sets of vertices or edges from a graph and the same notation is used.

These notions are illustrated in the following example:

Example 21.3.3.

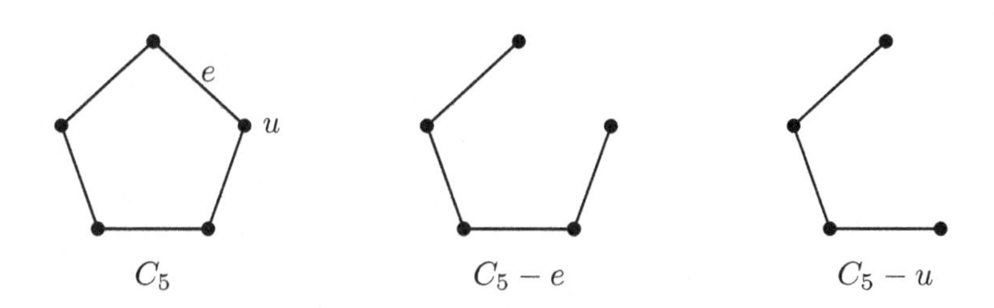

Now that we understand connectivity and removing vertices and edges from a graph, we may explore vertex and edge cuts.

21.3.3 Cut Vertices and Vertex Cuts

We will begin our exploration of cuts by considering when the removal of a single vertex disrupts a network.

Definition 21.3.4 (Cut Vertex). *A vertex v in a connected graph is called a* cut vertex *if $G - v$ is disconnected.*

In the following figure, v is a cut vertex of the connected graph G, and – in fact – is the only cut vertex of G.

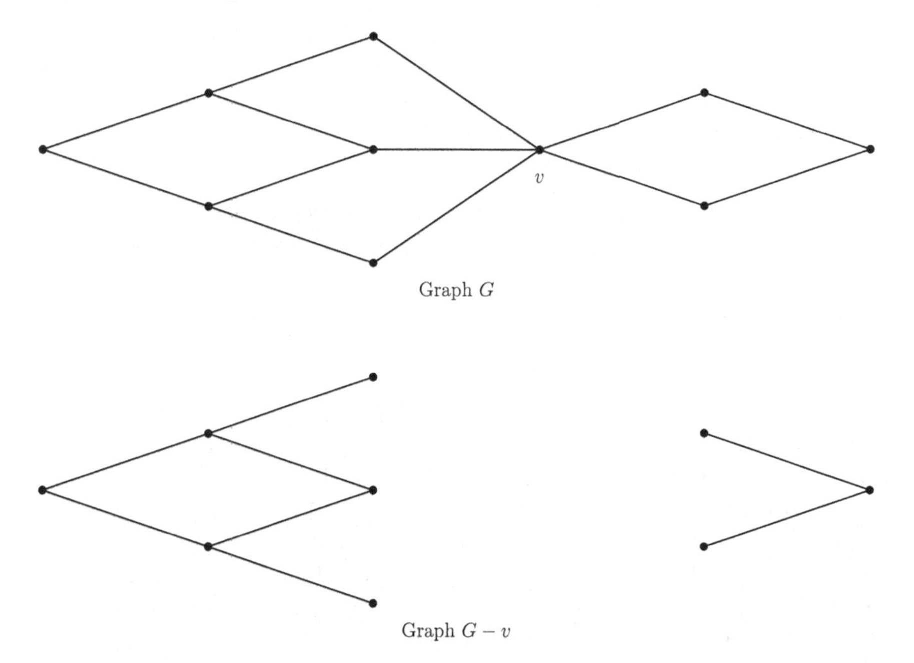

Graph G

Graph $G - v$

Note that any cycle C_n has no cut vertices. We also have the following useful result.

Theorem 21.3.5. *Every nontrivial connected graph contains at least two vertices that are not cut vertices.*

Proof. Suppose G is a nontrivial connected graph and let P be a longest path in G. Label the end vertices of P u and v so that P is a $u-v$ path. We claim that u and v are not cut vertices. To show this, suppose for contradiction that u is a cut vertex of G. Then $G-u$ is disconnected and therefore contains at least two components. Let w be the vertex adjacent to u in the path P. Since u is a cut vertex, the $w-v$ subpath P' of P must be contained in some component of $G-u$, say G_1. Let G_2 be another component of $G-v$ and since v was a cut vertex of G, v must be adjacent to some vertex x from G_2. This results in an $x-v$ path in G which is longer than P which is a contradiction. A similar argument also shows that v cannot be a cut vertex. □

The idea of the previous proof can be illustrated in the following diagram with the solid edges denoting path P. Note there may be other vertices and edges in G_1 and G_2.

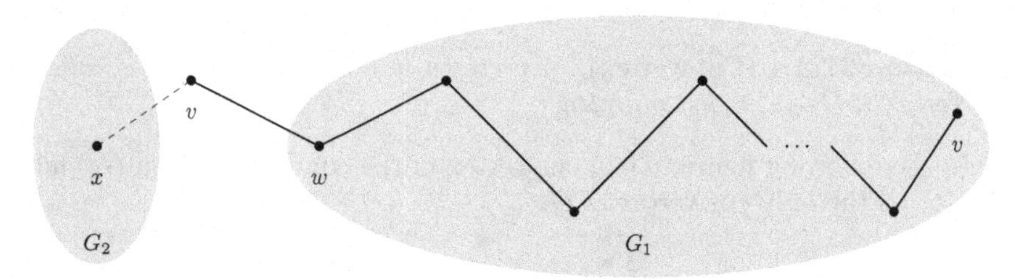

Moving beyond a cut vertex, it may be the case that to disrupt the connectivity of a graph we need to remove more than one vertex. We thus extend the definition to a collection of vertices.

Definition 21.3.6. *(Vertex Cut) Let G be a noncomplete graph with vertex set $V = V(G)$. Any $S \subseteq V$ such that $G - S$ is disconnected is called a* vertex-cut *of G.*

Note that in this definition we need the condition that G is not complete, else removing any vertices and their incident edges will result in a graph that is still connected (in fact, still complete!).

A vertex cut of minimum cardinality is known as a *minimum vertex cut*. The cardinality of a minimum vertex cut of a graph G is called the *vertex connectivity* (or just the *connectivity*) of G and is denoted $\kappa(G)$. $\kappa(G)$ is, thus, the smallest number of vertices of G whose removal results in a disconnected graph or the trivial graph. Since a complete graph of order n does not contain a vertex cut, we have $\kappa(K_n) = n - 1$. It follows that a trivial bound on the connectivity of a graph G of order n is $0 \leq \kappa(G) \leq n - 1$. We can do a little better for a bound, but need another definition first.

Definition 21.3.7. *(k-connected graph) A graph G is* k-connected *if* $\kappa(G) \geq k \geq 1$.

Thus a graph is k-connected if removing fewer than any of k vertices does not yield a graph that is disconnected or trivial. Clearly, if G is k-connected, then it is k^*-connected for each $k^* \in \{1, 2, \ldots, k\}$.

Many bounds involving k-connectivity are known but we will only present the following:

Theorem 21.3.8. *Let G be a graph of order n with k an integer such that $1 \leq k \leq n-1$. If*

$$\delta(G) \geq \left\lceil \frac{n+k-2}{2} \right\rceil,$$

then G is k-connected.

Proof. Suppose for contradiction that the theorem is false; that is, there exists a graph G of order n such that

$$\delta(G) \geq \left\lceil \frac{n+k-2}{2} \right\rceil$$

but G is not k-connected. Since it is possible to remove less than k vertices and have the resulting graph be disconnected or trivial, G cannot be complete. Moreover, since G is not k-connected, there exists a vertex-cut U of G where $|U| = l < k$, namely $l \leq k-1 \leq n-2$. Hence we are not removing enough vertices to make $G-U$ trivial. Further, since U is a vertex-cut, $G-U$ is disconnected and has order $n-l$.

Since $G-U$ is nontrivial and disconnected, it has at least two components. Let C^* be a component of $G-U$ or smallest order, say n^*. Since the $n-l$ vertices of $G-U$ are in at least two components and since C^* is a component of $G-U$, it follows from the Pigeonhole Principle that $n^* \leq \lfloor (n-l)/2 \rfloor$. Let v be any vertex of C^*. Since $G-U$ is disconnected, its only possible neighbors in G are the vertices in U and C^*. Thus

$$\begin{aligned}
\delta(G) \leq \deg v &\leq l + (n^* - 1) && (21.1)\\
&\leq l + \left\lfloor \frac{n-l}{2} \right\rfloor - 1 && (21.2)\\
&\leq \left\lfloor \frac{2l}{2} \right\rfloor + \left\lfloor \frac{n-l}{2} \right\rfloor - \left\lfloor \frac{2}{2} \right\rfloor && (21.3)\\
&\leq \left\lfloor \frac{2l}{2} + \frac{n-l}{2} - \frac{2}{2} \right\rfloor && (21.4)\\
&\leq \left\lfloor \frac{n+l-2}{2} \right\rfloor && (21.5)\\
&\leq \left\lfloor \frac{n+(k-1)-2}{2} \right\rfloor && (21.6)
\end{aligned}$$

$$\leq \left\lfloor \frac{n + (k-2) - 1}{2} \right\rfloor \tag{21.7}$$

$$< \left\lceil \frac{n + (k-2)}{2} \right\rceil \tag{21.8}$$

which contradicts the hypothesis. □

Suppose G is 2-connected. Thus removing a single vertex does not result in a disconnected or trivial graph, which is just another way to say that G does not have a cut vertex. With $k = 2$, we get a very nice result quickly from the above theorem:

Corollary 21.3.9. *Let G be a graph of order n. Then if $\delta(G) \geq \lceil \frac{n}{2} \rceil$, G does not have a cut vertex.*

21.3.4 Edge Cuts and Bridges

Since graphs consist of vertices and edges, it is natural to now explore edge cuts.

Definition 21.3.10. *(Edge Cut) Let G be a graph with vertex set $E = E(G)$. Any $X \subseteq E$ such that $G - X$ is disconnected is called a* edge cut *of G.*

A edge cut of minimum cardinality is known as a *minimum edge cut*. The cardinality of a minimum edge cut of a graph G is called the *edge connectivity* of G and is denoted $\lambda(G)$. $\lambda(G)$ is, thus, the smallest number of edges of G whose removal results in a disconnected graph or the trivial graph.

Since removing all the edges incident to a any vertex isolates the vertex, a set of edges incident to a given vertex forms an edge cut. It follows that a trivial bound on the edge connectivity of a graph G of order n is $0 \leq \lambda(G) \leq \delta(G) \leq n - 1$.

Let us also add that edge cuts that consist of a single edge have a special name. If e is an edge in a connected graph G such that $G - e$ is disconnected, then e is called a *bridge* of G. A bridge is illustrated in the following diagram.

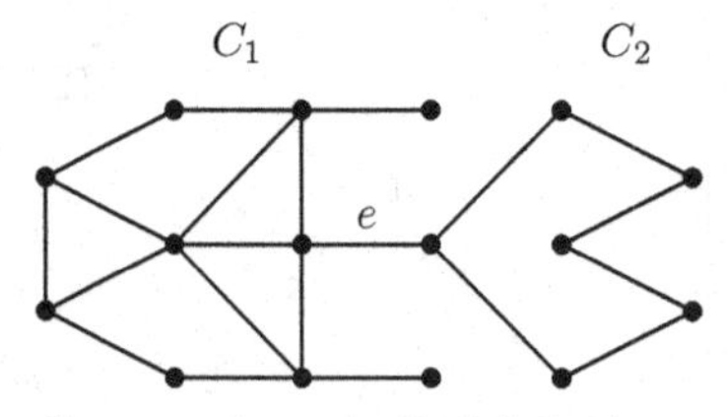

Connected graph G with bridge e

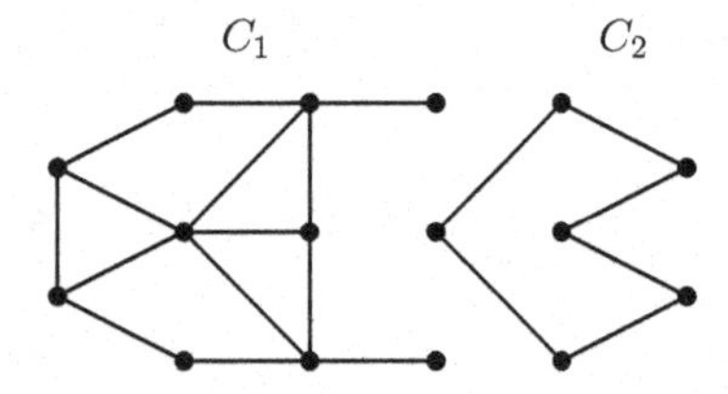

Disconnected graph $G - e$

We will close this section offering a result without proof that ties together the two types of connectivity of a graph. The result is due to Hassler Whitney [63] in 1932.

Theorem 21.3.11. *(Whitney's Inequalities) Let G be any graph of order n. Then*

$$0 \leq \kappa(G) \leq \lambda(G) \leq \delta(G) < n.$$

21.4 Some Useful and Interesting Results

We now offer results on the material that has been presented; most will be quite useful and the remaining illustrate an important mathematical reality.

We begin with the first theorem in almost every Graph Theory text.

Theorem 21.4.1 (The First Theorem of Graph Theory). *Let G be a graph with vertex set V and size m. Then*

$$\sum_{v \in V} \deg(v) = 2m.$$

Proof. As each edge is formed by two vertices, each edge will contribute 2 to the total degree of the graph. □

The First Theorem of Graph Theory is also often called *The Handshake Lemma.* This name arises from modeling the total number of handshakes among guests at a party by representing each guest as a vertex and an edge is present if a pair of guests shake hands.

It is an easy observation that the number of edges in a path of order n is $n-1$; that is the size of any P_n is always $n-1$. Moreover, since every edge in a graph is determined by its two vertices, the size of K_n is necessarily $\binom{n}{2} = \frac{n(n-1)}{2}$.

Since this is an Optimization text, let us now consider maximizing the number of edges in a complete bipartite graph of order n. We may suspect we get the most edges by making the partite sets as close in order as possible. This is indeed the case.

First, some notation. In Section 20.5 we were introduced to the ceiling function. We now introduce its kin.

Notation (Floor Function). *Let r be a real number. The* floor *of r is*

$$\lfloor r \rfloor = \textit{ the greatest integer } k \leq r.$$

For example, $\lfloor \pi \rfloor = 3$ and $\lfloor -\frac{5}{2} \rfloor = -3$. Now

Theorem 21.4.2. *The size of a bipartite graph of order n is at most $\lfloor n^2/4 \rfloor$.*

Proof. (Incorrect!) Let U and V be the partite sets and put $|U| = x$. Hence $|V| = n - x$. The maximum number of edges will occur when the graph is a complete bipartite graph which would then have size $m = x(n-x)$. n is fixed, but x is a variable, thus using Calculus, $m' = n - 2x$ and setting this to 0 we get a critical number when $x = \frac{n}{2}$. Since $m'' = -2 < 0$, this value is a maximizer. Thus $|U| = |V| = \frac{n}{2}$ yields the biggest size of $n^2/4$. □

Before reading further, stop and consider why this technique is incorrect.

We are counting and $n^2/4$ may be an integer. One may want to round down to the nearest integer, but recall Anna's Cozy Home Furniture in Example 8.1.1 from Chapter 8 and how rounding did not work. The problem here is that whatever $|U|$ is, it is an integer, and since Calculus is done over the reals (a continuous problem), its tools cannot be applied to discrete problems. As we saw with Integer Linear Programming in Chapter 8, discrete problems can be much more difficult than the continuous Linear Programming problems in Chapter 6.

Proof. (Correct) Suppose G is a bipartite graph with partite sets U and V and put $|U| = s$ where s is a positive integer less than n (if $s = 0$ or n, G would have no edges). Since[6]

$$0 \leq (n-2s)^2 = n^2 - 4ns + 4s^2 \tag{21.9}$$

we have $n^2 \geq 4s(n-s)$ giving $s(n-s) \leq n^2/4$. Thus

$$\text{size of } G \leq \text{ size of } K_{s,n-s} = s(n-s) \leq n^2/4. \tag{21.10}$$

But as the size of G is an integer,

$$\text{size of } G \leq \lfloor n^2/4 \rfloor.$$

□

The converse of Theorem 21.4.2 is quite useful. It tells us that if a graph has too many edges, then the graph cannot be bipartite. In particular.

Corollary 21.4.3. *(Converse of Theorem 21.4.2) If a graph G of order n has more than $\lfloor n^2/4 \rfloor$ edges, then G is not bipartite.*

Actually, we can say something even stronger than the converse.

Theorem 21.4.4. *Let G be a graph of order $n \geq 3$ and size $m \geq \lfloor n^2/4 \rfloor$. Then G contains a triangle (a C_3).*

Proof. First consider the case for G when $n = 3$. To meet the hypothesis of the theorem, $m > \lfloor 3^2/4 \rfloor = 2$. As the size of K_3 is $\binom{3}{2} = 3$, $m \leq 3$. Since $2 < m \leq 3$, $m = 3$ and thus $G = C_3$.

Next consider the case for G when $n = 4$. Thus, $m > \lfloor 4^2/4 \rfloor = 4$. As the size of K_4 is $\binom{4}{2} = 6$, $m \leq 6$. Since $4 < m \leq 6$, $m = 5$ or 6 and thus G is either K_4 or K_4 less an edge. In either case, G contains a triangle.

[6] We are secretly using the *Arithmetic Mean – Geometric Mean* (Theorem 18.5.1) here.

For the other cases, the proof will be by contradiction. Suppose that there exists a graph G of smallest order $n \geq 5$ and size $m > \lfloor n^2/4 \rfloor$ such that G does not contain a triangle. By hypothesis, G has size $m > \lfloor 5^2/4 \rfloor = 6$ (namely, its edge set is nonempty) and let uv be an edge in G. Since G has no triangles, there is no vertex in G adjacent to both u and v. Thus since u and v are neighbors and their neighborhoods are disjoint,

$$(\deg u - 1) + (\deg v - 1) \leq n - 2 \tag{21.11}$$

giving

$$\deg u + \deg v \leq n. \tag{21.12}$$

Put $G' = G - u - v$ (the graph G less vertices u and v and any edges incident to u and v). By assumption, G does not contain a triangle and as G' is a subgraph of G, neither does G'. Moreover, G' has order $n - 2 < n$ and size

$$\begin{aligned}
m' &= m - (\deg u + \deg v) + 1 && \text{(double counting removing edge } uv) && (21.13)\\
&\geq m - n + 1 && \text{(by 21.12)} && (21.14)\\
&> \lfloor n^2/4 \rfloor - n + 1 && \text{(by hypothesis)} && (21.15)\\
&\leq n^2/4 - n + 1 && \text{(since } m' \text{ is an integer)} && (21.16)\\
&= \frac{n^2 - 4n - 4}{4} && && (21.17)\\
&= \frac{(n-2)^2}{4} && && (21.18)\\
&\geq \lfloor (\text{order of } G')/4 \rfloor. && && (21.19)
\end{aligned}$$

Thus G' meets the conditions of the hypothesis having smaller order than G. Since G was a graph of the smallest order meeting the conditions but not having a triangle, G' must have a triangle; a contradiction. □

Putting together Theorem 21.4.2 and Theorem 21.4.4 we get quite a wonderful result: any graph that has order $n \geq 3$ with size $m > \lfloor n^2/4 \rfloor$ not only cannot be bipartite, it must contain a triangle.

21.5 Exercises

Exercise 21.1. *A* regular graph *is a graph G where every vertex has the same degree; that is, there is a nonnegative constant r such that* $\deg(u) = r$ *for every vertex u of G (such a graph is said to be r-*regular*). A graph in which every vertex has a different degree, that is* $\deg(u) \neq \deg(v)$ *for all distinct vertices u and v in G, is said to be* irregular. *Prove that there do not exist any irregular graphs (Hint: use the Pigeonhole Principle).*

Exercise 21.2. *Let G be a graph of order n and size $m \geq n^2/4$. Prove that G does not have a cut vertex.*

Exercise 21.3. *Determine the vertex-connectivity $\kappa(K_{s,t})$ and edge-connectivity $\lambda(K_{s,t})$ of an arbitrary complete bipartite graph $K_{s,t}$. (Hint: consider the order of the largest partite set.)*

Exercise 21.4. *Repeat the previous exercise but for an arbitrary complete k-partite graph $K_{n_1,n_2,\ldots,n_k}$.*

Exercise 21.5. *Using one of Whitney's inequalities (Theorem 21.3.11), prove that if G is a graph of order n and size m with $m \geq n-1$, then*

$$\kappa(G) \leq \left\lfloor \frac{2m}{n} \right\rfloor.$$

22

Network Flows

Network flow problems encompass a wide range of important applications including the flow of vital natural resources through a network of pipes, telecommunications, shipping oranges from orchards to processing facilities to distribution centers to stores, and so much more.

22.1 Basic Definitions

A *network* is a digraph D with two distinguished vertices s and t – the *source* and *sink* – together with a nonnegative real-valued function $c : E(D) \to \mathbb{R}_+$. The digraph D is called the *underlying digraph* of the network. Similarly, the graph G obtained by removing the orientation on the edges and compressing any two directed edges between vertices into a single edge is called the *underlying graph* of the network. The function c on the edges of D is called a *capacity function.* If $uv \in E(D)$, then the value $c(uv)$ is referred to as the *capacity* of edge uv. Vertices in a network that are not the source or the sink are often called *intermediate vertices.*

Example 22.1.1. *Figure 22.1 illustrates a network on 6 vertices with source s and sink t. This network has a, b, c, and d as intermediate vertices. Capacities of each edge written adjacent to the edge.*

It is possible to have multiple sources or sinks, but in these situations we can satisfy the definition of a network by introducing an artificial source which would then lead to each of the multiple sources. Likewise, an artificial sink may be introduced for a situation with multiple sinks.

Now that we have a network, we may introduce flow into the system.

Definition 22.1.2. *(Flow in a Network) Let N be a network with underlying digraph D, source s, sink t, and capacity function $c(e)$ defined on all $e \in E(D)$. A* flow *in N is a real-valued function f defined on $E(D)$ satisfying the conditions*

1. *$0 \leq f(e) \leq c(e)$ for every $e \in D$ and*
2. *$f^+(u) = f^-(u)$ for every intermediate vertex u in N*

DOI: 10.1201/9780367425517-22

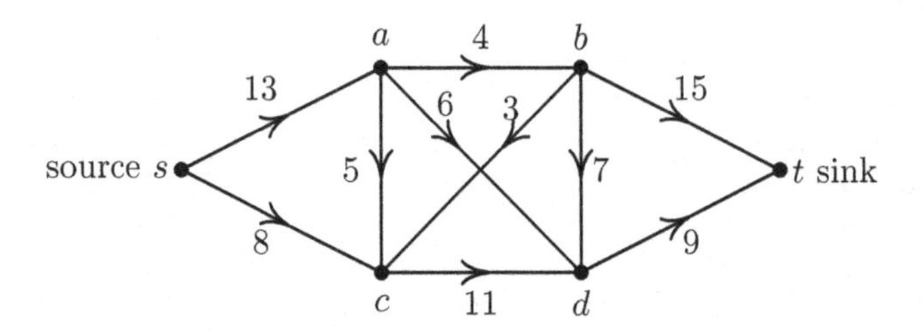

FIGURE 22.1
A network with a capacity function.

where

$$f^+(u) := \sum_{y \in N^+(u)} f(uy)$$

and

$$f^-(u) := \sum_{x \in N^-(u)} f(xu).$$

The $f^+(u)$ in the definition is the *total flow out of* u where $f^-(u)$ is referred to as the *total flow into* u. Thus the *net flow out of a vertex* u is

$$f^+(u) - f^-(u)$$

where the *net flow into a vertex* u is

$$f^-(u) - f^+(u).$$

Condition 2 states that the net flow into an intermediate vertex must equal to the net flow leaving the vertex. This is known as the *conservation equation* in network flows. The inequalities in condition 1 are referred to as the *capacity constraints*. Any edge e such that $f(e) = c(e)$ is said to be a *saturated edge*. Edges whose flow is strictly less than their capacity are called *unsaturated edges*. Figure 22.2 shows a network the capacity and flow of each edge included. Here, the notation a/b on edge e means $c(e) = b$ and $f(e) = a$. In this example, note that $f^+(a) = 4 + 3 + 3 = 10$ and $f^-(a) = 10$ as well, satisfying the conservation equation for vertex a.

By their definition, a source and sink are just distinguished vertices in a digraph. In practice, though, they are (as the names suggest) the place where the flow begins and the place where it ends. Usually the source only has directed edges exiting and the sink only has in-edges, though exceptions occur. Note that the definitions are still satisfied, though, if the sink has a positive in-degree and the source a positive out-degree. These possibilities are be considered in the following theorem, which begins to pave the way for the main result of this chapter.

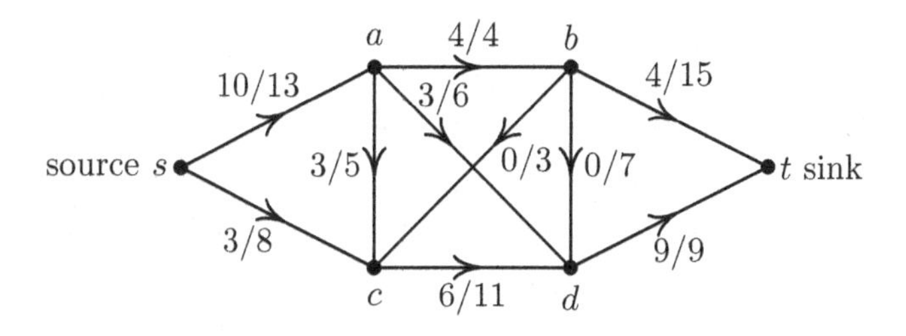

FIGURE 22.2
A network with edge capacities and flows.

Theorem 22.1.3. *Consider a network with source s, sink t, and flow f. Then*

$$f^+(s) - f^-(s) = f^+(t) - f^-(t).$$

Theorem 22.1.3 says that the net flow out of the source is the same as the net flow into the sink. This is obvious consequence of the conservation equations holding for each intermediate vertex and observe that $f^+(s) - f^-(s) = 13 = f^+(t) - f^-(t)$ in Figure 22.2.

22.2 Maximum Flows and Cuts

By Theorem 22.1.3, the net flow out of the source is the same as the net flow into the sink. It is natural to consider how large this net flow can be.

Definition 22.2.1 (Value of a Flow). *Let N be a network with source s and flow f. Then the* value of the flow *in N is*

$$\text{val}(f) := f^+(s) - f^-(s).$$

Thus $\text{val}(f)$ is just the net flow out of the source s which, by Theorem 22.1.3, is also the net flow into the sink. In Figure 22.2, $\text{val}(f) = 13$.

For a network N, let F be the collection of all possible flows on the network. Then the *maximum flow* on N is $\text{val}(f^*)$ such that $\text{val}(f^*) \geq \text{val}(f)$ for all $f \in F$. Of course, we are very interested in determining what the maximum flow is on a given network. To do this, we will need a very helpful tool.

Definition 22.2.2 (A Cut in a Network). *Let N be a network with source s, sink t, and underlying digraph D. For $U \subset V(D)$, define $\overline{U} := V(D) - U$ (the complement of U in $V(D)$). A* cut *in N, $[U, \overline{U}]$, where $s \in U$ and $t \in \overline{U}$, is the collection of all directed edges uv from $E(D)$ such that $u \in U$ and $v \in \overline{U}$ (i.e. all* forward edges from vertices in U to vertices in $\overline{U}$*).*

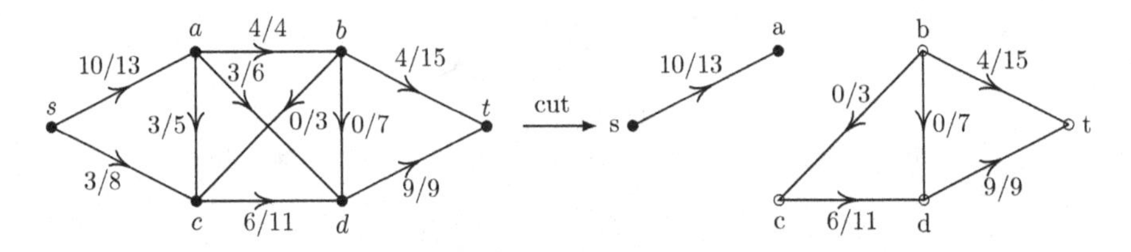

FIGURE 22.3
The cut $U = \{s, a\}$ in the network N.

Thus a cut in a network is an edge-cut but also necessarily a very special one. Here, exactly one of the source and sink end up in U and every edge of D that has its start vertex in U and end vertex in $\overline{U}$ is part of the cut. Figure 22.3 illustrates a cut $[U, \overline{U}]$ where $U = \{s, a\}$.

Now consider a network with a capacity function c. For a cut $K = [U, \overline{U}]$, we define the *capacity of the cut* to be

$$\text{cap}(K) := \sum_{(x,y)\in[U,\overline{U}]} c(xy).$$

The cut $K = [U, \overline{U}]$ in Figure 22.3 has capacity

$$\text{cap}(K) = c(sc) + c(ab) + c(ac) + c(ad) = 8 + 4 + 5 + 6 = 23.$$

For any $U \subset V(D)$, let us define $f^+(U) := \sum_{x\in U} f^+(x)$; that is, the total flow out of the set of vertices in U. Similarly, $f^+(U) := \sum_{x\in U} f^+(x)$; that is, the total flow into the set of vertices in U. For $U = \{s, a\}$ in Figure 22.3,

$$\begin{aligned} f^+(U) &= f^+(s) + f^+(a) = [f(sa) + f(sc)] + [f(ab) + f(ac) + f(ad)] \\ &= [10 + 3] + [4 + 3 + 3] = 23 \end{aligned}$$

and

$$f^-(U) = f^-(s) + f^-(a) = 0 + [f(sa)] = 10.$$

Notice that for this cut $f^+(U) - f^-(U) = 23 - 10 = 13$, which has been a reoccurring number in this example.

Theorem 22.2.3. *(Weak Duality in Flows) Let f be a flow in a network N with $K = [U, \overline{U}]$ a cut in N. Then*

$$\text{val}(f) = f^+(U) - f^-(U) \leq \text{cap}(K).$$

Notice that for the example in Figure 22.3, $\text{val}(f) = 13 = f^+(U) - f^-(u) \leq 23 = \text{cap}(K)$.

Proof. Let D be the underlying digraph of N with source s and a cut $K = [U, \overline{U}]$. Suppose $U = \{s, u_1, \ldots, u_k\}$. Then

$$f^+(U) - f^-(U)$$
$$:= f^+(s) + f^+(u_1) + \cdots f^+(u_k) - \big(f^-(s) + f^- + (u_1) + \cdots f^- + (u_k)\big) \quad (22.1)$$
$$= \big(f^+(s) - f^-(s)\big) + \big(f^+(u_1) - f^- + (u_1)\big) + \cdots \big(f^+(u_k) - f^- + (u_k)\big) \quad (22.2)$$
$$= f^+(s) - f^-(s) \text{ by the conservation equations} \quad (22.3)$$
$$=: \text{val}(f). \quad (22.4)$$

For the cut K, let U^* be the vertices of U incident with the edges $e_1, \ldots, e_l$ removed in the cut; i.e. $e_1, \ldots, e_l \in [U, \overline{U}]$. In Figure 22.3, we have $U^* = \{a\}$ and $U - U^* = \{s\}$. Edges incident with the vertices in u that are not involved in the cut have a net contribution of 0 to $f^+(U) - f^-(U)$ as in edge sa again in the example. Thus the only positive contributions to $f^+(U) - f^-(U)$ are the outflows on the vertices in U^*. Hence

$$f^+(U) - f^-(U) \leq f(e_1) + \cdots + f(e_l) \leq c(e_1) + \cdots c(e_l) = \text{cap}(K).$$

□

Theorem 22.2.3 is very important and states what so often occurs in optimization: one thing's maximum is another thing's minimum. In light of this, as we have seen maximum flow, it is now time to define a minimum cut. Consider a network N and let $\mathcal{K}$ be the collection of all possible cuts on N. K^* is a *minimum cut* on N if $\text{cap}(K*) \leq \text{cap}(K)$ for all $K \in \mathcal{K}$.

Two very nice results that fulfill our goal for the chapter follow from Theorem 22.2.3.

Corollary 22.2.4. *If f is a flow in a network N and K a cut where* $\text{val}(f) = \text{cap}(K)$, *then f is a maximum flow and K is a minimum cut.*

Proof. Suppose f^* is a maximum flow in N and K^* is a minimum cut. Thus $\text{val}(f) \leq \text{val}(f^*)$ and $\text{cap}(K^*) \leq \text{cap}(K)$. Together with Theorem 22.2.3 we then have

$$\text{val}(f) \leq \text{val}(f^*) \leq \text{cap}(K^*) \leq \text{cap}(K).$$

By assumption $\text{val}(f) = \text{cap}(K)$, thus $\text{val}(f^*) = \text{val}(f)$, making f^* a maximum flow. As well, the hypothesis leads to $\text{cap}(K^*) = \text{cap}(K)$, which means K^* is a minimum cut. □

Piecing together some of the details for proving the equality in Theorem 22.2.3 also gives us a sufficient condition for having a maximum flow and minimum cut.

Corollary 22.2.5 (Sufficient Conditions for a Max Flow and Min Cut)**.** *Suppose f is a flow in a network N and $K = [U, \overline{U}]$ a cut satisfying*

TABLE 22.1
Possible Edge Cuts in the Network N from Figure 22.2

U	$\overline{U}$	$K = [U, \overline{U}]$	cap(K)	Equal val(f) = $f^+(s) = 13$?	Max Flow/ Min Cut
s	a, b, c, d, t	sa, sc	21	No	No
s, a	b, c, d, t	sc, ab, ac, ad	23	No	No
s, b	a, c, d, t	sa, sc, bc, bd, bt	46	No	No
s, c	a, b, d, t	sa, cd	24	No	No
s, d	a, b, c, t	sa, sc, dt	30	No	No
s, a, b	c, d, t	sc, ac, ad, bd, bt	41	No	No
s, a, c	b, d, t	ab, ad, cd	21	No	No
s, a, d	b, c, t	sc, ab, ac, dt	26	No	No
s, b, c	a, d, t	sa, bd, bt, cd	46	No	No
s, b, d	a, c, t	sa, sc, bc, bt, dt	48	No	No
s, c, d	a, b, t	sa, dt	22	No	No
s, b, c, d	a, t	sa, bt, dt	37	No	No
s, a, c, d	b, t	ab, dt	13	Yes	Yes
s, a, b, c	d, t	ad, bd, bt, cd	39	No	No
s, a, b, d	c, t	sc, ac, bc, bt, dt	40	No	No
s, a, b, c, d	t	bt, dt	24	No	No

- *$f(uv) = c(uv)$ for all edges uv in the cut (thus $u \in U$ and $v \in \overline{U}$) and*
- *$f(vu) = 0$ for all edges vu in the cut with $v \in \overline{U}$ and $u \in U$,*

then f is maximum flow and K is a minimum cut.

Corollary 22.2.5 says that if a cut $[U, \overline{U}]$ has

- all *forward edges* in the cut (edges leaving the vertices in U and entering vertices in $\overline{U}$) are saturated as well as
- all *backwards edges* (those leaving the vertices in $\overline{U}$ and entering vertices in U) have no flow,

then f is a maximum flow and K a minimum cut.

We illustrate what we have discussed with the following example:

Example 22.2.6. *Let N the network shown in Figure 22.2 with the stated flows and capacities. As each of the 4 intermediate vertices can either be in U or not, there are 2^4 possible cuts in N. These 16 cuts, as well as whether each cut is a minimum cut, are listed in Table 22.1.*

Thus network N has a unique minimum cut. Notice that this cut, $U = \{s, a, c, d\}$ with removed edges ad and bt, has all forward edges saturated and any backwards edges (there are none) have no flow, as in Corollary 22.2.5.

We improve slightly the results above with the following important theorem. It was proven independently by Lester Randolf Ford, Jr and Delbert Roy Fulkerson [24] as well as Peter Elias, Alex Feinstein, and Claude E. Shannon [22] with both papers appearing in 1956. This result is regarded by many as one of the most beautiful theorems in mathematics.

Theorem 22.2.7 (Max-Flow Min-Cut Theorem). *In any network, the value of a maximum flow equals the capacity of a minimum cut.*

The proof of the Max-Flow Min-Cut Theorem is not difficult, but involves the idea of *augmenting paths* (which we will see shortly). We will exclude the proof and direct interested readers to the appropriate sections in either of the excellent texts [9] or [30].

22.3 The Dinitz-Edmonds-Karp-Ford-Fulkerson Algorithm

As often happens in mathematics, the Max-Flow Min-Cut Theorem provides important information, but the statement of the theorem does not describe how to obtain it. It is, of course, useful to know that the maximum flow is the same as the capacity of a minimum cut, but how do we determine exactly what this is?

Ford and Fulkerson's proof of their Max-Flow Min-Cut Theorem included elements that led to an algorithm for finding a maximum flow in a network; the well-known *Ford-Fulkerson Algorithm*. The algorithm does return the maximum flow in a network but also comes with some concerns:

- The algorithm may fail to terminate if some of the capacities are irrational (Ford and Fulkerson's example showing this can be found in [25]). This, of course, is not a concern in practice.
- It is possible for the algorithm to run very inefficiently.

As such, we will avoid introducing the original Ford-Fulkerson Algorithm and note that it is the essence of two similar algorithms that followed. Improved versions of Ford-Fulkerson were developed independently by Yefim A. Dinitz[1] in 1970 [18] and Jack Edmonds and Richard Karp in 1972 [37]. The algorithms both improve upon the Ford-Fulkerson Algorithm by establishing a search order for the augmenting path. This improvement removes the associated concerns. The algorithm is most often referred to as the *Edmonds-Karp Algorithm* but is also called *Dinitz's Algorithm*, *Dinic's Algorithm*, and even the *Dinitz (or Dinic)-Edmonds-Karp Algorithm*. As the improved algorithms

[1] Initially, Dinitz's name was transliterated from Russian to *Dinic* and this version of his name is often associated with the algorithm.

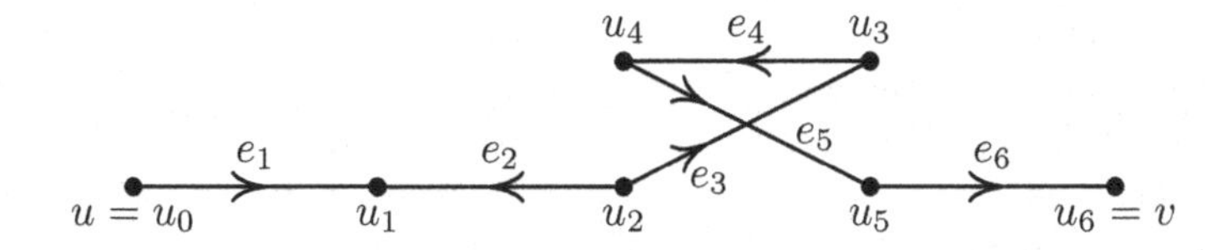

FIGURE 22.4
A $u-v$ semipath P from a network.

are the Ford and Fulkerson Algorithm with a breadth-first search introduced, it seems proper to honor all five of the algorithm's creators.

Before introducing the algorithm, we should establishing some of the nomenclature.

Definition 22.3.1 ($u-v$ semipath). *In a digraph D, a $u-v$* semipath *is a sequence $P = (u = u_0, e_1, u_1, e_2, u_2, \ldots, u_{k-1}, e_k, u_k = v)$ of distinct vertices and directed edges beginning with u and ending with v such that $e_i = u_{i-1}u_i$ or $e_i = u_iu_{i-1}$.*

Thus a semipath is a path but with directed edges and the condition that the edge e_i incident with vertices from the ordered list u_{i-1} and u_i is either be the directed edge $u_{i-1}u_i$ (called a *forward directed edge of P*) or the directed edge u_iu_{i-1} (a *backward directed edge of P*). Figure 22.4 shows a $u-v$ semipath P having forward edges e_1, e_3, e_4, e_5, and e_6 and backward edge e_2. Note, that as the example illustrates, whether an edge is regarded as a forward edge or backward edge in the semipath depends on the labeling of the vertices in the sequence giving the semipath.

Now that we understand semipaths,

Definition 22.3.2 (f-augmenting semipath). *Consider a network N with underlying digraph D and capacity function c. Let f be a flow in N. Then a semipath $P = (u_0, e_1, u_1, e_2, u_2, \ldots, u_{k-1}, e_k, u_k)$ in D is said to be* f-unsaturated *if for each i, $1 \leq i \leq k$,*

- *if e_i is a forward edge, then $f(e_i) < c(e_i)$ and*
- *if e_i is a backward edge, then $f(e_i) > 0$.*

A semipath from the source to the sink that is f-unsaturated is called an f-augmenting semipath.

Thus an f-augmenting semipath, or augmenting semipath, is a semipath from the source to the sink where each forward edge is unsaturated and the backward edges have positive flow.

Let f be a flow in a network N and suppose N has an f-augmenting semipath P. Since P is an f-augmenting semipath, it is possible to increase the flow across it and obtain a new flow f^* from the source to the sink such that $f^* > f$. Thus, if f-augmenting semipaths exist in N, f cannot be the

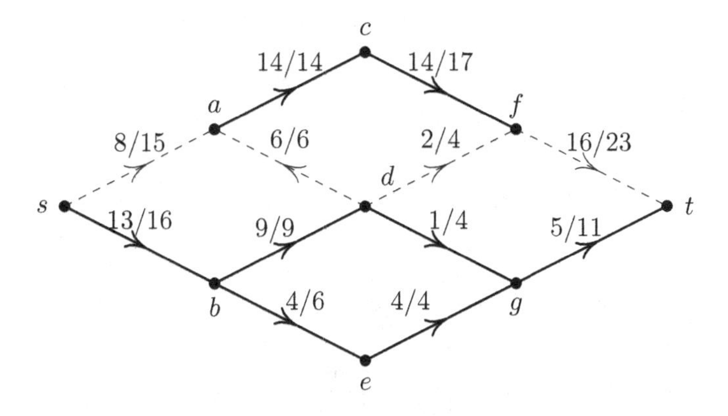

FIGURE 22.5
An augmenting semipath *sadft*.

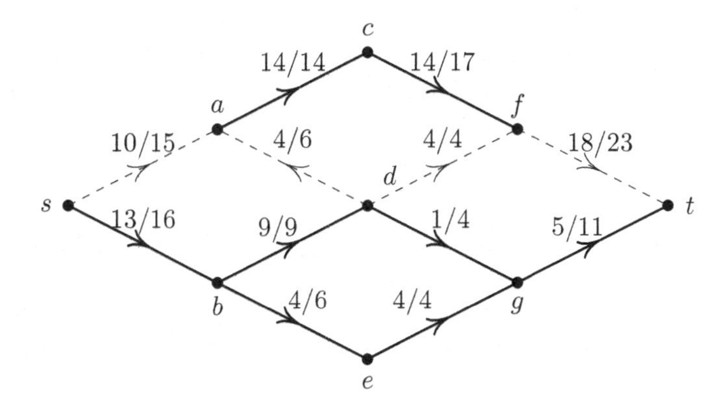

FIGURE 22.6
After augmenting the flow on semipath *sadft*.

maximum flow in N. Figure 22.5 shows a network with flow $f = 21$ and an augmenting semipath *sadft*. Decreasing the flow on edge *da* from 6 to 4 and moving that flow to *df* permits *sa* to have its flow increased (*augmented*) to 10 (but no more because of the capacity on *ac*). This augmentation results in a new network flow of $f^* = 23$ as shown in Figure 22.6. This is not the max flow as *sadgt* is also an augmenting path and a flow greater than 23 is possible.

The algorithm proceeds by first creating labels for the vertices in the network with the condition that a vertex v is labeled *only if* there exists an $s - v$ semipath (s is the source) that is unsaturated. Suppose such a semipath P exists and that u is the vertex immediately preceding v in the semipath. v is labeled with an ordered pair where

TABLE 22.2
Using DEKaFF to Find a First Augmenting Semipath in N in Figure 22.2

Iteration	v	Labeled	L	Labeled Vertices	Accepted Vertices
0		s labeled $(-, \infty)$	s	s	
1	s	s accepted and removed	$\emptyset$	s	s
		$sa \in E(D)$, a labeled $(s^+, 7)$	a	s,a	s
		$sb \in E(D)$, b labeled $(s^+, 3)$	a,b	s,a,b	s
2	a	a accepted and removed	b	s,a,b	s,a
		$ad \in E(D)$, d labeled $(a^-, 6)$	b, d	s,a,b,d	s,a
3	b	b accepted and removed	d	s,a,b,d	s,a,b
		$be \in E(D)$, e labeled $(b^+, 2)$	d,e	s,a,b,d,e	s,a,b
4	d	d accepted and removed	e	s,a,b,d,e	s,a,b,d
		$df \in E(D)$, f labeled $(d^+, 2)$	e,f	s,a,b,d,e,f	s,a,b,d
		$dg \in E(D)$, g labeled $(d^+, 3)$	e,f,g	s,a,b,d,e,f,g	s,a,b,d
5	e	e accepted and removed	f,g	s,a,b,d,e,f,g	s,a,b,d,e
6	f	f accepted and removed	g	s,a,b,d,e,f,g	s,a,b,d,e,f
		$ft \in E(D)$, t labeled $(f^+, 2)$	g,t	s,a,b,d,e,f,g,t	s,a,b,d,e,f

- the first component of the ordered pair is u^+ if uv is a directed edge in P or u^- if that direct edge is vu; and
- the second component is the positive number which gives the potential change in the flow f along P.

A list is then formed, conditions are checked to increase the flow, if possible, and vertices are accepted into the f-augmenting semipath. The details are stated in Algorithm 22.3.1. It is worthwhile to note that during the running of the algorithm, every vertex of the network will be either (i) unlabeled, (ii) labeled but unaccepted, or (iii) labeled and accepted.

Example 22.3.3. *We use Algorithm 22.3.1 to find the maximum flow on the network N in Figure 22.5.*

First Augmenting Semipath (Steps 3 – 33)

*Iterating through the **if** in steps 4 through 24 is a labeling procedure which produces the first f-augmenting semipath to increase the flow in N. This process is shown in Table 22.2 with notes on the iterations below. Steps 26 – 32 are then followed to produce an augmenting semipath and augment flows on the appropriate edges.*

1. *Since L is nonempty, the first vertex, s, in L is scanned and removed from the list. We then search the vertices of D and see that we have vertices a*

Algorithm 22.3.1 The Dinitz-Edmonds-Karp-Ford-Fulkerson (DEKaFF) Algorithm

Input: A network N with underlying digraph D, source s, sink t, and capacity function c.

1: let f be a given flow on D
2: assign each edge uv of D the values its flow $f(u, v)$ and capacity $c(u, v)$
3: label source s with $(-, \infty)$ and add s to sequence (ordered list) L of labeled, unaccepted vertices
4: **if** $L = \emptyset$ **then**
5: stop
6: **else** accept and remove the first vertex, v, of L which has as its label either $(u^+, A(x))$ or $(u^-, A(x))$.
7: **if** vertex x in D is unlabeled (break ties lexicographically) **then**
8: **if** $vx \in E(D)$ **then** ▷ a forward edge
9: **if** $f(v, x) < c(v, x)$ **then**
10: assign vertex x the label $(v^+, A(x))$ where $A(x) := \min\{A(v), c(v, x) - f(v, x)\}$
11: add x to the end of the list L
12: **end if**
13: **else if** $xv \in E(D)$ **then** ▷ a backward edge
14: **if** $f(x, v) > 0$ **then**
15: assign vertex x the label $(v^-, A(x))$ where $A(x) := \min\{A(v), f(x, v)\}$
16: add x to the end of the list L
17: **end if**
18: **end if**
19: **end if**
20: **end if**
21: **if** sink t is labeled **then**
22: go to step 25
23: **else** go to step 4
24: **end if**
25: proceed backwards from the sink t and following the first coordinate of the labels gives an f-augmenting semipath $P = (s, e_1, v_1, e_2, v_2, \ldots, e_k, t)$ where the edges are augmented by
26: **for** $i = 1$ to k **do**
27: **if** v_i is labeled $(v_{i-1}^+, A(v_i))$ (a forward arc) **then**
28: put $f(e_i) = f(v_{i-1}, v_i) := f(v_{i-1}, v_i) + A(t)$ ▷ t is the sink
29: **else if** v_i is labeled $(v_{i-1}^-, A(v_i))$ (a backward arc) **then**
30: put $f(e_i) = f(v_i, v_{i-1}) := f(v_i, v_{i-1}) - A(t)$ ▷ t is the sink
31: **end if**
32: **end for**
33: delete all labels and return to Step 3

Output: The max flow f in the network N.

and b incident with the selected scanned vertex s. As sa is an edge into vertex a, the first component of a's label is s+. Since this is an in-edge, by the algorithm, a has as its second component $A(a) := \min\{A(s), c(s,a) - f(s,a)\} = \min\{\infty, 15 - 8\} = 7$. *It is a similar process to obtain b's label with* $A(b) := \min\{A(s), c(s,b) - f(s,b)\} = \min\{\infty, 16 - 13\} = 3$.. *Note that at this moment we have three labeled vertices* $s = (-, \infty)$, $a = (s^+, 7)$, *and* $b = (s^+, 3)$ *with* $L = (a, b)$. *Since the sink t is not yet labeled, we go back to step 4.*

2. *Since* $L \neq \emptyset$, *we continue and scan a and remove it from L. Searching the vertices of D we have that vertices c and d are incident with the selected scanned vertex a. ac is an edge into vertex c, but as it is a forward edge at full capacity, it is not labeled by the algorithm. da is also at full capacity, but as it is a backwards edge, it is labeled. Thus the first component of d's label is* a^-. *Since this is an out-edge, by the algorithm, d has as its second component* $A(d) := \min\{A(a), c(d,a)\} = \min\{7, 6\} = 6$. *At this moment we have four labeled vertices s, a, b and d with* $L = (b, d)$. *Since the sink t is not yet labeled, we go back to step 4.*

3. *As b is the first vertex in the list, it is accepted and removed from L. Vertex d is already labeled, so nothing is done with it. be is an edge in D and as it is a forward edge its first component is* b^+ *and its second is* $A(e) := \min\{A(b), c(b,e) - f(b,e)\} = \min\{3, 6 - 4\} = 2$. *Since the sink t is not yet labeled, we go back to step 4.*

4. *As d is the first vertex in L and accepted and removed from the list. df is an edge in D and as it is a forward edge its first component is* d^+ *and its second is* $A(f) := \min\{A(d), c(d,f) - f(d,f)\} = \min\{6, 4 - 2\} = 2$. *Likewise for g with* $A(g) := \min\{A(d), c(d,g) - f(d,g)\} = \min\{6, 4 - 1\} = 2$. *Since the sink t is not yet labeled, we go back to step 4.*

5. *e is accepted and removed from L with* $eg \in E(D)$, *but g is labeled, hence nothing is done. Again, as t is not yet labeled, we return to step 4.*

6. *f is accepted and removed from L with* $ft \in E(D)$. *The first component of t's label is* f^+ *and its second is* $A(t) := \min\{A(f), c(f,t) - f(f,t)\} = \min\{2, 23 - 16\} = 2$. *Since the sink t is now labeled, we proceed to step 25.*

We now identify the f-augmenting semipath produced by these steps in the algorithm

The **for** *loop in steps 26 through 32 provide the augmentations on the stated edges to increase the flow in N. In this run through the algorithm the vertices have the labels given in Table 22.3.*

Reading backwards from the sink t and following the first components of the labels we have the augmenting semipath sadft. The edges sa, da (a backwards edge), df, ft are augmented by steps 26–32 with the results as in Table 22.4.

TABLE 22.3
Vertex Labels from the First Run through Algorithm 22.3.1

Vertex	Label
s	$(-,\infty)$
a	$(s^+,7)$
b	$(s^+,3)$
c	unlabeled
d	$(a^-,6)$
e	$(b^+,2)$
f	$(d^+,2)$
g	$(d^+,3)$
t	$(f^+,2)$

TABLE 22.4
Original Flow, Capacity and Augmentations on Edges in N from the First Run Through Algorithm 22.3.1

Edge	Original Flow	Capacity	Augmentation (Steps 26–32)	Augmented Flow
sa	8	15	+2	10
sb	13	16	none	13
ac	14	14	none	14
da	6	6	−2	4
bd	9	9	none	9
be	4	6	none	4
cf	14	17	none	14
df	2	4	+2	4
dg	1	4	none	1
eg	4	4	none	4
ft	16	23	+2	18
gt	5	11	none	5

The results of this run through DEKaFF are exactly what we obtained in Figure 22.6. As stated in the algorithm, all labels are now cleared and we repeat the process.

Second Augmenting Semipath (Steps 3 – 33)

The labeling done in steps 4 through 24 is stated in Table 22.5. The resulting labels are given in Table 22.6 and from this we see that sadgt is a new augmenting semipath.

TABLE 22.5
Using DEKaFF to Find a Second Augmenting Semipath in N in Figure 22.6

Iteration	v	Labeled	L	Labeled Vertices	Accepted Vertices
0		s labeled $(-,\infty)$	s	s	
1	s	s accepted and removed	$\emptyset$	s	s
		$sa \in E(D)$, a labeled $(s^+, 5)$	a	s,a	s
		$sb \in E(D)$, b labeled $(s^+, 3)$	a,b	s,a,b	s
2	a	a accepted and removed	b	s,a,b	s,a
		$ad \in E(D)$, d labeled $(a^-, 4)$	b, d	s,a,b,d	s,a
3	b	b accepted and removed	d	s,a,b,d	s,a,b
		$be \in E(D)$, e labeled $(b^+, 2)$	d,e	s,a,b,d,e	s,a,b
4	d	d accepted and removed	e	s,a,b,d,e	s,a,b,d
		$dg \in E(D)$, g labeled $(d^+, 3)$	e,g	s,a,b,d,e,g	s,a,b,d
5	e	e accepted and removed	g	s,a,b,d,e,g	s,a,b,d,e
6	g	g accepted and removed	$\emptyset$	s,a,b,d,e,g	s,a,b,d,e,g
		$gt \in E(D)$, t labeled $(g^+, 3)$	t	s,a,b,d,e,g,t	s,a,b,d,e,g

TABLE 22.6
Vertex Labels from the Second Run Through Algorithm 22.3.1

Vertex	Label
s	$(-,\infty)$
a	$(s^+, 5)$
b	$(s^+, 3)$
c	unlabeled
d	$(a^-, 4)$
e	$(b^+, 2)$
f	unlabeled
g	$(d^+, 3)$
t	$(g^+, 3)$

The results of the second augmentation are shown in Figure 22.7.

Search for Third Augmenting Semipath

The labeling process for the next search is given in Table 22.8.

Notice at the end of iteration 4, $L = \emptyset$. Thus DEKaFF terminates and the max flow has been found. Thus the max flow in this example is 26 and a flow yielding this flow value is given in Figure 22.7.

TABLE 22.7
Starting Flow, Capacity and Augmentations on Edges in N from the Second Run Through Algorithm 22.3.1

Edge	Start Flow	Capacity	Augmentation (Steps 26–32)	Augmented Flow
sa	10	15	+3	13
sb	13	16	none	13
ac	14	14	none	14
da	4	6	−3	1
bd	9	9	none	9
be	4	6	none	4
cf	14	17	none	14
df	4	4	none	4
dg	1	4	+3	4
eg	4	4	none	4
ft	18	23	none	18
gt	5	11	+3	8

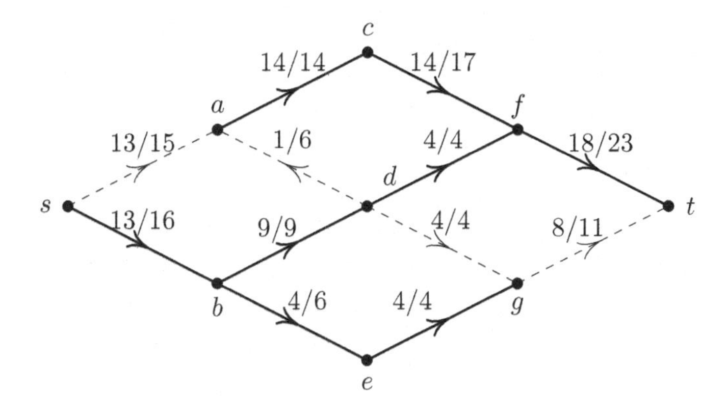

FIGURE 22.7
Second run of DEKaFF results through augmenting the flow on semipath *sadgt*.

It is worthwhile to note the we have something quite serendipitous occurring with DEKaFF in the example. Notice in Table 22.8 that when we reach the termination we have that s, a, b, and d are the labeled vertices. Putting $U := \{s, a, b, d\}$, we have a cut $[U, \overline{U}]$ which has a cut capacity of 26. This is also the max flow and is no accident. It is the case that the set of labeled vertices at the termination stage gives a min cut in the network. This remarkable fact is included in the following theorem.

TABLE 22.8
Third Search for an Augmenting Semipath in N in Figure 22.7

Iteration	v	Labeled	L	Labeled Vertices	Accepted Vertices
0		s labeled $(-,\infty)$	s	s	
1	s	s accepted and removed	$\emptyset$	s	s
		$sa \in E(D)$, a labeled $(s^+,2)$	a	s,a	s
		$sb \in E(D)$, b labeled $(s^+,3)$	a,b	s,a,b	s
2	a	a accepted and removed	b	s,a,b	s,a
		$da \in E(D)$, d labeled $(a^-,1)$	b, d	s,a,b,d	s,a
3	b	b accepted and removed	d	s,a,b,d	s,a,b
4	d	d accepted and removed	$\emptyset$	s,a,b,d	s,a,b,d

Theorem 22.3.4. *DEKaFF (Algorithm 22.3.1) terminates with a maximum flow f in a given network N. Moreover, Letting U be the set of labeled vertices at the termination stage of the algorithm gives a min cut $[U,\overline{U}]$ in the network N.*

A proof of the theorem appears in [9].

22.4 Max Flow as a Linear Programming Problem

It is possible to determine the maximum flow in a network via linear programming. For a network N with underlying digraph D, source s, and sink t with edge flows and capacities given by $f(u,v)$ and $c(u,v)$, then the maximum flow in N is found by

$$\begin{aligned} &\text{Maximize} && \text{val}(f) = f^+(s) - f^-(s) && (22.5)\\ &\text{Subject to} && f^+(u) = f^-(u) \text{ for all } u \in V(D), u \neq s,t \text{ and} && (22.6)\\ &&& f(u,v) \leq c(u,v) \text{ for all } u,v \in V(D). && (22.7)\end{aligned}$$

The decision variables will be, of course, the flow on each edge in N. The constraint equations in 22.6 are due the conservation of flow at each intermediate vertex and the inequality constraints in 22.7 are such that no edge flow exceeds its capacity.

Returning to Example 22.3.3 to find the maximum flow in the network in Figure 22.6, we have as the decision variables:

decision variable	initial value	capacity
$f(s,a) = X_{sa}$	8	15
$f(s,b) = X_{sb}$	13	16
$f(a,c) = X_{ac}$	14	14
$f(d,a) = X_{da}$	6	6
$f(b,d) = X_{bd}$	9	9
$f(b,e) = X_{be}$	4	6
$f(c,f) = X_{cf}$	14	17
$f(d,f) = X_{df}$	2	4
$f(d,g) = X_{dg}$	1	4
$f(e,g) = X_{eg}$	4	4
$f(f,t) = X_{ft}$	16	23
$f(g,t) = X_{gt}$	5	11

This yields the linear programming problem

Maximize $X_{sa} + X_{sb}$ (22.8)

Subject to $X_{ac} - X_{sa} - X_{da} = 0$ conservation of flow at vertex a (22.9)

$X_{bd} + X_{be} - X_{sb} = 0$ conservation of flow at vertex b (22.10)

$X_{cf} - X_{ac} = 0$ conservation of flow at vertex c (22.11)

$X_{da} + X_{df} + X_{dg} - X_{bd} = 0$ conservation of flow at vertex d (22.12)

$X_{eg} - X_{be} = 0$ conservation of flow at vertex e (22.13)

$X_{ft} - X_{cf} - X_{df} = 0$ conservation of flow at vertex f (22.14)

$X_{gt} - X_{dg} - X_{eg} = 0$ conservation of flow at vertex g (22.15)

$0 \leq X_{sa} \leq 15$ nonnegativity, capacity restriction edge sa (22.16)

$0 \leq X_{sb} \leq 16$ nonnegativity, capacity restriction edge sb (22.17)

$0 \leq X_{ac} \leq 14$ nonnegativity, capacity restriction edge ac (22.18)

$0 \leq X_{da} \leq 6$ nonnegativity, capacity restriction edge da (22.19)

$$0 \leq X_{bd} \leq 9 \quad \text{nonnegativity, capacity restriction edge } bd \tag{22.20}$$

$$0 \leq X_{be} \leq 6 \quad \text{nonnegativity, capacity restriction edge } be \tag{22.21}$$

$$0 \leq X_{cf} \leq 17 \quad \text{nonnegativity, capacity restriction edge } cf \tag{22.22}$$

$$0 \leq X_{df} \leq 4 \quad \text{nonnegativity, capacity restriction edge } df \tag{22.23}$$

$$0 \leq X_{dg} \leq 4 \quad \text{nonnegativity, capacity restriction edge } dg \tag{22.24}$$

$$0 \leq X_{eg} \leq 4 \quad \text{nonnegativity, capacity restriction edge } eg \tag{22.25}$$

$$0 \leq X_{ft} \leq 23 \quad \text{nonnegativity, capacity restriction edge } ft \tag{22.26}$$

$$0 \leq X_{gt} \leq 11 \quad \text{nonnegativity, capacity restriction edge } gt \tag{22.27}$$

As we can see, with an equality constraint for each intermediate vertex and inequality constraint for each edge, this approach is quite cumbersome and – it turns out – much less efficient than $DEKaFF$.

22.5 Application to a Major League Baseball Pennant Race

There are many wonderful applications of both maximum flows in a network as well as minimum cuts. We will explore an interesting one introduced by Benjamin Schwartz in 1966 [51].

After the completion of all games on Sunday, September 13, 1992, there were three weeks left in the regular season and the standings in the National League East were

Club	W	L	Pct	GB
Pittsburgh Pirates	82	60	.577	–
Montreal Expos	79	63	.556	3.0
Saint Louis Cardinals	71	69	.507	10.0
Chicago Cubs	70	71	.496	11.5
New York Mets	63	78	.447	18.5
Philadelphia Phillies	58	82	.414	23.0

As Major League Baseball plays a 162 game schedule, the Phillies at this point in the season have only 22 games remaining. The Pirates have 20 remaining, but as the Phillies are 23 games out, they are clearly eliminated from catching the Pirates and winning the division. The Mets, on the other hand, have 21 games remaining, and being 18.5 games out it seems they may have a shot at the division title. The problem is, though, that they would have to surpass both the Pirates and the Expos and these clubs have 6 games remaining against each other. We can incorporate the fact that those two clubs will have at least a total of 6 wins between them down the closing stretch of the season and determine if the Mets are eliminated by modeling this situation as a network flow.

This problem is referred to as the *baseball elimination problem* as this is how Schwartz framed the application, but it clearly applies to any tournament with n teams where winner is crowned based on tournament play.

First, we define the variables in the following table. What each variable represents will be written to the right of the variable:

n : number of teams in the league or tournament
w_i : the number wins team i has at the moment we are considering
r_i : the number of games remaining for team i
r_{ij} : the number of games remaining team i has with team j

We will consider a team eliminated if it has no chance to finish the season in first place.

The remaining games for the teams are (the number in parentheses in the club's column is the total number the team has remaining)[2]:

Club	Pirates	Expos	Cardinals	Cubs	Mets	Phillies	Others
Pittsburgh Pirates	(20)	4	4	3	6	3	0
Montreal Expos	4	(20)	3	6	3	4	0
Saint Louis Cardinals	4	3	(22)	4	4	7	0
Chicago Cubs	3	6	4	(21)	4	4	0
New York Mets	6	3	4	4	(21)	4	0
Philadelphia Phillies	3	4	7	4	4	(22)	0

To determine if team k is eliminated, a digraph in the network is formed by introducing a source s and sink t and two sets of intermediate vertices. One collection is the game or competition vertices: a vertex for each pair of

[2]The author cannot resist to point out the good old days of baseball: teams played 162 games that mattered and only the two division winners in each league advanced to the postseason. As well, down the stretch teams only played divisional opponents. There was nothing quite like a good ol' pennant race in September. As well, teams used to regularly play doubleheaders. During this three week period, the Phillies played doubleheaders on three consecutive days; two with the Cardinals and one against the Mets. That is six games in three days and they went 5-1.

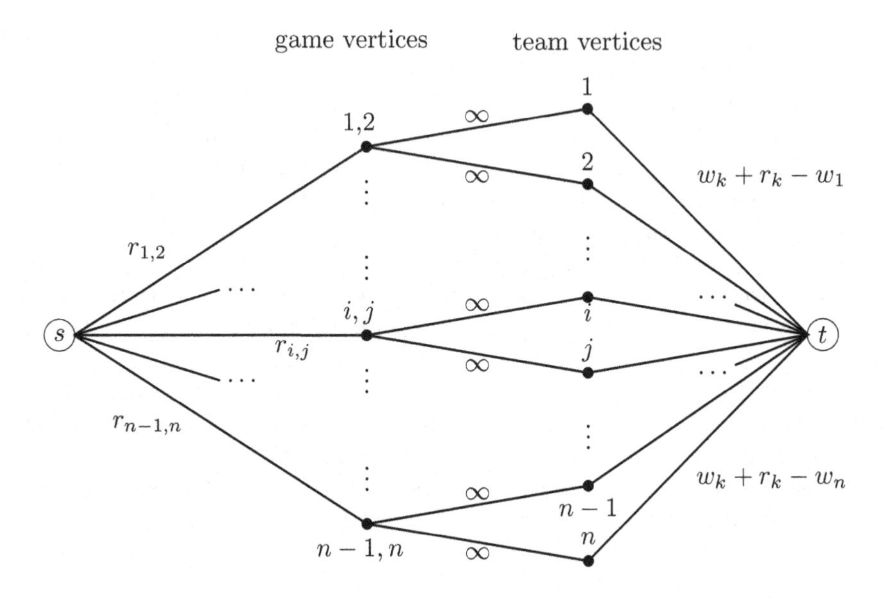

FIGURE 22.8
Design of the network for the baseball elimination problem (all edges are forward edges).

teams – for example, Pittsburgh versus Philadelphia. The second set is the team vertices; that is, one vertex for each team. A edge is introduce from the source to each competition vertex, then two edges will exit the competition vertex with an edge going into the respective team vertices. We then also introduce edges from the team vertices into the sink. To turn this into a network flow problem, each edge from the source to the competition vertex for the games between team i and j is assigned the capacity r_{ij}. No capacity needs to be assigned to the edges joining the intermediate vertices, but the edges leaving the team vertex i and entering the sink are given a capacity of $w_k + r_k - w_i$. Finally, let S bet the set of all teams in the league and define

$$R(k) := \sum_{\substack{i,j \in S - \{k\} \\ i<j}} r_{ij}.$$

Schwartz proved in [51] that team k is not eliminated if and only if there exists flow value of size $R(k)$ in the defined network. He further proved that if this flow exists, it is a max flow in the network.

A diagram of the construction of the network for an arbitrary problem is given in Figure 22.8.

For the 1992 National League Pennant race, we have the following:

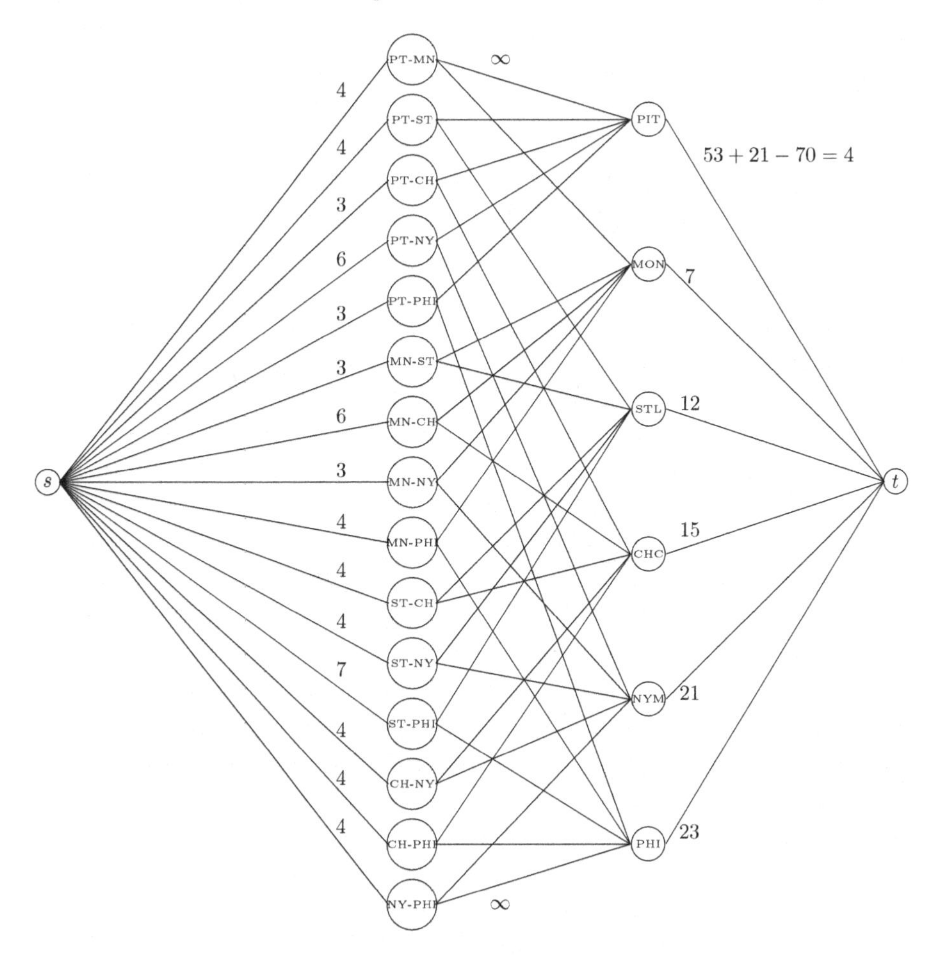

FIGURE 22.9
Determining if the Mets are eliminated in the 1992 NL East Pennant race.

22.6 Exercises

Exercise 22.1. *Make table similar to Table 22.1 of all cuts in the following network. Edges are labeled with their capacity. Also state the maximum flow through this network.*

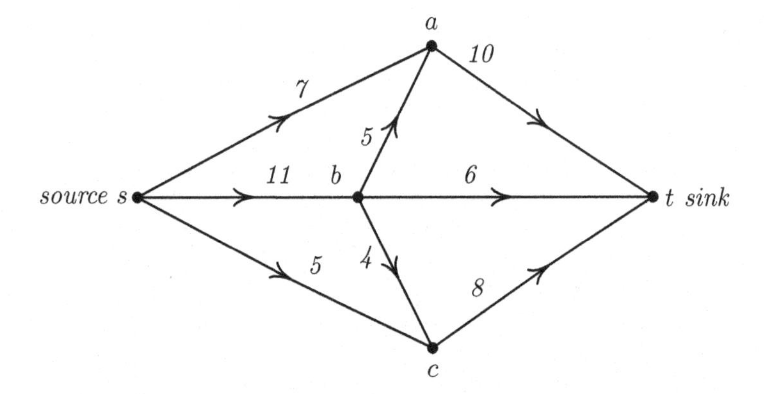

Exercise 22.2. *Make table similar to Table 22.1 of all cuts in the following network. Edges are labeled with their capacity. Also state the maximum flow through this network.*

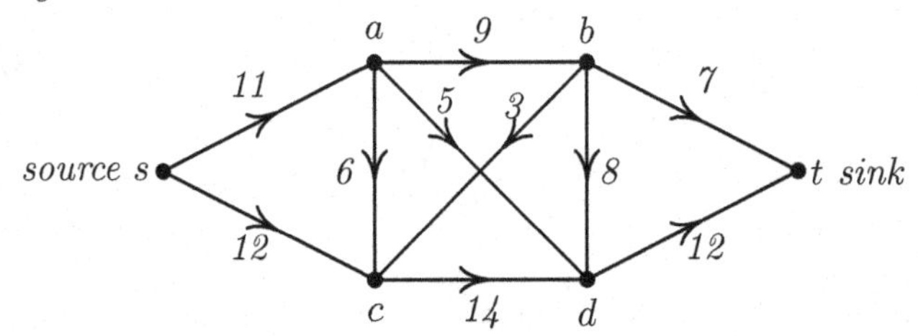

Exercise 22.3. *Make table similar to Table 22.1 of all cuts in the following network. Edges are labeled with their capacity. Also state the maximum flow through this network.*

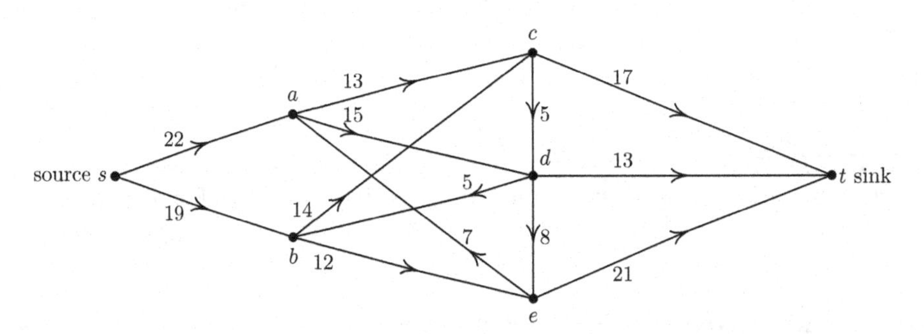

Exercise 22.4. *Use DEKaFF (Algorithm 22.3.1) to find the maximum flow on the network in Exercise 22.1.*

Exercise 22.5. *Use DEKaFF (Algorithm 22.3.1) to find the maximum flow on the network in Exercise 22.2.*

Exercise 22.6. *Use DEKaFF (Algorithm 22.3.1) to find the maximum flow on the network in Exercise 22.3.*

23

Minimum-Weight Spanning Trees and Shortest Paths

23.1 Weighted Graphs and Spanning Trees

In Chapter 21, we were introduced to trees. For the applications considered in this chapter, we will need to build on what we know of trees. The models of these problems will involve *weighted trees.* Since a tree is a special graph, we offer the following definition.

Definition 23.1.1 (Weighted Graph). *Let G be a connected graph. G is a* weighted graph *if there is a function $w : E(G) \to \mathbb{R}$.*

From the definition, we see that what makes a graph a weighted graph is that numerical values (weights) are assigned to each edge. In applications, the weights are most often some form of a cost.

Example 23.1.2. *A weighted graph.*

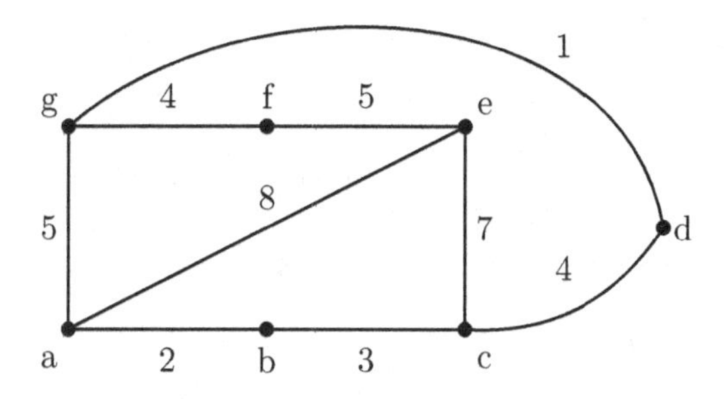

Spanning subgraphs were introduced in Section 21.1. Recall that the "span" in a spanning subgraph of G means that every vertex of G is reached; i.e. every vertex of the subgraph is incident with an edge. Since a tree is connected by definition, a *spanning tree* is a tree that reaches every vertex in the graph.

DOI: 10.1201/9780367425517-23

TABLE 23.1
Edges and Their Weights from Example 23.1.2

edge uv	ab	ae	ag	bc	cd	ce	dg	ef	fg
$w(uv)$	2	8	5	3	4	7	1	5	4

Example 23.1.3. *(Spanning Trees.)*

spanning tree

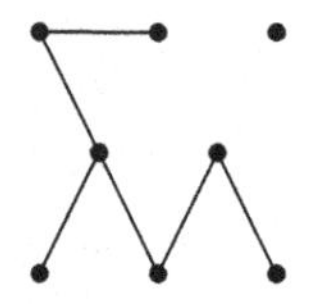

not a spanning tree

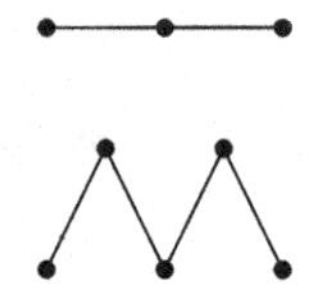

not a spanning tree

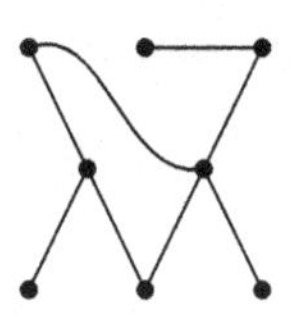

not a tree

23.2 Minimum-Weight Spanning Trees

Minimum-weight spanning trees have many applications, particularly when connections in a network are in consideration as in the following example:

Example 23.2.1. *The Delos Corporation is planning an amusement park with a central luxury resort and five separate themed locations: Western World, Medieval World, Roman World, Spa World ("where old age and pain have been eliminated"), and Future World*[1]*. Delos' plan is to have guests stay at the resort then visit the location of their choosing for a day of fun and excitement or relaxation. A sight for the resort and park has been chosen relatively close to a water source and Delos must decide where to lay the necessary pipe to get water from its source to the six locations. Distances between the source, resort, and locations are given in Table 23.2.*

Delos would like to minimize costs in getting water from its source to the park's locations. To minimize costs, Delos' approach is to connect each location to the source either directly or indirectly through a network of pipes with as few of miles of pipe in the network as possible. Delos chose a site very suitable to construction and there are no physical obstacles restricting the placement of pipe between any locations.

[1]This example borrows from the films *Westworld* [62] and its sequel *Futureworld* [27].

TABLE 23.2
Distances (in Miles) Between Sites at Delos Resort in Example 23.2.1

Site Vertex Label	Resort (R)	Western World (W)	Medieval World (M)	Roman World (O)	Spa World (P)	Future World (F)
Source (S)	7	16	9	11	1	14
Resort (R)	–	8	2	12	13	5
Western World (W)	–	–	8	5	13	4
Medieval World (M)	–	–	–	3	6	11
Roman World (O)	–	–	–	–	9	10
Spa World (P)	–	–	–	–	–	15

Solution. (Modeling Example 23.2.1.)
To create a model let the source, resort, and five themed locations each be represented by a vertex and there will be an edge between vertices if a pipeline is to be placed between the sites represented by the vertices. Recall that by Highlight 21.1.3 we do not need to be concerned with placing the vertices in our model in their exact physical locations. The only relevant aspects are that there are vertices and either an edge between distinct vertices or not. Thus the model will be a graph on 7 vertices and we consider pipelines between all the sites we have a K_7.

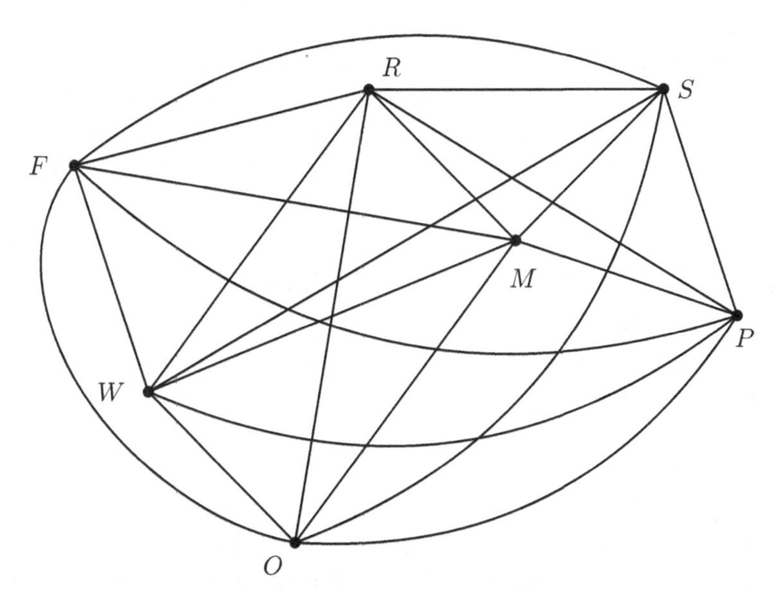

As discussed in Section 21.4, the size of K_7 is $\binom{7}{2} = 21$ and, of course, this is overkill. Delos does not need all these waterlines, and some thought reveals that a spanning tree would get the job done. Moreover, as Delos is seeking an optimal solution, a minimum-weight spanning tree is desired. ■

We will present two algorithms for finding minimum-weight spanning trees. Both are examples of a general class of algorithms known as greedy algorithms,

Definition 23.2.2 (Greedy Algorithm). *A* greedy algorithm *is an algorithm where a locally optimal decision is made at each iteration.*

Greedy algorithms are sometimes also called "unsophisticated locally optimal heuristics". Rather than remembering that mouthful is important is to realize that a greedy algorithm is exactly what the name says it is. During each run of the process the algorithm only considers what is happening at the moment; there is no consideration of anything regarding future decisions. We have already seen one example of a greedy algorithm: the Simplex Method from Section 6.2 for solving linear programming problems. At each iteration the algorithm decides which decision variable is best to optimize at that moment and then maximizes the value of only that particular decision variable. Recall that geometrically this meant that the algorithm chooses a decision variable that most increases the objective function and then moves as far as permitted by the feasible region in that direction.

23.2.1 Kruskal's Algorithm

The first algorithm we consider for finding minimum-weight spanning trees was developed by Joseph Bernard Kruskal [39] two years after he finished is Ph.D. dissertation. Kruskal built upon the work of the Austria-Hungarian (what would now be Czech) mathematician Otkar Borůvka. Borůvka's 1926 paper [4] presented and proved an algorithm that constructs minimum-weight spanning trees, a problem Borůvka worked on in designing efficient distribution networks of electricity while an employee of the West Moravian Power Company during World War I.

Algorithm 23.2.1 Kruskal's Algorithm (1956) for Finding a Minimum Weight Spanning Tree.

Objective: Find a minimum-weight spanning tree T of a weighted connected graph G on n vertices.

Input: A weighted undirected graph G.

1: List all the edges e_1, e_2, $\dots e_m$ of $G \backslash T$ in order of nondecreasing edge weight; i.e. $w(e_{i_1}) \leq w(e_{i_2}) \leq \cdots w(e_{i_m})$.

2: Starting with an edge of minimum weight, accept the edge into T as long as its acceptance does not create a cycle in T. Ties may be settled arbitrarily.

3: By Theorem 21.2.8, the algorithm terminates when T has $n-1$ edges.

Output: A minimum-weight spanning tree T of G. Though the total weight of T is the unique minimum weight, the minimum-weight spanning tree T need not be unique.

Let us now use *Kruksal's Algorithm* to determine the least expensive way to meet the water needs of Delos in Example 23.2.1.

Solution. (for Example 23.2.1 using Kruskal's Algorithm.) Following the algorithm, the edge weights in non-increasing order are

$$1, 2, 3, 4, 5, 5, 6, 7, 8, 8, 9, 9, 10, 11, 11, 12, 13, 13, 14, 15, 16$$

and since G has 7 vertices, we know the algorithm will terminate once we have selected 6 legal edges,

The first four iterations in building T of Kruskal's Algorithm add edges PS, MR, MO, and FW. There is a tie for the fifth iteration and we may arbitrarily choose between FR and OW and we will arbitrarily choose[2] FR. For the sixth and final iteration, adding edge OW creates a cycle, so it is not included, but the inclusion of MP creates an acyclic graph thus giving a minimum-weight spanning tree.

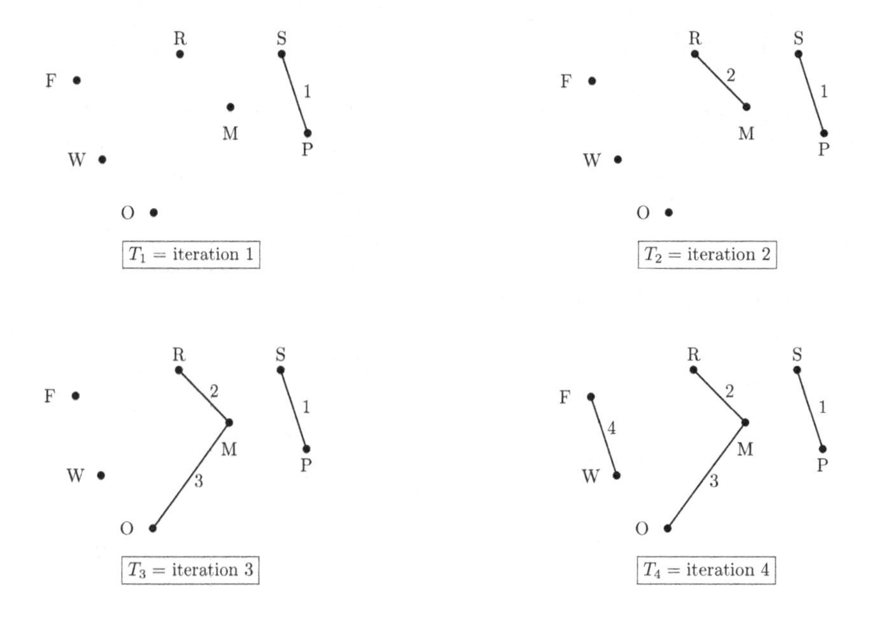

[2]Almost all implementations of Kruskal's Algorithm break ties lexiographically. With a runtime of $O(E \log V)$, Kruskal's Algorithm is very efficient, so there is insignificant computational advantage in considering which choice of tied edges is better.

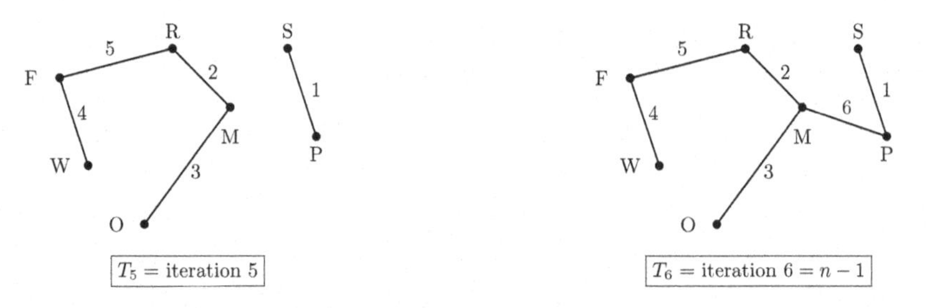

T_5 = iteration 5

T_6 = iteration $6 = n - 1$

■

By the design of the algorithm – namely, edges are not added to T if their inclusion creates a cycle – Kruskal's will produce a tree. Thus is remains to show that T is indeed a minimum-weight spanning tree of the given connected weighted graph.

Proof. (Kruskal's Algorithm produces a minimum-weight spanning tree.) Let G be a connected weighted graph of order n and let T be a spanning tree given by Kruskal's Algorithm. Hence T has $n-1$ edges $e_1, e_2, \ldots, e_{n-1}$ with $w(e_1) \leq w(e_2) \leq \cdots \leq w(e_{n-1})$.

We will show that T is a minimum-weight spanning tree of G. Assume for contradiction that T is not a minimum-weight spanning tree of G and let S be a a minimum-weight spanning tree of G that among all the minimum-weight spanning trees of G has the most edges in common with T. Since $S \neq T$, there exists at least one edge of T not is S. Let e_t be the first edge of T (the fixed ordering of the edges is by weight with ties settled arbitrarily) not in common with the edges of S. Note that if $t > 1$, then $e_1, e_2, \ldots, e_{t-1}$ is in $E(S) \cup E(T)$.

Put $S' := S + e_t$. Since S is a spanning tree and an edge has been added, S' must contain a cycle C. As T is acyclic, there is an edge e' in C that is not in T. Put $T' := S' - e' = S + e_t - e'$. Then T' spans G and is acyclic and thus a spanning tree of G. Moreover

$$w(T') = w(S) + w(e_t) - w(e') \tag{23.1}$$

but as S is a minimum-weight spanning tree, $w(S) \leq w(T')$. Thus by 23.1,

$$w(S) \leq w(T') = w(S) + w(e_t) - w(e') \tag{23.2}$$

giving

$$w(e') \leq w(e_t). \tag{23.3}$$

There are two case to be considered:

<u>Case $t = 1$</u>

Thus $e_t = e_1$ is an edge of least weight among all edges of G hence $w(e_t) \leq w(e')$. This together with (23.3) gives $w(e_t) = w(e')$ and thus by (23.1) $w(T') = w(S)$. Therefore T' is a minimum-weight spanning tree of G. Furthermore, as e_t is in T but e' is not in T, T' has more edges in common with T than S, contradicting the assumption that S is a minimum-weight spanning tree of G having the most edges in common with T.

Case $t > 1$

By Kruskal's Algorithm, e_t is an edge of G that can be added to $e_1, e_2, \ldots, e_{n-1}$ without introducing a cycle. But e' can also be added to T and not produce a cycle, thus $w(e_t) \leq w(e')$ by the process of the algorithm. This together with (23.3) gives $w(e_t) = w(e')$ and the argument follows as in *Case* $t = 1$. □

23.2.2 Prim's Method

We now consider an alternate algorithm for finding a minimum-weighted spanning tree for a connected graph G. As Kruskal's Algorithm focused on edges and as it is that graphs only consist of vertices and edges, it will not be a surprise that the alternate approach focuses on the vertices of G. This second approach is due to a 1957 paper by Robert Clay Prim [45] though the Czech mathematician Vojtěch Jarník had developed and rigorously proved the algorithm in 1930 [36].

Algorithm 23.2.2 Prim's Method (1957).

Objective: Find a minimum-weight spanning tree T of a weighted connected graph G on n vertices.

Input: An undirected weighted graph G.

1: Let $V = V(G)$ and $W = V(T)$. Select a start vertex from V to add to W.

2: Choose an edge out of W (i.e. an edge with one of its vertices from W) that has the least weight. Ties are broken arbitrarily. Add the vertex from $V \backslash W$ incident with this edge to W and the edge to T. The vertices in $V \backslash W$ are referred to as *not reached* where the vertices in W are said to be *reached.*

3: Repeat 2 until W has all n vertices.

Output: A minimum-weight spanning tree T of G. Though the total weight of T is the unique minimum weight, the minimum-weight spanning tree T need not be unique.

We will again determine the least expensive way to meet the water needs of Delos in Example 23.2.1, but this time using Prim's Method.

Solution. (for Example 23.2.1 using Prim's Method.)

The iterations of Prim's Method are given in the next table and illustrated in the two figures that follows. The first figure emphasizes the reached and unreached vertex sets through the first two iterations where the second figure shows the process without moving the vertices around. We will arbitrarily select S as our start vertex.

Iteration	Reached W	Unreached $V\backslash W$	Selected Edge	New Reached Vertex
Start	$\emptyset$	S,R,W,M,O,P,F	$--$	$--$
0	S	R,W,M,O,P,F	SP	P
1	S,P	R,W,M,O,F	PM	M
2	S,P,M	R,W,O,F	MR	R
3	S,P,M,R	W,O,F	RF	F
4	S,P,M,R,F	W,O	MO	O
5	S,P,M,R,F,O	W	FW	W
6	S,P,M,R,F,O,W	$\emptyset$	$--$	$--$

Reached and Unreached Vertices through Two Iterations

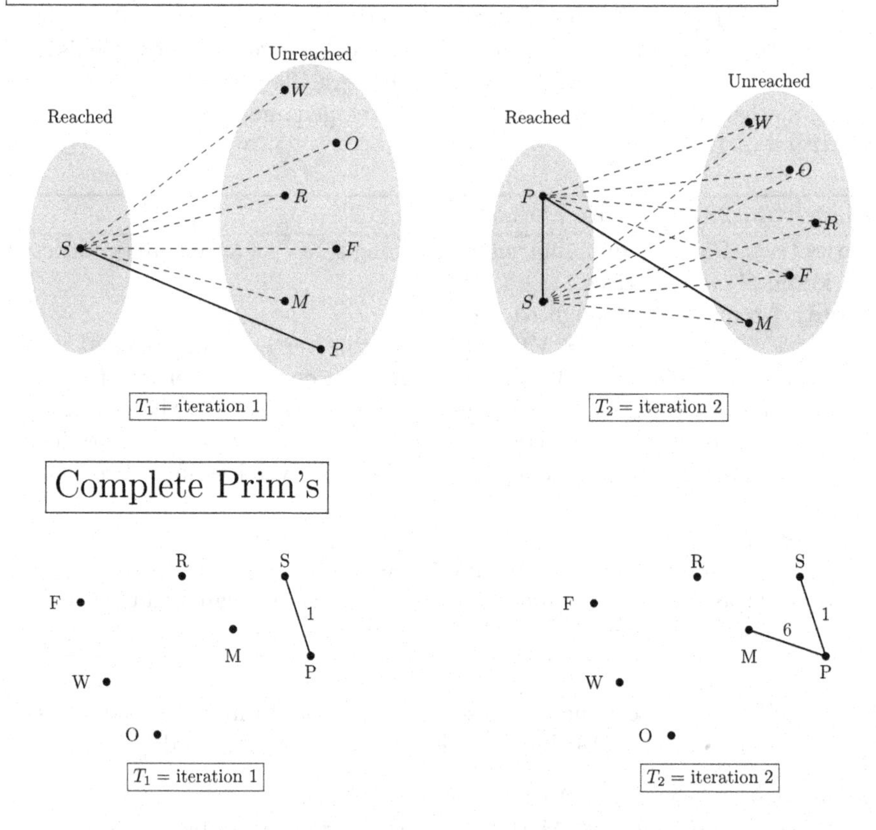

T_1 = iteration 1

T_2 = iteration 2

T_1 = iteration 1

T_2 = iteration 2

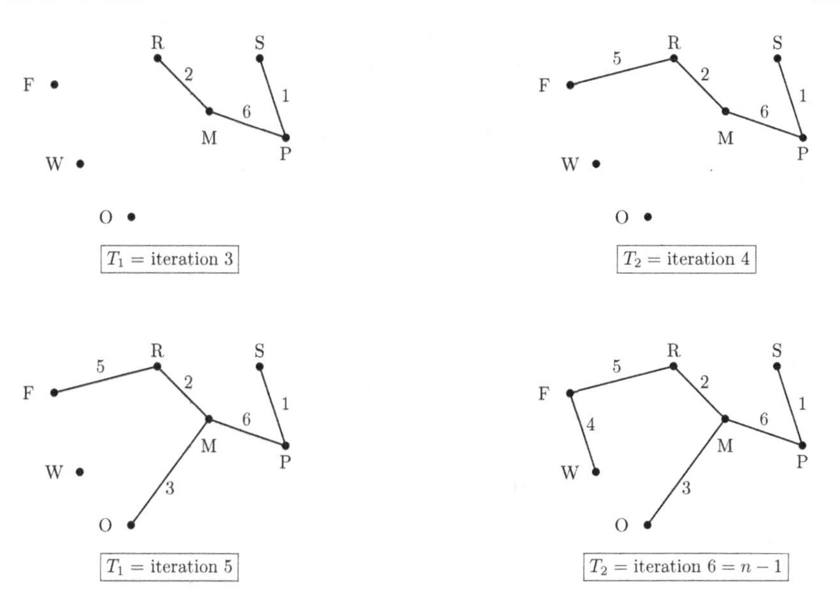

■

23.2.3 Kruskal's and Prim's Compared

For a graph of order n and size m, it can be shown that the runtime of Kruskal's is $O(m \log n)$ whereas Prim's runtime is $O(m+n \log n)$. Thus Kruskal's is best for a sparse graph and Prim's would be quicker for a dense graph.

23.3 Shortest Paths

Paths were introduced in Section 21.1 and are an essential part of the notion of distance in a graph.

Definition 23.3.1 (Distance in a Graph). *Let G be a weighted graph with vertex set $V = V(G)$. Then the* distance *between two vertices u, v of G is*

$$d_G(u,v) = d(u,v) := \textit{total weight of a shortest path in } G.$$

If there does not exist a u, v path in G, then $d(u,v) := \infty$.

Distance is defined in an unweighted graph G by counting the number of edges in each path (in other words, assigning each edge a weight of 1 and using the above definition). In a digraph, directed paths are considered.

Example 23.3.2 considers finding a shortest path in the weighted directed graph is shown in Figure 23.1. The following sections present different means to obtain a shortest path.

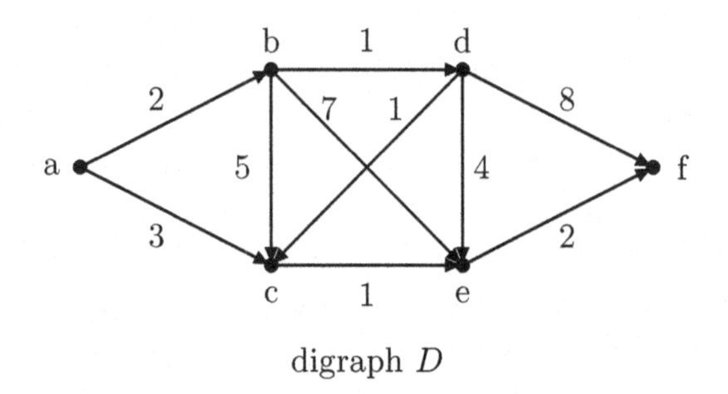

FIGURE 23.1
Finding a shortest path in D.

23.3.1 Dijkstra's Algorithm

One way to obtain a shortest path between vertices is by *Dijkstra's Algorithm.*

Algorithm 23.3.1 Dijkstra's Algorithm to Obtain Shortest Paths.

Objective: For a given vertex u in a weighted graph G, obtain the distances from u to v for all v in $V(G)$.

Input: A weighted directed or undirected graph G.

Require: Let $w(x, y)$ be the weight of the edge joining vertices x and y. If no edge xy exists, put $w(x, y) := \infty$.

1: Select u in $V(G)$ and put $S := \{u\}$ noting that $d(u, u) = 0$.
2: Put $t(u) := 0$ and $t(z) := w(u, z)$ for all $z \neq u$ in $V(G)$.
3: **while** $S \neq V(G)$ **do**
4: select a w in $V(G) \backslash S$ such that $t(w) = \min_{z \notin S}\{t(z)\}$ (ties are settled arbitrarily)
5: put $S := S \cup \{w\}$
6: for all z not in S, update $t(z) := \min\{t(z), t(w) + w(w, z)\}$
7: **end while**

Output: For all v in $V(G)$, $t(v) = d(u, v)$.

Note that the graph G can undirected or directed. What matters is if y can be reached directly from x.

Example 23.3.2. *Use Dijkstra's Algorithm to find the shortest $a - f$ path in the graph D shown in Figure 23.1.*

Solution. Iteration 0 $S = \{a\}$ and proceeding out of a:

$v \in V(D)$	Update $t(v)$	Selection	Shortest $a - v$ path
a	$t(a) = 0$	already in S	a
b	$t(b) = w(a,b) = 2$	select	a, b
c	$t(c) = w(a,c) = 3$		$a, +$
d	$t(d) = w(a,d) = \infty$		$a, +$
e	$t(e) = w(a,e) = \infty$		$a, +$
f	$t(f) = w(a,f) = \infty$		$a, +$

Update $S := \{a, b\}$. Since b has been selected, the algorithm has now returned $d(a, b)$.

Iteration 1 $S = \{a, b\}$ and proceeding out of b:

$v \in V(D)$	Update $t(v)$	Selection	Shortest $a - v$ path
a	$t(a) = 0$	already in S	a
b	$d(a,b) = t(b) = 2$	already in S	a, b
c	$t(c) = \min\{t(c), t(b) + w(b,c)\} = \min\{3, 2+5\} = 3$		$a, c, +$
d	$t(d) = \min\{t(d), t(b) + w(b,d)\} = \min\{\infty, 2+1\} = 3$	arbitrarily select	a, b, d
e	$t(e) = \min\{t(e), t(b) + w(b,e)\} = \min\{\infty, 2+7\} = 9$		$a, b, e, +$
f	$t(f) = \min\{t(f), t(b) + w(b,f)\} = \min\{\infty, 2+\infty\} = \infty$		$a, b, +$

Update $S := \{a, b, d\}$. Since d has been selected, the algorithm has now returned $d(a, d)$.

Iteration 2 $S = \{a, b, d\}$ and proceeding out of d:

$v \in V(D)$	Update $t(v)$	Selection	Shortest $a - v$ path
a	$t(a) = 0$	already in S	a
b	$d(a,b) = t(b) = 2$	already in S	a, b
c	$\mathrm{v}t(c) = \min\{t(c), t(d) + w(c,d)\} = \min\{3, 3+\infty\} = 3$	select	a, c
d	$d(a,d) = t(d) = 3$	already in S	a, b, d
e	$t(e) = \min\{t(e), t(d) + w(d,e)\} = \min\{9, 3+4\} = 7$		$a, b, d, e, +$
f	$t(f) = \min\{t(f), t(d) + w(d,f)\} = \min\{\infty, 3+8\} = 11$		$a, b, d, f, +$

Update $S := \{a, b, c, d\}$. Since c has been selected, the algorithm has now returned $d(a, c)$.

Iteration 3 $S = \{a, b, c, d\}$ and proceeding out of c:

$v \in V(D)$	Update $t(v)$	Selection	Shortest $a - v$ path
a	$t(a) = 0$	already in S	a
b	$d(a, b) = t(b) = 2$	already in S	a, b
c	$d(a, c) = t(c) = 3$	already in S	a, c
d	$d(a, d) = t(d) = 3$	already in S	a, b, d
e	$t(e) = \min\{t(e), t(c) + w(c, e)\} = \min\{7, 3 + 1\} = 4$	select	a, c, e
f	$t(f) = \min\{t(f), t(c) + w(c, f)\} = \min\{11, 3 + \infty\} = 11$		$a, b, d, f, +$

Update $S := \{a, b, c, d, e\}$. Since e has been selected, the algorithm has now returned $d(a, e)$.

Iteration 4 $S = \{a, b, c, d, e\}$ and proceeding out of e:

$v \in V(D)$	Update $t(v)$	Selection	Shortest $a - v$ path
a	$t(a) = 0$	already in S	a
b	$d(a, b) = t(b) = 2$	already in S	a, b
c	$d(a, c) = t(c) = 3$	already in S	a, c
d	$d(a, d) = t(d) = 3$	already in S	a, b, d
e	$d(a, e) = t(e) = 4$	already in S	a, c, e
f	$t(f) = \min\{t(f), t(e) + w(e, f)\} = \min\{11, 4 + 2\} = 6$	select	a, c, e, f

Update $S := \{a, b, c, d, e, f\}$. Since f has been selected, the algorithm has now returned $d(a, f)$.

Halt Since $S = V(D)$ the algorithm terminates and we have the following $d(a, v)$ and shortest a, v paths:

$v \in V(D)$	Distances	Shortest $a - v$ path
a	$d(a, a) = 0$	a
b	$d(a, b) = 2$	a, b
c	$d(a, c) = 3$	a, c
d	$d(a, d) = 3$	a, b, d
e	$d(a, e) = 4$	a, c, e
f	$d(a, f) = 6$	a, c, e, f

■

23.3.2 A Linear Programming Approach to Shortest Paths

We may also find a shortest path by solving a linear programming problem over a network flow. The approach is summarized in the following Highlight.

Highlight 23.3.3 (Shortest Path as a Linear Programming Problem). *Let G be a weighted directed or undirected graph. To introduce a linear programming problem to find a shortest $u-v$ path in G:*

1. *Introduce a decision variable X_{ij} for each edge ij in G.*
2. *Include the nonnegativity constraints $X_{ij} \geq 0$ for every edge's decision variable.*
3. *Introduce flow constrains for each vertex z in G by*

$$\begin{aligned}&(\textit{sum of in-edge variables } X_*) - (\textit{sum of out-edge variables } X_*)\\ &\quad = -1 \textit{ for source } u,\end{aligned} \tag{23.4}$$

$$\begin{aligned}&(\textit{sum of in-edge } X_*) - (\textit{sum of out-edge } X_*)\\ &\quad = 0 \textit{ for non-source/sink vertices },\end{aligned} \tag{23.5}$$

$$(\textit{sum of in-edge } X_*) - (\textit{sum of out-edge } X_*) = 1 \textit{ for sink } v. \tag{23.6}$$

4. *Lastly,*

$$\min \sum_{X_{ij}, ij \in E(G)} w_{ij} X_{ij} \tag{23.7}$$

where w_{ij} is the weight of edge ij in the graph G.

Example 23.3.4.
Rework Example 23.3.2 but by modeling it as a network flow problem.

Solution. The graph with flow constraints introduced is presented in Figure 23.2. The linear programming model of the problem is then

$$\min 2X_{ab} + 3X_{ac} + 5X_{bc} + 1X_{bd} + 7X_{be} + 1X_{ce} + 1X_{dc} + 4X_{de} + 8X_{df} + 2X_{ef} \tag{23.8}$$

subject to

$$\begin{aligned}
-X_{ab} - X_{ac} &= -1 \text{ (flow constraint for vertex } a) && (23.9)\\
X_{ab} - X_{bc} - X_{bd} - X_{be} &= 0 \text{ (flow constraint for vertex } b) && (23.10)\\
X_{ac} + X_{bc} + X_{dc} - X_{ce} &= 0 \text{ (flow constraint for vertex } c) && (23.11)\\
X_{bd} - X_{dc} - X_{de} - X_{df} &= 0 \text{ (flow constraint for vertex } d) && (23.12)\\
X_{be} + X_{ce} + X_{de} - X_{ef} &= 0 \text{ (flow constraint for vertex } e) && (23.13)\\
X_{df} + X_{ef} &= 1 \text{ (flow constraint for vertex } f) && (23.14)\\
X_{ij} &\geq 0 && (23.15)
\end{aligned}$$

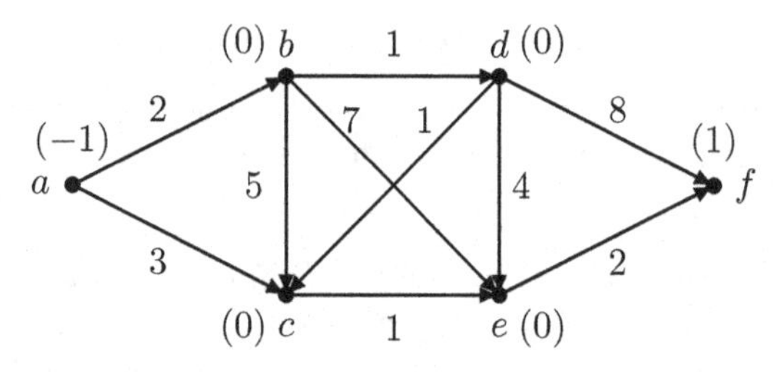

FIGURE 23.2
Flow constraints (f) for a shortest path in D.

C	D	E	F	G	H	I	J	K
		Linear Programming Model for Shortest						
Edge	Weight	Used?		Min	0	= SUMPRODUCT(D5:D14,E5:E14)		
ab	2							
ac	3			Subect to		(description)		Value
bc	5			flow constraint a	0	= -E5 - E6		-1
bd	1			flow constraint b	0	= E5 - E7 - E8 - E9		0
be	7			flow constraint c	0	= E6 + E7 + E11 - E10		0
ce	1			flow constraint d	0	= E8 - E11 - E12 - E13		0
dc	1			flow constraint e	0	= E9 + E10 + E12 - E14		0
de	4			flow constraint f	0	= E13 + E14		1
df	8							
ef	2							

FIGURE 23.3
The Excel setup for an LP solution for finding the distance in Example 23.3.4.

The Excel setup and solution are presented in Figures 23.3, 23.4, and 23.5. Note that the solution not only includes the minimum distance but also a shortest path.

■

Though Example 23.3.2 and Example 23.3.4 were both done on a directed graph, each technique is easily applicable to any undirected graph. Applying Dijkstra's Algorithm on an undirected graph is straight forward as what matters are the edges that reach the remaining vertices not in S from the vertex being considered. Regarding the linear programming approach, we would need to introduce two directed edges uv and vu for each undirected edge. Note further that as long as the edge weights are positive, there will be no need to return to the source or exit the sink. This would mean that if we are regarding vertex a as the source and, for example, a has the neighbor b, then the model only needs to introduce the directed edge ab and there is no need to introduce the directed edge ba.

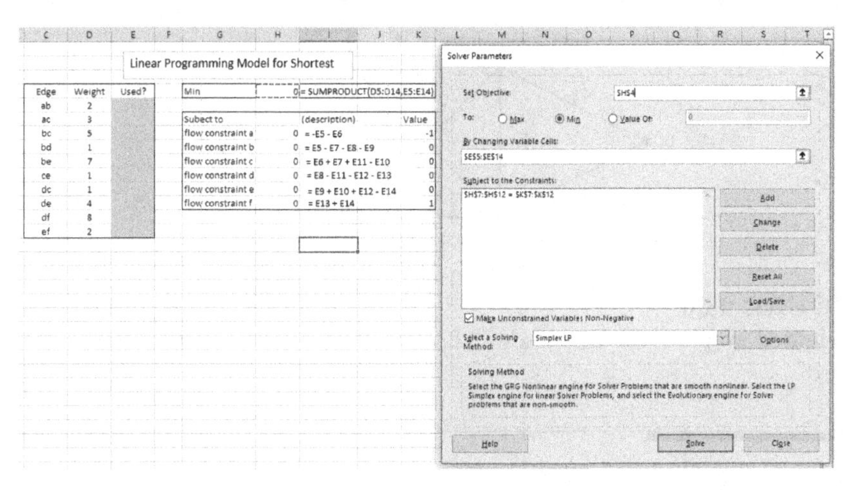

FIGURE 23.4
Using Solver for an LP solution for finding the distance in Example 23.3.4.

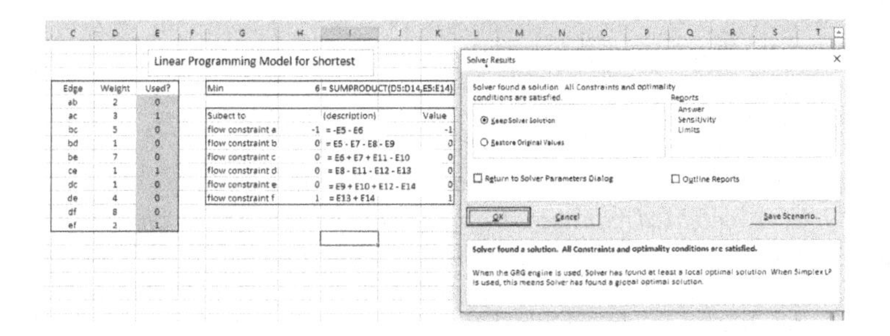

FIGURE 23.5
Excel's solution for an LP modeling of distance in Example 23.3.4.

Further note that as we will see in Section 24.2, as long as the coefficients involved in the network are integer[3], solving a linear programming problem over a network flow will always produce integer solutions. Thus, in Highlight 23.3.3, **there is no need to drastically complicate[4] the problem by introducing the requirement that the decision variables must be integer**.

[3]Since a collection of rational numbers can be scaled to a set of integers, "integer" can be replaced by "rational".

[4]Recall from Chapter 6 that the Simplex Method is used to solve LP problems over the reals will terminate with the global solution in a finite number of iterations. The existing heuristics used to solve ILP problems are not guaranteed to converge to a solution (see Chapter 8).

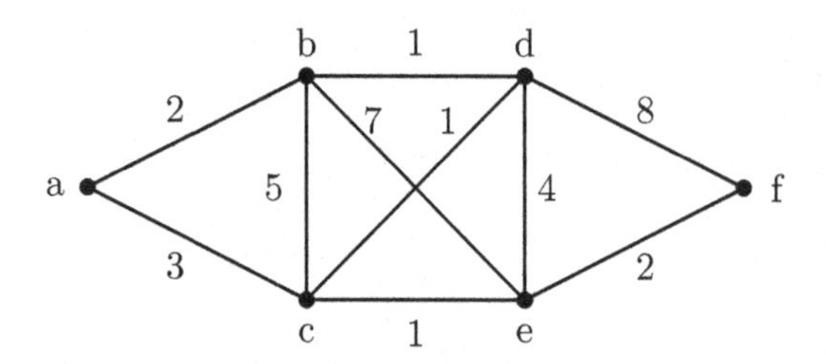

FIGURE 23.6
Weighted graph G for Exercises 23.2, 23.3, 23.4, 23.5.

23.4 Exercises

Exercise 23.1. *A* metric space *is a set S together with a function $d : S \to \mathbb{R}_+$*[5] *such that for any a, b, c in S*

1. $d(a, a) = 0$,
2. $d(a, b) > 0$ *whenever* $a \neq b$,
3. $d(a, b) = d(b, a)$, *and*
4. $d(a, c) \leq d(a, b) + d(b, c)$.

The function d in a metric space is known as a metric.

Let G be a weighted connected graph with vertex set V and let $d(u, v)$ be the distance between vertices a and b in G. Show that (V, d) is a metric space. Further, show that if G is not connected or is a digraph, that (V, d) may fail to be a metric space.

Exercise 23.2. *Use Kruskal's Algorithm to find a minimum weight spanning tree for the graph G Figure 23.6.*

Exercise 23.3. *Use Prim's Method to find a minimum weight spanning tree for the graph G Figure 23.6.*

Exercise 23.4. *Use Dijkstra's Algorithm to find the distance from a to f for the graph G in Figure 23.6. Include in your answer a shortest $a - f$ path.*

Exercise 23.5. *Find the distance from a to f for the graph G in Figure 23.6 by modeling the problem as a linear programming problem over a network flow. Include in your answer a shortest $a - f$ path.*

Exercise 23.6. *Use Kruskal's Algorithm to find a minimum weight spanning tree for the graph H Figure 23.7.*

[5] Recall that $\mathbb{R}_+$ is the set of *nonnegative real numbers*.

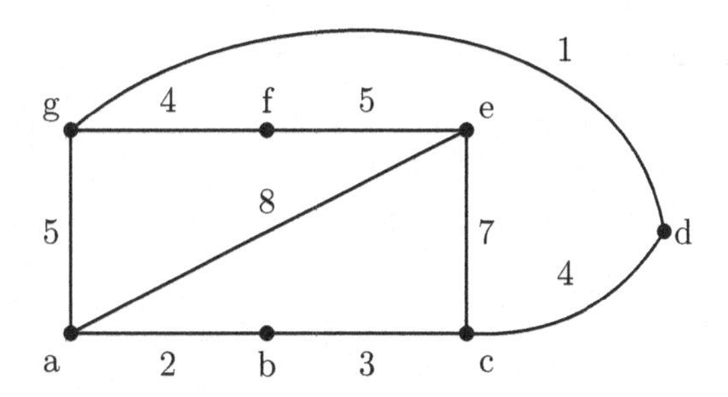

FIGURE 23.7
Weighted graph H for Exercises 23.6, 23.7, 23.8, 23.9.

Exercise 23.7. *Use Prim's Method to find a minimum weight spanning tree for the graph H Figure 23.7.*

Exercise 23.8. *Use Dijkstra's Algorithm to find the distance from b to f for the graph H in Figure 23.7. Include in your answer a shortest $b - f$ path.*

Exercise 23.9. *Find the distance from b to f for the graph H in Figure 23.7 by modeling the problem as a linear programming problem over a network flow. Include in your answer a shortest $b - f$ path.*

Exercise 23.10. *Show that Dijkstra's Algorithm does not unnecessarily find the shortest path if negative edge weights are allowed in a graph.*

Exercise 23.11. *What is the worst case runtime complexity of Dijkstra's Algorithm on complete graphs?*

24

Network Modeling and the Transshipment Problem

24.1 Introduction of the Problem

Numerous supply chain logistical problems can be modeled very nicely with networks. Determining an efficient way to ship a product from production facilities, to distribution centers, then to the retail stores is such an example. These scenarios, though, are neither shortest path problems, minimum weight spanning trees, nor maximum flows, and thus, require their own approach. We illustrate this with the following example:

Example 24.1.1. Jamie's SunLov'd Organic Oranges *grows oranges at five groves in Florida: Venice, Clearwater, Nokomis, Kissimmee, and Orange City. The oranges are then shipped to be cleaned, sorted, and packaged at processing plants in Bradenton and Sanford. From these facilities, the packaged, ready to sell oranges are taken to regional distribution centers in Tallahassee, Gainesville, and Fort Myers where they are available to regional local retailers. Bradenton and Sanford also double as regional distribution centers. Industry standards are that production is measured in the number of 90 pound boxes produced and to keep the problem simple, we will assume production = demand. Details are*

Daily Production		*Capacity*		*Daily Demand*	
Venice	*1400*			*Tallahassee*	*2500*
Clearwater	*900*	*Bradenton*	*4000*	*Bradenton*	*1000*
Nokomis	*1800*			*Gainesville*	*2000*
Kissimmee	*1700*	*Sanford*	*4500*	*Sanford*	*1100*
Orange City	*2200*			*Fort Meyers*	*1400*

DOI: 10.1201/9780367425517-24

Per box shipping costs (in cents) between the sites are

From	*Cost*	*To*
Venice	*35.6*	*Bradenton*
	170.4	*Sanford*
Clearwater	*44.8*	*Bradenton*
	133	*Sanford*
Nokomis	*34.1*	*Bradenton*
	169	*Sanford*
Kissimmee	*105.2*	*Bradenton*
	47.6	*Sanford*
Orange City	*146*	*Bradenton*
	12.7	*Sanford*
Bradenton	*314.1*	*Tallahassee*
	167.2	*Gainesville*
	413.6	*Fort Meyers*
Sanford	*285.6*	*Tallahassee*
	138.7	*Gainesville*
	186.6	*Fort Meyers*

What shipping schedule will meet demand but minimize shipping costs for Jamie's SunLov'd Organic Oranges*?*

We will begin solving this logistics problem for *Jamie's SunLov'd Organic Oranges* by first introducing a network modeling the situation. It is important to note that in Graph Theory, all we care about in that there are vertices and that they either are edges between vertices or not; namely, the vertices representing the cities in the problem need not be positioned to accurately represent geographical facts. A network modeling the problem is then (all edges are forward edges)

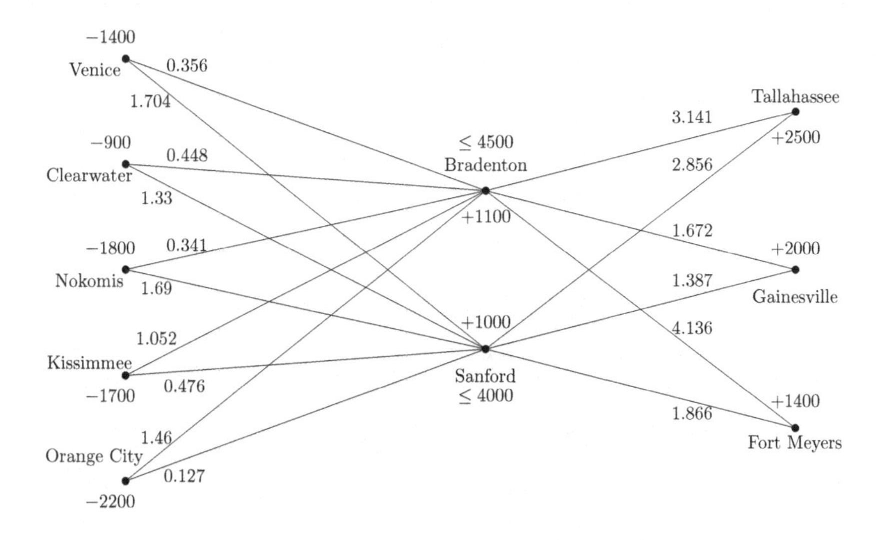

The decision variables in this problem will be how many boxes of oranges to ship along each route, namely

$$X_{ij} = \text{ the number of 90 pound boxes shipped from } i \text{ to } j$$

where i and j are facilities denoted by the first letter of the location. Notice in these problems both the edges and the vertices are weighted. The edges are weighed with the per unit shipping cost along the represented route whereas the vertices' weights have to do with production, capacity, and demand levels. We have followed convention and weighed the production vertices with negative values as those amounts are leaving the vertices and weighed the distribution centers with positive values as we need these net amounts to stay. The problem is then to

$$\begin{aligned}\text{Maximize} \quad C = {} & 0.356X_{VB} + 1.704X_{VS} + 0.448X_{CB} + 1.33x_{CS} \\ & + 0.341X_{NB} + 1.69X_{NS} + 1.052X_{KB} + 0.476X_{KS} \\ & + 1.46X_{OB} + 0.127X_{OS} + 3.141X_{BT} + 1.672X_{BG} \\ & + 4.316X_{BF} + 2.856X_{ST} + 1.387X_{SG} + 1.866X_{SF}\end{aligned}$$

Subject to (constraints at production)

$$-X_{VB} - X_{VS} \geq -1400 \tag{24.1}$$
$$-X_{CB} - X_{CS} \geq -900 \tag{24.2}$$
$$-X_{NB} - X_{NS} \geq -1800 \tag{24.3}$$
$$-X_{KB} - X_{KS} \geq -1700 \tag{24.4}$$
$$-X_{OB} - X_{OS} \geq -2200 \tag{24.5}$$

(constraints at processing)

$$X_{VB} + X_{CB} + X_{NB} + X_{KB} + X_{OB} - X_{BT} - X_{BG} - X_{BF} \geq 1100 \tag{24.6}$$
$$X_{VS} + X_{CS} + X_{NS} + X_{KS} + X_{OS} - X_{ST} - X_{SG} - X_{SF} \geq 1000 \tag{24.7}$$

(constraints at distribution)

$$X_{BT} + X_{ST} \geq 2500 \tag{24.8}$$
$$X_{BG} + X_{SG} \geq 2000 \tag{24.9}$$
$$X_{BF} + X_{SF} \geq 1400 \tag{24.10}$$

(and, lastly, capacity constraints)

$$X_{VB} + X_{CB} + X_{NB} + X_{KB} + X_{OB} \leq 4500 \tag{24.11}$$
$$X_{VS} + X_{CS} + X_{NS} + X_{KS} + X_{OS} \leq 4000 \tag{24.12}$$
$$\text{with } X_{ij} \geq 0.$$

Decision Variables

From	To	Value	Cost
Venice	Bradenton		0.356
Venice	Sanford		1.704
Clearwater	Bradenton		0.448
Clearwater	Sanford		1.33
Nokomis	Bradenton		0.341
Nokomis	Sanford		1.69
Kissimmee	Bradenton		1.052
Kissimmee	Sanford		0.476
Orange City	Bradenton		1.46
Orange City	Sanford		0.127
Bradenton	Tallahassee		3.141
Bradenton	Gainesville		1.672
Bradenton	Fort Meyers		4.136
Sanford	Tallahassee		2.856
Sanford	Gainesville		1.387
Sanford	Fort Meyers		1.866

Production

Facility	Shipped	Supply
Venice		1400
Clearwater		900
Nokomis		1800
Kissimmee		1700
Orange City		2200

Processing and Distribution

Facility	In	Out	Stay	Demand	Capacity
Bradenton				1100	4500
Sanford				1000	4000

Facility	In	Demand
Tallahassee		2500
Gainesville		2000
Fort Meyers		1400

Total Costs

FIGURE 24.1
The Excel setup for Jamie's SunLov'd Organic Oranges.

Of course, the X_{ij} need to be nonnegative and integer, but – for now – let us only worry about the nonnegativity requirement. We will address the integer constraints in the next section.

In establishing the constraints, we have followed the convention that edges flowing into a vertex are labeled with a + (they are bringing commodities in) but exiting edges have a – (those resources are leaving). It is, of course, the case that each of the production constraints $-X_{ij} - X_{kl} \geq -K$ is mathematically equivalent to $X_{ij} + X_{kl} \leq K$ and these inequalities hold because we can not ship more boxes than the groves are producing. Likewise, since we want to meet the given demand at the distribution centers, the flow into their vertices must be at least the level of demand (and hence these constraints are of the form $\geq$). The Excel set up of the problem is shown in Figure 24.1.

Since we are using Excel to solve the problem, forming the objective function is a good time to use Excel's SUMPRODUCT function. This is illustrated in Figure 24.2. Figure 24.3 show all the constraints for Jamie's SunLov'd Organic Oranges inclulding the (default in Excel) nonnegativity constraint.

Solver's optimal solution is given in Figure 24.4.

Notice that we obtained an integer solution without requiring the decision variables to be integer. This was not just a stroke of luck and is the subject of the next section.

=SUMPRODUCT(D5:D20,E5:E20)

Decision Variables

From	To	Value	Cost
Venice	Bradenton		0.356
Venice	Sanford		1.704
Clearwater	Bradenton		0.448
Clearwater	Sanford		1.33
Nokomis	Bradenton		0.341
Nokomis	Sanford		1.69
Kissimmee	Bradenton		1.052
Kissimmee	Sanford		0.476
Orange City	Bradenton		1.46
Orange City	Sanford		0.127
Bradenton	Tallahassee		3.141
Bradenton	Gainesville		1.672
Bradenton	Fort Meyers		4.136
Sanford	Tallahassee		2.856
Sanford	Gainesville		1.387
Sanford	Fort Meyers		1.866

Total Costs E5:E20)

Production

Facility	Shipped	Supply
Venice		1400
Clearwater		900
Nokomis		1800
Kissimmee		1700
Orange City		2200

Processing and Distribution

Facility	In	Out	Stay	Demand	Capacity
Bradenton				1100	4500
Sanford				1000	4000

Facility	In	Demand
Tallahassee		2500
Gainesville		2000
Fort Meyers		1400

FIGURE 24.2
Excel's SUMPRODUCT in Jamie's SunLov'd Organic Oranges.

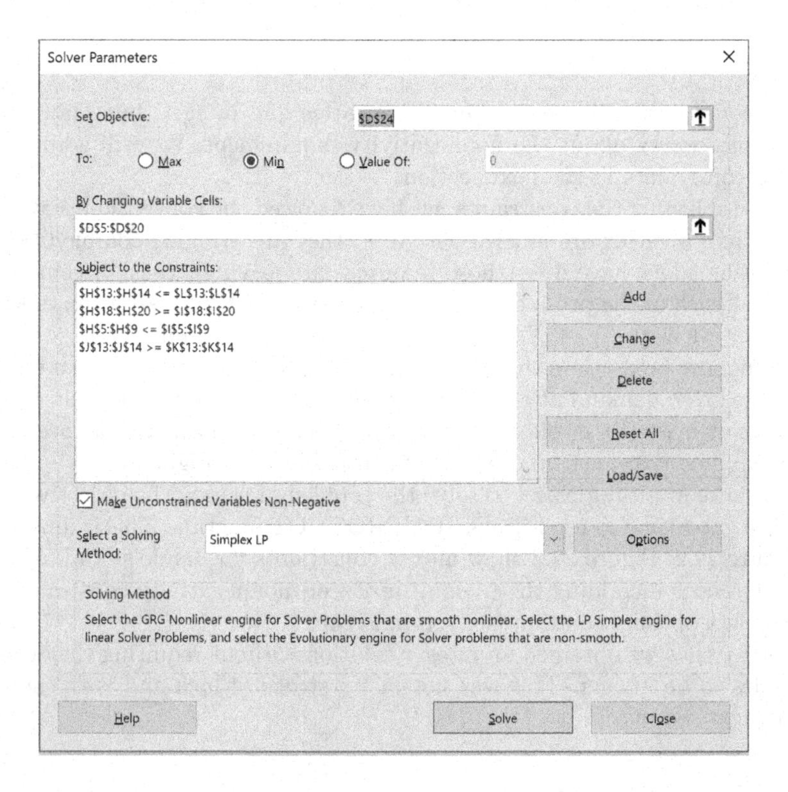

FIGURE 24.3
Constraints in Jamie's SunLov'd Organic Oranges.

Decision Variables

From	To	Value	Cost
Venice	Bradenton	1400	0.356
Venice	Sanford	0	1.704
Clearwater	Bradenton	900	0.448
Clearwater	Sanford	0	1.33
Nokomis	Bradenton	1800	0.341
Nokomis	Sanford	0	1.69
Kissimmee	Bradenton	0	1.052
Kissimmee	Sanford	1700	0.476
Orange City	Bradenton	0	1.46
Orange City	Sanford	2200	0.127
Bradenton	Tallahassee	1000	3.141
Bradenton	Gainesville	2000	1.672
Bradenton	Fort Meyers	0	4.136
Sanford	Tallahassee	1500	2.856
Sanford	Gainesville	0	1.387
Sanford	Fort Meyers	1400	1.866

Total Costs	15985.4

Production

Facility	Shipped	Supply
Venice	1400	1400
Clearwater	900	900
Nokomis	1800	1800
Kissimmee	1700	1700
Orange City	2200	2200

Processing and Distribution

Facility	In	Out	Stay	Demand	Capacity
Bradenton	4100	3000	1100	1100	4500
Sanford	3900	2900	1000	1000	4000

Facility	In	Demand
Tallahassee	2500	2500
Gainesville	2000	2000
Fort Meyers	1400	1400

FIGURE 24.4
Optimal distribution for Jamie's SunLov'd Organic Oranges.

24.2 The Guarantee of Integer Solutions in Network Flow Problems

We have seen that a transshipment problem is answered by introducing a linear programming model with inequality constraints. We know from the Fundamental Theorem of Linear Programming (Theorem 17.2.2) that a solution is a corner point of the feasible region. Fortunately, in the situation modeling transshipment problems (with the right conditions – which are always present in applications), we do get that all the corner points are integer valued. This comes from the fact that the matrix corresponding to the left-hand side of the constraints is *totally unimodular*, which was discussed in Section 4.4.2.

Let us walk through our example in building the theory behind integer solutions in transshipment problems. To match standard notation, we will start by rewriting all the constraints for *Jamie's Organic Sun Lov'd Oranges* to be of the form $\leq$, Thus we have as the model:

$$\begin{aligned}\text{Maximize} \quad C = {} & 0.356X_{VB} + 1.704X_{VS} + 0.448X_{CB} + 1.33x_{CS} \\ & + 0.341X_{NB} + 1.69X_{NS} + 1.052X_{KB} + 0.476X_{KS} \\ & + 1.46X_{OB} + 0.127X_{OS} + 3.141X_{BT} + 1.672X_{BG} \\ & + 4.316X_{BF} + 2.856X_{ST} + 1.387X_{SG} + 1.866X_{SF}\end{aligned} \tag{24.13}$$

Subject to
(constraints at production)

$$X_{VB} + X_{VS} \leq 1400 \tag{24.14}$$
$$X_{CB} + X_{CS} \leq 900 \tag{24.15}$$
$$X_{NB} + X_{NS} \leq 1800 \tag{24.16}$$
$$X_{KB} + X_{KS} \leq 1700 \tag{24.17}$$
$$X_{OB} + X_{OS} \leq 2200 \tag{24.18}$$

(constraints at processing)

$$-X_{VB} - X_{CB} - X_{NB} - X_{KB} - X_{OB} + X_{BT} + X_{BG} + X_{BF} \leq 1100 \tag{24.19}$$
$$-X_{VS} - X_{CS} - X_{NS} - X_{KS} - X_{OS} + X_{ST} + X_{SG} + X_{SF} \leq 1000 \tag{24.20}$$

(constraints at distribution)

$$-X_{BT} - X_{ST} \leq -2500 \tag{24.21}$$
$$-X_{BG} - X_{SG} \leq -2000 \tag{24.22}$$
$$-X_{BF} - X_{SF} \leq -1400 \tag{24.23}$$

(and, lastly, capacity constraints)

$$X_{VB} + X_{CB} + X_{NB} + X_{KB} + X_{OB} \leq 4500 \tag{24.24}$$
$$X_{VS} + X_{CS} + X_{NS} + X_{KS} + X_{OS} \leq 4000 \tag{24.25}$$
$$\text{with } X_{ij} \geq 0.$$

The constraints can be succinctly written in the form $A\mathbf{x} \leq \mathbf{b}$ where $\mathbf{b}$ is a column vector with the right hand side constants as it components. Suppose A is an $m \times n$ matrix; that is, it has m rows (one for each constraint inequality) and n columns (one for each decision variable; the flows on the edges), Then $A \in \mathbb{R}^{m \times n}$ and $\mathbf{b} \in \mathbb{R}^m$. Define

$$F = F(A, \mathbf{b}) := \{\mathbf{x} \in \mathbb{R}^n | A\mathbf{x} \leq \mathbf{b}, \mathbf{x} \geq 0\};$$

i.e. the feasible region of the linear programming problem.

$A =$

const.	x_{VB}	x_{VS}	x_{CB}	x_{CS}	x_{NB}	x_{NS}	x_{KB}	x_{KS}	x_{OB}	x_{OS}	x_{BT}	x_{BG}	x_{BF}	x_{ST}	x_{SG}	x_{SF}
(24.14)	1	1	0	0	0	0	0	0	0	0	0	0	0	0	0	0
(24.15)	0	0	1	1	0	0	0	0	0	0	0	0	0	0	0	0
(24.16)	0	0	0	0	1	1	0	0	0	0	0	0	0	0	0	0
(24.17)	0	0	0	0	0	0	1	1	0	0	0	0	0	0	0	0
(24.18)	0	0	0	0	0	0	0	0	1	1	0	0	0	0	0	0
(24.19)	−1	0	−1	0	−1	0	−1	0	−1	0	1	1	1	0	0	0
(24.20)	0	−1	0	−1	0	−1	0	−1	0	−1	1	1	1	0	0	0
(24.21)	0	0	0	0	0	0	0	0	0	0	−1	0	0	−1	0	0
(24.22)	0	0	0	0	0	0	0	0	0	0	0	−1	0	0	−1	0
(24.23)	0	0	0	0	0	0	0	0	0	0	0	0	−1	0	0	−1
(24.24)	1	0	1	0	1	0	1	0	1	0	0	0	0	0	0	0
(24.25)	0	1	0	1	0	1	0	1	0	1	0	0	0	0	0	0

Key to the result is understanding how transshipment problems are modeled. The decision variables in the model represent the flow on an edge between two vertices in the model's graph. Let A^* be the matrix of coefficients on the left-hand side of the constraints 24.14 through 24.23. As there is never more than one edge between any two vertices, every column of A will have nor more than two non-zero entries (if so, these entries are exactly $+1$ and -1).

Thus we have by Theorem 4.4.10 that A^* is totally unimodular. Let A be the matrix A^* with rows representing the capacity constraints from the distribution centers into A^*. The addition of these rows do not prevent the matrix from continuing to be a totally unimodular matrix (note that the reference theorem is sufficient conditions only and not an if-and-only-if).

As we saw in the Jamie's SunLov'd Organic Oranges example in the previous section, transshipment problems involve solving a linear programming problem. The corresponding system we solve will turn out to always have an integer solution provided the flow capacities at the vertices are integer valued. The keys in this result lies in the involved matrices being *totally unimodular* (introduced in Section 4.4.2).

Theorem 24.2.1. *Consider the linear system $A\mathbf{x} = \mathbf{b}$ where A is an $n \times n$ nonsingular square matrix with integer entries and* $\mathbf{b}$ *also having only integer components. Then* $\mathbf{x}$ *is integer if and only if A is totally unimodular.*

Proof. ($\Leftarrow$) Suppose A is totally unimodular. Since A is nonsingular, it has an inverse and $\det(A) \neq 0$, and thus $\mathbf{x} = A^{-1}\mathbf{b}$. By Cramer's Rule (Theorem 4.3.24), the ith entry of $\mathbf{x}$ is

$$x_i = \frac{\det(A_i)}{\det(A)}.$$

As the matrix A_i is the matrix A with its ith column replaced by $\mathbf{b}$, we can expand on the ith column A_i to determine $\det(A_i)$. The remaining matrices have determinant $0, \pm 1$ since A is totally unimodular and as the column we are using for expansion has integer entries, $\det(A_i)$ is clearly an integer. Since A is unimodular (it is totally unimodular, but as it as an inverse, 0 is ruled out as a value of its determinant), $\det(A) = \pm 1$ and, thus, the ratio giving x_i by Cramer's Rule must be an integer.

($\Rightarrow$) Suppose $\mathbf{x}$ has only integer components for all integer $\mathbf{b}$ with $A\mathbf{x} = \mathbf{b}$. Let $\mathbf{b} = [0, \ldots, 0, 1, 0, \ldots, 0]$; a vector with nonzero entries except for a 1 in the ith component. Since $\mathbf{x} = A^{-1}\mathbf{b}$ and $\mathbf{b} = [0, \ldots, 0, 1, 0, \ldots, 0]$, $\mathbf{x}$ is the ith column of A^{-1}. Thus the ith column of A^{-1} is integer. But this is true for any i, $0 \leq i \leq n$, therefore all entries of A^{-1} are integer, giving $\det(A^{-1})$ is integer. □

24.3 Exercises

Exercise 24.1. *Jackson Brewing makes three great beers: a stout, an IPA, and a bitter. The brewery has three locations at which it can brew: Pittsburgh, Knoxville, and Memphis. Given that each location can only make one kind of beer and that Neilan wants to continue making each of the three beers, Neilan is interested in assigning production in way that minimizes cost. The cost of brewing each beer at each location is given in the following table:*

	Stout	*IPA*	*Bitter*
Pittsburgh	*1.22*	*0.95*	*0.88*
Knoxville	*1.13*	*0.91*	*0.77*
Memphis	*1.29*	*0.99*	*0.92*

i. *Draw a network flow diagram representing the problem (be sure to clearly state what the vertices and edges represent).*

ii. *State the LP problem modeling this situation.*

iii. *For this problem, is it necessary to require the decision variables to be integer? Why or why not?*

iv. *Answer the question using any software of your choice. Submit a screenshot of your program's output.*

Exercise 24.2. *Rumpler Kayaks produces quality kayaks at three locations: Minneapolis, Pittsburgh, and Tucson. Additionally, the company has warehouses in Atlanta, Boston, Chicago, and Denver. The table below has the number of kayaks each warehouse demands as well as their respective capacities. As well, the table list the per kayak cost for shipping from a particular plant to each warehouse. Rumpler is interested in in fulfilling the warehouses' demands at a minimum cost.*

	Atlanta	*Boston*	*Chicago*	*Denver*	*Capacity*
Minneapolis	*6.00*	*5.60*	*2.20*	*4.00*	*10,000*
Pittsburgh	*3.60*	*3.00*	*2.80*	*5.80*	*15,000*
Tucson	*6.50*	*6.80*	*5.50*	*4.20*	*15,000*
Demand	*8000*	*10,000*	*12,000*	*9,000*	

i. *Draw a network flow diagram representing the problem (be sure to clearly state what the vertices and edges represent).*

ii. *State the LP problem modeling this situation.*

iii. *For this problem, is it necessary to require the decision variables to be integer? Why or why not?*

iv. *Answer the question using any software of your choice. Submit a screenshot of your program's output.*

25

The Traveling Salesperson Problem

Large portions of this chapter, including many of the figures, are taken from my student Corinne Brucato Bauman's Master Thesis [8].

25.1 History of the Traveling Salesperson Problem

The *Traveling Salesperson Problem* considers finding an optimal route in which a salesperson can visit the cities he or she needs to (each city once) then return home. This problem can be considered to be closely related to finding a closed *knight's tour* on a chessboard. That is, a sequence of legal moves by a knight that visits every square of a chessboard. A knight's tour is considered closed if the knight ends up one move away from where it started. In this sense, one can say the Traveling Salesperson Problem dates back to the 9th century when an example of such a tour appeared in the work of Kashmiri poet Rudrata [64]. It is very likely the chess players and many of the computer scientists reading this are very familiar with this problem. Note that Leonhard Euler presented a solution to the Knight's Tour problem in 1759, which he later published in 1766 [23].

The first example of such a problem appeared in the German handbook *Der Handlungsreisende – Von einem alten Commis – Voyageur* for salesman traveling through Germany and Switzerland in 1832 as explained in [10]. This handbook was merely a guide but did contribute to identifying the problem as an important one.

Much of the earliest work done on this problem was performed by Irish mathematician Sir William Rowan Hamilton and by the British mathematician Thomas Penyngton Kirkman in the 19th century. A good summary of the mens' work is presented in [2]. Hamilton even developed a board game based on the problem, his *Icosian Game*.

In Hamilton's Icosian Game, players are to find a tour visiting the 20 vertices along the available edges in in the dodecahedron graph as seen Figure 25.1. The game was played by having Player 1 place pegs 1 through 5 in order on adjacent vertices. A sample Player 1 move is shown in Figure 25.2. Player 2 would win if they could place pegs 6 through 20 in such a way to complete a Hamiltonian circuit. The game was evidently not very popular as

DOI: 10.1201/9780367425517-25

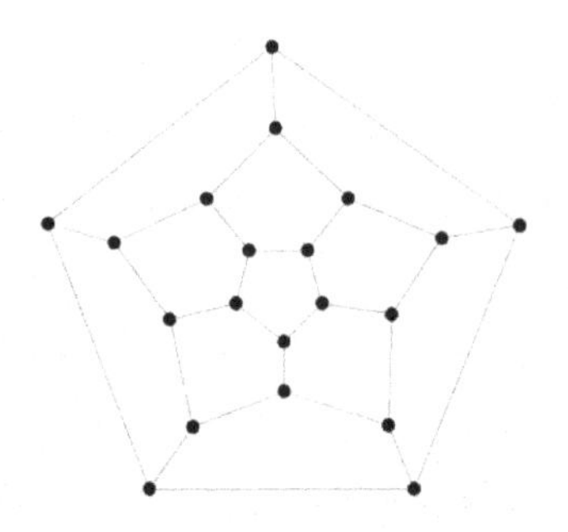

FIGURE 25.1
Hamilton's Icosian game.

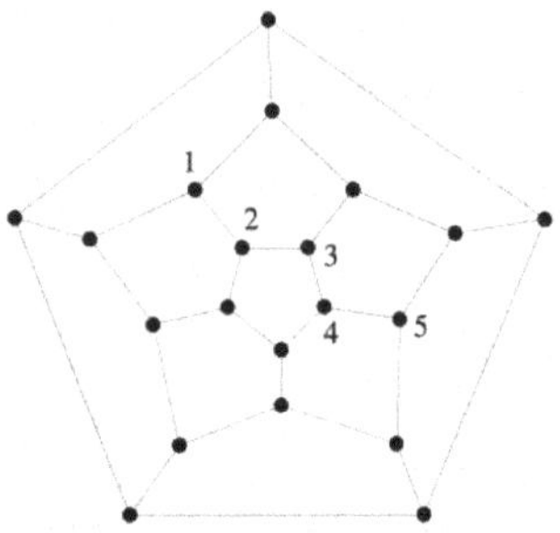

FIGURE 25.2
Sample start.

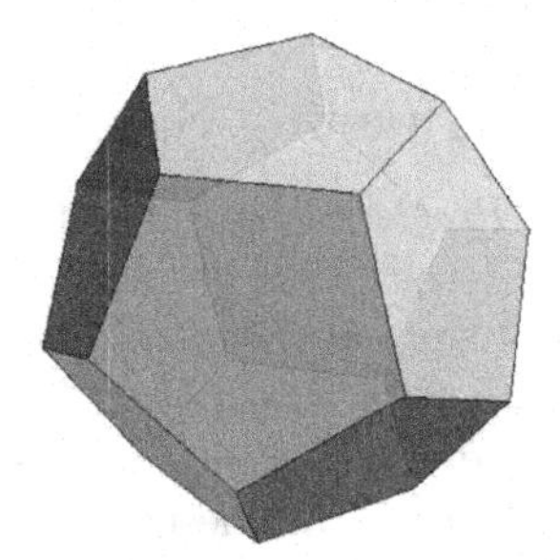

FIGURE 25.3
The platonic solid: Dodecahedron.

children complained that it was too easy [10] (it is the case that every starting position of the has a winning solution). *The Puzzle Museum* [46] has picture of a complete version of the original game.

It is worthwhile to explain the names used in describing Hamilton's game. As the graph in Figure 25.1 has 20 vertices, Hamilton referred to it as an *icosian* as *eikosi* (εικοσι) is an Ancient Greek word meaning "twenty". This graph is the *planar representation*[1] of the twelve-faced platonic solid dodecahedron from the Greek δοδεκαέδρον where δοδεκα means "twelve" and *hedra* (ἐδρα) can mean "face of a geometrical solid". A dodecahedron (the 12-sided die for my D&D friends) is pictured in Figure 25.3. The dodecahedron is one of the five Platonic solids and each of the solids has a representation as a planar graph[2].

The first paper related to the subject seems to be Hassler Whitney's 1931 paper in which the conclusion of the first theorem is "there exists a circuit

[1]By *planar graph* we mean that we can draw the graph in on a piece of paper or any other plane in some way and not have any edges cross.

[2]$V - E + F = 2$ anyone?

which passes through every vertex of the graph" [63]. The terms "Hamiltonian cycle" and "Traveling Salesperson" are not used. The first break-through on the problem was due to George Dantzig, Delbert Ray Fulkerson, and Selmer M. Johnson in 1954 while working at the RAND Corporation [14]. In this paper, the team found a provable solution to finding the shortest tour visiting Washington, D.C. and the 48 contiguous state capitals. The group also acknowledges Merrill Flood of Columbia University for "stimulating interest in the traveling-salesmen problem in many quarters" [14].

To date, no solution is known, but many good heuristics exist. The difficulty of the problem was placed on firm ground in Richard M. Karp's 1972 Paper [21] (this is the paper with Karp's 21 NP-complete problems). In this work, Karp established that finding a minimal weight Hamiltonian cycle is NP-complete, which implies the NP-hardness of the Traveling Salesperson Problem. In other words, Karp's paper formally explains why finding an optimal tour is computationally laborious.

25.2 The Problem

Suppose a salesperson must visit a single client in each of $n - 1$ other cities. Our salesperson would like to return home after the visits and, of course, it is highly desirable to determine a route that enables the salesperson to make these visits with the least possible amount of driving. In the language of mathematics, we seek a minimum weight Hamiltonian cycle (or tour) on a graph with n vertices.

Of course, a brute force approach will work in theory, but it will not take long until the problem becomes intractable. Consider the complete graph K_n on n vertices. An easy application of the Multiplication Principle (Theorem 20.2.1) gives that there are $(n-1)!$ Hamiltonian cycles on K_n, but if we regard going forward or backwards through the same sequence of cities as the same tour, then we can reduce this to $\frac{(n-1)!}{2}$. For various orders n, the number of Hamiltonian cycles on K_n is given in Table 25.1. Of course, if the graph is not complete, then we have the additional concern as to whether a tour is possible.

25.3 Heuristic Solutions

As stated when discussing the History of the problem, no technique yet exists to find a globally optimal solution in a reasonable amount of time (nor has it

TABLE 25.1
Number of Distinct Tours in a Complete Graph with n Vertices

n	Number of Distinct Hamiltionian Cycles on $K_n = (n-1)!/2$
3	1
4	3
5	12
6	60
7	360
8	2,520
9	20,160
10	181,440
*	2,852,592,966 (miles from Earth to Neptune on 1 February 2022)
15	43,589,145,600
*	$27,752,835,868,445.35 (U.S. National Debt, 1 February 2022)
20	60,822,550,204,416,000
*	4.35×10^{17} (estimated age of the universe in seconds)
30	4,420,880,996,869,850,977,271,808,000,000
*	8.7×10^{34} (number of floating point operations Oak Ridge National Laboratory's supercomputer Summit would have preformed if operating at peak since the Big Bang)
40	10,198,941,040,598,721,679,320,140,869,951,448,678,400,000,000
50	304,140,932,017,133,780,436,126,081,660,647,688,443,776,415,689, 605,120,000,000,000
*	80,658,175,170,943,878,571,660,636,856,403,766,975,289,505,440, 883,277,824,000,000,000,000 (number of unique ways to deal all 52 cards from a standard deck $= 52! \approx 8 \times 10^{64}$)
60	$\approx 1.3868 \times 10^{80}$
*	10^{82} (estimated number of atoms in the known universe)
75	$\approx 1.65394 \times 10^{107}$
100	$\approx 4.66631 \times 10^{155}$

been proven that one does not exist). In this section, we will present some of the tractable heuristics that attempt to find a tour of minimum weight.

25.3.1 Nearest Neighbor

The first technique we will consider is a greedy algorithm (see Definition 23.2.2). This technique is not to be confused with the *k-Nearest Neighbor* supervised learning algorithm used in Machine Learning.

TABLE 25.2
Distances (in Miles) Between Themed Parks in *Future World* in Example 25.3.1

Site	Western World	Medieval World	Roman World	Spa World	(F) Future World
(R) Resort	8	2	12	13	5
(W) Western World	–	8	5	13	4
(M) Medieval World	–	–	3	6	11
(O) Roman World	–	–	–	9	10
(P) Spa World	–	–	–	–	15

Algorithm 25.3.1 Nearest Neighbor Heuristic for Finding a (Local) Minimum Weight Tour.

Objective: Find a minimum-weight Hamiltonian cycle H in a weighted, connected graph or digraph G.

Input: A weighted, connected graph G.

1: Consider all vertices of G as unvisited or unreached.
2: Arbitrarily select a start vertex, u of G. Add u to H.
3: Of all the edges out of u to the unvisited vertices, select an edge of minimum weight, say edge uv and add it to H in an ordered manner. Settle ties arbitrarily.
4: Consider v as now visited and remove v from the set of unvisited vertices.
5: Repeat 3 with the last vertex selected until all vertices are visited.
6: Let w be the very last vertex visited. Add edge wu to H.

Output: A Hamiltonian cycle or tour H of locally minimum weight in G.

For an example, let us revisit Delos' *Future World* from Example 23.2.1.

Example 25.3.1. *The Delos Corporation has an amusement park with a central luxury resort and five separate themed locations: Western World, Medieval World, Roman World, Spa World ("where old age and pain have been eliminated"), and Future World*[3].

Suppose a guest would like to do a quick tour of the themed locations in one day before deciding how to plan their visits during their stay. We will assume a constant rate at which Future World *transports its guests, so minimizing tour distance will be our goal. Distances between the resort and locations are given in Table 25.2.*

Solution. We will model the problem with the following graph where the vertices represent the resort and themed locations and the edges are weighed by

[3]This example borrows from the films *Westworld* [62] and its sequel *Futureworld* [27].

their distance in miles. Note this graph and Table 25.2 are different from those in Example 23.2.1 because we are no longer concerned with the water source. Also note that this graph need only be a graph on the sense of Definition 21.1.2; that is, it has vertices and there is either an edge between distinct vertices or there is not (in this cast, all possible edges exist). For this algorithm, we do not need the graph to be drawn as the sites would appear on a map.

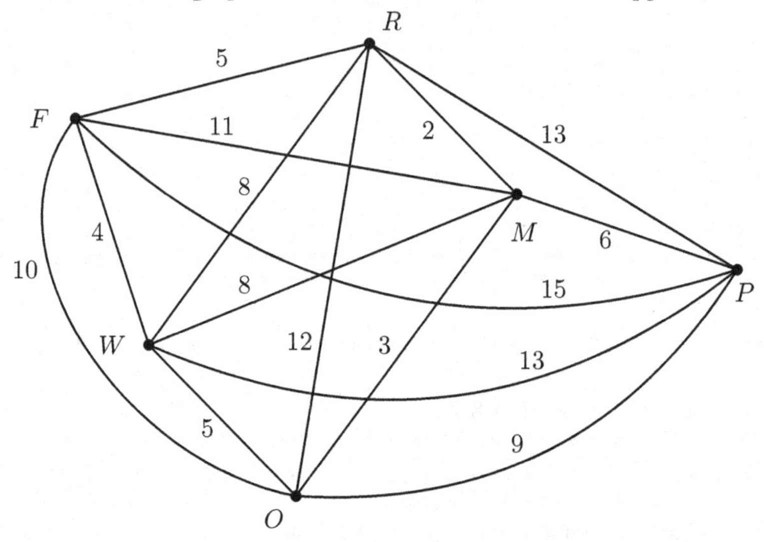

If the guest starts a the resort, then applying the Nearest Neighbor Algorithm gives the tour

$$R \xrightarrow{2} M \xrightarrow{3} O \xrightarrow{5} W \xrightarrow{4} F \xrightarrow{15} P \xrightarrow{13} R \tag{25.1}$$

which has a total weight (distance) of 42. ■

Example 25.3.1 illustrates a major drawback with the Nearest Neighbor Algorithm: we may get stuck at the end with large edges. It also turns out that we may get different tours depending on where we start, as shown in the next examples.

Example 25.3.2. *For this example, we will illustrate each step. Consider finding a minimum tour on the following weighted K_5 by starting at vertex A:*

Solution. Looking out of vertex A, we see that we can choose from four edges to start. These are illustrated with the darker dotted lines in Figure 25.5. Edge AD has the minimum weight of these four edges, so we accept it into our tour $H = A \xrightarrow{6} D$ and consider D as now being visited. The result of the first iteration of applying the Nearest Neighbor Algorithm to Example 25.3.2 is shown in Figure 25.6.

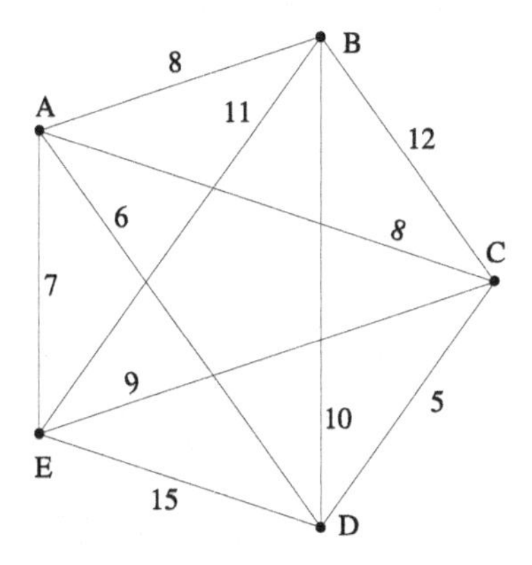

FIGURE 25.4
A weighted K_5.

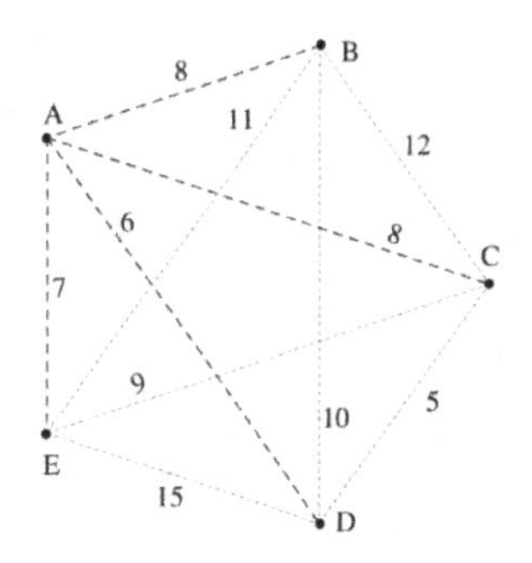

FIGURE 25.5
Choices leaving A.

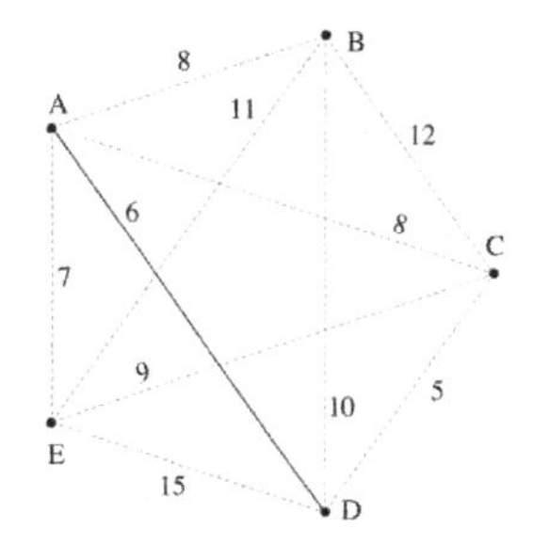

FIGURE 25.6
First edge selection.

For the next iteration, we consider all the edges out of D to the remaining unreached vertices. These are shown in Figure 25.7. Of the unvisited vertices, C is the closest, so we travel there on edge DC. The current tour is now updated to $H = A \xrightarrow{6} D \xrightarrow{5} C$ and C has now been reached.

The current tour is now at C and the edges out of C to unvisited vertices include CB and CE as edge CA would take us to a visited vertex. This is pictured in Figure 25.8. Of these two edges, CE is the shortest and the tour is now updated to $H = A \xrightarrow{6} D \xrightarrow{5} C \xrightarrow{9} E$ and E is now regarded as visited. This third iteration is illustrated in Figure 25.9.

Since only one vertex remains unvisited, the last two edges added to the tour are now determined. Visiting the remaining unreached vertex B and returning to the start produces the tour $H = A \xrightarrow{6} D \xrightarrow{5} C \xrightarrow{9} E \xrightarrow{11} B \xrightarrow{8} A$. The tour resulting from applying the Nearest Neighbor Algorithm and starting

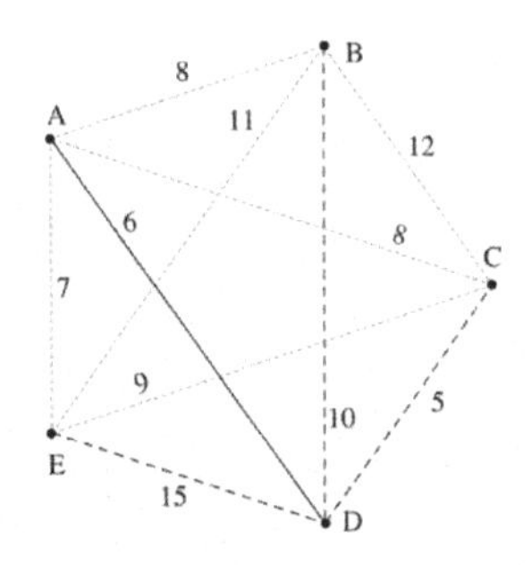

FIGURE 25.7
Choices leaving D.

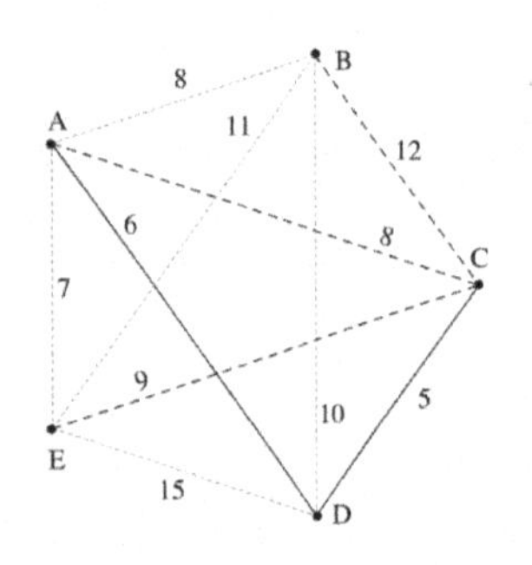

FIGURE 25.8
Second edge selection.

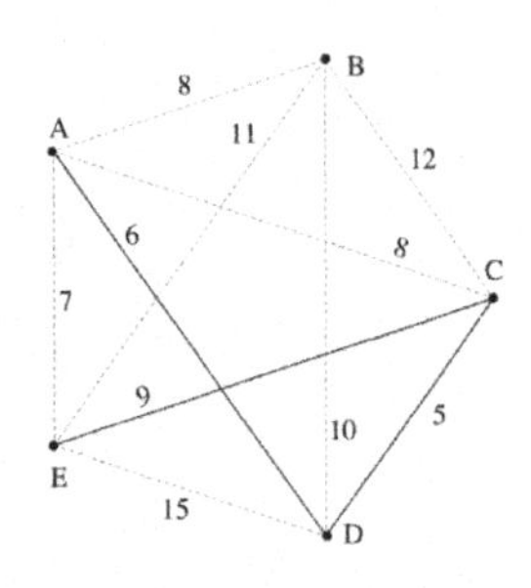

FIGURE 25.9
Reaching vertex E.

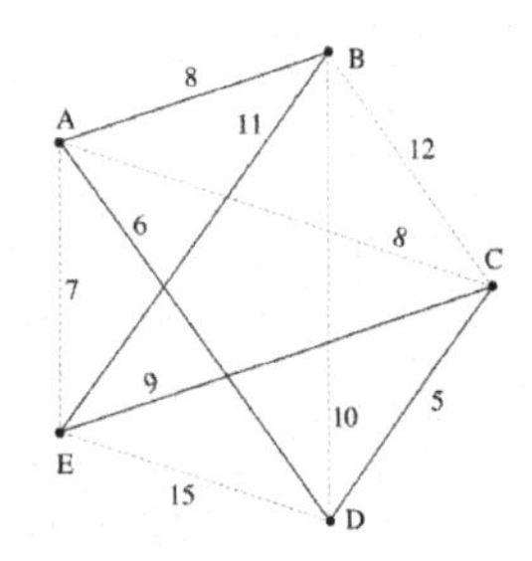

FIGURE 25.10
Resulting tour.

at A is illustrated in Figure 25.10. Note that the total weight of this tour is 39. ■

As stated, we may get a different tour of a different weight by choosing a different start vertex and applying the Nearest Neighbor Algorithm.

Solution. [Alternate solution to Example 25.3.2.]

For this application, let us suppose the traveling salesperson starts at vertex E. The first two iterations of applying the Nearest Neighbor Algorithm are to visit vertex A then vertex D. These steps are shown in Figures 25.11 and 25.12.

The resulting tour is $H = E \xrightarrow{7} A \xrightarrow{6} D \xrightarrow{5} C \xrightarrow{12} B \xrightarrow{11} E$. The tour resulting from applying the Nearest Neighbor Algorithm but starting at E is illustrated in Figure 25.13. Note that the total weight of this tour is 41, which differs from the previous tour.

■

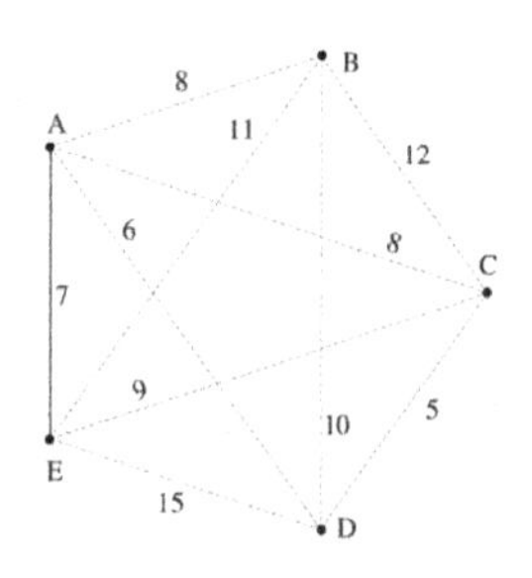

FIGURE 25.11
Leaving the start E.

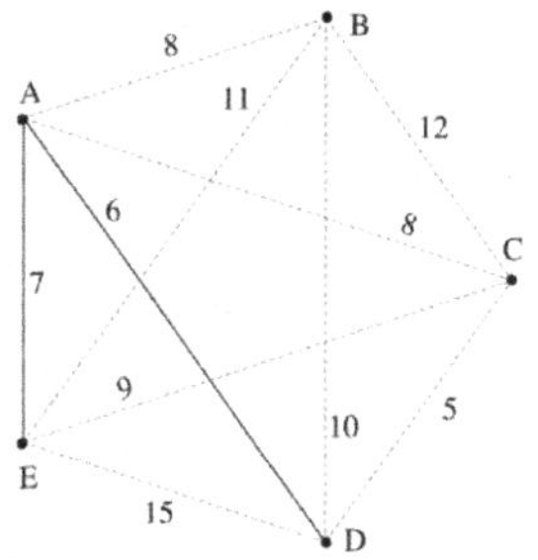

FIGURE 25.12
2nd step out of E.

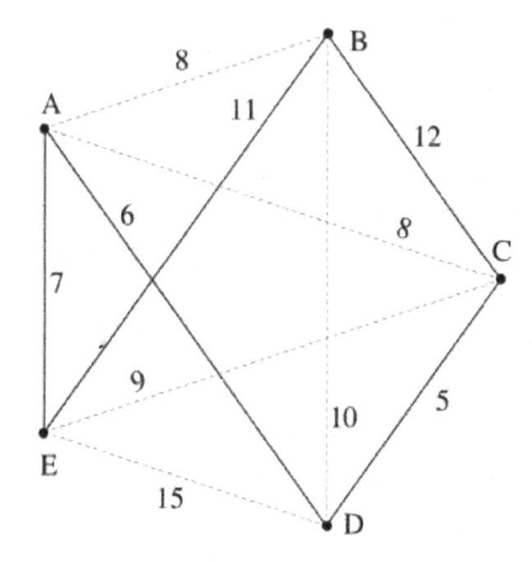

FIGURE 25.13
Nearest neighbor algorithm's tour for Example 25.3.2, but starting out of E.

25.3.2 Insertion Algorithms

We will now again employee a greedy algorithm, but this time from a different perspective. In the Nearest Neighbor Algorithm, our salesperson greedily chose to drive to the nearest unvisited city. Now, our salesperson will step back and take a bird's eye view. There are multiple version of this heuristic, but we will focus on two: our salesperson will still have a start city, but at each iteration she will build a tour by considering which unvisited city to add the existing tour by choosing the city 1) closest to the current tour or 2) choose a city whose inclusion increases the tour weight the least. In other words, the salesperson will greedily build tours in one of two ways until all cities are visited.

25.3.2.1 Closest Insertion

Our first insertion heuristic will involve selecting the city closest to the current existing tour. The technique is explained below.

Algorithm 25.3.2 Closest Insertion Heuristic for Finding a (Local) Minimum Weight Tour.

Objective: Find a minimum-weight Hamiltonian cycle H in a weighted, connected graph or digraph G.

Input: A weighted, connected graph G with $d_{i,j}$ as the weight of edge ij.

1: Consider all vertices of G as unvisited or unreached.

2: Arbitrarily select a start vertex, v of G. Add v to the tour H.

3: (Selection Step) From the unvisited vertices, select a vertex v to visit such that its distance to the current existing tour H is a minimum among all unvisited cities. Settle ties arbitrarily.

4: (Insertion Step) Insert into the tour H edge ij which minimizes $d_{i,v} + d_{v,j} - d_{i,j}$. Settle ties arbitrarily.

5: Consider v as now visited and remove v from the set of unvisited vertices.

6: Repeat 3 until all vertices are visited.

Output: A Hamiltonian cycle or tour H of locally minimum weight in G.

We illustrate the Closest Insertion Algorithm with a return to the *Future World* example.

Example 25.3.3. *Consider finding a minimum weight tour for the visitor to Future World whose visit starts from the Resort (see Example 25.3.1).*

Start and Iteration 1

We begin at R and currently have a tour empty of edges. As M is the vertex closest to the existing tour $H = R$, edge RM is added to H resulting in an updated $H = R \xrightarrow{2} M \xrightarrow{2} R$. Figure 25.14 illustrates the start from the resort R using the Closest Insertion Algorithm 25.3.2.

Iteration 2

Focusing on the visited vertices, the edges out of R to unvisited vertices have weight 5, 8, 12, and 13 while those out of M to unvisited vertices have weight 11, 8, 3, 6. Thus vertex closest to the existing tour $H = R \rightarrow M \rightarrow R$ is vertex O. Currently our only choice is to add edges MO and OP to the tour and there is no need to remove an edge from the existing tour H. Thus, at the end of this iteration, we have the updated tour $H = R \xrightarrow{2} M \xrightarrow{3} O \xrightarrow{12} R$. Figure 25.15 illustrates this second iteration with the dashed lines represent choices and the resulting tour illustrated with solid lines.

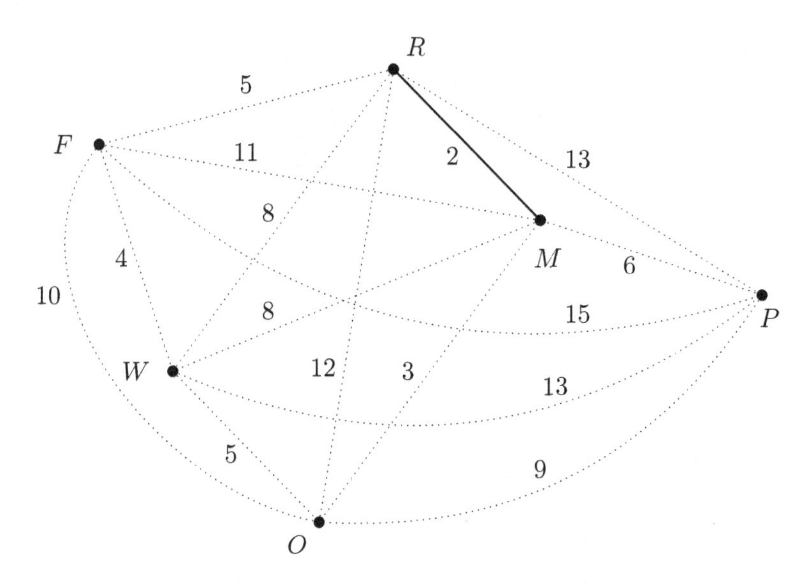

FIGURE 25.14
The weighted future world graph with start at R in Example 25.3.3.

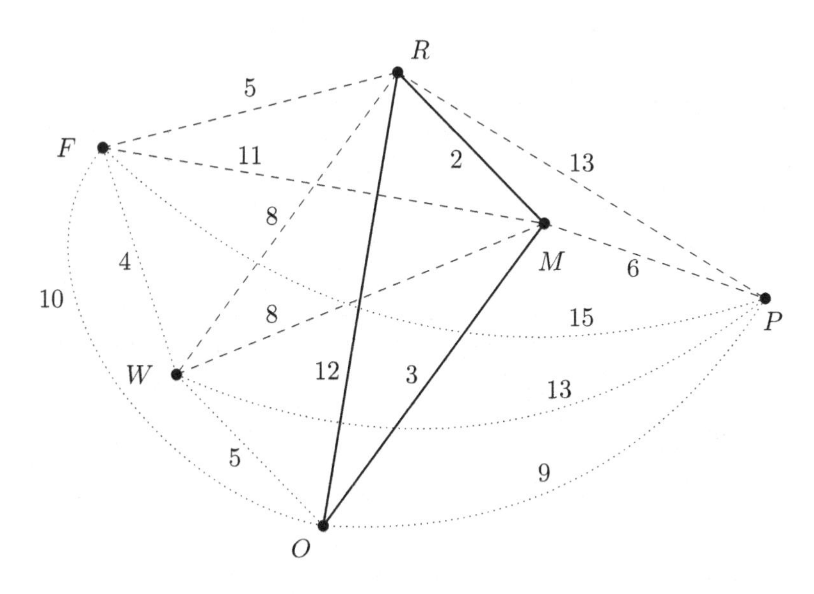

FIGURE 25.15
Future world closest insertion iteration 2 in Example 25.3.3.

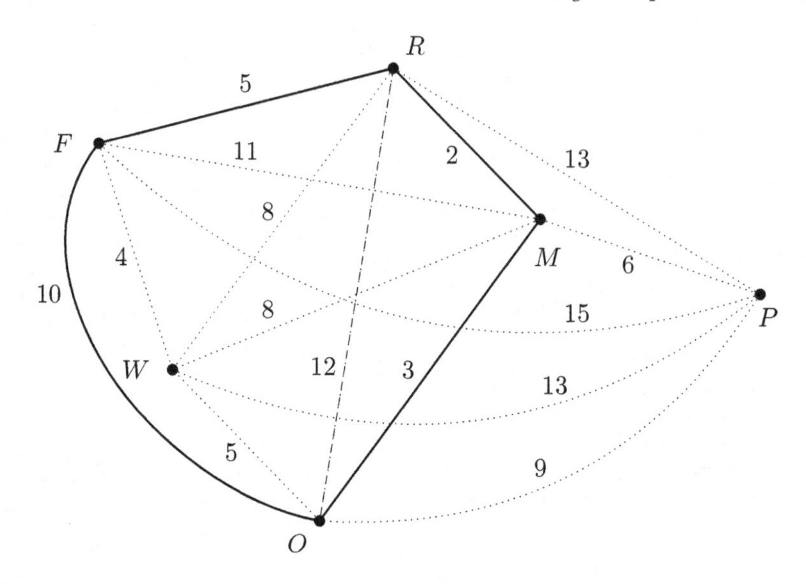

FIGURE 25.16
Future world closest insertion iteration 3 in Example 25.3.3.

Iteration 3

Of the unvisited vertices currently not on the tour, vertices F and W are the closest to H. We arbitrarily choose F and must now consider how to insert it into H.

Proposed Removed Edge	*Increase* $= d_{i,v} + d_{v,j} - d_{i,j}$	*Resulting Tour*	*Include?*
RM	$5 + 11 - 2 = 14$	$\langle R, F, M, O, R\rangle$	*No*
MO	$11 + 10 - 3 = 18$	$\langle R, M, F, O, R\rangle$	*No*
OR	$5 + 10 - 12 = 3$	$\langle R, M, O, F, R\rangle$	*Yes*

Thus, at the end of this iteration, we have the updated tour $H = R \xrightarrow{2} M \xrightarrow{3} O \xrightarrow{10} F \xrightarrow{5} R$*. Figure 25.16 illustrates the tour resulting from this third iteration with the dashed line representing the removed edge.*

Iteration 4

Of the unvisited vertices not yet on the tour, vertex W is closest to the current H. We now consider how to insert this into H.

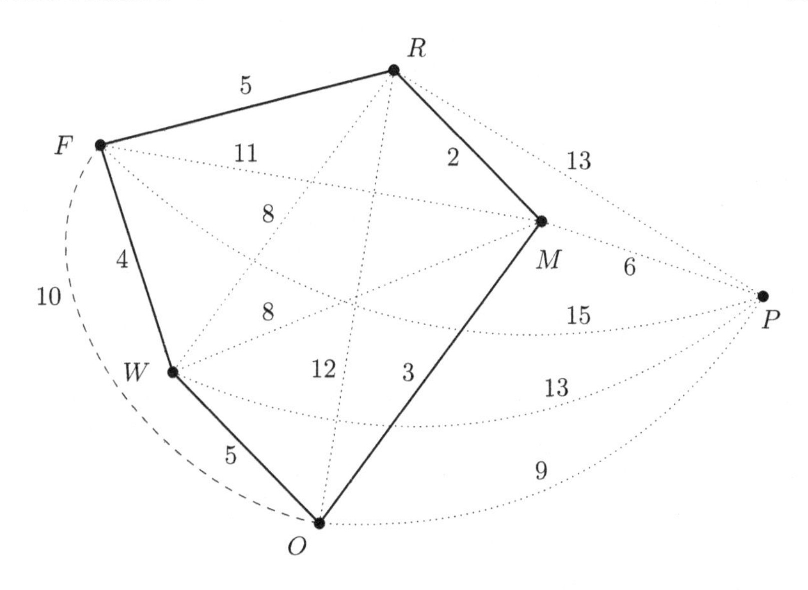

FIGURE 25.17
Future world closest insertion iteration 4 in Example 25.3.3.

Proposed Removed Edge	*Increase* $= d_{i,v} + d_{v,j} - d_{i,j}$	*Resulting Tour*	*Include?*
RM	$8+8-2=14$	$\langle R,W,M,O,F,R\rangle$	*No*
MO	$8+5-3=10$	$\langle R,M,W,O,F,R\rangle$	*No*
OF	$4+5-10=-1$	$\langle R,M,O,W,F,R\rangle$	*Yes*

Thus, at the end of this iteration, we have the updated tour $H = R \xrightarrow{2} M \xrightarrow{3} O \xrightarrow{5} W \xrightarrow{4} F \xrightarrow{5} R$. *Figure 25.17 illustrates the tour resulting from this fourth iteration.*

Iteration 5

One unvisited vertex remains and we must determine how the least expensively insert vertex P *into* H.

Proposed Removed Edge	*Increase* $= d_{i,v} + d_{v,j} - d_{i,j}$	*Resulting Tour*	*Include?*
RM	$13+6-2=17$	$\langle R,P,M,O,W,F,R\rangle$	*No*
MO	$6+9-3=12$	$\langle R,M,P,O,W,F,R\rangle$	*Yes*
OW	$9+13-5=17$	$\langle R,M,O,W,P.F,R\rangle$	*No*
WF	$13+15-4=24$	$\langle R,M,O,W,P,F,R\rangle$	*No*
FR	$15+13-5=23$	$\langle R,M,W,O,F,P,R\rangle$	*No*

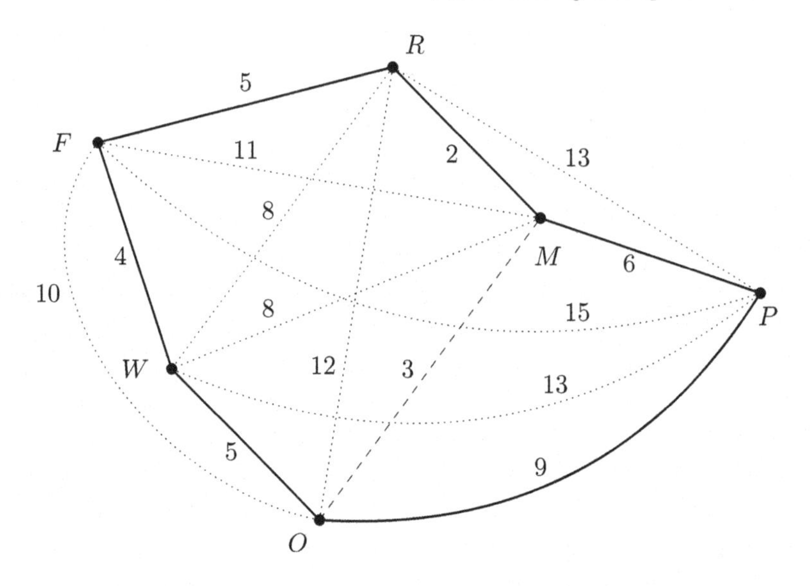

FIGURE 25.18
Future world closest insertion final tour in Example 25.3.3.

As there are no remaining unvisited vertices, we are finished and the algorithm has produced the tour $H = R \xrightarrow{2} M \xrightarrow{6} P \xrightarrow{9} O \xrightarrow{5} W \xrightarrow{4} F \xrightarrow{5} R$. *Figure 25.18 illustrates the tour resulting from this final iteration.*

25.3.2.2 Cheapest Insertion

Our second insertion heuristic is a slight variation of the first. This time, we forgo selecting to add the unvisited vertex closest to the tour but rather include the vertex whose insertion is the least costly.

Algorithm 25.3.3 Cheapest Insertion Heuristic for Finding a (Local) Minimum Weight Tour.

Objective: Find a minimum-weight Hamiltonian cycle H in a weighted, connected graph or digraph G.

Input: A weighted, connected graph G with edge weights $d_{i,j}$.

1: Consider all vertices of G as unvisited or unreached.
2: Arbitrarily select a start vertex, u of G. Add u to the tour H.
3: From the unvisited vertices, select a vertex v to visit such that its addition to the tour increases the total existing tour weight the least. Settle ties arbitrarily.
4: Consider v as now visited and remove v from the set of unvisited vertices.
5: Repeat 3 until all vertices are visited.

Output: A Hamiltonian cycle or tour H of locally minimum weight in G.

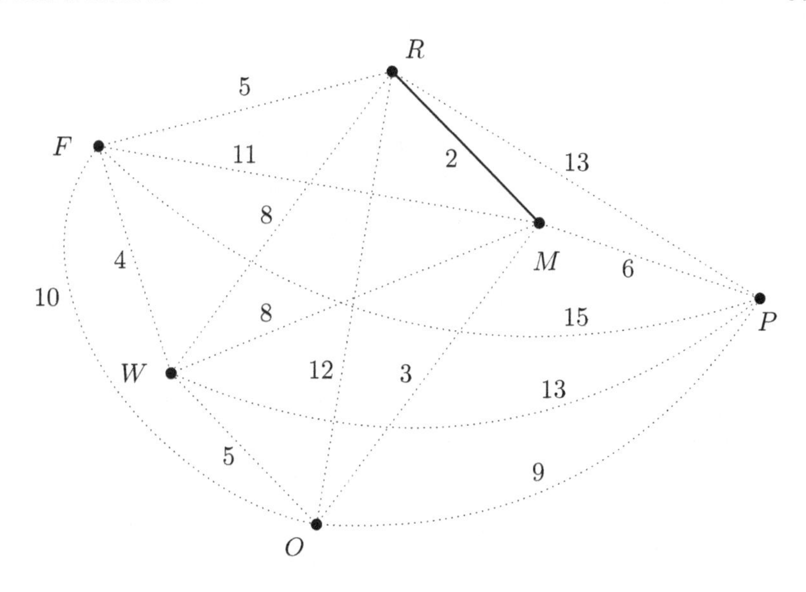

FIGURE 25.19
Weighted future world graph with start at R and 1st insertion in Example 25.3.4.

We illustrate the Closest Insertion Algorithm with a return to the *Future World* example.

Example 25.3.4. *Consider finding a minimum weight tour for the visitor to Future World whose visit starts from the Resort (see Example 25.3.1).*

Start and Iteration 1

We begin at R and currently have a tour empty of edges. The choices for insertions to the tour are considered in the following table:

Unvisited Vertex v	*Resulting Tour*	*Increase* $= 2 * d_{R,v}$	*Include?*
M	$\langle R, M, R\rangle$	4	*Yes*
P	$\langle R, P, R\rangle$	26	*No*
O	$\langle R, O, R\rangle$	24	*No*
W	$\langle R, W, R\rangle$	16	*No*
F	$\langle R, F, R\rangle$	10	*No*

As M is the vertex which is cheapest to insert into the existing tour $H = R$, we insert M into the tour, resulting in an updated $H = R \xrightarrow{2} M \xrightarrow{2} R$. Figure 25.19 illustrates the start from the resort R using the Closest Insertion Algorithm 25.3.3.

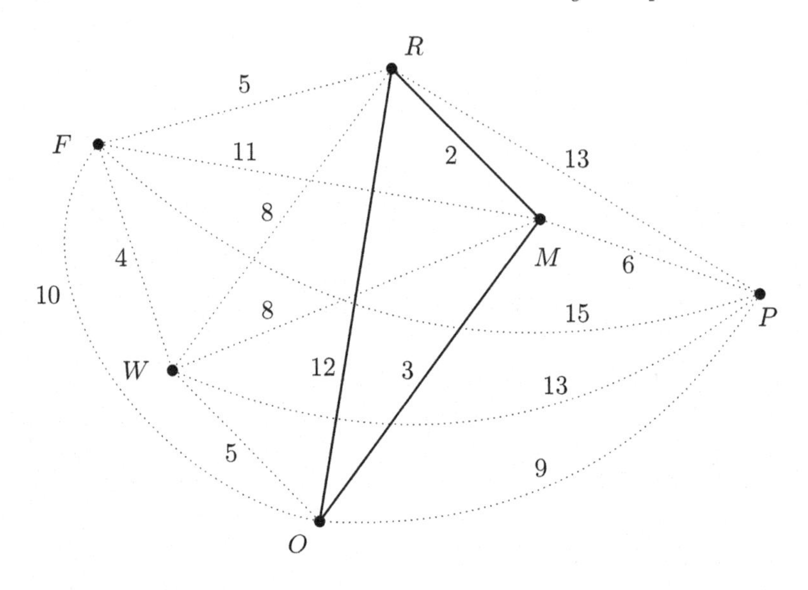

FIGURE 25.20
Future world cheapest insertion iteration 2 in Example 25.3.4.

Iteration 2

The unvisited vertices are now P, O, W, and F and note that we now have two choices where to inserted the new unvisited vertex (yes, they end up being the same tour but in a different direction this time, but for every other iteration this will matter). Possible insertions are:

Unvisited Vertex v	*Resulting Tour*	*Increase* $= d_{i,v} + d_{v,j} - d_{i,j}$	*Include?*
P	$\langle R, P, M, R\rangle$	$d_{R,P} + d_{P,M} - d_{R,M} = 13 + 6 - 2 = 17$	*No*
P	$\langle R, M, P, R\rangle$	$d_{M,P} + d_{P,R} - d_{R,M} = 6 + 13 - 2 = 17$	*No*
O	$\langle R, O, M, R\rangle$	$d_{R,O} + d_{O,M} - d_{R,M} = 12 + 3 - 2 = 13$	*Yes*
O	$\langle R, M, O, R\rangle$	$d_{M,O} + d_{O,R} - d_{R,M} = 3 + 12 - 2 = 13$	*Yes*
W	$\langle R, W, M, R\rangle$	$d_{R,W} + d_{W,M} - d_{R,M} = 8 + 8 - 2 = 14$	*No*
W	$\langle R, M, W, R\rangle$	$d_{M,W} + d_{W,R} - d_{R,M} = 8 + 8 - 2 = 14$	*No*
F	$\langle R, F, M, R\rangle$	$d_{R,F} + d_{F,M} - d_{R,M} = 5 + 11 - 2 = 14$	*No*
F	$\langle R, M, F, R\rangle$	$d_{M,F} + d_{F,R} - d_{R,M} = 11 + 5 - 2 = 14$	*No*

The cheapest insertion (settling the tie arbitrarily) will come by visiting previously unvisited vertex O first then proceeding to M. Thus, at the end of this iteration, we have the updated tour $H = R \xrightarrow{2} M \xrightarrow{3} O \xrightarrow{12} R$. Figure 25.20 illustrates this second iteration.

Iteration 3

The unvisited vertices are now P, W, *and* F *and note that we now have three choices where to inserted the new unvisited vertex into the current tour* $H = R \to M \to O \to R$. *Possible insertions are:*

Unvisited Vertex v	*Resulting Tour*	*Increase* $= d_{i,v} + d_{v,j} - d_{i,j}$	*Include?*
P	$\langle R, P, M, O, R\rangle$	$d_{R,P} + d_{P,M} - d_{R,M} = 13 + 6 - 2 = 17$	*No*
P	$\langle R, M, P, O, R\rangle$	$d_{M,P} + d_{P,O} - d_{M,O} = 6 + 9 - 3 = 12$	*No*
P	$\langle R, M, O, P, R\rangle$	$d_{O,P} + d_{P,R} - d_{O,R} = 9 + 13 - 12 = 10$	*No*
W	$\langle R, W, M, O, R\rangle$	$d_{R,W} + d_{W,M} - d_{R,M} = 8 + 8 - 2 = 14$	*No*
W	$\langle R, M, W, O, R\rangle$	$d_{M,W} + d_{W,O} - d_{M,O} = 8 + 5 - 3 = 10$	*No*
W	$\langle R, M, O, W, R\rangle$	$d_{O,W} + d_{W,R} - d_{O,R} = 5 + 8 - 12 = 1$	*Yes*
F	$\langle R, F, M, O, R\rangle$	$d_{R,F} + d_{F,M} - d_{R,M} = 5 + 11 - 2 = 14$	*No*
F	$\langle R, M, F, O, R\rangle$	$d_{M,F} + d_{F,O} - d_{M,O} = 11 + 10 - 3 = 18$	*No*
F	$\langle R, M, O, F, R\rangle$	$d_{O,F} + d_{F,R} - d_{O,R} = 10 + 5 - 12 = 3$	*No*

The cheapest insertion is obtained by visiting previously unvisited vertex W *first then out of* O *then proceeding home to* R. *Thus, at the end of this iteration, we have the updated tour* $H = R \xrightarrow{2} M \xrightarrow{3} O \xrightarrow{5} W \xrightarrow{8} R$. *Figure 25.21 illustrates this second iteration (previously included edge* OR *appears as a dashed edge).*

Iteration 4

The unvisited vertices are now P *and* F *and note that we now have four choices where to inserted the new unvisited vertex into the current tour* $H = R \to M \to O \to W \to R$. *Possible insertions are:*

Unvisited Vertex v	*Resulting Tour*	*Increase* $= d_{i,v} + d_{v,j} - d_{i,j}$	*Include?*
P	$\langle R, P, M, O, W, R\rangle$	$d_{R,P} + d_{P,M} - d_{R,M} = 13 + 6 - 2 = 17$	*No*
P	$\langle R, M, P, O, W, R\rangle$	$d_{M,P} + d_{P,O} - d_{M,O} = 6 + 9 - 3 = 12$	*No*
P	$\langle R, M, O, P, W, R\rangle$	$d_{O,P} + d_{P,W} - d_{O,W} = 9 + 13 - 5 = 17$	*No*
P	$\langle R, M, O, W, P, R\rangle$	$d_{W,P} + d_{P,R} - d_{W,R} = 13 + 13 - 8 = 18$	*No*
F	$\langle R, F, M, O, W, R\rangle$	$d_{R,F} + d_{F,M} - d_{R,M} = 5 + 11 - 2 = 14$	*No*
F	$\langle R, M, F, O, W, R\rangle$	$d_{M,F} + d_{F,O} - d_{M,O} = 11 + 10 - 3 = 18$	*No*
F	$\langle R, M, O, F, W, R\rangle$	$d_{O,F} + d_{F,W} - d_{O,W} = 10 + 4 - 5 = 9$	*No*
F	$\langle R, M, O, W, F, R\rangle$	$d_{W,F} + d_{F,R} - d_{W,R} = 4 + 5 - 8 = 1$	*Yes*

Hence the cheapest insertion is obtained by visiting previously unvisited vertex F *out of* W *then proceeding to* R. *At the end of this iteration, we have the updated tour* $H = R \xrightarrow{2} M \xrightarrow{3} O \xrightarrow{5} W \xrightarrow{4} F \xrightarrow{5} R$. *Figure 25.22 illustrates this second iteration (previously included edge* WR *appears as a dashed edge).*

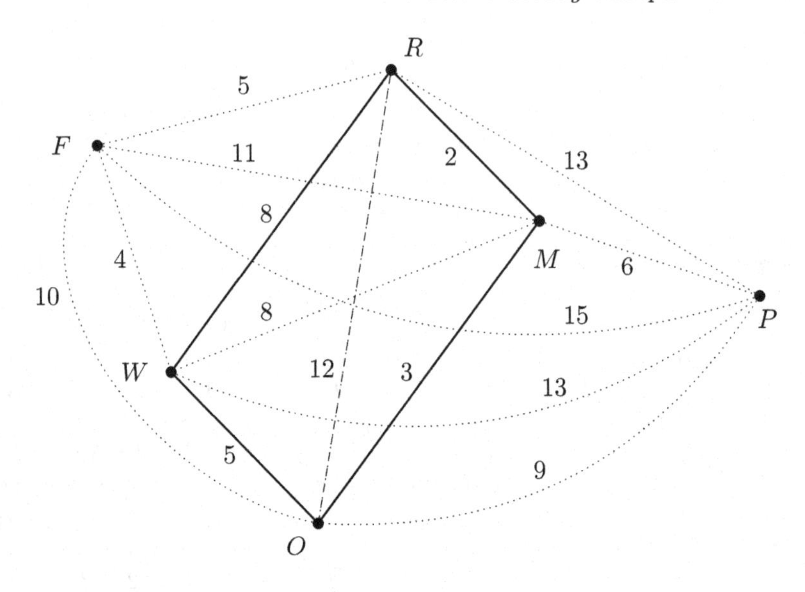

FIGURE 25.21
Future world cheapest insertion iteration 3 in Example 25.3.4.

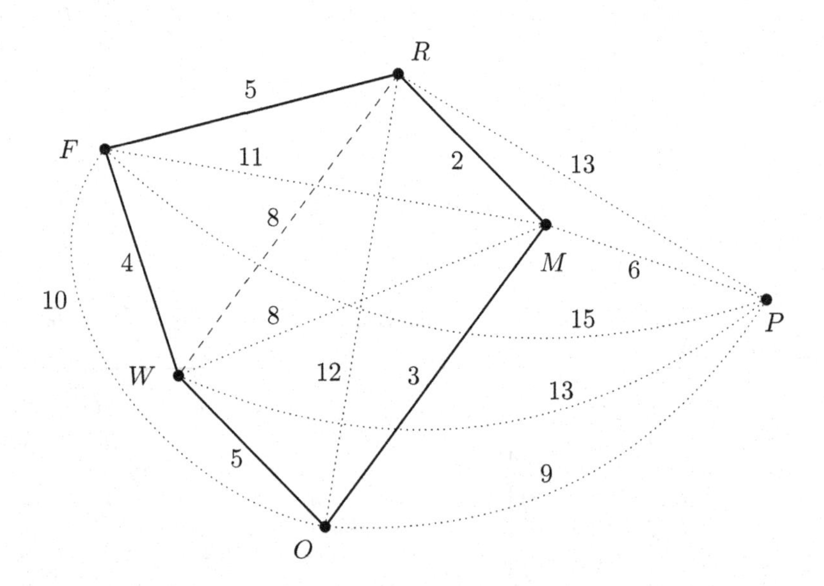

FIGURE 25.22
Future world cheapest insertion iteration 4 in Example 25.3.4.

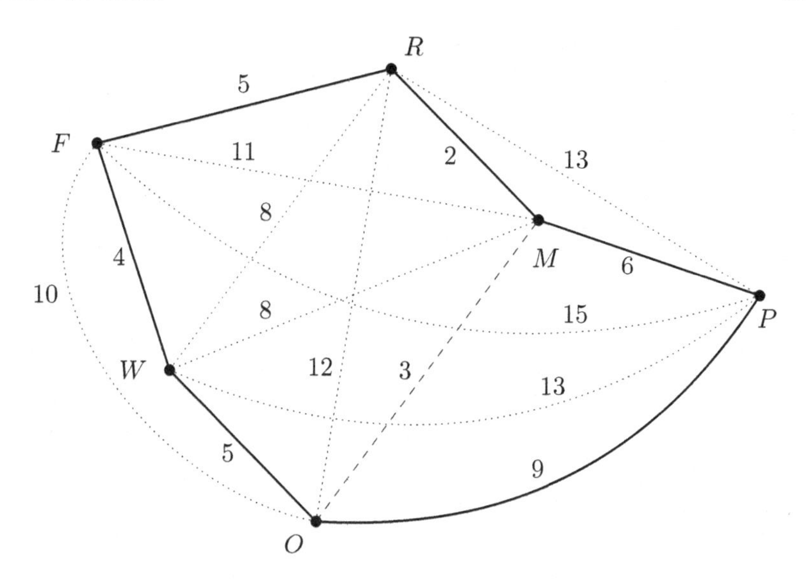

FIGURE 25.23
Future world cheapest insertion final tour in Example 25.3.4.

Iteration 5

One unvisited vertex remains and we must determine how to least expensively insert vertex P into $H = R \to M \to O \to W \to F \to R$.

Unvisited Vertex v	*Resulting Tour*	*Increase* $= d_{i,v} + d_{v,j} - d_{i,j}$	*Include?*
P	$\langle R, P, M, O, W, F, R\rangle$	$d_{R,P} + d_{P,M} - d_{R,M} = 13 + 6 - 2 = 17$	*No*
P	$\langle R, M, P, O, W, F, R\rangle$	$d_{M,P} + d_{P,O} - d_{M,O} = 6 + 9 - 3 = 12$	*Yes*
P	$\langle R, M, O, P, W, F, R\rangle$	$d_{O,P} + d_{P,W} - d_{O,W} = 9 + 13 - 5 = 17$	*No*
P	$\langle R, M, O, W, P, F, R\rangle$	$d_{W,P} + d_{P,F} - d_{W,F} = 13 + 15 - 4 = 24$	*No*
P	$\langle R, M, O, W, F, P, R\rangle$	$d_{F,P} + d_{P,R} - d_{F,R} = 15 + 13 - 5 = 23$	*No*

As there are no remaining unvisited vertices, we are finished and the algorithm has produced the tour $H = R \xrightarrow{2} M \xrightarrow{6} P \xrightarrow{9} O \xrightarrow{5} W \xrightarrow{4} F \xrightarrow{5} R$ (which happens to be the same tour we obtain using the Closest Insertion Algorithm for this example). Figure 25.23 illustrates the tour resulting from this final iteration (previously included edge MO appears as a dashed edge).

25.3.3 The Geometric Heuristic

Recall that, by definition, a (simple) graph has vertices and either an edge between distinct pairs of vertices or not. How the vertices are placed and the edges are drawn are immaterial. There are a few exceptions to this, and the current heuristic is one of them. Suppose the cities a salesperson is to visit are laid out as represented in the following graph; that is, their physical relative positions (in a 2-dimensional representation) are as the vertices appear in the following graph G.

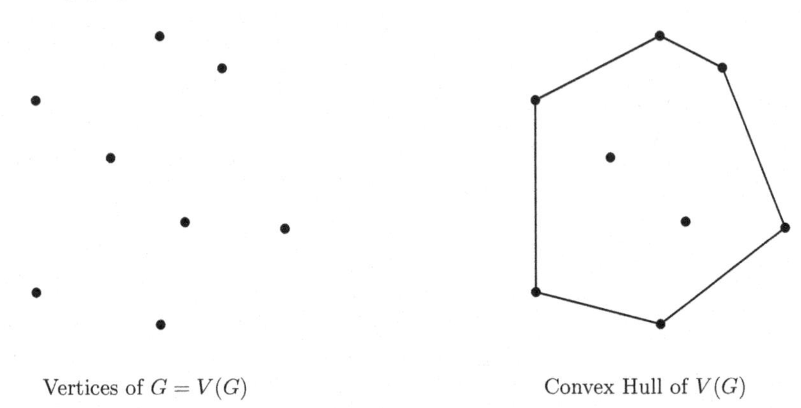

Vertices of $G = V(G)$ Convex Hull of $V(G)$

Suppose further that a direct route exists between every city; that is, that G is a complete graph (the edges have not been included in the figure).

The first step of the *Geometric Heuristic* is to form the convex hull (see part 4 of Definition 16.2.1) of the vertices representing the cities. A nice way to think of the convex hull of a collection of points to consider stretching a rubber band that is large enough to enclose all the points but not so large it loses its tension. Once all the points are in the interior of the rubber band, release it and let it tighten around the points. The convex hull of $V(G)$ is shown in the previous figure.

The next step will be to include the "free vertices" not yet part of the tour.

25.4 For Further Study

Bill Cook's book *In Pursuit of the Traveling Salesman: Mathematics at the Limits of Computation* [10] is an absolutely wonderful account of the problem including its history, applications, and some of the involved mathematics and computer science. It is not too technical and can be read by those outside these disciplines.

Reading both Wikipedia's [64] entry and Section 8.4 of [30] will provide the reader with good visuals a solid understanding of planar representations of the Platonic Solids.

Section 10.1 of [9] has a wonderful exposition of planar graphs and includes proofs of Euler's famous polyhedral equation $V - E + F = 2$ as well as why there are only five regular polyhedra.

25.5 Exercises

Exercise 25.1. *State all the Platonic solids and draw their planar representations. Show that each of the resulting planar graphs is Hamiltonian.*

Exercise 25.2. *Find tours for Example 25.3.2 via the Nearest Neighbor Algorithm starting at*

i. *vertex B*
ii. *vertex C*
iii. *vertex D*

Exercise 25.3. *Find tours for Example 25.3.1 via the Nearest Neighbor Algorithm starting at*

i. *Western World*
ii. *Medieval World*
iii. *Roman World*
iv. *Spa World*
v. *Future World*

Exercise 25.4. *Find tours for Example 25.3.3 via the Closest Insertion Algorithm starting at*

i. *Western World*
ii. *Medieval World*
iii. *Roman World*
iv. *Spa World*
v. *Future World*

Exercise 25.5. *Find tours for Example 25.3.4 via the Cheapest Insertion Algorithm starting at*

i. *Western World*
ii. *Medieval World*
iii. *Roman World*
iv. *Spa World*
v. *Future World*

Exercise 25.6. *How many total tours are considered when applying the Cheapest Insertion Algorithm on a complete graph on n vertices K_n?*

Exercise 25.7. *For the graph in Figure 25.24,*

i. *use the Nearest Neighbor heuristic six times to find an optimal tour by starting at each vertex,*

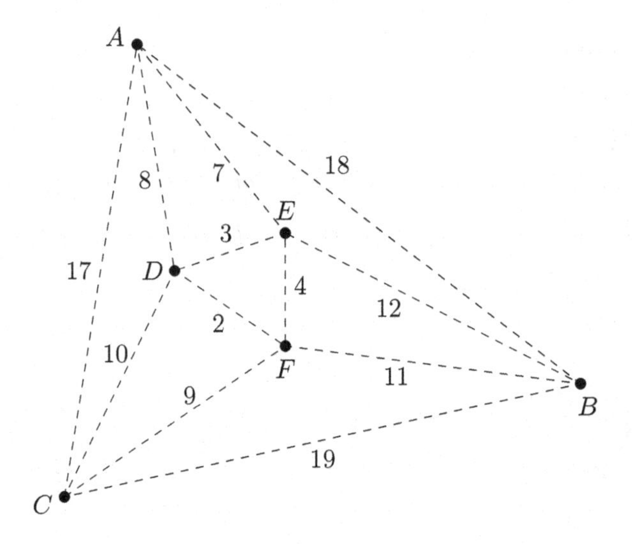

FIGURE 25.24
Weighted octahedral graph in Exercise 25.7.

ii. *use the Closest Insertion heuristic six times to find an optimal tour by starting at each vertex,*

iii. *use the Cheapest Insertion heuristic six times to find an optimal tour by starting at each vertex,*

iv. *and comment on any observations or difficulties that with the algorithms.*

Part VI

Optimization for Data Analytics and Machine Learning

26

Probability

26.1 Introduction

Probability is a common thread linking mathematics and statistics. A mastery of probability will grant someone a strategic command of many competitive games – including a deep, seething hatred of slot machines – as well as the means to analyse what is most likely to happen in more general situations. Some of the tools needed understand the examples and concepts in this chapter are covered elsewhere in this book. Specifically, Chapter 20 introduces the tools of basic counting and Chapter 9 reviews essential parts of calculus. Chapter 20 is especially helpful for understanding the important counting notions of *combinations* and *permutations.*

Since this is an introductory chapter to probability, we will avoid topics like *convolutions*, *joint* and *conditional distributions*, *moment-generating functions*, and many major concepts that are used in a deeper-level study of the subject. There are entire textbooks written on probability while the purpose of this chapter is to build a solid foundation of understanding so that when someone says "We expect [thing x] to happen" or "There is an $X\%$ chance of [thing y] happening", the reader will be equipped to know how those conclusions were reached. The reader will also be able to apply optimization methods learned in other chapters to the ideas presented in this chapter.

We will discuss probabilities in terms of their corresponding *sample space*, which is a special name for all possible *outcomes* (i.e. results) of an *experiment.* A simple example of an experiment is flipping a coin, which has the outcomes "heads" and "tails".

We can determine the *frequency* with which certain outcomes occur by performing an experiment multiple times. If we perform an experiment N times and some outcome of interest occurs X times, we can use the fraction $X/N = p$ to determine the frequency with which that outcome occurs. Often times, this relative frequency p can be used to represent the probability of an outcome occurring.

We will work within a *probability space*, (S, Z, P), which has sample space S, the σ-algebra Z created by that sample space, and a probability function P that takes as its input the outcome of an experiment and outputs the probability of that outcome.

DOI: 10.1201/9780367425517-26

But we are getting ahead of ourselves. It is first necessary to discuss sets and the basic concepts of probability that are defined in terms of them.

26.2 Set Theory

26.2.1 The Vocabulary of Sets and Sample Spaces

Mathematically, a *set* S is a collection of objects. The objects in S are *elements* of the set S and s being an element of S is denoted $s \in S$, which can be alternately be stated as "s is contained in S". There is no repetition of elements in a set, so $\{1, 1, 2, 3\}$ is the same set as $\{1, 2, 3\}$. In probability, the set S of all possible outcomes of an experiment is the *sample space* of the experiment. We illustrate this with a very simple experiment: flipping a coin once.

Example 26.2.1. *As the flip of a coin has two outcomes – heads or tails – we define the sample space of the experiment of a single flip of the coin flip, S, as*

$$S = \{H, T\}.$$

An *event* is any collection of outcomes of the experiment – that is, any subset of the sample space S (including S itself). Thus, an event can contain none, some, or all elements of a sample space. Mathematically, we say that some event A occurs if the outcome of our experiment is in A. This interpretation of events as subsets of the sample space will be useful in the coming sections.

Example 26.2.2. *You wake up and forget what day of the week it is. To uncover this mystery, you conduct an experiment by looking at your phone. We define the sample space as*

$$S = \{x \mid x = M, T, W, R, F, Sat, Sun\}$$

and our desired event E is defined as the day being a weekend; that is

$$E = \{x \mid x = Sat, Sun\}.$$

The outcome of our experiment reveals it is, in fact, Sunday. Sunday is an element in the subset E, and thus we can say that the event E, that today is a weekend, has occurred.

Sets give us a very useful way for us to describe, and later quantify, the outcomes of experiments.

26.2.2 The Algebra of Sets

To introduce how sets operate on each other, we first define some basic relationships.

Definition 26.2.3 (Set Containment, Subset). *A is contained in B if every element in A is also in B which is written as*

$$A \subseteq B$$

We also refer to A as a **subset** of B when it is entirely contained within B.

Definition 26.2.4 (Set Equality). *Sets A and B are equal if and only if they have the same elements. That is*

$$A = B \iff A \subseteq B \text{ and } B \subseteq A.$$

Definition 26.2.5 (Empty Set). *If $A = \{\}$, that is, if A is a set with no elements, then A is referred to as the* empty set *or* null set. *The symbol for the empty set is $\emptyset$; that is, $\{\} = \emptyset$.*

In the context of Probability, $\emptyset$ can be understood as an impossible outcome for an experiment. Note also that the empty set vacuously satisfies the definition of containment for *any* set; that is for any set S (including $\emptyset$) it is always true that $\emptyset \subseteq S$.

Example 26.2.6. *Consider the sets $S_1 = \{(x, y) \mid 0 \le x \le 1, 0 \le y \le 1\}$ and $S_2 = \{(x, y) \mid 0 \le x = y \le 1\}$. That is, S_1 represents all points in a square and S_2 represents all points on the diagonal of that square. In this case, $S_2 \subset S_1$. Also, the set of outcomes where $x + y > 2$ is empty.*

As this section is on the algebra of sets, we should consider operations that can be done to sets. There are three primary set operations: unions, intersections, and complements.

Definition 26.2.7 (Union of Sets). *Given two sets A and B the* union *of two sets is*

$$A \cup B := \{x \mid x \in A \text{ or } x \in B\}.$$

That is, the union of two sets A and B is a new set formed by putting together (without repetition) all the elements from the two individual sets. In particular, an element is in $A \cup B$ if and only if it is in A *or* it is in B.

Definition 26.2.8 (Intersection of Sets). *Given two sets A and B the* intersection *of two sets is*

$$A \cap B := \{x \mid x \in A \text{ and } x \in B\}.$$

That is, the intersection of two sets A and B is a new set formed by collecting (without repetition) all the elements that are common to both of the two individual sets. In particular, an element is in $A \cap B$ if and only if it is in A *and* also in B.

Example 26.2.9. *Suppose* $A = \{1, 3, 5, 7, 9\}$ *and* $B = \{2, 3, 5, 7\}$. *Then* $A \cup B = \{1, 2, 3, 5, 7, 9\}$ *and* $A \cap B = \{3, 5, 7\}$.

We can succinctly write the union or intersection of an arbitrary number of sets in a fashion familiar to calculus students:

$$A_1 \cup A_2 \cup ... \cup A_n = \bigcup_{i=1}^{n} A_i \text{ and}$$
$$A_1 \cap A_2 \cap ... \cap A_n = \bigcap_{i=1}^{n} A_i.$$

Definition 26.2.10 (Complement of a Set)**.** *Suppose S is a set in some* universe of discourse U. *Then the complement of S is the collection of all objects in the universe U that are not in S. That is,*

$$S^c = \bar{S} = \{s \in U \mid s \notin S\}.$$

Example 26.2.11. *Let $\mathbb{P}$ be the collection of primes and regard the universe of discourse to be $\mathbb{Z}^+$, the set of positive integers. Then $\mathbb{P}^c = \bar{\mathbb{P}}$ is the collection of positive composite numbers.*

Unsurprisingly, for any set S in any universe U, $(S^c)^c = S$. We also have that $U^c = \emptyset$ and $\emptyset^c = U$.

Example 26.2.12. *Consider a deck of cards as S. We could define the set of all red cards in terms of their suits as $Red = Hearts \cup Diamonds$. We could state the set of all black cards in terms of its complement as $Black = Red^C$. Notice that we have $Black \cap Red = \emptyset$.*

This last example gives us the opportunity to introduce an important term when working with sets. Sets that have nothing in common are said to be *disjoint.* That is

Definition 26.2.13 (Disjoint Sets)**.** *Sets A and B are* disjoint *if $A \cap B = \emptyset$.*

These disjoint sets can be thought of as events. Examples of disjoint events are easy to think of: e.g. "Today is a weekday" and "Today is Saturday" are two disjoint events. Disjoint events are also said to be *mutually exclusive.*

These operations can be combined to give us some useful laws in the algebra of sets. Each of these laws can be proven and can be found in [48].

Proposition 26.2.14. *For any three sets A, B, and C we have*

$$A \cup B = B \cup A, \quad A \cap B = B \cap A \quad \textit{(Commutativity Laws)}$$

$$A \cup (B \cup C) = (A \cup B) \cup C, \quad A \cap (B \cap C) = (A \cap B) \cap C \quad \textit{(Associativity Laws)}$$

$$A \cap (B \cup C) = (A \cap B) \cup (A \cap C), \quad A \cup (B \cap C) = (A \cup B) \cap (A \cup C) \quad \textit{(Distributivity Laws)}$$

$$(A \cup B)^c = A^c \cap B^c, \quad (A \cap B)^c = A^c \cup B^c \quad \textit{(DeMorgan's Laws)}$$

26.3 Foundations of Probability

Now that we've defined sets and how they operate on each other, we can define a probability function in terms of sets. We'll introduce a special kind of set that contains all possible combinations of outcomes for our experiments across multiple trials, then use that set to define a probability function.

26.3.1 Borel Sets

As a reminder, the overall goal we have is to obtain the relative frequency with which an event or combination of events occurs given a Sample Space S. In the latter case, we need to define a special set for each combination of events. This special set is called a Borel field (also a σ-field) on subsets of S.

Definition 26.3.1 (Borel Field). *A set $\mathcal{B}$ is a Borel field, or a σ-field, on a set S if $\mathcal{B}$ meets the following three conditions:*
1. $\emptyset \in \mathcal{B}$;
2. if $S \in \mathcal{B}$, then $S^c \in \mathcal{B}$; and
3. if a sequence of sets $A_1, A_2, \ldots, A_n \in \mathcal{B}$, then $\bigcup_{i=1}^{n} A_i \in \mathcal{B}$.

The first condition is just that $\mathcal{B}$ is not empty. The second condition requires that $\mathcal{B}$ is closed under countable intersections while the third condition is that $\mathcal{B}$ is closed under countable unions.

Example 26.3.2. *Recall that the sample space for flipping a coin is $S = \{x \mid x = H, T\}$. This sample space is the same whether we do one flip of the coin or two or 100 flips. The sample space for the case of two flips would be $\{x \mid x = (H,H), (H,T), (T,H), (T,T)\}$ which is a σ-field.*

If the reader is familiar with power sets, this definition should look familiar – the power set of an experiment's sample space is always a σ-field.

26.3.2 The Axioms and Basic Properties of Probability

Now that we have a sample space S and a collection of all possible outcomes $\mathcal{B}$, we can define the notion of a probability function, P. There are some common sense assumptions that are made about all probability functions.

Definition 26.3.3 (The Axioms of Probability/Definition of a Probability Function). *For a sample space S and a σ-field $\mathcal{B}$ on S with elements A, we can define a probability function $P : S \to [0,1]$ as a real-valued function on $\mathcal{B}$ where the following three axioms always hold:*

Axiom 1: *$P(A) \geq 0$ for all A in $\mathcal{B}$,*

Axiom 2: *$P(S) = 1$, and*

Axiom 3: *if $A_1, A_2, \ldots, A_n$ is a sequence of sets in $\mathcal{B}$ and $A_i \cap A_j = \emptyset$ for all $1 \leq i, j \leq n$, then $P(\bigcup_{i=1}^{n} A_i) = \sum_{i=1}^{n}(P(A_i))$.*

These axioms make intuitive sense – Axiom 1 states that there cannot be a negative probability. Axiom 2 implies that nothing outside of the sample space can happen (i.e. $P(\emptyset) = 0$) and, thusly, that the Sample Space is an exhaustive list of all possible outcomes. Axiom 3 says that the probability of the union of many disjoint events is equal to the sum of the individual probabilities of each disjoint event. Axiom 3 is listed for a finite amount of events, but it also applies to a theoretically infinite series of disjoint events as well.

The probability function tells us about the relative frequency of events in Z, where the relativity comes from the standardization of the function's output to the reals between 0 and 1. This idea of relative frequency will come up again in a couple sections when we discuss probability distributions.

There are some important consequences of the axioms that make up our next theorem.

Theorem 26.3.4 (Properties of a Probability Function). *Let A and B be events in S, $\mathcal{B}$ a Borel field on S, and $P : \mathcal{B} \to [0,1]$ be a probability function.*

i. $P(\emptyset) = 0$,
ii. $P(A^c) = 1 - P(A)$,
iii. *if* $A \subset B$, *then* $P(A) \leq P(B)$,
iv. $0 \leq P(A) \leq 1$, *and*
v. $P(A \cup B) = P(A) + P(B) - P(A \cap B)$.

Proof. It may seem silly that part i of Theorem 26.3.4 is necessary, but we need to it say that the probability of two disjoint events occurring is zero as in $P(A \cap A^c) = P(\emptyset)$ and intuitively the probability of an impossible event occuring should be zero. The result follows from part ii (proven next) with the fact that $\emptyset = S^c$.

Part ii follows from Axiom 2 of Probability (Theorem 26.3.3) and the relation that $S = A \cup A^c$. Thus, by Axiom 3, $P(S) = P(A) + P(A^c)$ and we have $1 = P(A) + P(A^c)$. The desired result then follows.

Part iii follows from Axioms 1 and 3 of Probability. If $A \subset B$, then $B = A \cup (A^c \cap B)$ thus by Axiom 3, $P(B) = P(A) + P(A^c \cap B)$. But Axiom 1 gives that $P(A^c \cap B) \geq 0$, giving the desired result.

Part iv follows from Axioms 1 and 2 and part iii of this theorem.

The last part is less obvious, but equally true. It follows from Axiom 3 in that we can rewrite $P(B)$ as the sum of two intersections involving A which then gives:

$$P(B) = P(A^c \cap B) + P(A \cap B) \tag{26.1}$$

Similarly, since $A \cup (A^c \cap B) = (A \cup A^C) \cap (A \cup B) = S \cap (A \cup B) = A \cup B$,

$$P(A \cup B) = P(A) + P(A^c \cap B) \tag{26.2}$$

We can solve 26.1 for the common term in both equation, which gives

$$P(A^c \cap B) = P(B) - P(A \cap B). \tag{26.3}$$

Substituting this into 26.2 establishes the desired result. □

Following the logic of last part of the previous theorem, we can even extend Axiom 3 to see that if we consider a sequence of increasing events, $A_1, \ldots, A_n$ (where we define "increasing" to mean that $A_i \subset A_{i+1}$ for all $i = 1, ..., n-1$) the probability of all events occurring is bounded by the sum of the probability of *any* these events occurring. If the A_i are decreasing, ($A_{i+1} \subset A_i, i = 1, ..., n-1$), the probability is bounded by the probability of *all* these events occurring. These ideas are formally stated in the following theorem.

Theorem 26.3.5 (The Continuity Theorem of Probability)**.** *1. If $A_1, A_2, \ldots$ is an increasing sequence of infinitely many events ($A_i \subset A_{i+1}$), then*

$$\lim_{n\to\infty} P(A_n) = P\left(\lim_{n\to\infty} A_n\right) = P\left(\bigcup_{n=1}^{\infty} A_n\right).$$

2. If $A_1, A_2, \ldots$ is a decreasing sequence of infinitely many events ($A_{i+1} \subset A_i$), then

$$\lim_{n\to\infty} P(A_n) = P\left(\lim_{n\to\infty} A_n\right) = P\left(\bigcap_{n=1}^{\infty} A_n\right).$$

The proof of this is a bit topological and is most important for extending the idea of a limit to probability functions. The key insight here is the summary statement that as a space increases through unions (is *an increasing sequence of events*), the probability of an event in that space is bounded by the *union* of all possible components of that space. Conversely, as a space decreases through intersections (a *decreasing sequence of events*) the probability of an event in that space is bounded by the *intersection* of all components in that space.

There is one more important matter fundamental to any study of probability. We could also think of subtracting the $P(A \cap B)$ in v in Theorem 26.3.4

as removing the instances in which we have *double-counted* the probability of the event $A \cap B$ occurring; that is, the probabilities of the results common to both A and B have been counted twice in $P(A) + P(B)$ and therefore must be subtracted once. This is the Probability version of the *Inclusion-Exclusion Principle* in Combinatorics (Theorem 20.2.5) for two sets.

There is a natural extension of this to determining the probability of three events A, B, and C occurring:

$$P(A \cup B \cup C) = P(A) + P(B) + P(C) - [P(A \cap B) + P(A \cap C) + P(B \cap C)] \\ + P(A \cap B \cap C).$$

This time we have removed the probability of occurrences common to all of A, B, and C twice and must make up for that by adding this probability back to the total.

This idea can be generalized to an arbitrary (finite) number of events, resulting in the probability version of *Inclusion-Exclusion Principle* in Combinatorics (20.2.7):

Theorem 26.3.6 (Inclusion-Exclusion Principle for Probability).

$$P(\bigcup_{i=1}^{n} A_i) = \sum_{i=1}^{n} P(A_i) - \sum_{1 \leq i < j \leq n} P(A_i \cap A_j) \\ + \sum_{1 \leq i < j < k \leq n} P(A_i \cap A_j \cap A_k) - \cdots + (-1)^{n-1} P(A_1 \cap \cdots \cap A_n).$$

In this formula, we remove the instances in which we have over-counted by subtracting all combinations of the intersections with an even number of events and add in all events in which we have over-removed the intersections with an odd number of events.

26.4 Conditional Probability

Now that we have defined a probability function, we can quantify probabilities. First, there is an important qualitative matter to consider – whether or not all outcomes of an experiment are equally likely.

26.4.1 Naive Probability

Naive probability is used to categorize probability where it is either the case that outcomes are equally likely or that this case is assumed. In this scenario, computing the probability of an event E given a sample space S is as simple as counting the number of objects in the set E (which is necessarily a subset of S) and dividing it by the number of objects in S. That is,

Definition 26.4.1 (Naive Calculation of Probability).

$$P(E) := \frac{|E|}{|S|}$$

where $|A|$ is the number of elements in the set A; that is, the cardinality *of the set A.*

Example 26.4.2. *A race has 42 runners $R_1, \ldots, R_{42}$ and will award its top prize to whomever finishes first. The sample space of potential outcomes of this experiment (i.e. a winner of the race) is $S = \{x \mid x = R_1, \ldots, R_{42}\}$. Since no further information is known, we will naively assume each runner has as good of a chance as winning the race as each of the rest of the entrants. Let i be a fixed integer such that $1 \leq i \leq 42$ and put $E = \{R_i\}$. Thus for each i, the probability of runner R_i winning the race is $P(E) = \frac{|E|}{|S|} = \frac{1}{42}$.*

Of course, the assumption of equally likely outcomes is often just not the case. Sometimes the different likelihoods chances of occurring may be the result of events depending on a condition that has already been established.

Example 26.4.3. *Suppose that runner A is an Olympian in the prime of their career, runner B works in an office but has a standing desk, and runners C through J are identical octuplet toddlers. Surely in this situation, we would want to re-weigh the probabilities in favor of runner A instead of assuming all race participants are equally likely to win this particular race.*

The above scenario shows why it is important to incorporate extra information (i.e. conditions) to our probability calculations. This is the motivation behind *conditional probability.*

26.4.2 Conditional Probability

We write conditional probabilities as $P(B|A)$, which is read as "the probability of B occurring given that A has already occurred". This value represents the chance that event B occurs given event A has already happened. To develop a way to quantify this, consider the following trivial identities:

$$P(A|A) = 1 \tag{26.4}$$

$$P(B|A) = P(A \cap B \mid A) \tag{26.5}$$

We would like to represent $P(B|A)$ in terms of non-conditional probabilities. From a relative frequency point-of-view, the ratio between $P(A)$ and $P(A \cap B)$ should not change based on whether or not A has occurred – the condition that A occurs should not affect the relative frequency between $P(A)$ and $P(A \cap B)$. Mathematically, that means:

$$\frac{P(A \cap B \mid A)}{P(A|A)} = \frac{P(A \cap B)}{P(A)} \tag{26.6}$$

Substituting into this 26.4.2 and 26.4.2 gives

Theorem 26.4.4 (Calculating Conditional Probability).

$$\frac{P(A \cap B)}{P(A)} = \frac{P(A \cap B \mid A)}{P(A|A)} = P(B|A).$$

It is worth explicitly noting that rearranging this result gives the important result

Theorem 26.4.5 (Multiplication Rule of Probability).

$$P(A \cap B) = P(B|A)P(A).$$

Example 26.4.6. *The 20 runners participating in a certain race have the following characteristics: 10 of the runners have red shirts while the other 10 runners have green shirts, 8 of the 10 runners in red shirts have brown hair, while 4 of the 10 runners in green shirts have brown hair. We know that the winner of the race is a runner that has brown hair.*

The chance that someone with a red shirt won is

$$P(red \mid brown) = \frac{P(red \cap brown)}{P(brown)} = \frac{8/20}{12/20} = \frac{8}{12} = \frac{2}{3}.$$

The chance that someone with a green shirt won is

$$P(red \mid brown) = \frac{P(green \cap brown)}{P(brown)} = \frac{4/20}{12/20} = \frac{4}{12} = \frac{1}{3}.$$

26.4.3 Bayes' Theorem

Conditional probabilities end up being quite useful and can be used to more accurately analyze the likelihood of something happening. For example, what is the probability that drivers on a certain road have an accident while driving that particular road could be useful information regarding a public safety concern, but knowing the probability that a driver on this road has an accident given that they were driving in the range of $50 - 59$ miles per hour will most likely help us develop a useful policy. Often in doing such analysis, we are interested in switching the order of the conditioning; that is, we may know $P(B|A)$ but be interested in $P(A|B)$. The means of reversing a condition is know as *Bayes' Theorem* and establishing it requires the following useful result in probability.

Suppose we have a sample space S, and event A, and a collection of nonempty events $B_1, B_2, .., B_n$ that partition S; that is $S = B_1 \cup \cdots \cup B_n$ with $B_i \cap B_j = \emptyset$ whenever $i \neq j$ (since the B_i satisfy these three conditions, we say they *partition* S). Then

$$A = A \cap S = A \cap \left(\bigcup_{i=1}^{n} B_i \right) \tag{26.7}$$

$$= \bigcup_{i=1}^{n} (A \cap B_i) \quad \text{(by the distributive law)} \tag{26.8}$$

and since the B_i are disjoint, by Axiom 3 of the Axioms of Probability, we have

$$P(A) = \sum_{i=1}^{n} P(A \cap B_i).$$

Thus by the Multiplication Rule of Probability, we have

Theorem 26.4.7 (The Law of Total Probability). *For a sample space S with an event A from S and a collection of events B_i that partition S,*

$$P(A) = \sum_{i=1}^{n} P(A \mid B_i)P(B_i).$$

This should make sense – it is somewhat similar to saying that $P(A) = P(A|S)$. The events B_i, $i = 1, ..., n$, encompass all of S by assumption, so the probability of any event A is the sum of all its conditional probabilities over the partition of its sample space.

Now that we are equipped with the Law of Total Probability, we can establish the very important *Bayes' Theorem.* We establish the simplest version of the theorem first.

Theorem 26.4.8 (Bayes' Theorem for Two Events). *Suppose that A and B are nonempty events in a finite sample space S. Then*

$$P(B|A) = \frac{P(A|B)P(B)}{P(A|B)P(B) + P(A|\bar{B})P(\bar{B})}$$

Proof. We have that $B \cup \bar{B} = S$. Thus by two applications of the Multiplication Rule of Probability as well as using the Law of Total Probability, we have

$$P(B|A) = \frac{P(A \cap B)}{P(A)} = \frac{P(B \cap A)}{P(A)} = \frac{P(A|B)P(B)}{P(A)}$$

$$= \frac{P(A|B)P(B)}{P(A|B)P(B) + P(A|\bar{B})P(\bar{B})}.$$

□

Example 26.4.9. *Suppose that a friend has recently attended a party and is concerned that she may have contracted COVID. To check this, your friend takes an at-home COVID test and tests negative all the while you suspect the test is of dubious quality. Let C represent the event that your friend has COVID, thus $\bar{C}$ will represent the event that your friend does not have COVID. Further, let N represent the event that any individual taking the test tests negative. After contact tracing, you estimate that $P(\bar{C}) = 0.6$ which means that $P(C) = 0.4$. You also have discovered that for this particular test, a negative result is only 80% reliable; that is, $P(N|\bar{C}) = 0.8$. As we are interested in the probability that your friend does not have COVID given her negative test result, we seek $P(\bar{C}|N)$. By Bayes' Theorem,*

$$P(\bar{C}|N) = \frac{P(N|\bar{C})P(\bar{C})}{P(N|\bar{C})P(\bar{C}) + P(N|C)P(C)} = \frac{(0.6)(0.8)}{(0.6)(0.8) + (0.4)(0.2)}$$
$$= 6/7 \doteq 0.8571.$$

Thus, if the given assumptions are reasonable, there is a good chance your friend does not have COVID but enough of a chance that she should be cautious.

Of course, Bayes' Theorem can be extended to situations involving many events.

Theorem 26.4.10 (Bayes' Theorem). *Suppose that A and $B_1, \ldots B_n$ are nonempty events in a finite sample space S and that $B_1, \ldots, B_n$ partition S. Then for each fixed $i = 1, \ldots, n$*

$$P(B_i|A) = \frac{P(A|B_i)P(B_i)}{P(A|B_1)P(B_1) + P(A|B_2)P(B_2) + \cdots + P(A|B_n)P(B_n)}$$
$$= \frac{P(A|B_i)P(B_i)}{\sum_{j=1}^{n} P(A|B_j)P(B_j)}.$$

Example 26.4.11. *Jackson Doyle had an amazing season his first full year playing AAA ball for the Hurdam Bulls. Among other successes, Jackson had a batting average of* 0.304 *and regularly saw only three different kinds of pitches (these can be subdivided further, but let's keep things simple for this example). Of the pitches he faced, he put 344 fastballs in play with 115 being hits, another 134 change ups in play with 37 hits, and 29 hits from the 118 balls he put in play from breaking ball pitches. Jackson had the game-winning walk-off hit in the series-clinching game five of the AAA Intercontinental League Championship Series. What is the probability that the pitch Jackson hit was a change up?*

Solution. Since all the pitches Jackson faced were one of these three pitches and since there is no overlap in the counts, this breakdown of pitches put in play partitions the set of all pitches Jackson put in play. Let's let F the event that a fastball was put in play, C the event that a change up was put in play, and B the event that a breaking ball was put in play. As well, we will let H the event that the batter earned a hit off the ball in play. We then have $P(H|F) = 115/344$, $P(H|C) = 37/134$, and $P(H|B) = 29/118$. Since Jackson faced a total of 596 pitches that he put in play, $P(F) = 344/596$, $P(C) = 134/596$, and $P(B) = 118/596$. Thus

$$\begin{aligned} P(C|H) &= \frac{P(H|C)P(C)}{P(H|F)P(F) + P(H|C)P(C) + P(H|B)P(B)} \\ &= \frac{(37/134)(134/596)}{(115/344)(344/596) + (37/134)(134/596) + (29/118)(118/596)} \\ &= \frac{37}{181} \doteq 0.2044. \end{aligned}$$

■

26.4.3.1 Bayesian Spam Filters

The initial programs to detect unwanted email were built using Bayes' Theorem and worked in the following way. First, a probability threshold is established as a cut-off for what email to let through and what to regard as unwanted. Care should be made in establishing this threshold as minimizing false positives in identifying spam should be prioritized since it is more important to let an unwanted email through than it is to quarantine a wanted email. We will illustrate the idea in the following example:

Example 26.4.12. *2500 emails known to be spam are collected as are 2500 known to not be spam. The word "prince" is appears in 313 of the emails known to be spam but only appears in 22 of the emails known to not be spam. According to the data site Statista.com, the five-year average of global spam volume 2015–2019 (we will leave out the COVID quarantine as it and the following two years may be anomalies) is* 55.85%; *that is, based on empirical data, an arbitrary email is more likely to be Spam than not Spam. We will set the threshold at* 92.5%; *that is, if an incoming message is* 92.5% *or higher likely to be spam, we will quarantine it.*

Suppose an incoming email contains the word "prince". Will the described filter accept or reject this message?

Solution. We will let R be the event that an incoming message contains the word "prince" and S be the event that an incoming message is spam. Thus we have $P(S) = 0.5585$, $P(R|S) = 313/2500$, and $P(R|\bar{S}) = 22/2500$. We seek $P(S|R)$ and by Bayes' Theorem

$$P(S|R) = \frac{P(R|S)P(S)}{P(R|S)P(S) + P(R|\bar{S})P(\bar{S})} \tag{26.9}$$

$$= \frac{\left(\frac{313}{2500}\right)(0.5585)}{\left(\frac{313}{2500}\right)(0.5585) + \left(\frac{22}{2500}\right)(0.4415)} \tag{26.10}$$

$$= 0.9474. \tag{26.11}$$

Since the probability is above our set threshold, we will regard the email as a unwanted and not permit its delviery. ■

There are two interesting things to note before we move on to new material. The usage of "spam" has everything to do with the hilarious Monty Python sketch where a man and woman seeking breakfast are in a small cafe full of Vikings. Every breakfast dish (even "the Lobster Thermidor au Crevettes with a mornay sauce served in a Provencale manner with shallots and aubergines garnished with truffle pate, brandy and with a fried egg on top") has Spam and the woman does not want the canned ham. To make matters worse, the Vikings flood the air when chants of "Spam! Spam! Spam! Spam!" whenever she mentions that she does not want any of it. Honoring the skit, individuals involved in the 1980s MUD community began using the term for unwanted emails and data [17].

Of course, the first Spam filters involved more than one trigger word and thus each email had a vector (e.g. $\langle 1,0,0,1,0\rangle$) associated with it instead of just a 0 or 1. Thus emails were represented by points in multidimensional space and the trick was to find a boundary curve to separate the wanted email from the unwanted email. The same idea is done in machine learning when training AI to identify whether a photo is a picture of a chihuahua or a blueberry muffin. The is the subject of support vector machines and is considered in Chapter 29.

26.4.4 Independence

Of course, in the real world many events are interconnected and whereas other sets of events may have outcomes that do not affect each other. This is such an important matter that we need a term for it.

If the outcomes of events A and B have no influence on each other, then we should expect that

$$P(B|A) = P(B) \text{ and } P(A|B) = P(A). \tag{26.12}$$

We could use this as the definition of when events are *independent*, but it is too much to check. By the definition of conditional probability together with the first part of 26.12:

$$P(B) = P(B|A) = \frac{P(A \cap B)}{P(A)} \tag{26.13}$$

which gives $P(B)P(A) = P(A \cap B)$. A similar calculation with the second part of 26.12 yields the same result. Thus we will use the following as a definition:

Definition 26.4.13 (Independent Events). *Two events A and B are* independent *if:*

$$P(A \cap B) = P(A)P(B).$$

Example 26.4.14. *Consider rolling a standard fair 10-sided die. Let A be the event that an odd number is rolled and B be the event that a prime number is rolled. Thus $A = \{1, 3, 5, 7, 9\}$, $B = \{2, 3, 5, 7\}$, and $A \cap B = \{3, 5, 7\}$. Thus we have*

$$P(A \cap B) = \frac{3}{10} \neq \frac{1}{5} = \left(\frac{1}{2}\right)\left(\frac{2}{5}\right) = P(A)P(B)$$

and that these two events are not independent. Since three out the possible five odd numbers are prime, we can see the lack of independence in a more intuitive way:

$$P(B|A) = \frac{3}{5} \neq \frac{2}{5} = P(B).$$

Example 26.4.15. *Consider the same die from the previous example, but this time we will have C be the event that a multiple of 5 is rolled and B will be the event that a roll results in an odd number. Thus we have*

$$P(C \cap B) = \frac{1}{10} = \left(\frac{1}{5}\right)\left(\frac{1}{2}\right) = P(C)P(B)$$

and that these two events are independent. As with the last example, we can see independence in a more intuitive way: since only one out the possible five odd numbers is a multiple of five:

$$P(C|B) = \frac{1}{5} = P(C).$$

Note in this example that $C \cap B \neq \emptyset$. Thus we have an example that shows that independent events need not be mutually exclusive.

26.5 Random Variables and Distributions

26.5.1 Random Variables

We will now consider the topic that gets the prize for the worst name among anything mathematical. A useful tool in probability is the notion of a *random variable.* Random variables are used to quantify the outcomes of random events and the values of these outcomes will vary. In this sense, we can see why "random" and "variable" are involved. Formally, though

Definition 26.5.1 (Random Variable). *A random variable is a function X that maps every element in a sample space S to a subset real number; that is $X : S \to \Omega$ where $\Omega \subset \mathbb{R}$.*

Thus, a random variable assigns a real number – usually a nonnegative integer – to the outcomes of a an experiment. Note that random variables are incredibly important, but they are not random and they are most certainly not variables.

Example 26.5.2. *Let S be the sample space for the experiment of flipping a coin three times. Let $X : S \to \mathbb{R}$ where X counts the number of heads of the event $s \in S$. Thus*

Event $s \in S$	*$X(s)$*
(T,T,T)	*0*
(H,T,T), (T,H,T), (T,T,H)	*1*
(T,H,H), (H,T,H), (H,H,T)	*2*
(H,H,H)	*3*

Example 26.5.3. *Consider the rolling of two standard six-sided dice. Define the random variable X as the sum of the two dice; that is, $X = x_1 + x_2$, where x_1 is the outcome of the first die roll and x_2 is the outcome of the second die roll. The sample space for this experiment is $S = \{(x_1, x_2) : 1 \leq x_1, x_2 \leq 6\}$ and the space created by the random variable X is $\Omega = \{2, ..., 12\}$.*

Event $s \in S$	*$X(s)$*
(1,1)	*2*
(1,2), (2,1)	*3*
(1,3), (2,2), (3,1)	*4*
(1,4), (2,3), (3,2), (4,1)	*5*
(1,5), (2,4), (3,3), (4,2), (5,1)	*6*
(1,6), (2,5), (3,4), (4,3), (5,2), (6,1)	*7*
(2,6), (3,5), (4,4), (5,3), (6,2)	*8*
(3,6), (4,5), (5,4), (6,3)	*9*
(4,6), (5,5), (6,4)	*10*
(5,6), (6,5)	*11*
(6,6)	*12*

Random variables can be discrete or continuous – that is, they can take either a finite number of values or occupy some interval on the real number line. The previous scenarios have been examples of discrete random variables – there are 2 possible outcomes of a coin flip and 11 possible outcomes of summing two dice rolls.

Observe that the creation of a random variable can simplify the sets on which we need to work. If we flipped a coin 100 times, there would be 2^{100} possible combinations of heads and tails in this experiment's sample space. But if we are only interested in the number of heads and we define a random variable X to be the number of Heads in the 100 coin flips, we get $\Omega = \{0, 1, ..., 100\}$, which is considerably easier to analyze.

26.5.2 Probability Mass and Probability Density Functions

Given a sample space S with a Borel field $\mathcal{B}$, we can define a random variable $X : S \to \Omega$ where Ω is the image of X. We will follow standard convention and use upper case letters such as X for random variables and lower case letters like x to denote particular values X may assume. Thus $x \in \Omega$. Hence an expression like $(X = x)$ will mean the collection of all events in S that are assigned the value x by the random variable X. Thus $P(X = x)$ is the probability that the random variable X takes on the value x. This is illustrated in the next example.

Example 26.5.4. *Consider the experiment of rolling a pair of standard six-sided dice as displayed in Example 26.5.3. Define the random variable X to be the sum of the values of the dice resulting from a roll. Thus $S = \{(a, b) | a, b = 1, 2, 3, 4, 5, 6\}$ and $\Omega = \{2, 3, \ldots, 12\}$. Thus $P(X = 7) = \frac{6}{36} = \frac{1}{6}$.*

We can summarize the probability values of the random variable X as

Event $s \in S$	$X(s) = x$	$P(X = x)$
(1,1)	2	$\frac{1}{36}$
(1,2), (2,1)	3	$\frac{1}{18}$
(1,3), (2,2), (3,1)	4	$\frac{1}{12}$
(1,4), (2,3), (3,2), (4,1)	5	$\frac{1}{9}$
(1,5), (2,4), (3,3), (4,2), (5,1)	6	$\frac{5}{36}$
(1,6), (2,5), (3,4), (4,3), (5,2), (6,1)	7	$\frac{1}{6}$
(2,6), (3,5), (4,4), (5,3), (6,2)	8	$\frac{5}{36}$
(3,6), (4,5), (5,4), (6,3)	9	$\frac{1}{9}$
(4,6), (5,5), (6,4)	10	$\frac{1}{12}$
(5,6), (6,5)	11	$\frac{1}{18}$
(6,6)	12	$\frac{1}{36}$

This table displays the probability of each possible value of the random variable X and is known as the *probability distribution of* X. When a probability distribution of a discrete random variable X can be expressed as a function f, this f is referred to as the *probability mass function* of the random variable X. If the random variable X is continuous, then the function f giving the probability distribution of X is called a *probability density function* of the random variable X.

We will explore some of the more prominent probability distributions after the next topic.

26.5.2.1 Expectation and Variance of a Discrete Random Variable

It is reasonable to consider what is an average outcome of a discrete random variable. This leads to the following definition.

Definition 26.5.5 (Expected Value of a Discrete Random Variable). *Let S be a sample space of an experiment with a random variable $X : S \to \Omega$ where X has the probability mass function f. Then the* expected value of X *is*

$$E[X] := \sum_{x \in \Omega} x f(X) = \sum_{x \in \Omega} x P(X = x) = \sum_{x \in \Omega} x P(x)$$

where $P(x)$ is an accepted representation of $P(X = x)$.

That is, the expected value of a random variable is a weighted average of all its outcomes and is often referred to as the *mean* of the random variable.

Example 26.5.6. *The expected value of the experiment of rolling a standard, fair six-sided die as in Example 26.5.4 is*

$$E[X] = 1\left(\frac{1}{6}\right) + 2\left(\frac{1}{6}\right) + 3\left(\frac{1}{6}\right) + 4\left(\frac{1}{6}\right) + 5\left(\frac{1}{6}\right) + 6\left(\frac{1}{6}\right) = \frac{7}{2}.$$

Knowing the expected value of a random variable can be quite useful.

Example 26.5.7. *Suppose a slot machine cost* \$1 *to play and has the following payouts with the stated probabilities:*

$$P(x) = \begin{cases} 0.0001, & x = \$100 \\ 0.0009, & x = \$10 \\ 0.999, & x = \$0 \end{cases}$$

Which gives the expected winnings as

$$E(X) = (\$100)(0.0001) + (\$10)(0.0009) + (\$0)(0.999) = \$0.019.$$

The expected value of the payouts (i.e. average player's reward) for playing this game on this slot machine is $\$0.019 - \$1 = -\$0.981$. *Thus if a person walked up to this slot machine with* \$10,000 *and played the game 10,000 times, they can expected to walk away with* \$19 *in hand.*

Note that the expectation of a discrete random variable[1] has the following useful property:

Theorem 26.5.8. *(Linearity of Expectation) Let S be a sample space of an experiment with a random variable $X : S \to \Omega$ where X has the probability mass function f giving $P(x)$ for all $x \in S$. Then for any real constants a and b*

$$E[aX + b] = aE[x] + b.$$

Proof.

$$\begin{aligned} E[aX + b] &:= \sum_{x\in\Omega} (ax + b)P(x) && (26.14) \\ &= \sum_{x\in\Omega} axP(x) + \sum_{x\in\Omega} bP(x) && (26.15) \\ &= a\sum_{x\in\Omega} xP(x) + b\sum_{x\in\Omega} P(x) && (26.16) \\ &= aE[X] + b. && (26.17) \end{aligned}$$

□

[1]The same is true for a continuous random variable, but the proof involves an integral instead of a sum and is conditioned upon the existence of the integral.

Let us now consider two experiments that both have five equally likely outcomes. The outcomes of Experiment A are $-2, -1, 0, 1, 2$ where the outcomes of Experiment B are $-100, -50, 0, 50, 100$. Both experiments have the same mean, but the outcomes of Experiment B vary much more greatly from the mean that do the outcomes of Experiment A. This leads to the following notion.

Definition 26.5.9 (Variance of a Discrete Random Variable). *Let X be a discrete random variable with expected value $E[x] = \mu$. Then the* variance *of X is*

$$Var(X) := E[(X-\mu)^2]. \tag{26.18}$$

Thus variance is a measurement of the expected difference of the values of a random variable from its mean. The $(X-\mu)$ is squared so that positive and negative terms do not cancel each other out. The positive square root of the variance of a random variable is called the *standard deviation*, σ, of the random variable. That is,

$$\sigma := +\sqrt{Var(X)}. \tag{26.19}$$

Example 26.5.10. *The probability distributions of the two experiments mentioned before the definition of variance of a discrete random variable are*

Experiment A		*Experiment B*	
x	$P(X=x)$	x	$P(X=x)$
-2	$\frac{1}{5}$	-100	$\frac{1}{5}$
-1	$\frac{1}{5}$	-50	$\frac{1}{5}$
0	$\frac{1}{5}$	0	$\frac{1}{5}$
1	$\frac{1}{5}$	50	$\frac{1}{5}$
2	$\frac{1}{5}$	100	$\frac{1}{5}$
$E[X]=0$		$E[X]=0$	

Though the experiments have the same mean, notice that for Experiment A,

$$\begin{aligned} Var(X) = (-2-0)^2 \cdot \left(\frac{1}{5}\right) + (-1-0)^2 \cdot \left(\frac{1}{5}\right) + (0-0)^2 \cdot \left(\frac{1}{5}\right) \\ + (1-0)^2 \cdot \left(\frac{1}{5}\right) + (2-0)^2 \cdot \left(\frac{1}{5}\right) = 2 \end{aligned}$$

and for Experiment B,

$$\begin{aligned} Var(X) = (-100-0)^2 \cdot \left(\frac{1}{5}\right) + (-50-0)^2 \cdot \left(\frac{1}{5}\right) + (0-0)^2 \cdot \left(\frac{1}{5}\right) \\ + (50-0)^2 \cdot \left(\frac{1}{5}\right) + (100-0)^2 \cdot \left(\frac{1}{5}\right) = 5000. \end{aligned}$$

We will close this subsection by observing some useful properties of variance.

By the linearity of the expectation of a discrete random variable, we have

$$\begin{aligned}
Var(X) &:= E[(X-\mu)^2] && (26.20)\\
&= E[X^2 - 2\mu X + \mu^2] && (26.21)\\
&= E[X^2] - 2\mu E[X] + \mu^2 \text{ by Theorem 26.5.8} && (26.22)\\
&= E[X^2] - \mu^2 \text{ since } E[X] = \mu. && (26.23)
\end{aligned}$$

Thus we have just shown:

Proposition 26.5.11. *Let X be a discrete random variable with expected value $E[x] = \mu$. Then*

$$Var(X) = E[X^2] - \mu^2.$$

We also have:

Theorem 26.5.12. *Let X be a discrete random variable and c be any constant. Then*

i. $Var(X) \geq 0$ *and*
ii. $Var(cX) = c^2 Var(X)$.

The proof of this theorem is Exercise 26.6.

In the context of Optimization, we often seek to maximize or minimize either the expectation of our random variable or the variance. In theoretical cases, this is done via methods such as Maximum Likelihood Estimation (which really just boils down to taking a derivative and setting it equal to zero). With real-world data, this is most often done with methods such as Steepest Descent.

26.5.3 Some Discrete Random Variable Probability Distributions

26.5.3.1 The Binomial Distribution

The *Binomial Distribution* is characterized by an experiment with n identical independent trials where each trial has two outcomes: either a success or a failure. If the probability of a success is p, then the probability of a failure is $q = 1 - p$. Let X be the random variable that counts the number of successes in n such trials. As there are $\binom{n}{k}$ ways to have k successes among the n trials, by the Multiplication Principle of counting (Theorem 20.2.1), we have

$$P(X = k) = \binom{n}{k} p^k (1-p)^{n-k}. \tag{26.24}$$

Equation 26.24 is the probability mass function of the Binomial Distribution. Note that $P(X = k)$ is often written as $P(k)$. The trials of a binomial distribution are often called *Bernoulli trials*.

Example 26.5.13. *You suspect a coin is unfair and devise a series of experiments to test your hypothesis. For the experiments, you have decided to flip the coin 100 times in each experiment and record the results. In the first experiment, the coin landed on heads 68 times out of the 100 flips and it has landed on heads 76 times the second time through. Since*

$$P(X = 68) = \binom{100}{68}(0.5)^{68}(1 - 0.5)^{(100-68)} \doteq 0.0001128 \text{ and} \tag{26.25}$$

$$P(X = 76) = \binom{100}{76}(0.5)^{76}(1 - 0.5)^{(100-76)} \doteq 0.000006293 \tag{26.26}$$

would be the respective probabilities of the results if the coin were fair, it is highly unlikely that this coin in fair.

26.5.3.2 The Geometric Distribution

Closely related to the Binomial Distribution is the Geometric Distribution.

The *Geometric Distribution* is also characterized by an experiment with identical independent trials where each trial has two outcomes: either a success or a failure. Again, if the probability of a success is p, then the probability of a failure is $q = 1 - p$. Let X be the random variable representing how many trials occur in this situation before the experiment reaches a success. For example, if we are again flipping a coin and regard a heads as a success and upon flipping the coin we get $TTTTH$, we would have $X = 5$. Thus the probability mass function of the Geometric Distribution is

$$P(X = k) = (1 - p)^{k-1}p \tag{26.27}$$

where k is any positive integer.

Example 26.5.14. *Suppose we toss a coin that is biased to heads with probability* 0.6. *What is the probability that the first head occurs on the fifth toss?*

Solution. Since the trials of this experiment are identical and independent, by 26.27

$$P(5) = (0.4)^4(0.6) = 0.01536.$$

■

26.5.3.3 The Negative Binomial Distribution

Our next distribution is very similar to the Geometric Distribution and we will explain it by considering the following example:

Example 26.5.15. *Bored one rainy day with a friend, the two of you create a game you call "Snake Eyes". For the rules of the game, the two of you decide to take a single, fair, standard six-sided die and roll repeatedly until you have accumulated two ones. The number of rolls to obtain two ones is counted and the winner is the player who obtained snake eyes in the fewest total rolls. What is the probability of obtaining two ones in exactly 12 rolls of the die?*

Let k be the number of desired successes in n identical independent trials where the successes have probability p. Thus there are $n-k$ failures in these n trials each with probability $1-p$. Since the game ends with the last success, there are $\binom{n-1}{k-1}$ many ways for the first $k-1$ success to occur among the first $n-1$ trials. By the Multiplication Principle of counting (Theorem 20.2.1), the probability of getting k success in the n trials is then

$$P(k) = \binom{n-1}{k-1} p^k (1-p)^{n-k}. \tag{26.28}$$

We thus have the pmf for what is know as the *Negative Binomial Distribution.*

Solution. Thus, the answer to the question posed in the example is

$$P(12, 2) = \binom{12-1}{2-1} \left(\frac{1}{6}\right)^2 \left(\frac{5}{6}\right)^{10} \doteq 0.04935.$$

■

26.5.3.4 The Hypergeometric Distribution

What is often referred to as the *Hypergeometric Distribution* is merely an exercise in using a basic counting principle.

Example 26.5.16. *For a youth baseball fielding practice, 6 baseballs are randomly selected from a ball bucket containing 40 baseballs. It turns out that the bucket contains 2 balls from Major League Baseball (MLB) while the rest are regular youth baseballs. What is the probability that at least one of the MLB balls are used in the practice?*

Solution. By the Multiplication Principle (Theorem 20.2.1), the number of ways that exactly one of the MLB balls is in the mix is $\binom{2}{1}\binom{38}{5}$. Thus the probability that exactly one MLB makes it to the field is

$$P(1\ MLB) = \frac{\binom{2}{1}\binom{38}{5}}{\binom{40}{6}} \doteq 0.26153.$$

Likewise,

$$P(2\ MLB) = \frac{\binom{2}{2}\binom{38}{4}}{\binom{40}{6}} \doteq 0.01923.$$

Thus, by the Addition Principle (Theorem 20.2.3) the probability that at least one MLB ball gets used for fielding practice is 0.2808. ■

Note that each of $P(1\ MLB)$ and $P(2\ MLB)$ is an example of a hypergeometric probability. Thus given N items consisting of M items of type 1 and thus $N - M$ items of type 2, the probability mass function of a hypergeometric probability of selecting n items and having k many of them being type 1 is

$$P(n,k) = \frac{\binom{M}{k}\binom{N-M}{n-k}}{\binom{N}{n}}.$$

As continuous random variables are not considered in this text, we will not present any common distributions of continuous random variables and refer interested readers to any introductory probability textbook.

26.6 Exercises

Exercise 26.1. *Determine the following probabilities where the various experiments involve rolling a standard, fair, six-sided die.*

i) For rolling two dice, let E be the event that the sum of the rolls is even. What is $P(E)$?
ii) When six dice are rolled, what is the probability that all of their numbers are different?
iii) When four dice are rolled, what is the probability that all of their numbers are different?

Exercise 26.2. *If $A \subset B$ and $P(B) > 0$, then what is $P(A|B)$?*

Exercise 26.3. *Use Bayes' Theorem to find the probability that Jackson Doyle's game winning hit in Example 26.4.11 came off a*

i) fastball.
ii) breaking ball.

Exercise 26.4. *In a large high school, there are 600 juniors and 800 seniors. 60% of the juniors play a sport, while 75% of the seniors play a sport. Given that a student picked at random does not play a sport, what is the chance they are a senior?*

Exercise 26.5. *If A and B are independent events, prove that A^c and B^c are also independent.*

Exercise 26.6. *Prove both parts of Theorem 26.5.12.*

Exercise 26.7. *Show that the expectation a binomial random variable with n trials and p probability of success is equal to np. For the same random variable, show that the variance is $np(1-p)$.*

Exercise 26.8. *Show that $E[X(X-1)] = \mu(\mu-1) + \sigma^2$, where $E(X) = \mu$ and $Var(X) = \sigma^2$.*

27

Regression Analysis via Least Squares

27.1 Introduction

One of the most important applications for optimization is statistical modeling, or the analysis of relationship between several variables based on observations. This has wide applicability from stock forecasting, to observations of biological processes, to general data science and analysis. Regression analysis determines the expected probability of observing a dependent variable Y given independent variables X and model parameters (represented as a vector β), and is the most common type of statistical modeling used. Any student or employee who works with data should become comfortable with regression analysis in their tool set. The goal of this chapter is to introduce students to the formulation of regression problems, both linear and nonlinear, and methods of least-squares, including linear, regularized, and nonlinear, which is commonly used to solve them.

27.2 Formulation

Let $X \in \mathbb{R}^{N \times M}$ be a set of data observations or measurements, i.e. each row X_i of the matrix is a data vector containing M points. Let $Y \in \mathbb{R}^P$ be a vector of dependent variables which are related to X by some unknown process. It is not always clear what the relationship between Y and X is, thus we require a **statistical model** to determine the most likely relationship given some modeling assumptions. This is commonly written as $E(Y|X) = f(X, \beta)$ where E is the expectation (in probability) of observing Y given X, f is an unknown statistical model for this expectation which takes in input X, and also has a vector of parameters $\beta = (\beta_1, \beta_2, \dots, \beta_D) \in \mathbb{R}^D$.

In general, it is difficult to ascertain the structure of f or the values of the parameter vector β for real-world problems. Problems like predicting the stock exchange given previous day's numbers do not readily have a statistical model explicitly given. Instead, data analysts usually estimate a candidate f and parameter vector β which practically performs the best on a given dataset

DOI: 10.1201/9780367425517-27

or application space. The remainder of this book will deal with the two main types of f that are considered: linear and nonlinear models.

Linear Regression:

The linear model is the simplest case, but in practice is the most widely used model for regression analysis. This is because the linear assumption is in the parameter coordinate β_i, not in the independent variables x_i. For instance, the expression $y = \beta_1 x_1 + \beta_2 x_1^2$ is still linear in β_i, even though the independent variable x_1 is nonlinear in the second term. Thus most data fitting problems can be expressed as linear regression, with the independent variables x_i being any basis functionals such as polynomials or sinusoids or exponentials. A "best fit" chooses the appropriate parameters β_i to sum up these basis elements to get a final function $f(X, \beta)$ which is linear in β, but can be highly nonlinear in X.

Let $Y \in \mathbb{R}^P$ have same dimension as the number of rows in the data matrix X, i.e. $P = N$, and $\beta \in \mathbb{R}^M$ have the same dimension as the number of points per data row. Then the linear regression problem can be formulated as follows:

$$Y = X\beta + \epsilon \tag{27.1}$$

where $X(i,j) = x_{ij}, i = 1, \ldots, N, j = 1, \ldots, M$ is a matrix of the independent variables. Each row of the matrix (denoted X_i) is the set of independent variables for each observed dependent variable y_i. Note by the definition of matrix multiplication that each $y_i =< X_i, \beta >$ is an inner product of the row of the data matrix with the parameter vector. Finally, ϵ is a noise or distortion vector which corrupts the linear relationship between Y and β. Note that in applications, each y_i is an observation of a process governed by the unknown β and the known independent variables or basis functionals X_i.

27.3 Linear Least Squares

Given the linear regression model expressed by 27.1, the problem lies with solving for the coefficients contained in β. We fit a linear model for the relationship between our independent and dependent variables:

$$y_i = f(X_i; \beta) = \beta_1 x_{i1} + \beta_2 x_{i2} + \ldots \beta_M x_{iM}. \tag{27.2}$$

We want to choose β so that the distances between the regression model and each data point is minimized. Let R be the aggregate of residual errors defined as

$$R = \sum_{i=1}^{N} r_i = \sum_{i=1}^{N} |y_i - f(X_i; \beta)|^2 = ||y - f(X; \beta)||_2^2. \tag{27.3}$$

Minimizing R is one common way of fitting β to our data. The reader may be curious as to why we construct R the way we do. A simple answer is that R aggregates all the Euclidean distances between y and our model $f(X;\beta)$. Why do we have a squared norm? One answer is that it makes our calculation easier (we get rid of a square root) with no impact on the optimal value of β. It is also true that the square comes from a probabilistic interpretation of the regression problem that we discuss later.

In general, we can construct R given Y and X and we want to use it to solve for β. Based on our linear model, we can write

$$y = X\beta. \tag{27.4}$$

Note that sometimes the first column of X can be sometimes augmented with all 1's which allows β_1 to act as an intercept parameter for the regression.

The common approach would be to take the inverse of X and solve for β, right? Not exactly, X is likely to be not square. We need to use the *pseudo-inverse*. We detail the particulars of the pseudo-inverse in Section 27.3.1, but for right now assume we can construct the *right pseudo-inverse* $X^{+} = (X^T X)^{-1} X^T$, then we can solve

$$\beta = X^{+} y$$

and we are done. Easy, right? Unfortunately, our noisy data is unlikely to follow a trend line so nicely as to hold the equality naturally, so we have to turn to a choice of β that minimizes the error. Let us consider the residuals R and choose β in the framework of optimization. We want β such that R is minimized. If we had scalars instead of matrices and vectors, we would do this by taking the equation for R, taking the derivative with respect to β, finding the critical values, and calculating the values for β that give us the global minimum. The thing is: we can still do that here! First, let's rewrite R:

$$\begin{aligned} R &= ||y - X\beta||_2^2 \\ &= (y - X\beta)^T (y - X\beta) \\ &= y^T y - y^T X\beta - \beta^T X^T y + \beta^T X^T X\beta \\ &= y^T y - 2\beta^T X^T y + \beta^T X^T X\beta. \end{aligned} \tag{27.5}$$

Now, like we would with a scalar, let us take the derivative of R with respect to the variable we want to optimize, β:

$$\frac{dR}{d\beta} = -2X^T y + 2X^T X\beta. \tag{27.6}$$

To arrive at the derivative, we need to use the two relationships:

- $d/d\beta(\beta^T X^t y) = X^T y$
- $d/d\beta(\beta^T X^T X\beta) = 2X^T X\beta.$

We leave it to the reader to verify these two derivatives. Now we can find the critical points as the solution to

$$X^T X\beta = X^T y \tag{27.7}$$

which leads to the solution

$$\beta = (X^T X)^{-1} X^T y = X^+ y \tag{27.8}$$

(note that we discuss the invertibility of $X^T X$ in Section 27.3.1; we assume it is invertible for now).

Of course, there is one last question: is this solution for β the global minimum? Our construction for R is convex with respect to β, so the answer is yes! Therefore, we have a solution for β that gives us the guaranteed minimum residual distances between all the points on aggregate. We focused on the trend line case, but notice that this pertains to any number of independent variables.

27.3.1 Pseudo-Inverse

In Chapter 4 we discussed matrix properties in depth. Of that, we covered if a square matrix could be inverted and how that could be done. Now we will provide a brief overview of what we can do when we do not have a square matrix, like in the regression problem.

Consider some matrix $B \in \mathbb{R}^{N,M}$ where $N \geq M$ (this is typically the case for regression). The *right pseudo-inverse* is defined as:

$$B^+ = (B^T B)^{-1} B^T. \tag{27.9}$$

This construction allows us to solve the equation $B^+ B = I$ much like a "normal" inverse. The *left pseudo-inverse* is defined similarly

$$B^+ = B^T (BB^T)^{-1} \tag{27.10}$$

and it lets us solve $BB^+ = I$. For the purposes of this chapter, we will use the right pseudo-inverse exclusively but the rest of this section applies in a similar fashion (basically, instead of working with $B^T B$, the left pseudo-inverse has requirements on BB^T).

The derivation of the pseudo-inverse is straightforward. We are trying to solve the equation

$$VB = I \tag{27.11}$$

for some matrix V. If B is square and full rank, this is easy: V is the inverse B^{-1}. What if B is not, i.e. $N \neq M$? Notice that $B^T B$ *is* square. In fact, $B^T B \in \mathbb{R}^{M,M}$. Thus, let's now try to solve

$$VB^T B = I. \tag{27.12}$$

Again, from Chapter 4, if B^TB is full rank, then this is simple: we let $V = (B^TB)^{-1}$ and we are done. What does this mean for B? Recall that the rank of a matrix is preserved by the transpose operation, i.e.

$$rank(B^T) = rank(B).$$

Next, we need to derive some theory regarding the rank of a product of matrices.

Theorem 27.3.1. *Given matrices $V \in \mathbb{R}^{M,N}$ and $B^{N,M}$, then $rank(VB) \leq \min\{rank(V), rank(B)\}$.*

Proof: Recall that the rank of a matrix is equivalent to the dimension of the space spanned by its columns. Therefore, for some $v \in \mathbb{R}^M$, we can define $u = Bv \in \mathbb{R}^N$ and, by definition, Vu cannot span a space of dimension greater than $rank(V)$.

Similarly, the rank is equal to the dimension of the space spanned by a matrix's row. Since $rank(VB) = rank(B^TV^T)$, we see that for $u = V^Tv$, then, again, B^Tu cannot span a space greater of dimension greater than $rank(B^T) = rank(B)$. These two inequalities give us our desired result.

Corollary 27.3.2. *The rank of B^TB is equal to the rank of B.*

Proof Left to the reader as an exercise.

Putting our knowledge of matrix inversion and these theorems together, we arrive to:

Theorem 27.3.3. *For a matrix $B \in \mathbb{N}, \mathbb{M}$ with $N \geq M$, the right pseudo-inverse $B^+ = (B^TB)^{-1}B^T$ exists if $rank(B) = M$.*

27.3.2 Brief Discussion of Probabilistic Interpretation

As we saw, if we use R as the least square difference between the data and model, we arrive to $\beta = X^+y$. Our argument lies with R being a convenient choice based on the Euclidean norm. There is another way to derive this solution for β that reveals it to be the best *probabilistic* choice if we make some simple assumptions.

To start, let us rewrite our relationship between our regression and dependent variable as

$$y = X\beta + \varepsilon \tag{27.13}$$

where $\varepsilon \sim \mathcal{N}(0, \sigma^2 I)$. That is, assume that we can model the residuals of our model (the error) with *white Gaussian noise*. Errors are typically modeled in this way. Adding this noise means that we are then modeling y as

$$y \sim \mathcal{N}(X\beta, \sigma^2 I). \tag{27.14}$$

It is outside the scope of this book to go into further details, but note that by solving for β in this way – i.e. fitting a normal distribution to the errors of the data – provides the *exact* same solution for β. That is, $\beta = X^+y$ is the *maximal likelihood solution.*

27.4 Regularized Linear Least Squares

As noted in the section above, linear least squares commonly results in trying to solve the following optimization problem:

$$\beta^* = \text{argmin}_\beta ||y - X\beta||_2^2. \tag{27.15}$$

This solution is known analytically as the least squares solution given by the pseudo-inverse $\beta^* = X^+ y$, and minimizes the residual error on average in the data.

However, there are several instances in data analysis and optimization where the minimizing the error on average is not the ideal solution. For instance, if β is sparse or contains only a few non-zero entries, then we want the solution to the optimization to preserve this sparsity condition. Thus, we turn to *regularization* as a tool to help us solve these optimization problems.

Generally, a regularization to the linear least squares problem can be expressed as follows:

$$\beta^* = \text{argmin}_\beta ||y - X\beta||_2^2 + \Gamma(\beta). \tag{27.16}$$

Γ here can be any function, but common choices include: (1) Tikhonov regularization with $\Gamma(\beta) = ||\beta||_2$, (2) sparse ℓ_1 regularization with $\Gamma(\beta) = ||\beta||_1$, and (3) total variation regularization with $\Gamma(\beta) = ||\nabla\beta||_2$. Note that sometimes the regularization term is not differentiable such as the ℓ_1 norm, and thus closed form solutions for these problems do not necessarily exist.

Solving regularized least squares problems is outside the scope of this book, so we recommend interested readers refer to the Convex Optimization book by Boyd and Vandenberghe for more details. Typically such solutions require numerical or iterative methods to solve the optimization efficiently.

28

Forecasting

Forecasting is one of the broadest applications of Statistics. With it, one can predict the sales of a product, the temperature on a certain day, or a disease recurrence rate. It can be used to guess the probability that it will rain, that a person has an irregular heartbeat, or whether or not the global economy will crash in the next 5 years. We can look to predict at a specific point in time, or predict how a trend will change over time.

The data we will be looking at in this chapter are called time series – the same data measured at various points in time. The methods discussed in the previous chapter can be used to build regression models from time series data (where the covariates have measures at the same time as the dependent variables – e.g. yearly a GDP forecast modeled by yearly unemployment, yearly median home value, etc.). Forecasts can be obtained by running regressions over the data set.

But what if we lack reliable covariates? What if we lack any covariates? What if our worldview predetermines that everything is chaos, all correlations are spurious, and that processes can't be predicted using outside measures?

It is in this setting that this chapter exists: we have a dependent variable y with no covariates.

28.1 Smoothing

Suppose you are a little child, incapable of observing anything other than the reading of a thermostat. You would like to know what the temperature will be today. The most natural way to do this would be to look at yesterday's temperature use it as your prediction for today. That is, $\hat{y}_t = y_{t-1}$, where your prediction for today is $\hat{y}_t$ (A qucik note on notation – anything wearing a hat is a prediction.)

Perhaps your lack of observational prowess is offset by diligent bookkeeping. If you have the temperature from the last 7 days, you could also take the average of the last 7 days as your forecast for tomorrow. More generally, you

DOI: 10.1201/9780367425517-28

could take the average of the last n days. This process is known as **smoothing**.

$$\hat{y}_{t+1} = s_t = \frac{\sum_{i=1}^{n} y_t}{n} \tag{28.1}$$

where s_t is the smoothed value at time t, as your estimate for tomorrow's temperature. This is obviously not rocket science (that's the Kalman Smoother), but it illustrates the basic idea behind smoothing – use some average of multiple past data points to predict the future. The simplest way to do that is with the arithmetic average of the last n days.

28.1.1 Exponential Smoothing

Exponential Smoothing (also called Simple Exponential Smoothing by people who like to show off) is similar in concept, but the past data is weighted so that more recent data points are given a larger influence on the prediction. Formula 28.1's numerator changes to $\Sigma c_t y_t$, where the c_t values are typically user-defined and $\Sigma c_t = 1$. A more common way of writing this formula is

$$\hat{y}_{t+1} = s_t = \alpha y_t + (1 - \alpha)s_{t-1} = \alpha y_t + (1 - \alpha)\hat{y}_t \tag{28.2}$$

where s_{t-1} is the smoothed value at time $t - 1$ (so it's just the smoothed value that we computed "yesterday"). We call α a smoothing parameter. All smoothing parameters are bounded between 0 and 1.

It's useful to note here that the formula can be thought of as a weighted average between our newly observed value and our past prediction, where the weighting is done by α (where α is just a re-write the c_ts from the previous formula).

The smoothing parameter can obviously be tuned for precision (model evaluation comes in a later section), but right now, this is almost certainly a waste of time. One of the many reasons for this is trends.

28.1.2 Trends

It is highly reasonable to suggest that things change over time. This can be quantified as a trend. It is equally reasonable to embrace trends and incorporate them into our forecasts. Thus, we need a more complicated model for more accurate forecasting.

Adding a trend into our model makes it look like this:

$$\hat{y}_{t+n} = s_t + nb_t \tag{28.3}$$

where n is the number of days and

$$s_t = \alpha y_t + (1 - \alpha)(s_{t-1} + b_{t-1})$$

and

$$b_t = \beta(s_t - s_{t-1}) + (1 - \beta)b_{t-1}$$

α and β are smoothing parameters, and the trend is determined by the difference in smoothed values $s_t - s_{t-1}$ along with the previous trend. Also note that the b term is scaled by how far in the future you wish to predict.

Note that the two terms are just weighted averages. For the smoothed term, s_t, the average is between our observed value at time t y_t and our predicted value at time t if we take our trend into account $(s_{t-1} + b_{t-1})$. For the trend term, the average is between our newly observed trend value $(s_t - s_{t-1})$ and our old trend b_{t-1}. So this continues the idea of using weighted averages between a newly observed value and our past prediction, but now it's broken up into two components.

It is helpful to think of the s_t term as a baseline expectation for time t, with the adjustment for trend acting as a modifier to make the forecast more precise.

So now we can account for trends! Note that some trends are nonlinear, and must be reduced over time. We call this damping, and it's done by replacing all of the b terms with ϕb where ϕ acts as a damping parameter. As with the smoothing parameter, the damping parameter is bounded between 0 and 1. For clarity, we will assume that all trends are linear for the rest of this chapter.

Others may be concerned with something else: seasonality.

28.1.3 Seasonality

When variations are predictably recurrent in some regular interval of time, we refer to these variations as **seasonal**. Seasonality, then, is the presence of seasonal variations. We denote seasonality with the term a_t. We use m to denote the number of seasons (i.e. intervals) in a given period of time. If you treat each week as a different "season" and are predicting sales based on the previous 28 days, $m = 4$. The term a_{t-m} will be used to denote the seasonal adjustment from m intervals ago.

Suppose you have grown from being a little child and are now in high school. You've loved, lost, and secured a job waiting tables at a local restaurant. Your boss doesn't trust outside help, so they ask the waitstaff to forecast sales for the next month. They know that you're a star student, so they come to you first.

As a good employee, you know that the restaurant is busiest on Friday, Saturday, and Tuesday (due to the extremely popular Taco Tuesday promotion). This is an example of seasonality – during a period of time (7 days, in this case) the values have a predictable pattern of variability (Tuesday's sales are always higher than the days immediately before and after). Adjusting the model for seasonality (and continuing to account for trends), we get

$$\hat{y}_{t+1} = s_t + b_t + a_{t-m} \tag{28.4}$$

where

$$s_t = \alpha(y_t - a_{t-m}) + (1-\alpha)(s_{t-1} + b_{t-1})$$
$$b_t = \beta(s_t - s_{t-1}) + (1-\beta)b_{t-1}$$
$$a_t = \gamma(y_t - s_{t-1} - b_{t-1}) + (1-\gamma)a_{t-m}$$

All we're doing here is adding another weighted average term to the forecast. The term $(y_t - s_{t-1} - b_{t-1})$ is just our observed value (y_t) minus yesterday's baseline forecast s_{t-1} and trend b_{t-1} adjustment The difference between these terms represents our observed value for the amount added due to the season. And a_{t-m} is just a previously predicted value for the seasonal component. So again, just a weighted average term between an observed value and a previously predicted value.

Adjusting for seasonality changes our estimate for our smoothed term s_t by accounting for the value of a_t in some past interval. The trend term, b_t is independent of our seasonality term a_t. This makes intuitive sense – our estimate for an overall trend shouldn't be impacted by recurrent variations.

Your predictions will now take the day of week into account. If you only used exponential smoothing, your forecasts for Wednesday and Friday would be off (too high and too low, respectively).

At this point, the reader should probably be wondering where optimization comes in. The smoothing discussed here is not really optimizable in a statistical sense. The goodness-of-fit is judged by one-number summaries (discussed in a later section) meant to be optimized, but it's done by iterative work rather than maximizing a likelihood function.

Smoothing requires no assumptions, making it an excellent first approach to forecasting. However, making some assumptions about the properties of the data will allow us to extend our forecasting toolkit.

28.2 Stationary Data and Differencing

Stationary time series data has properties that don't vary with time. This is a key assumption of the models in the next section – that the properties of our data are constant with respect to time. When data has this property, specifically a constant mean and variance, is called **stationary** data. We formally define the property of stationarity as:

Definition 28.2.1 (Stationary). *A set of data with finite variance, denoted by y, is said to be stationary if*
1. $\forall t, E(y_t) = \mu$
2. $\forall t, i, j, cov(y_t, y_{t-i}) = cov(y_{t-j}, y_{t-i-j})$

Of course the means and variances will almost never be completely constant – we can perform hypothesis tests to test these assumptions and determine whether any changes over time are significant enough to violate our assumptions. It is also worth noting that this definition of stationarity is referred to as "weak" stationarity – a "strong" version of stationary requires us to confirm that the CDF of every possible subset of y is equivalent to every other possible subset of y. This property is much more difficult to achieve, whereas we have much simpler methods of achieving weak stationarity. Non-stationary data can be transformed into weakly stationary data, most popularly through differencing.

Differencing is as simple as subtracting the value of interest from one period to the next. We will denote the differenced series of data as y'_t, and it is calculated as

$$y'_t = y_t - y_{t-1} \tag{28.5}$$

We also refer to y_{t-1} as the first **lag** of y_t. y_{t-2} would be the second lag of y_t, etc.

Differencing primarily helps stabilize the mean of the time series – that is, it removes the trend's effect on the mean. Consider a product that always sells 100,000 more units than it sold the previous month (i.e. a product with a trend of increasing by 100,000 units sold). The differenced data will have a mean of 100,000 (because every data point in the differenced series is 100,000), whereas the moving average of the un-differenced sales will vary based on which months you look at.

To stabilize the variance, transformations can be used. Logarithms are a popular choice for this.

Note that the stationarity of data doesn't affect how forecastable it is. Consider the product from a couple sentences ago that always sells 100,000 more units than it did the previous month – we could use our most basic smoothing techniques to estimate sales for the next few months with no problem, even though it doesn't have a constant mean over time and is thus non-stationary.

28.2.1 Autocorrelation

To determine stationarity (and thus whether or not differencing is necessary), we look at the autocorrelation function (ACF) of the data. The value of the ACF at time t compared to lag i is equal to

$$r_t = \frac{\Sigma(y_t - \bar{y})(y_{t-i} - \bar{y})}{\Sigma(y_t - \bar{y})^2} = \frac{cov(y_{t-i}, y_t)}{var(y_t)} \tag{28.6}$$

The word autocorrelation can be intimidating, but it's really the same as correlation. Correlation is the relative amount to which some variables e.g. x and y vary together: autocorrelation is the relative amount to which e.g. y_t and y_{t-i} vary together. The "auto" part of autocorrelation just means that the action is performed on itself. So then autocorrelation is just the correlation of a set of data with a past version of itself.

We can determine the whether or not data is stationary with hypothesis tests. One widely used way of checking is to use the ACF and the Ljung-Box test. The test statistic, Q, is used to see if autocorrelation is significantly different from 0 at any time, and $Q \sim {\chi^2}_m$. The formula is:

$$Q = n(n+2) \sum_{i=1}^{h} \frac{(\rho_k)^2}{n-k}$$

Where n is the sample size, ρ_k is autocorrelation at lag k (which means that it is k days away from time t, and h is the number of lags being tested. Basically the sums of the autocorrelation (with more importance given to smaller lag times) and a scaling factor.

There are other tests (such as the Dickey-Fuller, KPSS, and the unit root test) but Ljung-Box is the most widely used. It has the somewhat desirable property of a loosely defined alternative hypothesis (which means that the test has a moderate amount of power in a large number of situations rather than a large amount of power in a moderate number of situations), making it a generally useful first test for stationarity.

Why is a high level of autocorrelation (also known as serial autocorrelation) bad? Technically speaking, autocorrelation violates the assumption that all of our data is independent and identically distributed (which is an assumption of basically every statistical model you will ever encounter). But there is an intuitive understanding as well – When we use optimization methods (such as least squares estimation), we're essentially trying to find a "signal" to the data, some kind of underlying structure. If everything is just a function of what came before it (rather than the manifestation of an underlying pattern/structure), then the signal is much harder to parse. This results in unstable estimates.

There is also a Partial Autocorrelation Function (PACF), which measure the relationship between y_t and y_{t-k} after removing the effects of $y_{t-1}, y_{t-2}, \cdots, y_{t-(k-1)}$. Both the ACF and PACF should be checked to make sure serial autocorrelation doesn't exist in the data. The ACF and PACF are guaranteed to be equal at time $t-1$.

28.3 ARIMA Models

If we assume our data is stationary, we can use other statistical techniques to generate a forecast. A basic technique for this is an **ARIMA** model. A formal definition for an ARIMA will follow after all the component parts are explained.

ARIMA stands for AutoRegressive Integrated Moving Average. ARIMA models represent a different approach to forecasting than smoothing – they capture autocorrelation in the data rather than just trends.

ARIMA models have 3 components: the number of AutoRegressive coefficients (denoted as p), the value of differencing (d, corresponds to the "Integrated" part of the model name), and the number of terms used in the Moving Average (q). The model is written like ARIMA(p,d,q).

It will be useful to compare these models to well-known "classical" time series. The two big classical time series are **white noise** and **random walks**.

Definition 28.3.1 (White Noise). *A set of time series data with finite variance is said to be white noise if 1.* $\forall t, E(y_t) = 0$
2. $\forall t, i, j, cov(y_t, y_{t+i}) = 0$

The values can adhere to a distribution (denoted WN), but are completely random. That is, a sequence of data Z_t is denoted $Z_t \sim WN(0, \sigma_Z)^2$. White noise is basically an unpredictable sequence of values – note the complete lack of covariance among terms implied by condition 3. We generally assume that white noise plays a part in every set of time series data.

A random walk builds on the idea of white noise by incorporating serial autocorrelation. Rather than just an unpredictable sequence of values, a random walk is a sequence where y_t is heavily correlated with past values $y_{t-1}, y_{t-2}, \cdots$ plus a white noise component. A random walk looks like

$$y_t = y_{t-1} + Z_t = \sum_{i=1}^{t} Z_i \tag{28.7}$$

We can also add a term for "drift", which is very similar to accounting for an overall trend in the data. The formula becomes

$$y_t = \delta + y_{t-1} + Z_t = \sum_{i=1}^{t} (\delta + Z_i) = t\delta + \sum_{i=1}^{t} Z_i \tag{28.8}$$

Random walks (even without drift) are not stationary.

28.3.1 Autoregressive Models

Autoregression (AR) is simply regressing a dependent variable on past values of itself (rather than values of independent variables). The value of a series with a mean of zero would be:

$$y_t = \mu + \phi_1 y_{t-1} + \phi_2 y_{t-2} + \cdots + \phi_p y_{t-p} + \epsilon_t = \sum_{i=1}^{p} \phi_i y_{t-i} + \epsilon_t \tag{28.9}$$

where ϵ_t is white noise. This model is called an AR(p) model (an autoregressive model of order p) because it uses p lagged values in the model equation.

Recall that we assume our series is fundamentally just white noise. We can express this mathematically as:

$$y_t - \sum_{i=1}^{p} \phi_i y_{t-i} = \mu + \epsilon_t \tag{28.10}$$

Which is a nice way of showing that our method is just taking out the components that aren't white noise from the value of y_t via the autoregressive components. That is, after accounting for the autoregressive component, our series is just white noise centered around μ.

In an AR(1) model, $-1 < \phi_1 < 1$. If $\phi_1 = 0$ in an AR(1) model, then it implies our data is all white noise (no correlation to the past). If $|\phi_1| = 1$ in an AR(1) model, then it implies our data is a random walk (perfect correlation to the past).

You might remember that in a linear regression, a regression coefficient b can be estimated by simply multiplying the ratio of standard deviations between x and y along with the correlation between x and y, or $b = r\frac{s_y}{s_x}$. It's not quite so simple for autoregressions (though the partial autocorrelation plays a large role in determining ϕ), and a more detailed explanation of how these are calculated can be found in the 4th edition of Shumway and Stoffer (cited, along with other useful resources, at the end of the chapter)

28.3.2 Moving Average Models

Moving Average (MA) models are not the same as obtaining forecasts using smoothing. Moving average models uses past errors from another forecasting model (rather than the past values of the dependent variable, as in an AR model), and looks like

$$y_t = \mu + \theta_1 e_{t-1} + \theta_2 e_{t-2} + \cdots + \theta_q e_{t-q} + \epsilon_t = \sum_{i=1}^{q} \theta_i e_{t-i} + \epsilon_t \tag{28.11}$$

where e_t is equal to $y_t - \hat{y}_t$. This model is called an MA(q) model (a moving average model of order q) because it uses q lagged errors in the model equation.

28.3.3 ARIMA Model Structure

Putting the last few sections together, we can now formally define an ARIMA model.

Definition 28.3.2 (ARIMA). *An ARIMA model of order (p,d,q) is the combination of an autoregressive model of order p and a moving average model of order q, possibly on differenced data. The model is written as*

$$y_t = \mu + \sum_{i=1}^{p} \phi_i y_{t-i} + \sum_{i=1}^{q} \theta_i e_{t-i} + \epsilon_t \tag{28.12}$$

Where $y_t = y'_t$ if the data is differenced. Our biggest concern when fitting ARIMA models is optimizing the value of p,d,and q. Luckily, they're integers! So it's just a matter of brute force and checking some goodness-of-fit statistics! Thus again, we're optimizating using iterative methods.

We'll consider the special (and frequently used) case where d=1.

We can connect these notions of ARIMA models to our smoothers from earlier. Check this out - We can rewrite our simple exponential smoothing formula to be

$$\hat{y}_{t+1} = \alpha y_t + (1-\alpha)\hat{y}_t = \hat{y}_t + \alpha(y_t - \hat{y}_t) = \hat{y}_t + \alpha e_t$$

which can be rewritten to say

$$\hat{y}_t = \hat{y}_{t-1} + \alpha e_{t-1}$$

using $e_{t-1} = y_{t-1} - \hat{y}_{t-1}$, this can be rewritten as

$$\hat{y}_t = y_{t-1} - (1-\alpha)e_{t-1}$$

Now, an ARIMA(0,1,1) model prediction looks like:

$$\hat{y}_t = y_t - y_{t-1} = \theta_1 e_{t-1}$$

which is actually just

$$\hat{y}_t = y_{t-1} + \theta_1 e_{t-1}$$

So they are the same! An ARIMA(0,1,1) model gives the same value as Simple Exponential Smoothing when $\theta = -(1-\alpha)$!

The connection is hard to see, and is not really useful in a practical sense. But is worth showing the overlaps in our methods to show how ARIMA models represent a conceptual extension to our previous methods, even if they don't really look anything alike. It should make sense that a simple exponental smoothing model can be equivalent to a differenced MA model. The AR component simply allows for additional information (specifically, the autocorrelation) to be utilized as well.

28.4 Partial

The reader almost certainly would like to know whether they should use Smoothing or ARIMA techniques. Luckily, there is no definitive answer! Both have their uses, and the effectiveness of a model can be quantified using goodness-of-fit metrics. However, caution and common sense should be use when interpreting models. Smoothing has its uses outside of forecasting as well – many data imputation methods are just some form of exponential smoothing.

28.4.1 Goodness-of-Fit Metrics

Choosing which model is not especially difficult – there are a few metrics we use to measure the accuracy of our forecasting models. We will focus on 2: **BIC** (Bayesian Information Criterion) and **MSE** (Mean Squared Error).

The BIC is similar to AIC (Alkaline Information Criterion), but it adjusts for sample size. Many statistical packages include both AIC and BIC, though BIC is the more useful metric for comparing models with different sample sizes. AICc (basically AIC but with an adjustment to limit the number of parameters) also exists. The formulas for all 3 are

$$AIC = 2k - 2\ell$$

$$AICc = AIC + \frac{2k^2 + 2k}{n - k - 1}$$

$$BIC = k \ln n - 2\ell$$

Where ℓ is the maximized value of the log-likelihood function, k is the number of parameters to be estimated, and n is the sample size.

Lower values indicate a better fit (note that we aim to maximize the likelihood function, which is multiplied by a negative number to compute BIC), and all 3 of these statistics can be negative.

The MSE is covered elsewhere in the book, but the formula is

$$MSE = \frac{\sum_{i=1}^{n} (y_t - \hat{y}_t)^2}{n}$$

Where n is the number of datapoints. We also want to minimize this, but it cannot be negative.

Also note that transforming your data will change your goodness-of-fit metrics – make sure that all models being compared are being run on the same data!

28.5 Exercises

1. What are the benefits of including more terms in an exponential smoother? What are the downsides?

2. Explain what would happen if you used a linear transformation to stabilize the variance of a non-stationary time series. Would it solve the problem of heteroskedasticity?

3. Explain why ACFs that go to zero quickly are more likely to be stationary than ACFs that go to zero asymptotically.

4. Show that the random walk with no drift is nonstationary (Hint: $Var(Y_t) = Var(\sum_{i=1}^{t} Z_i)$).

5. Show that an ARIMA(0,1,0) model is a random walk.

29

Introduction to Machine Learning

29.1 Introduction

In this chapter, we provide a brief introduction to the field of machine learning, a subfield of artificial intelligence and pattern recognition. This area has experienced a lot of growth in the past several years, with companies such as Google, YouTube, and Facebook pioneering the use of this technology in their products and services. Machine learning techniques are becoming ubiquitous in modern technology.

For students in optimization, machine learning becomes a major application area for the techniques taught in this textbook. Most machine learning involves learning from data or examples, which typically involves data analysis in the form of model fitting and optimizing the parameters of a model to best describe (or generate) the data. In this chapter, we briefly survey the field of machine learning to highlight how optimization is used in the central algorithms of the field. We encourage the reader to refer to dedicated sources for machine learning to learn about more advanced applications.

29.2 Nearest Neighbors

One of the simplest tasks for machine learning is for a computer to determine if an example data point is **similar** to another data point in a set of data points. The key word here is similar, and can apply to many contexts. For instance, two images of two different dogs may be similar, even though they are two completely different dogs.

To understand this, its important to understand that data can be **encoded** into features, which are represented as vectors $x \in \mathbb{R}^N$. Sometimes the signal itself can act as its own feature vector (i.e. the audio signal in time), but most often feature vectors are constructed from the original signal or data point.

Thus, we formulate the problem as follows: Given a new test datapoint with feature vector $x \in \mathbb{R}^N$, what is the most similar data point x^* in the training set $[x_1, \ldots, x_M]$? To calculate this, one can solve the following optimization

DOI: 10.1201/9780367425517-29

algorithm:

$$x^* = \text{argmin}_{x_i} d(x, x_i), \tag{29.1}$$

where $d : \mathbb{R}^N \times \mathbb{R}^N \to \mathbb{R}$ is typically a distance function. This simple optimization algorithm is typically solved via brute force search, namely checking all pairwise distances between the new data point and all training datapoints to find the closest match. While simple, this algorithm has seen widespread use in systems such as recommendation systems (e.g. find the movie most like the one the user just watched), and pattern recognition.

29.3 Support Vector Machines

One of the main tasks in machine learning is that of classification: dividing the data into distinct categories. For instance, this could include separating pictures into those taken during the day and those taken at night, or categorizing the breed of dog present in a given picture. Of these classification tasks, one of the most important ones is that of binary classification. This is particularly important for questions about the data which have yes or no answers (e.g. is this a dog in the picture? Is this a spam email?).

Formally, we write that data x_i has a corresponding label $y_i = \pm 1$ where $+1$ is one class, and -1 is the opposite class. The task thus becomes to find a function f such that $f(x) = y$ for all (x, y) pairs in the world. This would be equivalent to the perfect binary classifier.

In this section, we introduce the idea of support vector machines as classifiers that can learn from training data in order to achieve high performance in binary classification. We note that multi-class classification is more difficult, and requires either training separate one-against-all binary classifiers for each class or more advanced joint optimization schemes.

Support Vector Machine: For a given dataset (x_i, y_i), we define a hypothesis as the following function: $h(x_i) = sign(w^T x_i + b)$ where $w \in \mathbb{R}^N$ is a weight vector, b is a scalar bias. Thus h makes a decision on its input on whether it belongs to ± 1. Thus our problem reduces to how to assign weights w and bias b such that our classifier evaluates correctly.

Hyperplanes: We first note a natural underlying geometry to the problem of support vector machines. The equation $w^T x + b$ defines a hyperplane in $\mathbb{R}^N$. This means that this binary classification problem is equivalent to finding a hyperplane which "shatters" or separates the $+1$ and -1 classes. For a given point $a \in \mathbb{R}^N$, we calculate the distance between a and hyperplane in the direction of the weight vector w as $\frac{1}{||w||}(w^T a + b)$.

Not only do we want to find a w, b which separate the points, but we want a hyperplane that most separates the two classes equally. If we consider the distance between two hyperplanes as $1/||w||$, we then seek to minimize the following quadratic programming problem:

$$\text{argmin}_{w,b}\frac{1}{2}w^T w \text{ s.t. } y_i(w^T x_i + b) \geq 1, \quad \forall i. \tag{29.2}$$

What this optimization basically returns is the weight vector w and bias b such that it achieves no mistakes on the training data $y_i(w^T x_i + b) \geq 1$. While this is the most straightforward linear classifier formulation of the SVM, it shows how optimization can be used to perform a binary classification task.

29.4 Neural Networks

Currently, the most common type of machine learning algorithm deployed are neural networks. The networks are a nested composition of linear and nonlinear functions, which transform the input data (or features) into an output that can either be a classification vector or directly regress another set of points. Neural networks are roughly modeled after the electrical signaling that occurs for our brain neurons, where inputs are weighted, aggregated, and cause the neuron to spike after a certain nonlinear threshold is reached. While this is a highly simplistic model of neurons in our brain, assembling multiple of these neurons has resulted in state-of-the-art in pattern recognition tasks.

Optimization is primarily used for the training of the weights/parameter sets of neural networks. Typically, these nonlinear functions are trained using stochastic gradient descent, where the gradient is computed over batches of training data. More explicitly, each weight parameter $w_{new} = w_{old} - \lambda\frac{\partial L}{\partial w}$ is updated with respect to the gradient of a loss function L which typically encodes how far the output of the neural network is with respect to the ground truth annotated label.

29.4.1 Artificial Neural Networks

The simplest model for the neural network is typically called an artificial neural network (ANN) or a perceptron. Given an input $x \in \mathbb{R}^N$, the perceptron computes the following:

$$y = f(\sum_i^N w_i x_i + b) \tag{29.3}$$

where f is typically a non-linear function such as the Heaviside step function or a sigmoidal function, and w is a vector of weights, and b is a bias term. Note this is very similar to the SVM, except with the non-linearity attached.

To train this perceptron, usually the following algorithm is deployed:

Method 29.4.1 (Perceptron Training Algorithm). *Initialize $w = 0$. Then perform the following steps for $i = 1, \dots, T$ or until a desired accuracy rate is achieved:*

1. *Calculate the estimate of the perceptron:* $\hat{y}^t = sign(\sum_i^N w_i^{t-1} x_i^t)$

2. *Query the ground truth value* y^t

3. *If* $y^t \neq \hat{y}^t$*, update* $w^t = w^{t-1} + \lambda y^t \cdot x^t$ *with some weight* λ.

Usually the bias term is folded into the weight term by appending an additional coordinate of 1 to all data points. This method learns to update the weights w to learn the correct classification on all the data points. There have been some proofs of convergence for the perceptron algorithm, although this is out of scope for this book, and requires strong assumptions on the datasets (such as linear separability similar to the linear SVM case).

Further advances in neural networks have introduced multiple layers of perceptrons, or multi-layer perceptrons (MLPs) to form artificial neural networks. These are usually trained using backpropagation, an update rule that uses a loss function to calculate the gradient of the loss function with respect to the weights of the multilayer perceptron. Even more advanced deep learning techniques have introduced more types of neural networks such as convolutional neural networks and recurrent neural networks among others. We advise the reader to read a contemporary text on machine learning and pattern recognition to get a full picture of the state-of-the-art in neural networks.

29.4.2 Exercises

Exercise 29.1. 1. *Let* D *be the following dataset consisting of points* $x_i = (x_1, x_2) \in \mathbb{R}^2$ *with associated label* $y_i = \pm 1$*, written in form* $[(x_1, x_2), y]$. *Let*

$$D = ([(1,1),+1],[(2,-2),-1],[(-1,-1.5),-1],[(-2,-1),-1],\\ [(-2,1),+1],[(1.5,-0.5),+1]).$$

Perform all steps of the perceptron algorithm by hand, going through each point in order (left to right, do not randomize the order). Show all calculations, and plot the hyperplane and all data points (with their labels) for every iteration. Note: you will have to come up with a way to take care of the bias term. Report the final converged solution, and plot it.

2. *Now write a code to perform the perceptron algorithm given any set of data points. Show that you can reproduce the result of the dataset given in the step above.*

3. *Add the minimum number of additional points to make this particular dataset not linearly separable. Show that your code keeps looping and doesn't attain convergence (it suffices to measure the change in* w *from iteration to iteration, and plot this to show that it never converges to zero).*

Exercise 29.2. *1. Using the same dataset as the previous problem,*

$$D = ([(1,1),+1],[(2,-2),-1],[(-1,-1.5),-1],[(-2,-1),-1],\\ [(-2,1),+1],[(1.5,-0.5),+1]),$$

let's solve this problem using SVMs. Write a code to find a SVM which linearly separates the above data. Plot this support vector machine and its (hard) margin. We recommend using pre-existing packages for SVMs in a language like Python or Matlab to solve this problem.

A

Techniques of Proof

The sciences come to what is regarded as "true" by employing the *Scientific Method.* Developed in the 17th century and held as the standard since, the Scientific Method is "a method of research in which a problem is identified, relevant data are gathered, a hypothesis is formulated from these data, and the hypothesis is empirically tested" [16]. This works well for the sciences, but not for Mathematics. Rather than observing phenomena, Mathematics is built on definitions and using logic to derive results from those definitions and other results; i.e. the only way to acquire truth in Mathematics is to employ logic and rules of inference[1].

A.1 Introduction to Propositional Logic

We will work towards proof techniques after first establishing the principles on which they are based. Fundamental to this is the notion of a proposition.

Definition A.1.1 (Proposition). *A* proposition *is a declarative statement that is either true or false.*

Examples of propositions include "Today is Monday", "$3 + 4 = 8$", "Julie is taking Optimization", and "The moon is made of green cheese". The statements "Optimization is a great class" and "The Steelers are the greatest NFL franchise" are not regarded as propositions as they are matters of opinion and not true or false (though I am pretty sure reasonable people would agree with both).

Propositions are usually denoted by lowercase letters p, q, r, etc., and it is possible to perform operations on them. Possible operations are best defined using a *truth table* where all combinations of possible truth values of the included propositions are considered.

Definition A.1.2 (Negation of a Proposition.). *Let p be a proposition. Then the* negation of p *is*

[1]We will not introduce the rules of inference here but rather assume the basics are understood. The curious reader is encouraged to see Section 1.6 of [48] or any logic text.

TABLE A.1
Truth Table for Negation of a Proposition

p	$\neg p$
T	F
F	T

It is acceptable to use to alternate notation $-p$ or $\overline{p}$ for the negation of p.

We may also combine two or more propositions, as in the conjunction and disjunction operators. Conjunctions correspond to the word "and" as in "Take out the garbage and do the dishes". A conjunction is true exactly when both propositions are true and false otherwise.

Definition A.1.3 (Conjunction of Propositions). *Let p and q be propositions. Then the* conjunction of p and q *is*

TABLE A.2
Truth Table for Conjunction Operator on Propositions

p	q	$p \wedge q$
T	T	T
T	F	F
F	T	F
F	F	F

Likewise, a disjunction corresponds to "or" as in "Take out the garbage or do the dishes". A disjunction is false when both propositions are false and true otherwise.

Definition A.1.4 (Disjunction of Propositions). *Let p and q be propositions. Then the* disjunction of p and q *is*

TABLE A.3
Truth Table for the Disjunction Operator on Propositions

p	q	$p \vee q$
T	T	T
T	F	T
F	T	T
F	F	F

An important operator on propositions is the implication operator $\rightarrow$ which is usually referred to as the conditional statement $p \rightarrow q$. This can be read "p implies q" or, more common in spoken language, "If p, then q". The "if" part is called the *hypothesis* or *antecedent* where the "then" part is referred to as the *conclusion* or *consequent*. Implications are also called conditional statements and do not always appear in "if-then" form. For instance, a classic example of a conditional statement is "If it is raining, then it is wet". On of many other ways to say this is "It is wet whenever it rains".

Before we state the truth table for conditional statements let us consider an important scenario. Suppose Optimization professor has on the class' syllabus that the course is only offered Pass/Fail and any semester average that is 70% or higher will be assigned a pass. If a student has a semester average that is at least 70% and the student is awarded with passing, then the professor has honored the conditions in the syllabus. If a student has an average that is less than 70% and does not get a pass, then certainly the professor has not violated the syllabus. If, on the other hand, a student has an average that is less than 70% but the professor assigns a pass, one cannot say that the professor has violated the syllabus. The only condition where the professor would violate this particular agreement is if a student has an average that is at least 70% but is assigned a fail. Thus the truth table for conditional statements is

Definition A.1.5 (Conditional Statement). *Let p and q be propositions. Then the* conditional statement $p \rightarrow q$ *is*

TABLE A.4
Truth Table for Conditional Statements

p	q	$p \rightarrow q$
T	T	T
T	F	F
F	T	T
F	F	T

Other conditional statements are related to a given conditional statement. The *converse* of $p \rightarrow q$ is the conditional statement $q \rightarrow p$. The converse of "If it is raining, then it is wet" is "If it is wet, then it is raining". As well, the *inverse* of $p \rightarrow q$ is the conditional statement $\neg p \rightarrow \neg q$. The inverse of "If it is raining, then it is wet" is "If it is not raining, then it is not wet". Since it can be wet without it currently raining we see that just because a conditional statement is true does not mean its converse or inverse are true. Lastly and important to our work in this chapter, the *contrapositive* of $p \rightarrow q$ is the conditional statement $\neg q \rightarrow \neg p$. The contrapositive of "If it is raining, then it is wet" is "If it is not wet, then it is not raining" which, of course, has to be true since it gets wet every time it rains. This example is not an exception as it is the case that $p \to q$ is logically equivalent to $\neg q \to \neg p$ as we now prove by showing they have the same truth values.

TABLE A.5
Proof That a Conditional Statement Is Logically Equivalent to Its Contrapositive

p	q	$p \to q$	$\neg q$	$\neg p$	$\neg q \to \neg p$
T	T	T	F	F	T
T	F	F	T	F	F
F	T	T	F	T	T
F	F	T	T	T	T

A special situation arrives when a conditional statement is joined with its converse by a conjunction. The result is called a *biconditional* which is usually read "if and only it".

Definition A.1.6 (Biconditional).
The biconditional *statement on two propositions p and q is defined to be*

$$p \iff q := (p \to q) \wedge (q \to p).$$

The truth table for the biconditional is given in Table A.6:

TABLE A.6
Truth Table for the Biconditional

p	q	$p \to q$	$q \to p$	$(p \to q) \wedge (q \to p)$	$p \iff q$
T	T	T	T	T	T
T	F	F	T	F	F
F	T	T	F	F	F
F	F	T	T	T	T

From the table we see that a biconditional on two propositions is true precisely when the propositions have the same truth values and false otherwise.

We conclude this section with introducing language often associated with conditional statements. Given a conditional statement $S \to N$, one can say that S *is sufficient for* N, meaning that if S is true then it must follow that N is true. S does not need to be true for N to be true (see row 3 of Table A.4); hence the term *sufficient.* To be wet outside, it is sufficient that it is raining; though it may also be wet for another reason. Similarly, one can say that given $S \to N$, that N being true is a *necessary condition* for S to be true; meaning S cannot be true without N also being true. Being wet outside is a necessary condition for it raining; it cannot rain and not get wet outside, Another example would be that if Brenna wants to roast marshmellows, then it is necessary for Brenna to have a fire. Roasting marshmellows is only a sufficient condition for having a fire as Brenna may have a fire for some other reason.

A.2 Direct Proofs

As discussed at the beginning of this appendix, mathematical results do not follow from the Scientific Method but are proven. These results go by many names – Theorem, Lemma, Corollary, Observation, or Remark – and have one thing in common: they are all conditional statements. In a direct proof of $p \to q$ an argument is made by starting with the hypothesis then using rules of inference to build to q. This will be demonstrated in the following examples:

Observation A.2.1. *If n is even, then n^2 is also even.*

Proof. The hypothesis of this statement is the condition that n is even. To build a direct proof we start with the hypothesis: n even means that exists some integer k such that $n = 2k$. Thus $n^2 = (2k)^2 = 4k^2 = 2(2k^2)$. Since $2k^2$ is an integer, we have shown that n^2 is twice an integer thus establishing that the conclusion does follow from the hypothesis. □

Of course, it is not necessary for a proof to be so wordy as we show with the next example.

Observation A.2.2. *The product of two odd numbers is an odd number.*

Proof. Suppose m and n are odd; that is, $m = 2s + 1$ and $n = 2t + 1$ for some integers s and t. Then $mn = (2s+1)(2t+1) = 4st+2s+2t+1 = 2(2st+s+t)+1$ which is thus odd. □

A.3 Indirect Proofs

Sometimes a direct proof is quite cumbersome or even impossible. In these situations, it is very helpful to try a different, less direct, approach.

A.3.1 Proving the Contrapositive

Consider proving that of n^2 is odd then n must be odd. A direct approach would start with $n^2 = 2k + 1$ for some integer k but from here it is impossible to conclude that n is of the same form. In this situation it will be helpful to use the result proven in Table A.5: that $p \to q$ is logically equivalent to $\neg q \to \neg p$.

Observation A.3.1. *If n^2 is odd, then n is odd.*

Proof. We instead prove the contrapositive, namely if n is even then n^2 is even. This was established in Observation A.2.1. □

For the next example, recall that a number r is *rational* if there exist integers a and b with $b \neq 0$ such that $r = a/b$. A number that is not rational is said to be *irrational.*

Lemma A.3.2. *If x is irrational, then $1/x$ is irrational.*

Proof. We instead prove the contrapositive: if $1/x$ is rational, then x is rational. If $1/x$ is rational, then there exist integers a and b, $b \neq 0$, such that $1/x = a/b$. As $1/x \neq 0$, we have $a \neq 0$. Hence $x = b/a$ for integers a and b. □

A.3.2 Proof by Contradiction

Sometimes both a direct proof and proving the contrapositive are incredibly difficult or even impossible. When this happens, it will be helpful to have another tool in the toolbox.

Recall from Table A.4 that the only time the conditional statement $p \to q$ is false is when p is true and q is false. A proof by contradiction establishes a result by ruling out this possibility. In particular, a proof by contradiction begins by assuming that the conclusion is false then establishes that under this assumption the hypothesis cannot be true. Consider the following example:

Theorem A.3.3. *Let n be a positive integer. If $n = ab$, then either $a \leq \sqrt{n}$ or $b \leq \sqrt{n}$.*

Proof. Assume for contradiction that the conclusion is false; namely that $a > \sqrt{n}$ and $b > \sqrt{n}$ (this is DeMorgan's Law; see Exercise A.2). Then $n = ab > \sqrt{n} \cdot \sqrt{n} = n$; a contradiction since n cannot be greater than itself. □

The next example is a classic and something taught to elementary school children without proof. It depends on the fact that every positive integer can be written as a product of primes which is also something taught known to elementary school children (we will prove this result in the next section). The statement and proof of the example were known to Euclid 2300 years ago.

Theorem A.3.4. *There are infinitely many primes.*

Proof. Suppose for contradiction that there are only finitely many primes: $p_1, p_2, \ldots, p_k$. Put $M = p_1 p_2 \cdots p_k$ and consider the number $M + 1$. Either $M + 1$ is prime or it can be written as a product of primes. If it is prime, we have a contradiction as it is too big to be any of the primes in our list. If it is not prime, it can be written as a product of primes, but as none of p_1, p_2, ..., p_k divide $M+1$, its prime factors are also not in our list. □

This technique is important enough to merit a third example.

Observation A.3.5. *$\sqrt{2}$ is an irrational number.*

Proof. Suppose for contradiction that $\sqrt{2}$ is rational. Then there exist integers a and $b \neq 0$ such that $\sqrt{2} = a/b$. Without loss of generality[2] suppose a and b have no common factor; i.e. a/b has been reduced to lowest terms. Squaring both sides gives $2 = a^2/b^2$ which leads to $a^2 = 2b^2$. Thus a^2 must be even and by Exercise A.5 a is even, so $a = 2m$ for some integer m. Substituting this value yields $2b^2 = 4m^2$. Thus $b^2 = 2m^2$ and again using Exercise A.5 b must be even. Hence $b = 2n$ for some integer n which contradicts that a/b was in reduced form. □

A.4 The Principle of Mathematical Induction

Imagine standing dominoes in such a way that if one falls over, a chain reaction begins and all the dominoes after the one that falls are also knocked over. This imagery illustrates the process of using the Principle of Mathematical Induction induction as a proof technique.

Theorem A.4.1 (The Principle of Mathematical Induction)**.** *Let $P(n)$ be a statement about a positive integer n. If*

1. *$P(1)$ is true and*

2. *$P(k)$ true $\rightarrow$ $P(k+1)$ true for all positive integers k,*

then $P(n)$ is true for all positive integers[3].

Part 1. of the Principle of Mathematical Induction is referred to as the *base case* or *basis step*. Part 2. is called the inductive step with the assumption that $P(k)$ is true known as the *induction hypothesis*. Our first example is of an proof by induction is a result which is quite useful and was discovered by Gauss as a young schoolboy.

[2]It is important to realize that WLOG can only be used when the assumption does not remove generality. Saying "WLOG suppose $x = 3$" is absurd as assigning a specific value to x removes all generality.

[3]This is a theorem by Exercise A.16. It is the case that neither the Principle of Mathematical Induction nor the Well Ordering Principle can be derived from first principles. One must be accepted as an axiom then the other can be proven.

Theorem A.4.2. *Let n be a postive integer. Then*

$$1 + 2 + 4 + \cdots + n = \frac{n(n+1)}{2}.$$

Proof. Let $P(n)$ be the proposition that $1 + \cdots + n = \frac{n(n+1)}{n}$. Since $1 = 1$, the $P(1)$ base case is established. Now suppose for some positive integer k that $1 + \cdots + k = \frac{k(k+1)}{n}$ (this is the induction hypothesis). Then

$$\begin{aligned}
&1 + 2 + 3 + \cdots + k + (k+1) \\
&\quad = \frac{k(k+1)}{2} + (k+1) \quad \text{(by the induction hypothesis)} && \text{(A.1)} \\
&\quad = \frac{k^2 + k}{2} + \frac{2k+2}{2} && \text{(A.2)} \\
&\quad = \frac{k^2 + 3k + 2}{2} && \text{(A.3)} \\
&\quad = \frac{(k+1)(k+2)}{2}. && \text{(A.4)}
\end{aligned}$$

Since $P(k+1)$ is true when $P(k)$ is true for any positive integer k, the result holds by the Principle of Mathematical Induction. □

From this point on we will refer to the Principle of Mathematical Induction as simply "induction". Induction is a very important proof technique, but as we can see from the proof of Theorem A.4.2, induction is good for establishing that a result is true, but gives no insight as to why it is true. In induction proofs, the result must be discovered by some other means.

For our second example of a proof by induction, we will consider an important sum in Mathematics and Computer Science. This example will be important for a second reason in that though the statement for the Principle of Mathematical Induction states that n is a positive integer, it also holds if n is any nonnegative integer (in other words, we may start at 0 instead of 1 if necessary).

Theorem A.4.3 (Sum of a Geometric Series)**.** *Let a and r be real numbers with $r \neq 1$ and n a nonnegative integer. Then*

$$a + ar + ar^2 + \cdots + ar^n = \frac{ar^{n+1} - a}{r - 1}.$$

Proof. For $n = 0$,

$$\frac{ar^{0+1} - a}{r - 1} = \frac{a(r-1)}{r-1} = a$$

thus establishing the base case. Now suppose that for some nonnegative integer k,

$$a + ar + ar^2 + \cdots + ar^k = \frac{ar^{k+1} - a}{r - 1}.$$

Then

$$\begin{aligned} & a + ar + ar^2 + \cdots + ar^k + ar^{k+1} \\ &= \frac{ar^{k+1} - a}{r - 1} + ar^{k+1} \quad \text{(by the induction hypothesis)} && \text{(A.5)} \\ &= \frac{ar^{k+1} - a}{r - 1} + \frac{ar^{k+2} - ar^{k+1}}{r - 1} && \text{(A.6)} \\ &= \frac{ar^{k+2} - a}{r - 1} && \text{(A.7)} \\ &= \frac{ar^{(k+1)+1} - a}{r - 1}. && \text{(A.8)} \end{aligned}$$

□

There is second form of induction, known as *strong induction*, which can be rather useful. It can be shown that regular induction and strong induction are logically equivalent.

Theorem A.4.4 (Strong Induction). *Let $P(n)$ be a statement about a positive (or nonnegative) integer n. If*

1. *$P(1)$ is true and*
2. *$P(1) \wedge p(2) \wedge \cdots \wedge P(k)$ is true $\rightarrow P(k+1)$ true for all positive (or nonnegative) integers k,*

then $P(n)$ is true for all positive (or nonnegative) integers.

Strong induction is useful when it is not enough to just assume that $P(k)$ is true, as the next example shows.

Theorem A.4.5. *Let $n > 1$ be a positive integer. Then n can be expressed as a product of primes*[4].

Proof. As 2 and 3 are prime, they are trivially a product of primes. Moreover $4 = 2 \cdot 2$. Let $k \geq 4$ and suppose that k can be written as a product of primes. If $k+1$ is prime, we are done. If it is not prime, then there exist integers a and b with $2 \leq a, b < k$ such that $k+1 = ab$. By the (strong) induction hypothesis, a and b can be written as the product of primes, making $k+1$ expressible as a product of primes. □

[4]This is the *Fundamental Theorem of Arithmetic* when it is included in the conclusion that the factorization is unique. As this example is to illustrate strong induction, we will not clutter the proof with a uniqueness argument.

A.5 Exercises

Exercise A.1. *Use a truth table to show that* $\neg(p \wedge q)$ *is logically equivalent to* $\neg p \vee \neg q$ *(This is one of DeMorgan's Laws).*

Exercise A.2. *Use a truth table to show that* $\neg(p \vee q)$ *is logically equivalent to* $\neg p \wedge \neg q$ *(This is the other DeMorgan's Law).*

Exercise A.3. *Use a truth table to show that* $p \to q$ *is logically equivalent to* $\neg p \vee q$.

Exercise A.4. *Prove that if* n *is odd, then* n^2 *is odd.*

Exercise A.5. *Show that* n^2 *even implies* n *must be even.*

Exercise A.6. *Prove that if* $n^2 + 4$ *is odd, then* n *is odd.*

Exercise A.7. *Prove that if* $n^2 + 3$ *is odd, then* n *is even.*

Exercise A.8. *Show that the square root of a positive irrational number is irrational; that is, if* $x > 0$ *is irrational, then* $\sqrt{x}$ *is also irrational.*

Exercise A.9. *Prove that the sum of an irrational number and a rational number must be irrational.*

Exercise A.10. *Show that for a positive integer* n, $n < 2^n$.

Exercise A.11. *Prove for a positive integer* n *that* $2^n < n!$ *where* $n! := n \cdot (n-1) \cdot (n-2) \cdots 3 \cdot 2 \cdot 1$.

Exercise A.12. *Show that* $n^3 - n$ *is divisible by 3 whenever* n *is a positive integer.*

The next three exercises illustrate that for induction arguments, all that is really needed is a starting point.

Exercise A.13. *Prove for a positive integer* $n > 1$ *that* $n! < n^n$.

Exercise A.14. *Show that for an integer* $n > 4$, $n^2 < 2^n$.

Exercise A.15. *Prove for a positive integer* $n > 6$ *that* $3^n < n!$.

Exercise A.16. *The* Well Ordering Principle *states that any nonempty subset of positive integers has a least element. Prove that the Principle of Mathematical Induction is logically equivalent to the Well Ordering Principle; that is* $PMI \iff WOP$.

Exercise A.17. *Prove that (regular) induction is logically equivalent to strong induction; that is,* $PMI \iff$ *strong induction.*

B

Useful Tools from Analysis and Topology

B.1 Definitions

Definition B.1.1 (Euclidean Distance). *Let $x = (x_1, x_2, \ldots, x_n)$ and $y = (y_1, y_2, \ldots, y_n)$ be points in $\mathbb{R}^n$. Then the* Euclidean distance *between x and y, denoted $||x - y||$, is*

$$||x - y|| := \sqrt{\sum_{i=1}^{n}(x_i - y_i)^2}.$$

Many distances (metrics) exist in Mathematics, and the Euclidean Distance is the name given to the distance based upon applying the Pythagorean Theorem (Theorem B.2.1) to determining how far apart points are in $\mathbb{R}^n$.

Definition B.1.2 (ϵ-neighborhood). *Let x be a point in space and ϵ a positive real number. Then an* ϵ-neighborhood, *written $N_\epsilon(x)$ is*

$$N_\epsilon(x) := \{y \in \mathbb{R}^n \mid ||x - y|| < \epsilon\}.$$

It is now natural to define global and local extrema, but we note that these definitions were stated in Section 2.1. We know introduce some important topological ideas.

Definition B.1.3 (Interior, Exterior, and Boundary). *Let R be a subset of a universal set Ω.*

- *The Guarantee of Integer Solutions in Network Flow Problems The* interior *of R, denoted $int(R)$, is*

$$int(R) := \{x \in R \mid \exists \epsilon > 0 \text{ such that } N_\epsilon(x) \subseteq R\}.$$

- *The Guarantee of Integer Solutions in Network Flow Problems The* exterior *of R, denoted $ext(R)$, is*

$$ext(R) := \{x \in \Omega \mid \exists \epsilon > 0 \text{ such that } N_\epsilon(x) \subseteq \Omega \backslash R\}.$$

DOI: 10.1201/9780367425517-B

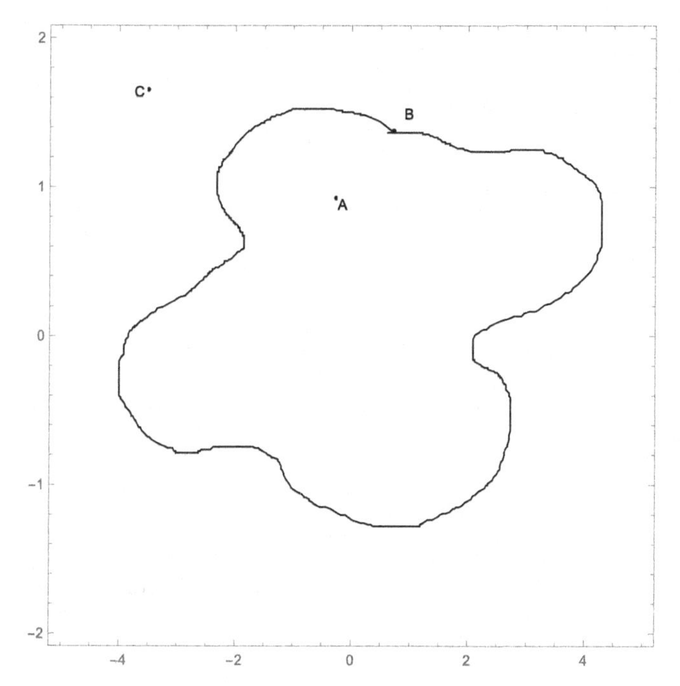

FIGURE B.1
Point A inside a region R, C outside R, and B on the boundary of R.

- *The Guarantee of Integer Solutions in Network Flow Problems x is a* boundary point *of R if and only if for all $\epsilon > 0$ there exist $y_1, y_2 \in N_\epsilon(x)$ such that $y_1 \in R$ but $y_2 \in \Omega \backslash R$. The* boundary *of R is the collection of all boundary points of R.*

In Figure B.1, A is an interior point of R, C an exterior point of R, and B a boundary point of R.

Definition B.1.4 (Metric). *A* metric *on a set X is a function*

$$d : X \times X \to \mathbb{R}$$

such that

1. *$d(x, y) \geq 0$ for all $x, y \in X$ with $d(x, y) = 0$ iff and only if $x = y$ (definiteness),*
2. *$d(x, y) = d(y, x)$ for all $x, y \in R$ (homogeneity), and*
3. *$d(x, z) \leq d(x, y) + d(y, z)$ for all $x, y, z \in X$ (the **triangle inequality**).*

B.2 Useful Theorems

From high school Geometry, we have

Theorem B.2.1 (Pythagorean Theorem). *Let a and b be the lengths of the legs of a right triangle with its hypotenuse having length c. Then*

$$a^2 + b^2 = c^2.$$

A more generalized version of the Pythagorean Theorem is

Theorem B.2.2 (The Law of Cosines). *Let a, b, and c be the lengths of the sides of a triangle with c opposite angle θ. Then*

$$c^2 = a^2 + b^2 - 2ab\cos\theta.$$

Note that if c is the length of the hypotenuse of a right triangle, then $\cos\theta = 0$ and the Law of Cosines reduces to the Pythagorean Theorem.

Theorem B.2.3 (The Triangle Inequality). *Let $\mathbf{x}$ and $\mathbf{y}$ be in $\mathbb{R}^n$. Then*

$$||\mathbf{x} + \mathbf{y}|| \leq ||\mathbf{x}|| + ||\mathbf{y}||. \tag{B.1}$$

Note that if $n = 1$, i.e. $\mathbf{x}$ and $\mathbf{y}$ are real numbers, then B.1 reduces to

$$|x + y| \leq |x| + |y|.$$

Theorem B.2.4 (Cauchy-Schwarz Inequality). *For all vectors u and v of an inner product space*

$$|\langle \mathbf{u}, \mathbf{v}\rangle|^2 \leq \langle \mathbf{u}, \mathbf{u}\rangle \cdot \langle \mathbf{v}, \mathbf{v}\rangle \tag{B.2}$$

where $\langle\cdot,\cdot\rangle\langle\cdot,\cdot\rangle$ is the inner product.

Bibliography

[1] Kendall Atkinson and Weimin Han; *Elementary Numberical Analysis,* 3rd edition, John Wiley and Sons (2004).

[2] N.L. Biggs, E.K. Lloyd, and R.J. Wilson; *Graph Theory 1736–1936*, Clarendon Press, Oxford, 1976.

[3] Robert G. Bland, *New Finite Pivoting Rules for the Simplex Method*, Mathematics of Operations Research, vol. 2, no. 2 (1977), 103–107.

[4] Otakar Borůvka, *O jistém problému minimélném (About a certain minimal problem).* Préce Mor. Přérodověd. Spol. V Brně III (in Czech and German), 3 (1926), 37–58.

[5] Miklós Bóna, *Introduction to Enumerative and Analytic Combinatorics*, 2nd edition, CRC Press (2016).

[6] Anthony Brabazon, Michael O'Neill, and Seán McGarraghy, *Natural Computing Algorithms*, 1st edition, Springer Publishing Company (2015).

[7] Anthony Brabazon and Seán McGarraghy, *Foraging-Inspired Optimisation Algorithms*, 1st edition, Springer Publishing Company (2018).

[8] Corinne Brucato Bauman, *The Traveling Salesman Problem*, Master's Thesis, University of Pittsburgh, D-Scholarship@Pitt, `http://d-scholarship.pitt.edu/id/eprint/18770` (2013).

[9] Gary Chartrand, Linda Lesniak, and Ping Zhang, *Graphs and Digraphs*, 6th edition, CRC Press (2016).

[10] William J. Cook, *In Pursuit of the Traveling Salesman: Mathematics at the Limit of Computation*, Princeton University Press, 2012.

[11] Thomas H. Cormen, Charles E. Leiserson, Ronald L. Rivest, and Clifford Stein; *An Introduction to Algorithms,* 3rd edition, MIT Press (2009).

[12] Gérard Cornuéjols, *Revival of the Gomory Cuts in the 1990's*, Gérard's notes, Carnegie Mellon University, "`https://www.andrew.cmu.edu/user/gc0v/webpub/gomory.pdf`"

[13] R.J. Dakin, *A Tree-Search Algorithm for Mixed Integer Programming Problems*, Comp. J., 8, 250–255 (1964).

[14] G.B. Dantzig, D.R. Fulkerson, and S.M. Johnson, *Solution of a Large-Scale Traveling-Salesman Problem*, Journal of the Operations Research Society of America, vol. 2, no. 4 (1954).

[15] Dictionary.com `https://www.dictionary.com/browse/heuristic?s=t`

[16] The Little Oxford Dictionary, 6th edition, Oxford University Press (1986).

[17] Simon Hill, *Why do we call it spam? Blame spiced ham shoulder, Monty Python, and Usenet*, digitaltrends, *https://www.digitaltrends.com/computing/why-junk-email-is-spam/*, February 8, 2015.

[18] Yefim A. Dinitz, *An Algorithm for the Solution of the Problem of Maximal Flow in a Network with Power Estimation*, Doklady Akademii Nauk SSSR (Proceedings of the USSR Academy of Sciences), vol. 11 (1970), 1277–1280.

[19] Marco Dorigo, Vittorio Maniezzo, and Alberto Colorni, *Ant System: Optimization by a Colony of Cooperating Agents*, IEEE Transactions on Systems, Man, and Cybernetics—Part B, vol. 26, no. 1 (1996), pp. 29–41.

[20] Mario Dorigo and Luca Maria Gambardella, *Ant Colony System: A cooperative Learning Approach to the Travelling Salesman Problem*, IEEE Transactions on Evolutionary Computation, vol. 1 (1997), pp. 53–66.

[21] Jack Edmonds and Richard Karp, *Theoretical Improvements in Algorithmic Efficiency for Network Flow Problems*, Journal of ACM, vol. 19 (1972), 248–264.

[22] Peter Elias, Alex Feinstein, Claude E. Shannon, *A Note on the Maximum Flow Through a Network*, IRE Transactions on Information Theory, vol. 2, no. 4 (December 1956), 117–119.

[23] Leonhard Euler, *Solution of a Curious Question which does not seem to have been Subjected to any Analysis*, Mémoires de l'Academie Royale des Sciences et Belles Lettres, Annêe 1759, vol. 15, Berlin (1766), 310–337.

[24] Lester Randolf Ford, Jr. and Delbert Ray Fulkerson, *Maximum Flow Through a Network*, Canadian Journal of Mathematics, vol. 8 (1956), 399–404.

[25] Lester Randolf Ford, Jr. and Delbert Ray Fulkerson, *A Simple Algorithm for Finding Maximal Network Flows and an Application to the Hitchcock Problem*, Canadian Journal of Mathematics, vol. 9 (1957), 210–218.

[26] L.R. Foulds, *Optimization Techniques*, Springer-Verlag (1981).

[27] Futureworld, Director: Richard T. Heffron, American International Pictures (1976), Film.

[28] Rivka Galchen, *The Mysterious Disappearance of a Revolutionary Mathematician*, New Yorker, May 9, 2022.

[29] R.E. Gomory, *Outline of an Algorithm for Integer Solutions to Linear Programs*, Bull. Am. Math. Society 14 (1958), 275–278.

[30] Jonathan L. Gross and Jay Yellen, *Graph Theory and its Applications*, 2nd edition, CRC Press (2006).

[31] Jonathan L. Gross, Jay Yellen, and Ping Zhang; *The Handbook of Graph Theory,* 2nd edition, CRC Press (2013).

[32] R.J. Hyndman and G. Athanasopoulos, *Forecasting: Principles and Practice*, 2nd edition, OTexts: Melbourne, Australia (2018). `OTexts.com/fpp2` Accessed on 05/13/20

[33] Alan J. Hoffman and David Gale, *Appendix* to I Heller and CB Tompkin, *An Extension of a Theorem of Dantzig's*, in HW Kuhn and AW Tucker's (editors) *Linear Inequalities and Related Systems*, Annals of Mathematics Studies, 38, Princeton (NJ): Princeton University Press, (1956) pp. 247-–254.

[34] Alan J Hoffman and Joseph B Kruskal, *Integral Boundary Points of Convex Polyhedra*, in HW Kuhn and AW Tucker's (editors) *Linear Inequalities and Related Systems*, Annals of Mathematics Studies, 38, Princeton (NJ): Princeton University Press, (1956) pp. 223–246.

[35] E.E. Holmes, M.D. Scheuerell, and E.J. Ward, *Applied Time Series Analysis for Fisheries and Environmental Data.* NOAA Fisheries, Northwest Fisheries Science Center, 2725 Montlake Blvd E., Seattle, WA 98112. `https://atsa-es.github.io/atsa-labs/`

[36] Vojtěch Jarník, *O jistém problému minimálním (About a minimal problem)*, Práce Mor. Přírodvěd. Spol. v Brně, Acta Societ. Scient. Natur. Moravicae 6 (1930), 57–63.

[37] Richard M. Karp, *Reducibility Among Combinatorial Problems*; in R.E. Miller, J.W. Thatcher, and J.D. Bohlinger (eds.) *Complexity of Computer Computations*, Plenum Press, New York, pp. 85–103 (1972).

[38] William Karush, *Minima of Functions of Several Variables with Inequalities as Side Constraints* (M.Sc. thesis). Dept. of Mathematics, Univ. of Chicago, Chicago, Illinois (1939).

[39] J.B. Kruskal, *On the Shortest Spanning Tree of a Graph and the Traveling Salesman Problem*, Proceedings of the American Mathematical Society, vol. 7 (1956), pp. 48–50.

[40] Harold W. Kuhn and Albert W. Tucker, *Nonlinear Programming.* Proceedings of 2nd Berkeley Symposium. University of California Press, Berkeley, pp. 481–492 (1951).

[41] A.H. Land and A.G. Doig; *An Automatic Method of Solving Discrete Programming Problems*, Econometrica, vol. 28, no. 3 (July, 1960), pp. 497–520.

[42] Robert Nau, *Introduction to ARIMA models*, Duke University. Published electronically at `https://people.duke.edu/~rnau/411arim.htm`.

[43] A.L. Peressini, F.E. Sullivan, J.J. Uhl Jr., *The Mathematics of Nonlinear Programming*, Springer (1991).

[44] David Poole, *Linear Algebra, a Modern Introduction*, 3rd edition, Brooks/Cole, Cengage Learning (2011).

[45] R.C. Prim, *Shortest Connection Networks and Some Generalizations*, Bell System Technical Journal, vol. 36 (1957), 1389–1401.

[46] The Puzzle Museum `https://www.puzzlemuseum.com/`

[47] Cliff Ragsdale, *Spreadsheet Modeling & Decision Analysis: A Practical Introduction to Business Analytics*, 8th edition, Brooks/Cole, Cengage Learning (2015).

[48] Kenneth H. Rosen, *Discrete Mathematics and Its Applications,* 8th edition, McGraw-Hill (2019).

[49] Conor Ryan, Michael O'Neill, and J.J. Collins, eds., *Handbook of Grammatical Evolution*, 1st edition, Springer Publishing Company (2018).

[50] Alexander Schrijver; *Theory of Linear and Integer Programming*, John Wiley & Sons (1999).

[51] Benjamin L. Schwartz, *Possible Winners in Partially Completed Tournaments* SIAM Review, vol. 8, no. 3 (1966) pp. 302–308.

[52] P.D. Seymour, *Decomposition of Regular Matroids* in Linear Inequalities and Related Systems, Journal of Combinatorial Theory (B), vol. 28, Elsevier, (1980) pp. 305–359.

[53] N.J.A. Sloane, editor, *The On-Line Encyclopedia of Integer Sequences*, published electronically at `https://oeis.org`.

[54] Shumway, Robert H. and Stoffer, David S. Time Series Analysis and Its Applications, 4th edition, Springer Statistics Series. `https://www.stat.pitt.edu/stoffer/tsa4/`

[55] Krzysztof Socha and Marco Dorigo, *Ant Colony Optimization for Continuous Domains*, European Journal of Operational Research, vol. 185, no. 3 (2008), pp. 1155–1173.

[56] Thomas Stützle and Holger H. Hoos, *MAX-MIN Ant System*, Future Generation Computer Systems, vol. 16, no. 8 (2000), pp. 889–914.

[57] Luara Taalman and Peter Kohn, *Calculus*, W.H. Freeman and Company (2014).

[58] Lloyd N. Trefethen and David Bau III, *Numerical Linear Algebra*, SIAM (1997).

[59] William R. Wade, *An Introduction to Analysis*, 4th edition, Pearson (2010).

[60] Hassler Whitney, *Congruent Graphs and the Connectivity of Graphs*, American Journal of Mathematics, vol. 54 (1932), pp. 150–168.

[61] Eric W. Weisstein, *Unimodular Matrix*, MathWorld–A Wolfram Web Resource, `https://mathworld.wolfram.com/UnimodularMatrix.html`.

[62] Westworld, Director: Michael Crichton, Metro-Goldwyn-Mayer (1973), Film.

[63] Hassler Whitney, *A Theorem of Graphs*, Annals of Mathematics, Apr., 1931, Second Series, vol. 32, no. 2 (Apr., 1931), pp. 378–390.

[64] Wikipedia, `https://en.wikipedia.org/wiki/Main_Page`.

Index

Printed in the United States
by Baker & Taylor Publisher Services